Genetic Organization

A Comprehensive Treatise
Volume I

CONTRIBUTORS

Walter F. Bodmer
Andrew J. Darlington
L. C. Dunn
Sterling Emerson
Rhoda F. Grell
Henry M. Sobell
J. Herbert Taylor

GENETIC ORGANIZATION

A Comprehensive Treatise

Edited by

ERNST W. CASPARI
Department of Biology
The University of Rochester
Rochester, New York

ARNOLD W. RAVIN
Department of Biology
University of Chicago
Chicago, Illinois

Volume I

ACADEMIC PRESS New York and London 1969

ACADEMIC PRESS, INC.
111 Fifth Avenue, New York, New York 10003

United Kingdom Edition published by
ACADEMIC PRESS, INC. (LONDON) LTD.
Berkeley Square House, London W1X 6BA

Library of Congress Catalog Card Number: 68-8429

PRINTED IN THE UNITED STATES OF AMERICA

LIST OF CONTRIBUTORS

Numbers in parentheses indicate the pages on which the authors' contributions begin.

WALTER F. BODMER (223), Department of Genetics, Stanford University School of Medicine, Palo Alto, California

ANDREW J. DARLINGTON (223), Department of Genetics, Stanford University School of Medicine, Palo Alto, California

L. C. DUNN (1), Emeritus Professor of Zoology, Columbia University, Nevis Biological Station, Irvington-on-Hudson, New York

STERLING EMERSON (267), Division of Biology, California Institute of Technology, Pasadena, California

RHODA F. GRELL (361), Biology Division, Oak Ridge National Laboratory, Oak Ridge, Tennessee

HENRY M. SOBELL (91), Department of Chemistry, The University of Rochester, Rochester, New York

J. HERBERT TAYLOR (16), Institute of Molecular Biophysics, Florida State University, Tallahassee, Florida

PREFACE

At first we wondered what a multivolume treatise dealing with the current state of knowledge of genetics could offer that available publications did not. *Advances in Genetics* periodically brought certain problems and fields up-to-date. Several excellent textbooks and paperbacks were appearing regularly; these dealt with everything from the mechanics of genetic transmission to gene action to ecological genetics. In recent years, the *Annual Reviews of Genetics* has been added to the literature. Indeed, the growing list of publications dealing with aspects of genetics attested to the impressive development that the biological discipline concerned with heredity and variation has enjoyed since its foundation at the turn of the present century. The discipline, in fact, appeared to be fragmenting. The molecular geneticist, caught up in the fascinating and rapid advances concerning the physicochemical basis of heredity, was losing communication with the developmental and the evolutionary geneticist whose problems became no less complex as the structure of the gene was elucidated. Genetics, which played such a vital role in the unity of biology, was threatened with a deterioration of the kind of discourse among its practitioners that assured full recognition of the frontiers of the science. We realized that a treatise of this type might meet this challenge.

The title for the work, "Genetic Organization," was chosen deliberately. Continuity and change are seen to occur at every level of biological analysis, from the molecular through the cellular and organismic to the populational. At any level continuity and change are assured by an organization of elements, by their spatial and temporal coordination. The interrelation between the organizations of various levels, however incomplete our knowledge of it may be at the present time, is the foundation on which the unity of genetics is based. We offer this treatise as a medium in which the organization of continuity and change may be viewed at its various levels, as a medium by which the students of the specialized subdisciplines of genetics may attempt to integrate what has been learned thus far into a single coherent structure. In this effort at integration we shall be compelled to assess in a critical manner what we believe we know of genetic organization, to distinguish between models and hypotheses, on the one hand, and experimental findings and observations, on the other, to reveal, where they become apparent, the gaps in our understanding and the lack of bridges within our theoretical structure. Perhaps because of the brilliant successes of genetic research, we do need to take stock at this time and evaluate how future studies may be most effective. And in the effort we trust the scope of genetics will be viewed as a whole, and a common language will bring together all of those who study genetic organization in its various forms. The arrangement of this treatise is intended to help achieve this goal.

This volume opens with a chapter setting the historical perspectives of our science and indicates the themes of unity within it. The remainder of the volume is devoted to the structure and transmission of genetic material. The next volume will be devoted to the action of the gene and the genome on the development of the phenotype. A final volume will be concerned with mutation and evolution. Because the chapters are written by different authors, a certain amount of overlap is unavoidable. For our present purpose, some overlap is desirable, and we have sought the optimal level by promoting communication between the authors. For any shortcomings, we, the editors, take full responsibility.

We dedicate this treatise to the memory of Kurt Jacoby laboring in the good cause of intellectual communication and to the future students of genetics laboring in the good cause of intellectual unity.

August, 1969

Ernst W. Caspari
Arnold W. Ravin

CONTENTS

CONTENTS OF OTHER VOLUMES

Genetic Organization

A Comprehensive Treatise
Volume I

I GENETICS IN HISTORICAL PERSPECTIVE

L. C. DUNN

I. Introduction

There is some truth in the naive view that modern science is often influenced in its early stages by disciplined curiosity about practical questions. The boy who takes the clock apart in his effort to find out what makes it tick inevitably silences the tick. The satisfaction of discovery comes when he has found out how to resuscitate the tick by restoring the original order to the dismembered parts. That represents to him, when he first accomplishes it, a real act of creation. The power and confidence this confers will enable him to do it again and to show another child how it is done. The realization that his act was one of re-creation of a pattern that had evolved from a succession of mechanical inventions will come later, fortunately too late to inhibit his first impulse to take the clock apart.

If the boy retains this pristine interest as he matures, he will probably come to wonder about the steps by which the clock, as mechanism, came into being. Some grown-up boys will assemble, in chronological order, all the kinds of clockwork mechanisms they can find. Others will seek to introduce further improvements and economies in the making of time machines. A few will search for a general principle underlying both clocks and other moving mechanisms. Very few will be moved to reflect on the process of discovering and of knowing, and on the nature of knowledge and reality. Only an occasional boy will attain the pinnacle of maturity from which he will look backward over the panorama of the development of mechanisms and of the ideas connected with them. At any rate that has seemed to be the myth in the United States: Many young men investigate; only a few old ones turn to history.

If this was ever true, it is certainly changing now. A goodly segment of the history of science and technology has taken place during the lifetime of young scientists now living. That part of history is a part of their own experience. They have become familiar, as a matter of course, with some of the events that gave rise to the questions in which their own interests have come to center. The fact that such events were recent ones does not make them any less historical. Judgments about historical importance will be rendered rather by the degree to which ideas underlying events in a science determine its future course. Sometimes this becomes quickly apparent, as in the case of the DNA model of 1953. Knowledge only of recent events, however, and opinions based upon them are seldom sufficient for making sound judgments of future developments. Interest in the recent changes in science may act as a kind of decoy, directing attention toward more remote sources of present ideas, and thus opening the prospect of another kind of satisfaction to students of natural science, quite beyond the practical uses of recent history in setting guidelines for today's studies. That prospect is the attainment of what one might call a certain composure, as when parts have fallen into such relations that the attention shifts to the composition itself. If that simile is borrowed from aesthetics, its application to science is well justified, both now and in the past. And if composure implies that the mind comes to rest, at least while it contemplates some aspects of a particular science which embody an underlying unity and harmony, that too can be seen as one of the satisfactions, if not the goal, of scientific study.

Such reflections seem especially relevant at a time when the accumulation of facts in science is advancing at a feverish and accelerating pace. The contrast between the present and the earlier states of science is especially marked in biology. There was a time when a man might leave on a research voyage in 1831, begin six years later to collect facts to test some ideas acquired during the journey, and yet in 1856 could write to a friend who was urging speed ". . . am doing my work as completely as my present materials allow and without waiting to perfect them. And this much acceleration I owe to you."* In the same year in which this letter was written, another man began some experiments to test an idea which had come to him about heredity. Ten years later the results of the

* C. Darwin to C. Lyell, Nov. 10, 1856 (vide F. Darwin, 1887).

experiments appeared in a 41-page paper which at the time (Mendel, 1866) no one seems to have understood. By such leisurely means biology was put upon a new course. Today, of course, only a few months may intervene between the moment of conception of an idea and its birth in print, and speed is, in fact, of the essence for the author since he knows that others are racing to the same goal.

The application of this preamble to a discussion of genetics in historical perspective will become obvious to those who read the contributions in this volume. Genetics, which has now become an important scientific activity, was strongly influenced in its beginnings in the eighteenth and nineteenth centuries by a concern with practical problems, many of them connected with agriculture and especially with plant breeding. These interests, as they culminated in the published records of experiments that supplemented and made more durable the improved varieties of plants which had been sought, turned more and more toward development of methods, toward learning how for its own sake, and thus toward the quantification and abstraction by which Mendel helped to change an art into the beginnings of a science.

If the first source of interest in questions of heredity and variation, which led to modern genetics, came from agricultural pursuits such as animal and plant breeding, horticulture, and gardening, these fields were also the first to benefit from the scientific attitudes which came to supplement the arts of husbandry. Strong influences upon human economy have come from such advances as the controlled use of hybrids in the breeding of corn and other crop plants, in poultry and swine breeding, and the development of disease-resistant strains of plants and animals. The latest field to feel the effect of ideas and methods from genetics has been human medicine. Although specific disease entities like hemophilia were recognized as hereditary in the early nineteenth century, and the method of inheritance of polydactyly was known as early as 1753 (Maupertuis), these remained as isolated and rather curious facts and no useful generalizations arose from them.

The best gifts of science to practice have generally been ways of thinking and methods of general applicability; the development of biometry provides one good example. Although the use of probability methods and of statistics applied to biology antedated the rediscovery of Mendel's principles, they received a great impetus from the discovery of the gene. What Galton had begun and Pearson had worked out in great detail was developed by R. A. Fisher and others into methods of inference, computation, and design of experiment which were most useful in plant breeding and agriculture generally. These methods have most often served to purge genetics of false hypotheses, and that is as essential a step as inventing new ones.

A. Darwin and Mendel

A great influence, of both practical and theoretical importance since it provided a central unifying idea, was the development of Darwin's methods of careful observation, often accompanied by counting and measuring, which affected animal and plant breeding and agricultural research before Mendel's did. "I have no faith," wrote Darwin, "in anything short of actual measurement and the Rule of Three." The foundations for this kind of quantitative analytical study applied to variation and heredity had been laid by Galton in response to stimuli from Darwin. They appeared in definitive form in Galton's "Natural Inheritance" of 1889.

Mendel's contribution was of a different sort. It was the application of probability methods to the study of alternative inheritance. It was the logical rather than the computational features of Mendel's method that formed the basis for a science of genetics and gave it great practical value.

It is an interesting reflection that the methods of Darwin and Mendel, which developed in isolation from each other, have led, in part through the quantitative form now assumed by their basic ideas, to so close an interdependence that each has now become the touchstone by which the other is to be judged. They are firmly and profitably joined in both the theory and practice of plant and animal breeding. They form the basis of evolutionary biology and have provided a rationale for much of agriculture.

B. Diversification of Genetics

Genetics today seems to have become an extremely diversified field. Before it acquired its present name (about 1908), its central doctrine was referred to as "Mendelism," and a question of real interest was whether Mendel's laws had general validity, or at least whether they applied to animals, including man, and to other organisms in which they had not been experimentally demonstrated. Now, of course, one hears of nearly as many kinds of genetics as there are kinds of organisms: plant, animal, and human genetics, with their counterparts in practice—plant breeding, animal breeding (limited, usually, to "useful" plants and animals), and medical genetics. There are treatises on poultry genetics, cattle genetics, rice genetics, corn genetics, and, of course, *Drosophila* genetics, bacterial, fungal, and yeast genetics, protozoan and viral genetics, and many more. But this spawning of special names does not mean essential diversification. Quite the contrary: they are all genetics using one common, symbolic language and one set of logical canons when describing the outcome of breeding experiments.

Special fields dealing with processes, however, by about 1950 had come to express the diversification actually needed for dealing with problems of evolution and individual development as well as those of the structure and function of the genetic material. Well-marked fields such as population genetics, developmental and physiological genetics, cytogenetics, and others had all taken form before the advent of molecular genetics. The older fields and many other diversifications based on process are subject to the unifying influences of the logical and symbolic methods used in statistical or biometrical genetics, as discussed above.

However, the chief evidences of diversification within genetics, as within biology generally, have appeared during the last fifteen to twenty years with the use of microorganisms and the rise of methods for analyzing genetic processes and structure at a molecular level. There has resulted a great expansion of the ranges of phenomena dealt with by genetic ideas, not only in breadth but in depth of penetration. Conceptions of gene structure and function have taken on new dimensions not generally foreseen since they were not required to explain the results obtained by the older methods.

The diversity which genetics now encompasses extends from the bonds that hold atoms together, at one level, to the relations that integrate societies and ecosystems at another. At the same time, no aspect of recent genetics is more impressive than the concrete documentation it has brought to the view that all life is one.

Biology came to be recognized as a science when, in the early years of the nineteenth century, it began to appear that living organisms might share a history, a physical organization, and capacity for reproduction common to all. The great revelation of the continuous history and evolution of life, derived from the arguments of Lamarck, the evidence of Darwin and Wallace, and the persuasiveness of T. H. Huxley, can now be said to be securely based on the material substrates of molecular organization that are common to all living matter. The effects of the realization of what this means for man, now in the 1960's, may well be greater than those that followed the establishment of the natural origins of man and other living creatures a century earlier.

The prospect now is that a new biology, transformed by ideas and methods from a new physics and a new biochemistry, and acting through the new genetics, will greatly expand the range of science itself. If the mechanisms that control the reproduction of nucleic acid molecules and the synthesis of specific proteins are known, those processes themselves can be subjected to deliberate control. If old limitations which prevented the control of functions at the molecular level are removable, then why not also those that restricted progress at other levels, for example, those concerned with the biology of man, with human relations, and with guided changes in human populations?

It is at least a tenable hypothesis that it was the ideas, ways of thinking, energies, and confidence generated during the first century of the development of modern genetics that set off the transformation of biology. To what extent this was true and why it was genetics that provided so strong a stimulus will be discussed at the end of this essay. It will in any case be admitted that the lure of the future, held out above, can best be appreciated when seen in the context of the past.

C. Histories of Genetics

It is notable that a recent rather sudden burst of interest in the history of genetics coincided with the emergence of a new level of genetics—the molecular one marked by the deciphering of the genetic code. Several books, special articles, and symposia devoted to the history of genetics appeared during 1965 and 1966. These publication dates coincided with the centenary of the delivery (1865) and publication (1866) of Mendel's analysis of his hybridization experiments with peas. The books of Sturtevant (1965), Dunn (1965), Olby (1966), Stubbe (1965), the reprinting of documents concerning Mendel and his rediscoverers and critics, one edited by Křiženecký (1965a), one by Bennett (1965), and one by C. Stern and Sherwood (1966), would doubtless have been published whether or not the birth of molecular genetics was to be celebrated as well as Mendel's anniversary. But the fact is that we have just been provided with some of the materials needed for getting the historical perspective which it is the purpose of this chapter to provide. Needless to say, these sources have been drawn upon together with the excellent source books of Roberts (1929), Iltis (1924), and Barthelmess (1952). Brief outline accounts of the development of theories of heredity by Babcock (1949) and Grant (1956) have also been helpful. The studies of Glass (1947, 1953, 1963, and others) have already provided views in depth of the development of essential ideas of genetics, and have related them to the broader questions of the history of science.

For the recent period inaugurated by the DNA model of Watson and

Crick, I have found Bresch's "Klassische und molekulare Genetik" (1964) excellent while Watson's "Molecular Biology of the Gene" (1965) relates the new developments with the old in a remarkably clear way. The latter was not intended for the use of historians and thus contains no bibliography or, in general, specific citations documenting the events described. These books cannot, of course, repair the deficiencies in biochemical training and experience with bacterial genetics which I share with some of those who, like myself, work with the methods and viewpoints referred to as "classic genetics." But at least they made me aware of the fundamental continuity of ideas and problems that the new genetics inherited from the old, or in some cases rediscovered.

One book of outstanding interest for those wishing to gain perspective in genetics appeared too late to contribute to the planning or writing of this article. This is "Heritage from Mendel," edited by R. A. Brink from the papers contributed by the 27 scientists from all fields of genetics who participated in the Mendel Centennial Symposium held at Fort Collins, Colorado, in September, 1965. It was published in February 1967 by the University of Wisconsin Press. The detailed literature lists in the individual articles are a most valuable and useful feature. The valuable book of Carlson (1966) "The Gene: A Critical History" was not available during the writing of this chapter.

Nearly all of the works mentioned above, including those devoted to historical surveys, have been written by persons whose main interests and occupation have been in research and teaching of genetics. The fact that genetics has not yet been dealt with comprehensively by professional historians should not occasion surprise. What a practicing scientist thinks about the development of his field, his attitude toward documentation, what problems he chooses as the significant ones, how they relate to the scientific, cultural, and social matrix in which the developments took place—these may well differ from the attitudes inculcated by historical scholarship. That is why I have chosen to speak of historical perspective rather than history. Moreover, I believe that perspective is what is most useful for students of genetics at its present juncture when most of its practitioners are recent adherents.

II. Genetics Today

One revealing indication of what genetics is like is that today, one hundred years after the discovery of the first of its first principles and in the midst of a period of feverish research activity, there are still people who try to reduce the whole subject to a few simple statements. Many writers of textbooks and of popular expositions, speakers at symposia and international congresses behave as though there were a unity underlying the processes of heredity and change and an order of which glimpses can be obtained from the study of any kind of living matter. This is not a matter of faith, although faith had to sustain hope in the early stages of the search for what now has become a demonstrated fact.

How some of these principles came to be established will be discussed in more detail later on in this chapter. Just now we want to know what they are, so that our attempts to view them in perspective may be selective and not dissipated in the byways of detail. During the last few years the rapid pace of discovery of new rules about the structure and the functioning of the genetic material have created a widespread impression that what is most important is the newest. It does not reduce the significance of the new to realize that it rests upon a foundation laid one hundred years ago.

A. First Principles

The first principles of heredity were discovered by Mendel as a result of questions restricted to the means of transmission of characters differentiating the existing varieties of a cultivated plant. The outcome of his work was the view of heredity as particulate, consisting of the transmission of elements (later called genes) which could take two alternative (allelic) forms of which only one member could enter any one reproductive cell. Such cells consequently could be conceived as containing one member of each of many pairs of elements in all possible combinations. The demonstration of assortment or recombination of segregating elements was an essential part of the proof of the particulate nature of heredity and immediately gave an insight into the sources of variety on which natural or artificial selection could act in forming new varieties or species. Since the existence of particles was inferred from numerical relations among the offspring of hybrids, the particles were in fact symbols and the laws governing their behavior were statistical in nature. This by itself put them in a category which, being subject to rules of probability, had a general character.

Later, after others had rediscovered these rules in the same and other plants, the question arose as to whether they were in fact general. In the first decade of this century, in a burst of activity participated in by both botanists and zoologists, these rules were shown to apply wherever reproduction involved a reduction division in the production of gametes. Their application proved to be as wide as the existence of meiosis.

B. Additional Principles

This connection with the chromosome mechanism first suggested that the particles were located in the chromosomes. The first additions to Mendel's principles were thus those concerned with linkage, crossing-over, and the linear order of particles in the linkage groups, each of which was assumed to correspond to one chromosome. By 1926, these principles, known collectively as the theory of the gene, could be claimed to be general although strict proof extended only to diploid animals and plants.

In the following decade the growing points of genetics were in the experimental study of mutation and in the analysis of the correspondence between gene order and chromosome structure. These studies produced rules governing the changes in the particles (gene mutation) as random errors in the process of gene-copying, while the analysis of rearrangements induced in chromosomes provided confirmation in physical terms of the rules deduced from linkage and recombination. Thus the statistical or logical order was shown to be based upon a physical order. The gene-containing bands of the salivary chromosomes of *Drosophila* appeared to consist of nucleoproteins while the wavelengths of ultraviolet energy effective in producing mutations appeared to be those absorbed by nucleic acid.

Beginning about 1941 the combination of biochemical and genetic methods in the analysis of gene action in microorganisms led to the rule that enzymes are gene-controlled and in 1947 a mutant gene was found to control the structure of a protein by changing the order of the amino acids of which it was composed. A gene thus came to be regarded as a unit of function, the function being to specify the structure of a protein.

The proof in 1943 that mutations occur in bacteria by changes in genes governing functions like those in other organisms was an extension of the theory of the gene to nondiploid organisms.

C. Molecular Genetics

In 1944 came the first proof that the material basis of heredity was in fact DNA and in 1953 a model of the structure and mode of replication of DNA paved the way for the statement of genetic rules in molecular terms. The degree to which this has been accomplished is a question of current investigation which will be dealt with by other authors in this treatise.

The chief alterations in views of the transmission system required by discoveries in genetics after about 1945 concerned the concept of the gene as the fundamental unit of organization of the genetic material. Today genes appear to be complex linear sequences of a hundred or more separately mutable sites along the polynucleotide chain of DNA. Each site probably consists of a single pair of nucleotides. It is this pair of small molecules, or the bonds which hold them together, that constitute the elementary structures. Micromaps of genes, attained by classic methods of recombination, reveal an order at a level below that of the gene as a unit of function, and research chiefly with microorganisms has revealed that genes are also elements in an organization of higher order. They participate as structural genes, that is, genes specifying protein structure, and along with regulator and contiguous operator genes are organized into regulatory circuits. The activity of the structural genes is modulated and controlled by the regulatory circuit. Regulation, where best understood as in bacteria, usually takes place by selective repression. The regulatory substance from the regulator gene passes through the cytoplasm where it may be activated or inhibited en route to the structural genes in the chromosome.

Finally, the manner in which a structural gene specifies the structure of the protein for which it is responsible is embodied in a set of rules known as the genetic code. The chief feature of this is that the nucleotide pattern of one of the strands in a segment of DNA acts as a template upon which is formed a complementary strand of ribonucleic acid in which U (uracil) takes the place of the thymine of DNA. The DNA order is thus transcribed into a corresponding RNA order. This RNA copy acts as a template upon which the amino acids from the pool in the cell will be assembled in the sequence specified by the order of bases (nucleotides) in the RNA. The message in the RNA is in a four-letter code, the letters representing the bases A(adenine), U(uracil), G(guanine), and C (cytosine). This is to be decoded or translated into the twenty-letter language of the amino acids found commonly in proteins. The code consists of triplets or codons of three successive bases such as AGU upon which a specific amino acid (e.g., methionine) is attached by a specific enzyme. The message transcribed from the DNA order of nucleotides is translated, by means of the code, into a sequence of amino acids assembled in polypeptide chains to form a specific protein.

Experimental studies with microorganisms, especially bacteria and viruses, have provided material for much of the new body of principles in genetics that have arisen since 1943. Before that time it was not apparent that bacteria have a genetic system comparable to that of higher organisms. Since then there have been profitable feedback relations between work with microorganisms and with higher forms. The classic rules of segregation and recombination of alleles have been shown to hold in bacteria, but recombination can also occur in mitotic division and with means such as transformation and transduction in addition to those found in diploid organisms. Rules discovered in microorganisms, such as those concerned with gene complexity, the correspondence between the sequence of genes along the bacterial chromosome and the sequence of biosynthesis, the groupings of elements that control gene function, and the genetic code itself are for the most part novel gifts from research with bacteria. If the direction of influence from genetics was first from research on the higher plants and animals, it is now, with some exceptions, reversed as far as discovery of the finer mechanisms of transmission and control of gene function are concerned.

The great legacy of classic genetics can now be seen to have been (1) the statement of questions about heredity and variation in forms amenable to experimental analysis, and (2) the development of methods of reasoning from experimental breeding data and of criteria for testing hypotheses. This might be called the logic of genetics and one could say that the early geneticists learned that it was better to define elements by their behavior than by physical description. A third contribution of the classic period was the discovery in the genetic material of order of a particular linear kind which has proved to have general application to biological systems.

A fourth contribution to genetics today was a cooperative spirit by which groups, even when physically separated, pooled their resources and efforts. Scientists from other disciplines have often remarked upon this spirit which is so strongly in evidence in genetics and responsible in part for the rapid progress of genetics today.

III. The Main Channel of Genetics

The decisive impulse which set off the development of genetics came from the rediscovery of Mendel's principles in 1900. That the principles do not derive exclusively from Mendel is shown by the fact that they were independently rediscovered. There can be little doubt, however, that it was the stimulus of Mendel's sharp and persuasive experimental demonstration of segregation and recombination, and the bold generalization of these in a symbolic system that got the new science off to a running start. It was as much the effect of Mendel's model as of his facts upon Bateson, Correns, and others which led them to proclaim, test, verify, and extend a theory which they never called anything but Mendel's.

A. Sex in Plants

Although the central theory, that heredity is particulate, had not been stated before Mendel published his report in 1866, many of the facts discovered and some of the methods used by Mendel and his successors resulted from the work of botanists and plant breeders of the two centuries preceding Mendel's scientific lifetime. This earlier work was motivated by interest on the part of both scientists and practical horticulturalists in the mechanism of reproduction of flowering plants. The story of the working-out of the sexual system in higher plants has been told many times, with some good accounts just published (Crew, 1966; Nemeč, 1965; Olby, 1966; Sturtevant, 1965; and especially Stubbe, 1965) although the main sources were set forth in the earlier work of Roberts (1929) and Zirkle (1935).

In essence, morphological descriptions of male and female floral parts, e.g., as summarized by Zaluzansky (1592) and Grew (1682), proved to be insufficient as guides to practice, and the understanding of the function of pollen and ovule had to be derived from experimental observations on crossing of varieties differing in visible characters. The results of some such experiments were published by Camerarius in Tübingen (1694).* Some good observations on sex and the effect of crossbreeding in maize were reported from the English colonies in America by Cotton Mather (1716), Paul Dudley (1724), and especially James Logan (1735). What part these early experiments played in the breeding of Indian corn has been well described by Henry A. Wallace and William L. Brown in "Corn and Its Early Fathers" (1956). In England, Thomas Fairchild (1718) produced by artificial pollination one of the first plant hybrids, showing that he understood the mode of reproducing well enough to control it.† The chief advance, however, came from Kölreuter, the German botanist, who carried out hundreds of hybridizations (1761–1766) and showed that hybrids were usually intermediate between the varieties or species which supplied pollen and ovule respectively. It was he who started rational plant breeding and bequeathed the methods, the knowl-

* References to Zaluzansky, Grew, and Camerarius will be found in Nemeč (1965).
† References to works cited in the balance of this section are in Stubbe (1965).

edge, and especially the confidence to his successors, one of whom, 100 years later, was Mendel, a self-trained amateur.

Plant breeders were very active during the first half of the nineteenth century, especially in England, Germany, France, and Italy. Some of the observations from which Mendel fashioned his theory had already been made by some of them. Thus T. A. Knight, President of the Horticultural Society of London, had by 1823 recognized in the pea plant the dominance of gray over white seed color of peas and segregation of the two colors without intermediates. Since he did not count the proportions nor progeny-test those with the dominant color, he missed making the discovery made by Mendel 40 years later. The Italian pomologist, Count Giorgio Gallesio, in addition to important breeding work with citrus fruits, in 1816 published a book, "Teoria della Riproduzione Vegetale," in which, in describing the results of crossing carnations of different colors, he noted that the colors appeared in descendants unequally according to which color was "dominant." According to Stubbe (1965) (who credits Martini with first pointing out this passage) this use of the concept and the word "dominant" antedated by 10 years its use by Sageret in 1826. John Goss (1824) and Alexander Seton (1824) both observed segregation and dominance of seed coat color and form in peas and this was later confirmed by Laxton. In France, the great plant breeders of the de Vilmorin family, Louis (1816–1860) and his son Henry (1843–1899), found 3:1 ratios in flower colors in lupines, but failed to interpret them as Mendel did, nor did they derive a rule for segregation in pea hybrids in which their results resembled those of Goss. These observations antedated those of Mendel, although Mendel seems not to have known of them.

One set of observations by Mendel's contemporary, the French plant breeder, Naudin, was also unknown to Mendel, but could readily be interpreted on Mendel's theory. Naudin also did not count the different types of progeny, although he did recognize disjunction of opposed "essences" as accounting for the diversity of F_2 plants following a cross. These examples of results repeatedly obtained, resembling in one or several ways those of Mendel, from which, nevertheless, rules like Mendel's were not derived, could be extended to observations on breeding of house mice (Haacke, von Guaita,* Coladon), wheat (Rimpau, Spillman), and barley (Tedin) among others. But enough has been said

* In fact, Castle (1903) used von Guaita's data, obtained in ignorance of Mendel's rules, to prove that these results on inheritance of albinism could be predicted from Mendel's first principle.

to show that what in his time was unique about Mendel's work was not his material but his quantitative methods of observation and analysis.

B. Charles Darwin

One of the most influential biological books of the nineteenth century was Charles Darwin's "Variation of Animals and Plants under Domestication," first published in 1868. This contained the evidence by which Darwin proposed to support his theory of evolution by natural selection, which he had published in the "Origin of Species" in 1859. The book on variation was a great compendium of the experience of breeders, horticulturalists, field naturalists, and zoologists and botanists of all kinds. It also contained the first statement of Darwin's theory of heredity which he presented in the last chapter of the second volume as a "Provisional Hypothesis of Pangenesis." This was the theory, seriously entertained by Darwin in spite of its "provisional" nature, that inheritance occurs by the transmission of representative units, gemmules, which are generated in the cells, increase by self-division, circulate in the tissues, and, assembling in the germ cells, are transmitted to offspring where they cause the development of the parts from which they were derived in the parents. Although the hypothesis was disproved both by Galton and by Romanes (1895, p. 143), it exercised a fascination even upon those who did not accept it. Among these was Hugo de Vries, who constructed a different speculative theory of heredity in his "Intracellular Pangenesis" of 1889. It is also to be noted that the theory was revived under a different name by the Soviet agronomist, Lysenko, who in the mid-twentieth century employed it for the same reason used by Darwin, namely, to explain the inheritance of acquired characters. By the time of its revival, however, there were no longer any grounds for regarding inheritance of acquired characters as a fact.

What made Darwin's book so useful and important, both in the nineteenth century and today, was its careful marshalling of solid facts about variation and inheritance in crop plants and domestic animals. It also contained accounts of some of Darwin's own extensive observations and breeding experiments on plants, in one of which Darwin noted a clear example of an F_2 segregation from a cross of two varieties of snapdragon. He noted the proportions because of his habit of making exact observations, not to test a theory, since segregation of alternatives was not a part of his pangenesis hypothesis.

This book was published after Mendel's experiments with peas had been completed and published, so it could not have influenced Mendel.

Mendel's continuing interest in Darwin's work is attested (Richter, 1940) by his annotations in Darwin's later book "Effects of Cross and Self-Fertilization" (1877). Darwin strongly influenced the generation of biologists immediately following Mendel: Weismann, de Vries, Bateson, Johannsen, and others.

IV. The Discovery and Rediscovery of the First Principles

There was great interest in theories of heredity in the second half of the nineteenth century, but most of them were speculative and were not based on observations (Dunn, 1965). The way toward experimental tests of ideas about the transmission of hereditary traits was prepared by the work of the plant hybridists, described above, and by the discovery of the essential features of fertilization. By 1865 it was known that in plants the new individual took origin from the union of two single cells, egg and pollen grain. This provided a rationale for experimental crossing experiments. A number of botanists engaged in such work, but only one set of observations, those of Mendel, led to a consistent theory which could be generalized and confirmed in animals and plants of all kinds.

A. Mendel

Mendel* began the preparations for such experiments in 1854 after his return from two years of scientific study at the University of Vienna to which he had been sent by his monastery. He was then 32 years old, a priest and member of the Augustinian order, and a substitute teacher of physics and natural history in the *Realschule* (Modern School) at Brünn, then a part of Austria, now Brno, Czechoslovakia. He first collected from seedsmen some 34 varieties of edible peas, differing in such readily observable traits as height, flower color, seed form and color. He made a 2-year test of these by raising offspring derived from each variety by self-fertilization. After observing that each variety bred true, he chose 22 varieties for crossing experiments and maintained these in his trial garden as controls, examining their offspring for constancy of characters throughout the period of experimental breeding analysis which extended from 1856 to 1863.

Mendel analyzed seven "differentiating characters," as he called them. It is significant that when he described these in his report of 1865 he

* For an autobiographical account see Mendel (1965).

arranged them in pairs. The form of the ripe seeds for example, was either round or wrinkled, and the words *entweder–oder* (either–or) occur in the description of each pair. This made possible the basic distinction of alternative traits which led to the discovery of allelism, a fundamental feature of heredity.

He then crossed, by carefully guarded artificial cross-pollination, plants showing alternative traits such as round or wrinkled seeds, and observed the generation grown from the hybrid seeds (now called F_1); he allowed these to self-fertilize, and planted such seeds to produce an F_2 which he grew in considerable numbers in order to count the numbers of plants with each of the alternative traits. He allowed such plants to self-fertilize and produce an F_3 generation, in which he classified and counted the numbers of individuals with the alternative traits from each type of F_2 parent. He did the same with combinations of traits, such as plants with round and yellow seeds crossed with plants with wrinkled and green seeds. In one such experiment he crossed parents that differed in three traits.

I have been at some pains to emphasize the method that was carried out for the first time by Mendel. It was deliberately designed to carry out a purpose that Mendel was also the first to recognize. It was stated in the introduction to his report of 1865:

> Whoever surveys the work done in this field will reach the conviction that among all the numerous researches, not one has been carried out to such an extent and in such a way as to make it possible to determine the number of different forms under which the offspring of hybrids appear, or to arrange these forms with certainty according to their separate generations, or definitely to ascertain their numerical relations to each other.

By carrying out this plan Mendel established the two principles that bear his name: First, the principle of segregation or splitting into different gametes of the members of a pair of elements responsible for the observed alternative characters such as round and wrinkled. This rule was often referred to by Mendel's successors as the "law of the purity of gametes" since its essence was that each gamete could contain only one member of an alternative pair, either *R* (round) or *r* (wrinkled). Mendel's second rule was the principle of independent assortment or recombination by which, in the formation of gametes by a hybrid, elements belonging to different pairs of alternatives such as *A, a,* and *B, b,* enter into all possible combinations with each other, viz., *AB, aB, Ab, ab.* The two rules together convey Mendel's view of heredity as particulate. For him this meant that intact elements, treated by Mendel as symbols, such as *A* or *a,* are transmitted through the reproductive cells. It was this view, and not the other nineteenth century idea of material units (Spencer's physiological units, Darwin's gemmules, Weismann's biophores, de Vries' pangenes) which underlay the development of genetics and of modern particle biology.

The whole tenor of Mendel's paper was to base conclusions only on statistical observations from experiments designed to test specific hypotheses. Units or elements established by such observations can have the status of symbols only; hence we may say that Mendel proved the existence of a statistical gene concept, applicable to his own material and to some other plants in which he had tested the rules. He made it quite clear that the "law formulated for Pisum required still to be confirmed"; and left it to be decided "experimentally" whether hybrids of other plant species would conform to the same rules. He evidently thought they would, for he wrote: "In the meantime we may assume that in material points an essential difference can scarcely occur, since the unity in the developmental plan of organic life is beyond question." That was a bold and prophetic statement to make in 1865. The now well-demonstrated universal character of the organization of living matter, based on self-replicating elements in the pattern of Mendel's units, is his best justification.

B. Fisher's Criticism of Mendel

The originality of Mendel's view, which was unique in its time, is emphasized by the evidence in his paper that he had the theories of segregation and recombination in mind before he carried out his experiments. This is the conclusion required by the fact that Mendel's numerical data fitted the expectations based on his theory to a degree that can hardly be accounted for by chance. This excessive "goodness-of-fit" was first pointed out by Fisher in 1936. Fisher's paper has been reprinted together with a translation of Mendel's paper, and with notes and comments by Fisher (see Bennett, 1965). It also is reprinted in C. Stern and Sherwood (1966) with comments by Sewall Wright. Fisher's most telling point was that one set of Mendel's observations shows too close a fit to an expectation which was wrongly calculated by Mendel. In testing F_2 plants with the dominant character, Mendel's theory would predict that these should occur in the ratio 1*AA* to 2*Aa*. In fact Mendel grew a set of only ten plants from each F_2 dominant plant which could be *AA* or *Aa*. Those sets that segregated, i.e., produced *A* and *aa* offspring, obviously came from *Aa* plants, but those that did not contain one recessive still had a chance of $(3/4)^{10}$ ($= 0.0563$) of being *Aa* but by chance failed to produce 1 recessive in 10. Thus the corrected expectation from testing samples of 10 would be 1.887 segregating:1.113 nonsegregating. But Mendel's actual data fitted the erroneous 2:1 expectation. Mendel clearly knew what to expect in F_2, F_3, backcross, and recombination experiments and he obtained it almost precisely, perhaps by unconscious bias in classifying plants or by stopping the counts when a close fit had been obtained. Wright has confirmed Fisher's finding of excessive goodness of fit in Mendel's results and reaches this opinion (C. Stern and Sherwood, 1966, p. 175): "I am afraid it must be concluded that he [Mendel] made occasional subconscious errors in favor of expectation, especially in this case. Taking everything into account, I am confident, however, that there was no deliberate effort at falsification."

It is not evident just when Mendel thought out the theory for which the observations provided such startling confirmation. It may have come from his first (1858) results, although Fisher thinks it possible that Mendel (and perhaps other mid-century biologists) could have invented a particulate view of heredity *ad hoc* and then set out to prove it. No direct evidence of this has been found.

C. The Rediscovery of the First Principles

Mendel's arguments from his pea experiments, persuasive as they now seem to us, had no influence in their own time—not until 34 years had passed. In the spring of 1900 they were restated as a result of similar experiments carried out independently by three European botanists. This story has been so frequently told that it need not be repeated here. It is set forth in several recent histories of genetics (Crew, 1966; Dunn, 1965; Sturtevant, 1965; Olby, 1965). All of the original publications bearing on the discovery and rediscoveries have recently been reprinted with commentary by Křiženecký (1965b). C. Stern and Sherwood (1966) have provided a corrected English translation of Mendel's papers on peas and hawkweed together with the "rediscovery" papers of de Vries and Correns, considering that only these deserve that title.

The result of the spate of publication during 1900–1903 by Correns, de Vries, Tschermak, Bateson, Castle, Cuénot, Boveri, Wilson, Sutton, and McClung was to establish firmly Mendel's principles as governing normal alternative inheritance in many species of plants and several species of animals. Mendel's rules were shown not only to be widely applicable, but features not worked out by Mendel were added. Dominance of one alternative (called allelomorph by Bateson, 1902) over its opposite was found not to be a rule and in fact not to have anything to do with the transmission system. Failure of independent assortment of two pairs of characters, the first suggestion of linked or coupled genes, was found by Correns in 1900.* Interaction between different pairs of allelomorphs in producing what at first were called compound characters was found to affect the expression but not the segregation or recombination of the alleles (Bateson, 1902). Correns (1902) even found an exception to the rule that the union of gametes bearing different alleles (e.g., sperms *A* and *a* with eggs *A* and *a*) occurs purely at random as called for by Mendel's theory. He showed that departures from expected ratios in this case were due to selective fertilization, i.e., pollen tubes carrying one allele, e.g., *s* (sugary carbohydrate in maize seed) less often fertilize ovules with the same allele than those with the opposite allele S (starchy carbohydrate). The theory of segregation nevertheless surmounted this and all other exceptions and emerged as the cornerstone of what had come to be called Mendelian heredity, or simply Mendelism.

V. The Chromosome Theory and the Theory of the Gene

At the time when Mendel's theory was published (1866) there were not known any other biological processes to which assumptions like those of Mendel were applicable, that is, assumptions of paired alternatives like *A* or *a*, *B* or *b* which formed combinations at random.

Such an idea was independently expressed by Francis Galton in a letter to Charles Darwin in 1875. Galton suggested that the character of hybrid offspring as intermediate between the parents might be explained by supposing them to contain equal numbers of two contrasted elements, such as a mixture of black and white gemmules which if combining in pairs by chance would produce one-quarter black, one-half gray, one-quarter white. This has a superficial resemblance to Mendel's explanation, but Galton did not apply it and did not further develop the idea (cf. Olby, 1965, pp. 70–76). There is, however, a strong suggestion that Galton at that time would have understood Mendel's theory if he had encountered it.

A. Weismann's Prediction, 1887

In the interval between the appearance of Mendel's paper of 1866 and the rediscovery of the principle in 1900, another biological process

* Original papers reprinted in Křiženecký (1965b); bibliography of papers relating to Mendel in Jakubiček and Kubíček (1965).

operating by the same rules was being worked out, namely, the behavior of the chromosomes during meiosis. It was in the year of Mendel's death (1883–1884) that Van Beneden enunciated his important law, viz., that "the chromosomes of the offspring are derived in equal numbers from the nuclei of the two conjugating germ-cells, and hence equally from the two parents" (Wilson, 1928, p. 14). The addition of those in the sperm to those in the egg produces the diploid number characteristic of the species: $n + n = 2n$. This was the cytological parallel of Mendel's assumption concerning the elements in gametes and zygote, but no one noted it at the time.

At the same time Roux proposed that there was a linear alignment of different qualities along each threadlike chromosome. On these foundations Weismann built his ingenious theory of vital units or biophores which he supposed to be grouped into ids aligned along the chromosome. Each chromosome was thus viewed as a linear organization of different qualities representing the species characteristics, that is, the ancestral germ-plasms. But in Weismann's scheme, unlike that arrived at later by Boveri, each chromosome carries the whole set of germ plasms. Weismann's brilliant prediction in 1887* of the reduction division was based on the need to counteract the effect of doubling the numbers of these chromosomes in fertilization. He stated his prediction as follows: ". . . there must be a form of nuclear division in which the ancestral germ-plasms contained in the nucleus are distributed to the daughter nuclei in such a way that each of them receives only half the number contained in the original nucleus." Although the prediction was confirmed quantitatively, by 1900 it had still not been proved that each chromosome of the haploid group is qualitatively different and has a homolog in the haploid group of the opposite sex. Synapsis of pairs prior to meiosis was suggested by Henking in 1891, and Montgomery in 1900 proposed that one member of the pair was paternal and one of maternal origin.

B. The Sutton–Boveri Hypothesis, 1903

But the full meaning of meiosis as the segregation and recombination of pairs of homologous chromosomes, one from each parent, was not reached until 1903 when W. S. Sutton and Theodor Boveri provided cytological evidence for this interpretation. Sutton (1903)† said that

* For references in this paragraph and quotation from Weismann see Wilson, 1928, p. 500.

† The original papers of Sutton (1903) and of Boveri (1904) have been reprinted in Křiženecký (1965b).

he had reached his interpretation "from purely cytological data before the author had knowledge of the Mendelian principles." Boveri (1902) got his idea of individual differences among the chromosomes from observations of multipolar mitosis and noted in a footnote (p. 81) "that these were like the behavior of hybrids and their descendants."

By 1903 the cytological observations had reached a stage at which an exact parallelism could be seen between the behavior of the chromosomes at meiosis and Mendel's laws of the distribution of genes to offspring. Sutton was the first to point this out clearly in January 1903, but several others had recognized at least the central idea that the chromosomes provided a basis for the operation of Mendel's principles. The proof of the parallelism and the operation of the meiotic mechanism was most convincingly given by Sutton (1903)* and Boveri (1904)* whence this first statement of the chromosome theory of heredity is referred to as the Sutton–Boveri hypothesis.

We can see now that this hypothesis was a dramatic climax made possible by the intensive observations on the chromosomes during mitosis and meiosis extending over the previous twenty years. In this earlier work there was the same fruitful association of observation and experiment, of interest in the functional significance of morphological facts as we find in the complementary relations of the contributions of Sutton and of Boveri. As Boveri wrote in his paper of 1904: "Sutton from the morphological side came to the same conclusions about the individuality [*Verschiedenwertigkeit*] of the chromosomes and on the nature of reduction as I from the physiological side" (p. 114).* In a similar way Weismann took the observations of O. Hertwig, Henking, Moore, and others and wrought them into a functional theory of the reduction division and the meaning of the tetrads. Although Weismann's theory contained a major misconception, it still portrayed an essential truth, that the meaning of the maneuvers of the chromosomes was to be sought in their genetic content.

Mendel's system was based on breeding experiments without reference to cytological observations, and nineteenth-century cytology operated without knowledge of Mendel's work. The conception of purely symbolic elements led the way in the elucidation of the meaning of the behavior of the visible bodies, the chromosomes.

The state of these questions and the intensity of the disputes about them on the eve of the revelation of the Mendelian system in 1900 were well set forth in the great summary of Yves Delage (1895) and more

* Original paper reprinted in Křiženecký (1965b).

succinctly by Wilson (1900). The initial steps toward the further development of the chromosome theory were also provided by genetic experiments which disclosed exceptions to Mendel's principle of independent assortment. Correns had already called attention to one such case in 1900,* in which flower color and hairiness in a flowering plant (stock) were inherited together, without recombination. He guessed, correctly, that the arrangement of the genes (*Anordnung der Anlagen*) before the segregation division might be decisive for the outcome in such cases, but since he found no exceptions to the association he did not discover incomplete linkage. Bateson and Punnett discovered this in 1905 by observing partially coupled inheritance of two different characters in the sweet pea, although their interpretation (reduplication before meiosis of the gametes with the coupled characters) proved to be wrong. Lock (1906) suggested that partial linkage might be due to location of the partially coupled genes in the same chromosome, but no proof of this was offered. The first evidence that this was the correct interpretation came from T. H. Morgan in 1910 (Morgan, 1914) and it led to the general interpretation. Morgan's theory was usually referred to as the "chromosome theory of heredity" although Weismann's, Boveri's, Sutton's, and several others equally merited the designation. Morgan referred to the developed form of the theory as the "theory of the gene" and this became the title of his definitive discussion of it in 1926.

C. Sex Chromosomes and Sex Linkage

An important preparation for the development of the chromosome theory was the discovery that one chromosome, the X, is normally responsible for sex determination in animals. This began with Henking's observation (1891)† of an unpaired element in a male bug. He was not certain that it was a chromosome, hence the designation "X." McClung later (1901)† showed it to be so, referred to it as the "accessory" chromosome, and suggested that it was sex-determining. The relation of the X chromosome to sex determination was clearly pointed out in 1905 almost simultaneously by Nettie M. Stevens and E. B. Wilson.‡ Both found cytological evidence, Stevens in a beetle and Wilson in a bug, that half of the sperm had an X chromosome and half either lacked it (in a bug) or had an unlike partner, the Y chromosome (in the beetle). There was no doubt, after that, that in insects sex was inherited in a Mendelian way

* Original paper reprinted in Křiženecký (1965b).

† In Křiženecký (1965b).

‡ Cf. Wilson (1928, p. 768).

by the segregation of an X chromosome. The idea that the sex difference depends on the genotypic constitution of the gametes which unite to form the new individual was expressed by Correns in 1903. In 1907 he gave the first experimental proof of it in his book on the inheritance and determination of sex. He showed that in dioecious species of the plant genus *Bryonia* male plants form two kinds of pollen with respect to sex transmission, female plants only one kind of gamete. He gave the first thorough account of the inheritance of sex as a Mendelian character. Later, sex chromosomes, like the X and Y of insects, were found to be responsible for sex determination in a liverwort, *Sphaerocarpus* (Allen, 1917) and subsequently in some dioecious angiosperms.

Sex-linked inheritance, the association of a genetic trait with the segregation of a sex-determining element, was discovered by Doncaster and Raynor in 1906* in a moth. Here, however, it appeared that two kinds of eggs were formed, one transmitting the sex-linked trait, the other not, while all sperm were alike. Some confusion was caused at first by the fact that opposite forms of sex linkage and sex determination were found in different orders of insects. This was cleared up by a combination of cytological and genetic investigations which showed that in birds (Durham and Marryatt, 1908; Goodale, 1909)*, as well as in Lepidoptera, males transmit sex-linked traits to both sons and daughters, but females only to sons. Males could thus be homozygous for sex-linked traits and "homogametic," all sperm being alike, in respect to sex determination; females could only be heterozygous (hemizygous) for such traits and "heterogametic" in respect to sex determination. Doncaster (1910)* and Seiler (1913)* later showed that in Lepidoptera the female produces two kinds of eggs, one with and one without an X chromosome (at first called a Z chromosome) while all sperm are alike.

The opposite type of sex determination and sex linkage was proved first in *Drosophila* and later in several other animals including man. This was first revealed in *Drosophila* by the behavior in inheritance of a mutant character, white eyes, recessive to the normal red eye color. White-eyed females mated to red-eyed males transmitted the mutant trait only to their sons, while red-eyed females crossed to white males transmitted red to both sons and daughters in F_1, and in F_2 a ratio of 75% red to 25% white was found, the white-eyed flies being males. In explaining this, Morgan (1910)* assumed that males were heterozygous for a sex-determining "factor" X; ♀ XX, ♂ XO (later shown to be XY); and

* Literature in Morgan (1914).

that the gene determining the red-white difference was completely linked with the sex factor. The regularity of sex-linked inheritance in this case, as in that of moths and birds, left no doubt that sex in both cases was determined by the genetic constitution of the gametes, by the same mechanism in different species, in some of which it was the males which were heterogametic, in others, the females.

Soon other mutant genes showing the sex-linked pattern of transmission were studied in *Drosophila*. These made possible proof of these important points: first, that sex-linked genes are linked with each other, from which Morgan concluded that all are located in the X chromosome; second, they also show recombination with each other, indicating that linkage between sex-linked genes is not complete. For example, it was shown that females heterozygous for the two recessive sex-linked genes, white eye and yellow body (having received white and yellow together from the same parent) form four kinds of eggs, in unequal but symmetrical proportions: 49.5% (yellow body and white eye), 49.5% wild type (nonyellow, nonwhite), 0.5% yellow but not white, 0.5% not yellow but white; that is, 99% in the original or parental combinations, 1% in the new combination. Other pairs of sex-linked genes show different proportions of parental and recombinant types.

D. Crossing-Over

In explanation of these facts, Morgan invoked the observations of the Belgian cytologist, Janssens, who in 1909 had described the meiotic chromosomes of two amphibian species. He had found what appeared to be homologous chromosomes paired side by side in synapsis, each homolog split into two. Morgan wrote (1914, p. 94):

> Thus there are formed four parallel strands each equivalent to a chromosome—the tetrad group. At this time Janssens has found that cross unions between the strands are sometimes present. In consequence a strand is made up of a part of one chromosome and part of another. . . . A consequence of this condition is that the chromosomes that come out of the tetrad may represent different combinations of those that united to form the group. On the basis of this observation we can explain the results of associated inheritance. For, to the same extent to which the chromosomes that unite remain intact, the factors are linked, and to the extent to which crossings occur the exchange of factors takes place. On the basis of the assumption of the linear arrangement of the factors in the chromosomes the distance apart of the factors is of importance. . . . If the two factors lie near together, the chance of a break occurring between them is small in proportion to their nearness.

These simple mechanical assumptions became the basis for the chief major advance in genetics since Mendel. It is interesting that it was hints or suggestions from Janssens rather than proof of his chiasmatype hypothesis which gave the impetus to Morgan's theory of linkage and crossing-over. This in turn underlay the whole development of the theory of the gene. Janssen's hypothesis of exchanges between homologous chromosomes during meiosis must, by its nature, have remained without decisive proof until members of a pair of homologous chromosomes could be found, which by reason of a different structure or chromosome aberration in one member could be distinguished under the microscope.

Such heteromorphic pairs were used by Carothers (1913)* to provide cytological evidence of independent assortment of chromosome pairs.

However, proof of the association of cytological exchange with genetic crossing-over was not obtained until 1931. Nevertheless, in Morgan's hands chiasmatype became a "working hypotheses" in the most useful sense, since although never strictly proved, it "worked" as a guide in the planning of experiments and the formation of a general theory of the transmission mechanism of heredity which turned out to fit the facts from breeding analysis.

E. Mapping the *Drosophila* Chromosomes

The first breeding experiments with *Drosophila* had been carried out by W. E. Castle and his students between 1901 and 1905. They had worked out methods of rearing the small, rapidly breeding vinegar fly in quantity, chiefly for studying the effects of inbreeding. Morgan began his research in 1909, quickly found several mutant genes in *Drosophila,* and had the good fortune to be joined in 1910 by three Columbia University students, C. B. Bridges, A. H. Sturtevant, and H. J. Muller. Together they found and analyzed large numbers of mutant genes. By late 1911 data on recombination among several sex-linked genes were available; using recombination values for closely linked genes as estimates of distance, Sturtevant (1913) produced the first chromosome map. Mapping of the X chromosome was followed by the second linkage group (1912)†, the third (1913), and the fourth (1914)† thus quickly accounting for the four pairs of chromosomes of *Drosophila melanogaster.*

In 1913 Bridges showed that an exception to the rule of criss-cross inheritance of sex-linked genes (mother to son following the X chromosome in transmission) was due to exceptional behavior of the sex chromosomes. Mothers showing the exceptional behavior were found to have two X chromosomes and a Y, caused by nondisjunction of the X's, two

* Cf. Wilson (1928, p. 933).

† Cf. Sturtevant (1965).

having entered a single egg which was then fertilized by a Y-bearing sperm. All the exceptional facts could be accounted for by assuming that sex-linked genes were physically located in the X chromosome. The exact correspondence of cytological and genetic behavior provided "Proof of the Chromosome Theory of Heredity" the title of Bridges' definitive paper of 1916. Similarly, anomalous genetic behavior of groups of linked genes was explained in some cases by translocation of a piece of chromosome with its contained genes to a nonhomologous chromosome with which the translocated genes thereafter showed linkage; in others the order of genes in a segment of chromosome had become inverted, such as the order *a d c b e f* derived from *a b c d e f*. This changed the pairing pattern determined by affinity between homologous genes, so that gametes containing chromosome segments derived from crossing-over in the inverted region were incomplete or otherwise abnormal and failed to give rise to offspring. This loss of recombinant offspring had the appearance of crossover suppression associated with inverted regions. In these and other anomalies, such as duplication and deficiency, the breeding results were found to be accompanied by the chromosomal anomalies predicted by the theory that each chromosome consisted of a linear sequence of physical sites occupied by discrete genes which could change separately by mutation.

By 1915 the evidence provided by breeding experiments and cytological study of *Drosophila* was such that the group (Morgan, Sturtevant, Muller, and Bridges) chose as the title of the book they then published "The Mechanism of Mendelian Heredity." Their claim to have worked out the formal mechanism of the genes in the chromosomes was justified by subsequent work both with *Drosophila* and with other diploid animals and plants.

F. The Theory of the Gene

Morgan, in his final book on the transmission mechanism of heredity, published in 1926 as "The Theory of the Gene," summarized as follows the new principles added to those derived from Mendel (p. 25):

> We are now in a position to formulate the theory of the gene. The theory states that the characters of the individual are referable to paired elements [genes] in the germinal material that are held together in a definite number of linkage groups; it states that the members of each pair of genes separate when the germ-cells mature in accordance with Mendel's first law, and in consequence each germ-cell comes to contain one set only; it states that the members belonging to different linkage groups assort independently in accordance with Mendel's second law; it states that an orderly interchange—crossing-over—also takes place, at times, between elements in correspond-

ing linkage groups; and it states that the frequency of crossing-over furnishes evidence of the linear order of the elements in each linkage group and of the relative position of the elements with respect to each other.

In his reformulation of the principles of transmission genetics, Morgan made no mention of the location of genes in chromosomes; indeed, the principles were reflections only of the numerical results of breeding experiments. The theory was concerned with genes which could be dealt with as abstract units just as Mendel had conceived them, and Morgan's chief interest was in the construction of a logical system, consistent within itself and to be tested or justified by genetic analysis. Although Morgan, Sturtevant, Muller, and Bridges laid the groundwork for what came to be called cytogenetics, cytological observations always seemed to them to be ancillary operations, necessary to complete the description of the system in which the genes are organized, but not by themselves throwing new light on the essential nature of the gene. Those studying *Drosophila,* which was the chief source of support of the new theory from 1910–1925, utilized to the full the information on chromosome structure and behavior being gathered by others at that time. Any biologist would have been blind indeed who could not see that the behavior of the chromosomes in meiosis provided just the mechanism required for the operation of the principles stated above. In fact Morgan said something to this effect in the preface to the book of 1915 cited above:

We have made no assumption concerning heredity that cannot be made abstractly without the chromosomes as bearers of the postulated factors. Why then, we are often asked, do you drag in the chromosomes? Our answer is that since the chromosomes furnish exactly the kind of mechanism that the Mendelian laws call for; and since there is an ever-increasing body of information that points clearly to the chromosomes as the bearers of the Mendelian factors, it would be folly to close one's eyes to so patent a relation. Moreover, as biologists, we are interested in heredity not primarily as a mathematical formulation but rather as a problem concerning the cell, the egg, and the sperm (Morgan *et al.*, 1915, p. IX).

The wisdom of such an attitude has been witnessed time and again in the subsequent development of genetics. Bacteria, viruses, even filamentous fungi, which contributed so profoundly to the maturation of genetics after 1940, had to be studied without reference to a chromosomal mechanism. In these organisms the study of the physical basis was preceded by study of the method of reproduction and by the abstract analysis of the genetic system.

There was, of course, another reason why, in the work with the *Drosophila* species, genetic analysis always led the way. The meiotic chromosomes of *Drosophila* were notoriously difficult to study, especially in the female. It was only after the study of linkage began that the chromosome groups of *Drosophila melanogaster* were shown to be ♀, XX + 3 pairs of autosomes; ♂, XY + 3A. It was the painstaking work of Bridges (1916), in proving the agreement between the physical presence or absence of a specific chromosome with the presence or absence of specific genes, deduced from breeding evidence, which brought convincing proof of the location of specific genes in specific chromosomes. This also showed that *Drosophila* cytology was possible and led to proofs comparable to the above in cases of duplication, deficiency, triploidy, and the analysis of triploid intersexuality, and in the hands of Metz, Huettner, and others to the establishment of chromosome number, morphology, and behavior in several other species of *Drosophila*. Stevens (1908) and Metz (1914)* had shown that in *Drosophila* species and some other Diptera the chromosomes normally show "somatic pairing," the homologs lying near together and side by side. This was not only an aid to the identification and structural comparison of homologs, but assumed considerable importance when later applied to the interpretation of the giant salivary gland chromosomes.

G. Cytogenetics

1. *Salivary Chromosomes of Drosophila*

Drosophila's disadvantages, that is, small size and lack of detail in the meiotic chromosomes, quickly changed to great advantages as soon as the giant chromosomes of the larval salivary glands came to be studied. It was this study which put the capstone on the chromosome theory, by identifying the actual physical sites and arrangements of groups of genes.

The banded nature of the huge chromosomes of a dipteran larva had been mentioned by Balbiani in 1881 but it was not until Heitz and Bauer in 1933† showed that the long, wormlike banded filaments in the nuclei of the Malpighian tube cells of a dipteran larva corresponded to the haploid chromosome number that their significance for genetics was realized. Painter (1933) opened a new era in cytogenetics by showing that, in *Drosophila melanogaster*, the banding sequence in the salivary gland chromosomes corresponds to the sequence of the genes as determined by breeding methods. As Painter said (1933, p. 585):

> This discovery places in our hands, for the first time, a qualitative method of chromosome analysis, and once the normal morphology of any given element is known,

* Cf. Wilson (1928, p. 575).

† Cf. Sturtevant (1965, p. 75).

by studying chromosome rearrangements of known genetic character, we can give morphological positions to gene loci and construct chromosome maps with far greater exactness than has been heretofore possible.

The mapping of the salivary chromosomes of *Drosophila melanogaster* was soon taken up and extended by Bridges (1935), who had been chiefly responsible for the genetic maps (based on recombination) of this species. By this time the maps contained hundreds of genes with specified locations. In his first "salivary" maps Bridges recognized over 3540 bands in the four chromosomes, and in a few years this was nearly doubled. The concept of a chromosome thus became that of a highly differentiated linear sequence of thousands of smaller elements, chromomeres. This picture was quite different from that suggested by the condensed meiotic chromosomes with their few landmarks. The conclusion was not to be avoided that the bands were or contained the genes. In fact, the correspondence extended beyond that between sequence of genes and of bands in the genetic and salivary maps and led to the close pinpointing of single loci within one or two bands. It was at this time that the true meaning of chromosomes began to be expressed in new synonyms for the term "chromonema," which Wilson had introduced in 1896, to designate the chromosome thread. Now the terms "genonema" or "gene-string" were being applied to it. Koltzoff, the Russian biologist, was speaking of "les molécules héréditaires" (1939), meaning the genonemata.

2. *Cytogenetics of Maize*

While this picture of the chromosome as a complex linear structure of thousands of elements separable and recombinable by crossing-over was emerging from the analysis of the *Drosophila* chromosomes, parallel developments were occurring elsewhere. The plant most comparable to *Drosophila* in numbers of genes analyzed and located on linkage maps was Indian corn (*Zea mays*). Breeding observations on this plant are on record from the eighteenth century, and it had been domesticated and grown by American Indians in many genetically different races before the arrival of Europeans in the New World. Mendelian analysis of its characters was begun by de Vries (1900)* and by Correns (1899)* in connection with his studies of xenia, the effect of pollen on the endosperm of the seed that it has fertilized. The last sentence in Correns' paper of 1899 contained the first reference to Mendel by someone applying his theory to new results. In Correns' first paper on Mendel's rules (1900)† he said (p. 159): "The varietal hybrids of maize show identical

* Original paper reprinted in Křiženecký (1965b).

† Cf. Křiženecký (1965b).

behavior in all essential points." In de Vries' first paper reporting the rediscovery of Mendel's rules (1900)* he reported crosses made in 1898, showing that sugary kernels (sweet corn type) behaved as a Mendelian recessive to starchy kernels (field corn type).

Maize was actively studied by many geneticists and plant breeders in the first two decades of this century. The simultaneous analysis of the genetic system and the chromosomal mechanism came to center in the group gathered about R. E. Emerson at Cornell University in the early 1920's. In 1935 Emerson, Beadle, and Fraser published a summary of linkage studies in maize while Rhoades and McClintock (1935) summarized the cytogenetic studies. Over 400 gene loci had been identified and most of them had been assigned to one of the ten chromosomes, each one of which was shown to carry one of the ten linkage groups. Each chromosome was found to be morphologically distinguishable at the pachytene stage by a variety of topographical landmarks and chromomere sequences which made the maize pachytene chromosomes comparable to those then being described in the salivary gland nuclei of *Drosophila,* although with less detail. It was significant that proof of the concomitance of chromosomal and genetic crossing-over was given simultaneously in maize by Creighton and McClintock (1931) and in *Drosophila* by Curt Stern (1931).

Work with *Drosophila* and maize provided the chief proofs of the chromosome theory in all of its details. In this the two kinds of material complemented each other in important ways. In maize, major categories of gene effects could be analyzed which were not identifiable at all in insects. Thus by 1935, over 100 genes out of 400 then known in maize had been shown to be concerned with chlorophyll development. There was no counterpart to this demonstrated relation of genes to the synthesis of products of primary importance until the gene-controlled antigens and hemoglobins of mammals, beginning with man, began to be worked out some 10 to 15 years later. In maize the discovery and mapping of genes affecting the haploid cells of the gametophyte generation showed that such processes as pollen tube growth (often affecting selective fertilization) or the synthesis of specific carbohydrate reserves by one gene in single dose supplied clues to gene activities not available at all in animals, as did also the analyzed effects of many genes on the functioning of the chromosome mechanism itself. The known width of control of genes over biological processes was greatly extended by work with maize and other plants; and the basic similarity between the plant and the animal in the

* Cf. Křiženecký (1965b).

manner in which the controlling factors were arranged in the genome reinforced the conclusions drawn from different kinds of organisms.

Thus in both *Drosophila* and maize the primary principle of limitation of the linkage groups first became evident, that is, the correspondence of each group of linked loci with one visible chromosome pair. In maize this was extended largely by the work of McClintock to ten linkage groups, each associated with a chromosome pair cytologically identifiable in its extended pachytene state by landmarks such as knobs, succession of chromomeres, arm length, etc. When such landmarks were transposed, made deficient, or otherwise visibly altered by mutational events, the association of the cytological with the genetic (linkage) changes served to identify the segments containing specific loci. In maize, the association of a gene group with a whole visible chromosome was made rapidly and precisely by the occurrence of viable plants or tissues which lacked one member of one chromosome pair, in the monosomic, $n-1$, condition which Bridges had recognized in *Drosophila,* but which, when it affected the major autosomes II and III in *Drosophila,* failed to survive, as is usually the case in other animals.

3. *Datura*

The same kind of proof of the occurrence of specific groups of genes in specific chromosome pairs had also been provided by the work of Blakeslee and Belling on the Jimson weed, *Datura stramonium* (review in Blakeslee, 1936). In this plant there could be recognized by phenotype and by corresponding chromosome complement: haploid ($n=12$), diploid (2n), triploid (3n) and tetraploid (4n) plants, as well as plants monosomic ($2n-1$) and trisomic ($2n+1$) for each of the twelve chromosomes, and other more complex combinations for some of the chromosomes, e.g., $4n-1$, $4n-2$, etc. In *Datura* the first cases of "segmental interchange," that is, reciprocal translocations between arms of different chromosome pairs, were analyzed cytologically by Belling (1926) and the phenotypic effect of dosage changes due to addition or subtraction of specific chromosome segments could thus be recognized. In *Datura,* very few point mutations had been discovered, so that associations between chromosome arms and genes had to be inferred from changes in gross or microscopic morphology. Changed phenotypes appeared to be due to changes in the balanced relations of many genes in different specific chromosomes. In maize, however, the known location of many mutated genes in each chromosome pair made the proof of association much more precise and convincing.

A major effect of cytogenetic analysis in both maize and *Drosophila* was the proof that the genetic system was essentially a linear arrangement of particles identifiable both as abstract units and as parts of a physical system. The principle of specific pairing affinities between homologous particles derived from father and mother turned out to be a reliable guide not only to the interpretation of the behavior of chromosomal aberrations of all kinds but to better understanding of the processes of mitosis and meiosis. The key to solving problems of chromosomal behavior was to envisage them as strings of genes.

Cytogenetic analysis, which came to fruition in the 1930's, led off in two main directions. One was to a view of some of the mechanisms by which the system of genes and chromosomes had evolved; the other led toward what Stadler (1954), a leader in this field, referred to as the central problem of biology, namely, the physical nature of the living substance.

Some of the ideas on evolution derived directly from cytogenetical observations will be briefly pointed out here; those bearing on the physical nature of genic material will be discussed in a later chapter.

4. *Theoretical Consequences of Cytogenetic Analysis*

Bridges' (1919) discovery of duplication of a part of a linkage group in *Drosophila* was accompanied by the suggestion that this might offer a method for the increase during evolution of the lengths of chromosomes by addition of another set of the same genes which in the new position could mutate separately and thus acquire new effects or properties. Cytological proof of the occurrence of small duplications was obtained in 1935 by Bridges and by Muller (1951) from the examination of the banding patterns of the *Drosophila* salivary chromosomes. Similar sequences of bands found in different chromosomes often showed somatic pairing behavior like that between identical sections of homologous chromosomes. Bridges appealed to these "repeats" as evidence of a mode of increase in number of genes and of diversification among them. Muller's view (1951, p. 94) was that, "By a succession of such duplications followed inevitably by the differentiations, the genotype can become not only more compound but more complex."

A proved duplication of one section of bands at the locus of Bar-eye mutation in *Drosophila melanogaster* was shown by Bridges (1936) to be responsible for the "position effect" which Sturtevant had discovered by genetical methods. The juxtaposition of repeated sections in the same chromosome had been found to have a greater restricting effect on the size of the eye than the same sequences distributed on opposite homologous chromosomes. The new phenotypes Bar-eye (one extra set of bands) and Ultra-Bar (two extra sets) had been acquired not by change in a gene but by rearrangement of previously existing genes. The view of mosaic discontinuity reflected by the linear distribution of gene loci, like beads on a string, had been disturbed by a glimpse of a new kind of continuity in the chromosome and of functional interdependence among neighboring parts. Longer views of "the evolution of the genetic system" by a variety of rearrangements of the chromosome complement were outlined by Darlington in a book of that title first published in 1939 and in White's "Animal Cytology and Evolution" (1945). These provided both summaries of cytogenetics at its peak of development and outlines of applications of its principles to problems of evolution.

VI. Mutations and the Origins of Genetic Variety

The idea of mutation as abrupt change in organisms is much older than genetics. Stubbe (1965) traces the first careful description of a sudden heritable change in the form of a plant to the end of the sixteenth century. The Heidelberg apothecary Sprenger found in his experimental garden in 1590 a new cut-leaved form of the greater celandine, *Chelidonium majus*. To the species name he then added the varietal designation of the mutant: *foliis et floribus incisis*. It was described by other gardeners during the next century from specimens all of which appear to have been descended from Sprenger's sport. The variant strain apparently bred true from the beginning. This experience was repeated and recorded for other ornamental plants by gardeners and horticulturalists many times over the next three centuries.

By the time Darwin wrote his "Variation of Animals and Plants under Domestication" (1868) hundreds of instances had been noted in both plants and animals of what Darwin referred to as "single variations" which often became the starting points for new domesticated varieties. He distinguished these from the more extreme monstrosities which would have little chance to contribute to species formation in nature, although some sports appearing under domestication, such as a crest or feathered legs or bizarre feather forms in pigeons, had given rise to well-marked varieties.

Stubbe (1965) has in a sense brought Darwin's book on variation up-to-date, and in enlarging his first historical accounts of 1938 and 1963 has added new material on the origins of domesticated animals and plants in prehistoric times. The subject has clearly been of great interest to gardeners and husbandmen as well as to scientists and has generated an enormous literature.

Glass (1947) has pointed out that Maupertuis in 1751 had proposed an explanation for the origin of the hereditary variations found in domestic animals and in man. This came very close to the modern view of mutation as fortuitous or spontaneous unit changes which produce new combinations.

Much of the older discussion of the causes of the origin of variation, although not that of Maupertuis, tacitly assumed it to be due to the inheritance of acquired characters. The explanation commonly accepted by both amateurs and scientists (including Charles Darwin) until late in the nineteenth century, was that new hereditary characters were likely to arise through the direct influence of the environment (Zirkle, 1935, 1946). The undermining of this general belief, largely by the arguments of August Weismann, was an essential preparation for the investigation of other theories.

A. History of the Mutation Concept

As for the use of the term "mutation," that too has a long and, as Mayr (1963, p. 168) says, "a tortuous history which has not yet been properly presented." Mayr, however, made a good beginning to remedy this defect and traced some of the changes in the usage of this word. It was in literary use in European languages, through its Latin meaning of change of any sort, before its employment in the seventeenth century to denote any drastic change in form or animals or plants. Robert Hooke used the term about 1675 in describing sudden changes in fossil lines. It appears repeatedly in both zoology and paleontology over the next 200 years. In 1838 it was used by Spring as a change from the type of the species.* In his autobiography Charles Darwin wrote (p. 75), "I had become, in the year 1837 or 1838, convinced that species were mutable productions . . . ." Mutation was used more precisely and quite commonly in paleontology, following Waagen (1869), to denote sudden changes in phylogenetic lineages (see Mayr, 1963). Steinmann (1908) pointed out (unjustly, according to Stubbe, 1965, p. 188): "Nothing illustrates better the disregard which historical research has received at the hands of biologists, than the sad fact that the concept of mutation which Waagen set up in 1867 . . . and which is illustrated in every textbook of paleontology, has recently been used by a botanist for an essentially different phenome-

* See Mayr, 1963, p. 168.

non." The offending botanist was, of course, Hugo de Vries, who adopted the term for sudden genetic changes, typified by the heritable departures from type which he had observed in the evening primrose, *Oenothera lamarckiana.*

1. *The Mutation Theory of Hugo de Vries*

De Vries' usage was quite in line with the sense in which biologists had used the term, and he could hardly be accused of neglect for failing to follow the paleontologists, although he never did refer to their usage. On the other hand, as it turned out, de Vries introduced a further confusion of usage, since the mutations each of which he regarded as a starting point of a new species proved to be a heterogeneous collection of several kinds of genetic differences from the standard type. Nevertheless, de Vries' great book, "The Mutation Theory" (1901–1903) was based on years of careful observation and experiment. Its basic idea of sudden change in the hereditary constitution by discontinuous variation proved to be valid and extremely useful. This established the term "mutation" in genetics even though his first examples were not, strictly speaking, mutations at all.

De Vries began his observations on *Oenothera lamarckiana* in 1886 when among a stand of this American species near Hilversum, Holland, he noted striking variations in the organs of certain plants. In 1887 he found two distinct forms, one with a short pistil (*O. brevistylis*) and another with narrow petals (*O. laevifolia*). When bred in his Amsterdam garden both forms bred true; and in the succeeding years many other true-breeding variant forms were found. This led de Vries into experimental plant breeding, to the invention of pedigree methods like those used by Mendel, and eventually to the rediscovery of Mendel's principles. Thus at the very beginning of the experimental study of mutation, it served purposes wider than the investigation of sudden hereditary change, just as later in the hands of Muller, Stadler, Timoféef-Ressovsky, Demerec, and others, it became a powerful tool for analyzing the nature of the gene.

Before 1889, when he wrote his prophetic book on "Intracellular Pangenesis," Hugo de Vries had reached the view that species characters are composed of more or less independent unit factors. It was abrupt change in one of these factors, with resultant change in combinations of them, which he supposed to occur by mutation and thus to give rise to a new species.

Needless to say, this view was anathema to those biologists who had accepted Darwin's theory of evolution by natural selection acting on small continuous variations. De Vries did not convince his opponents, but the failure of his chief evolutionary thesis is of less importance now than the success of the efforts which he initiated to analyze the results of mutation.

What set the study of mutation in motion as a central problem of genetics was de Vries' recognition that the origin of variations could be subjected to systematic study. It was no accident that Mendelian genetics and the scientific study of mutation both date to the same year 1900, nor that one man was so largely responsible for both innovations.

In the event, most of the *Oenothera* variants analyzed by de Vries and his many followers proved to be due not to a new change in a gene, which was to become the commonly understood meaning of mutation, but to one of several different genetic processes. The elucidation of these by Renner, Cleland, Belling, and others during the clearing up of the "*Oenothera* puzzle" between 1900 and about 1930 was one of the chief legacies of the work which de Vries had begun.*

Only two examples of this will be given here. One of the peculiarities of several of de Vries' *Oenothera* mutants was that they reappeared in low frequency season after season. This raised the question whether they might be due to segregation or recombination from a hybrid, kept rare because of factors interfering with viability or with the frequency of recombination. Otto Renner, beginning in 1917,* showed that most of the *Oenothera* species from which mutants arose are very complex hybrids. Usually, however, they produced only two types of gametes; and since in self-fertilization only one phenotype, that of the heterozygote, survives, each type of gamete could be assumed to carry a gene which was lethal when homozygous. An essential clue was Muller's discovery (1918) of the first balanced lethal system in *Drosophila*. In this case two different nonallelic lethal genes on homologous chromosomes were prevented from crossing-over by an inversion. Consequently homozygotes for each lethal die, leaving only heterozygotes to survive plus an occasional recombinant which had lost one of the lethals by a rarely occurring crossover. The latter showed a different phenotype and therefore appeared as rare exceptions. Muller suggested that *Oenothera lamarckiana* would behave like a permanent hybrid breeding true as a balanced lethal system, if crossing-over of all genes were prevented. Cleland in 1922 showed that complete linkage of all characters in "complexes" was due to their location in a complement of seven chromosomes which formed rings at meiosis rather than associating as bivalents in synapsis.

* For a more recent review of this work and references to original literature see Cleland (1962).

Crossing-over between homologs in the ring was thus usually prevented and the complex hybrid bred true except for the rare cases in which a recombinant, which de Vries had called a mutant, survived. Belling's hypothesis (1922) of segmental interchange or reciprocal translocation of chromosome arms, which he had shown to lead to formation of rings at meiosis in *Datura,* suggested the mechanism by which ring formation and suppression of recombination in *Oenothera* had come about (cf. Cleland, 1962).

Others of de Vries' mutants proved to be due to duplication of whole chromosome sets as in *Oenothera gigas,* a tetraploid, or to be heteroploids such as $n + 1$ or $n - 1$. But one of the original mutants, *brevistylis,* turned out to differ by a single gene from the type form from which it had arisen. It was a true gene mutation of the kind recognized subsequently in most other animals and plants when they were inbred under observation. It was these to which appeal was increasingly made as the source of the new variations on which evolutionary change would chiefly depend. It is important to recognize that de Vries used the term "mutation theory" to describe his views of the relation of suddenly occurring, discontinuous, heritable variations to the origin of species, and thus to evolution. The theory was not intended as an explanation of the mechanism of origin of mutations, apart from the general notion with which he had begun. This had been described in his book of 1889 "Intracellular Pangenesis," which developed the thesis, in a speculative way, that species were composed of assortments of individual units, pangenes, which replicated at each division so that each kind of pangene as a rule was transmitted to each descendant cell and organism. Heritable variations were seen as due to "the formation of new kinds of pangenes" and to "altered numerical relation of the pangenes already present" (p. 74, English translation of "Intracellular Pangenesis," 1910). "The pangenes are not chemical molecules but morphological structures, each built up of numerous molecules. They are life units, the characters of which can be explained in an historical way only" (p. 70). De Vries was satisfied that when he had proved the existence of units which segregated according to Mendel's rules, he had thereby verified the hypothesis of pangenesis from which he had originally derived his mutation theory. Apart from this general view he had no special theory of the physical mechanism by which mutations occurred.

The important idea contributed by de Vries was that changes occur in hereditary elements and that these changed elements may be studied by means of breeding experiments. It was this view that influenced later research on mutation and not his demonstration of the discontinuous nature of variations. Bateson (1894) had emphasized discontinuous variation and Korshinsky ("Heterogenesis und Evolution," 1899) had based a general theory on it (vide Stubbe, 1965).

2. *The Pure Line Theory*

After 1900, the study of mutation as begun by de Vries became an essential part of genetics. A more precise meaning was given to it by Johannsen's method of distinguishing experimentally between nonheritable modifications and variations due to genotypic differences. This was made clear in his concept of the "pure line" (1903), meaning the descendants by self-fertilization of a single individual, hence all genetically identical except when a new mutation should arise. Normal transgressive variation among members of a pure line was shown to be due to nongenetic causes. Jennings soon duplicated Johannsen's results with a protozoan as did Woltereck with a crustacean.

Erwin Baur further developed this view in his experimental studies of modifications (nonheritable) and mutations, many of them with small but discontinuous effects in the snapdragon *Antirrhinum majus*. Baur's extensive analyses of mutations in this plant were published in his textbook of 1911, and in the revised editions (1914, 1919). A chief point was that mutations in both germinal and somatic tissues could be recognized and proved to be due to changes in genes producing new alleles; and that large size or conspicuous effect was not, as de Vries had thought, an essential feature of a mutation since small ones also occurred.

B. Mutation in *Drosophila* and Maize

Many mutations were found and analyzed in *Drosophila* (several species). Indian corn (*Zea mais*) provided many mutants including the first example of a mutable gene, analyzed by Emerson (1914). This led the way toward the proof by Emerson, Stadler, Demerec, Rhoades, and others that different genes have different normal rates of mutation. In the extensive populations observed from controlled matings in these and other species, the usual cause of the appearance of unusual variants, namely, segregation and recombination of hidden recessives already present could be excluded by previous long-continued inbreeding. In the absence of overt environmental agents that could be associated with the changes, the newly arisen gene forms came to be referred to as "spontaneous" in origin.

In addition to this central category of allele or "point" mutations which could be assigned to individual loci on the linkage maps, the breeding experiments, especially with *Drosophila* and maize in which both linkage maps and chromosome maps were first developed, revealed many other kinds of changes in the genotype. These were also referred to as mutations in the broad sense. They included newly arisen changes in the chromosome complement from 2n to 3n, 4n, etc., to 2n — 1 or 2n + 1, etc., as described earlier (p. 32); changes within chromosomes, such as inversions of gene order; deficiencies of a part of a chromosome; translocations between chromosomes; duplications and other forms of chromosomal aberrations. It became evident, however, especially from the extensive observations on the *Drosophila* species, that the essential event which produced new hereditary variations was the origin of a new gene and attention was centered on this category.

C. Induction of Mutations

As soon as it seemed probable that mutation was the ultimate cause of hereditary variety, investigations began concerning the causes of sudden changes in the genotype; they took the form of attempts to induce mutations by physical and chemical means. De Vries had proposed the mutation theory at a time when the penetrating character of radium emanation (discovered in 1898) was under active discussion; it was perhaps not surprising that he should predict in 1901 that a planned artificial induction of mutations might produce new varieties of cultivated plants and animals. This served at least to indicate de Vries's view of mutation as an event subject to external physical influences. He became more definite in 1904 in suggesting that "the rays of Roentgen and Curie" might be used in such attempts. C. S. Gager* took up the suggestion in 1908,† treating plant ovaries with radium salts, and D. T. MacDougall did the same in 1911.* Although changes in genotype appeared to follow the treatments, the possibility could not be excluded that the material treated was heterozygous. Similarly, heritable abnormalities in the descendants of mice treated with X rays were reported in 1923 by Little and Bagg, but the causal connection between genotype change and the external agent could not be proved.†

1. *Muller's Analyses*

The problem was placed on a new basis by H. J. Muller's convincing

* Literature and references in Gustafsson (1963).

† This was not the first use of X rays in attempts to induce changes. In 1901 Aschkinass and Caspari had already found an induced heritable change in a bacterium.

proof (1927) of increases of many hundredfold in the mutation rate of *Drosophila melanogaster* following treatment with X rays. This proof, which marked a major turning point of genetics toward analysis of the molecular basis of heredity and variation, had been made possible by the application by Muller and Altenburg of methods of detecting a numerous class of mutations, those with recessive lethal effects. Their experimental results gave the first evidence that the frequency of mutations increased with increase in the temperature at which the flies were maintained.

It was observed also that when a mutation occurred in a gene, the other gene of the same type, its allele or homolog, closely adjacent in the same cell, did not mutate. This suggested, in Muller's words, that "mutations ordinarily result from submicroscopic accidents, that is, from caprices of thermal agitation that occur on a molecular or submolecular scale" (1947, p. 261).

Shortly after Muller (1927) had described his first proof of mutation induction by X rays, Stadler (1928) provided similar proof of the induction by X rays of mutations in barley, maize, and other plants. In the case both of Muller's and of Stadler's experiments, success was due to exclusion of causes other than radiation in the mass production of mutations, to the invention of special breeding methods, and to previous measurements of the natural or spontaneous mutation rate in the species treated. This had been done with especial thoroughness and ingenuity by Muller in experiments begun in 1918. His first method, devised for large-scale observations, was to identify a chromosome such as the X in *Drosophila* by a dominant marker gene accompanied by a crossover suppressor and a lethal gene. This chromosome was then brought into combination (by crossing) with a homologous X from a male. Mating such a heterozygote to any male results in the death of half of the sons, those that got the marked X chromosome with its lethal from the mother. If the other maternal X had acquired a lethal gene by mutation, sons receiving that X would also die and in such cultures no sons at all would be found. Thus from such tests the frequency of cultures containing no sons was a measure of the frequency of lethal mutations in the X chromosome. When X chromosomes from untreated males were compared in a large series of such tests with those from fathers exposed to a measured dose of radiation, the latter showed a mutation rate drastically increased over that of the controls. Subsequently, the increase in mutation rate was found to be roughly proportional to the X-ray dosage to which the father had been exposed.

2. *Stadler's Work*

In barley, Stadler used a special method which depended on the fact that each of three stalks (tillers) bearing a flower and seeds arises from a separate primordium in the seed. A recessive mutation occurring in one primordium will appear only in the progeny from self-fertilization within that flower head; progenies from other heads served as controls. Stadler recorded some 800 mutations in seedlings descended from heads treated by X rays. The mutation rate was many times higher from these than from control heads.

Similar evidence was later provided by experiments with maize and other plants treated with X rays, radium emanation, ultraviolet light, and heat. Work with insects, mammals, and eventually with microorganisms confirmed Muller's original proof. By the early 1930's there was no longer room for doubt that mutation was a process occurring at some determinable rate under given normal conditions and that this rate could be greatly increased by application of additional energy of several different kinds. Mutation had all the earmarks of being itself a physical process.

Both Muller and Stadler stressed that the mutations obtained after treatment resembled those that had been observed to arise without treatment. Like the spontaneous mutations they belonged to several categories, both lethal and "visible" changes, usually recessive but some dominant, point mutations localized in gene loci that had previously been mapped in *Drosophila* and maize, chromosomal aberrations such as translocations, inversions, deletions, and other changes involving more than one break in a chromosome. By and large a rough proportionality was found by Timofeeff-Ressovsky and others between amount of radiation applied and response by production of mutations. This proportionality appeared to hold for all ionizing radiations—X rays, γ rays from radium, or from other radioactive elements such as cobalt-60—and to be independent of both time and intensity of treatment provided the total ionization effect (measured in r units) was the same. N. W. Timofeeff-Ressovsky, Zimmer, and Delbrück (1935) derived from these relations the so-called target or hit theory of induction of gene mutation by high-energy radiations. The change in the gene was assumed to be caused by a single ionization within a target or sensitive volume of molecular order of size. This was a refinement in physical terms of Muller's interpretation of gene mutation as an extremely localized accident in an elementary particle of gene substance, which thereafter replicated in the new form.

A well-documented history of mutation research including the early work on induction has been provided by Timoféeff-Ressovsky (1937) and by Stubbe (1938). This has been brought up-to-date in a succinct review by Gustafsson (1963). Induced chromosomal changes have been reviewed in the book edited by Wolff (1963).

D. Toward the Molecular Basis of Mutation

Important influences in the direction of understanding the molecular basis of mutation and of the genetic material were provided by Stadler and Sprague's (1936) study of the effects of ultraviolet radiation in inducing mutational changes in the pollen of maize. When monochromatic ultraviolet was used later, the effective part of the spectrum was found to be that in which lay the specific absorption of nucleic acid (Stadler and Uber, 1942).

In the 1930's Caspersson and his colleagues at Stockholm had been developing cytochemical methods for measuring the relative quantities of nucleic acid in different cell structures. Caspersson and Schultz (1938) applied these methods to the salivary gland chromosomes of *Drosophila* mutants in which rearrangements had caused juxtaposition of normal and heterochromatic regions. They thereby found the first evidence of a connection between nucleic acid synthesis and gene reproduction. These beginnings gave indications and prepared the way for the later identification of DNA as the genetic material.

In fact, the clue provided by the discovery of the action of ultraviolet light as a mutagen pointed in the direction of the present view that mutation is essentially a chemical change in a segment of a nucleic acid molecule. It was not responsible for the discovery that mutation involves substitutions or similar localized changes in the nucleotide base pairs within a gene, but it is a curious fact that it was more coincident with this view than results obtained from the intensive research on chemical mutagenesis. This was opened by Auerbach's work (Brink, 1967) on nitrogen mustards and other mutagens done during World War II. The research of this sort during the next few years showed that the sensitive volume refinement of the target theory could be used also in interpreting effects of chemical mutagens and of other agencies affecting the presence and local action of free radicals in causing change in a gene. But it did not lead to the understanding of the chemical structure of the genetic material which was reached in a few short years of research with microorganisms, guided by a model of the structure of DNA.

The experimental study of mutation induction had been undertaken by Muller as a means of studying the nature of the genetic material. It clearly changed the course of the growth of genetics in the direction of physical and chemical methods and ideas and in this sense was a cause of the development which culminated in present views of the basis of the gene in DNA structure. Without the new facilities provided by combined biochemical and genetic research with microorganisms, mutation research by itself probably would not have solved the problem.

VII. Mendelism in Populations

The recognition from the results of breeding experiments that heredity is transmitted by discrete particles—genes—brought with it the first glimpse of the dynamics of gene distribution among the descendants. Mendel realized that certain automatic consequences flowed from the operation of the segregation principle by which members of pairs of alleles such as *A* and *a* always entered different gametes. He worked out these consequences for the special condition of reproduction by self-fertilization found in peas. Since *Aa* plants would always produce 25% *AA*, 50% *Aa*, and 25% *aa*, of which *AA* and *aa* would produce only their likes, it was easy to see that the proportion of *Aa* would be halved in each generation. Mendel showed that the descendants of a plant *Aa* should consist of $(2^n - 1)AA + 2Aa + (2^n - 1)aa$ in which n was the number of generations of self-fertilization from *Aa*. Consequently such a hybrid would automatically give rise to two pure lines *AA* and *aa* with a proportion of hybrids rapidly reducing to very low frequency. Mendel cited as example the frequency of *Aa* in the tenth generation of self-fertilization as 1/1024 or less than 0.001, 0.999 being homozygous. This became the basis of the later interpretation of the effects of inbreeding. It should be noted that, although the proportion of genotypes changed as inbreeding proceeded, the proportion of alleles remained constant—0.5*A* and 0.5*a*.

The automatic and symmetrical character of this system of segregation implied the important corollary of genetic equilibrium in cross-fertilizing populations such as those of most higher animals including man. The results to be expected from the operation of Mendelian heredity in interbreeding populations were first adumbrated in 1902 by the English statistician G. Udny Yule, who showed that if an F_2 population consisting of 25% *AA*, 50% *Aa*, and 25% *aa* interbreeds at random, the three genotypes will be present in the same proportions in subsequent generations. Castle (1903) (vide Li, 1967), in correcting an error in Yule's paper concerning the effects of selection (elimination of *aa*), confirmed the predicted constancy of genotypes and showed that when selection against *aa* ceases, the proportions of genotypes which then exist will continue in stable equilibrium. This statement was derived in algebraic form by Karl Pearson in 1904 for the special case in which *A* and *a* are present in equal frequency as in an F_2 population.

A. Hardy-Weinberg Relation

The full generalization, which we now know as the Hardy-Weinberg formula, was reached independently in 1908 by the English mathematician G. H. Hardy and the German physician Wilhelm Weinberg. They showed, as extension of Pearson's formulation, that in a large (infinite) population interbreeding at random with respect to genotype, the proportions of genotypes should be constant. With frequencies p for A and q for a and $p + q = 1$, the relation is:

$$p^2\ AA + 2\ pq\ Aa + q^2\ aa = 1$$

In addition to large population size and random mating, additional conditions for stability of genotype proportions (allele frequencies) were given by: no preferential mutation, i.e., if there is mutation, then changes from *A* to *a* must equal those from *a* to *A*; no selective advantage of any of the three genotypes, i.e., effect of *A* and *a* on fertility, viability, etc., should be equal; and no differential migration of genotypes into or out of the population.

Since the above are conditions for a steady state, it is apparent that a departure of any one of them in a given population will open the possibility for gene frequencies to change. For this reason accidents of sampling in small populations (random drift), preferential mutation, selection, and differential migration have been referred to as evolutionary forces.

The Hardy-Weinberg relation has often been referred to as Mendelian equilibrium, but it should be observed that what it expresses is not equilibrium as used in physics, but rather the random state of a system without implications concerning the action of forces upon it.

B. Application to Evolution Theory

The theoretical effects of variations in the parameters cited above were studied during the 1920's by J. B. S. Haldane (cf 1932) on the effect of selection, by R. A. Fisher in his book, "The Genetical Theory of Natural Selection" (1930), and by Sewall Wright (1931), who brought together into one formula mutation rate, natural selection, migration, and population size as determinants of change in gene frequency and hence of evolution.

While these theoretical foundations of population genetics were being developed, a beginning was made in the comparison of gene frequencies actually found in natural populations with those predicted by earlier models such as that of Pearson. This early observational work on populations of *Drosophila* in nature was carried out in the U.S.S.R. by S. S. Chetverikov (1927). His work was continued by Dobzhansky, Serebrovsky, Timoféeff-Ressovsky, Dubinin, and others. The effect of these studies was to reveal that natural populations of *Drosophila* species regularly retain a great store of hidden variability originating by mutation and maintained by the integrity of the genes responsible and the tendency toward the equilibrium emphasized by Hardy and Weinberg.

C. Polymorphisms

Drosophila species were not exceptional in this respect. Ludwik and Hanka Hirszfeld had already reported in 1918 (see Boyer, 1963, p. 32) that the blood of all human races examined revealed the presence of the four gene-determined human blood groups A, B, AB, and O, the genes *A* and *B* being present in different proportions in different races. Later studies, summarized by Mourant (1954), have shown that this type of genetic polymorphism, the maintenance of several genotypes within the same population, is characteristic of all the numerous human blood-group genes and indeed of most normal human genes.

The presence of such genetic polymorphisms in most of the cross-fertilizing species examined was emphasized by E. B. Ford, who restricted the term polymorphism to describe the mutual occurrence in the same habitat of two distinct forms of the same species in such proportions that the rarest of them cannot be maintained by recurrent mutation (Ford, 1953).* The term "morphism," introduced by Huxley (1955),* described the more general situation in which alleles of major genes (like the blood-group genes) are common in a population maintained by natural selection. Further discussions of polymorphism as an evolutionary mechanism have been given by Ford (1964), Sheppard (1958), and Dobzhansky (1962).*

---

* Literature references in Ford (1964).

D. Population Genetics

The concepts of gene frequency and of the gene pool representing the collection of alleles which circulates through a Mendelian population made it possible to conceive of evolution as changes in the relative frequency of genes due to interactions between the forces tending to upset the Hardy-Weinberg relation. Dobzhansky brought these views to a focus in his book of 1937, "Genetics and the Origin of Species." This book, following those of Fisher (1930), Haldane (1932), and Wright's "Evolution in Mendelian Populations" (1931), marked the reconciliation of Mendelian heredity and Darwin's theory of natural selection and thus the emergence of modern evolutionary biology.

The Hardy–Weinberg principle also made it possible to test genetic hypotheses in populations like those of man, which could not be subjected to experimental control. It was first used for this purpose by Wright (1917) and by Bernstein (1924). They argued that if the proportions of phenotypes in a population interbreeding at random were those called for by the relation:

$$p^2 + 2\,pq + q^2 = 1$$

then it could be assumed that they were generated by gene segregation. In this way Wright disproved an hypothesis about eye color inheritance while Bernstein first showed that a three-allele system responsible for the four phenotypes of the human ABO blood group fitted the population distribution better than the previous assumption of two independent pairs of alleles.

The discovery that most interbreeding populations are polymorphic for a variety of gene and chromosome variants made it possible to investigate the causes of maintenance of such intrapopulation variety. Muller's analysis (1918) of a case of balanced lethal factors in *Drosophila* pointed to an important cause, namely, the superior fitness of the heterozygote. If *Aa* is superior to *AA* or the mutant homozygote *aa*, natural selection will cause the mutant allele to be retained in the population, even though *aa* is poorly adapted or even lethal. Dobzhansky (1951) has reviewed cases of this sort in which the superiority of the heterozygote has been proved experimentally and has been found to be responsible for the retention of inferior homozygotes in populations bred in confinement.

An important result of the study of Mendelian equilibria has been the recognition that genetic organization is a property of populations and species as well as of individuals. The number of possible combinations of gene forms that can be maintained in Mendelian populations is much larger than the number found in any population at any one time. This has been taken to mean (vide Dobzhansky, 1962, pp. 295–296) that there exist in normal populations a great variety of genotypes, coadapted to each other and to the environment, and that this balance or equilibrium is both the result of evolution and the condition for further evolution.

VIII. Cytoplasmic Inheritance (Non-Mendelian or Nonchromosomal Heredity)

Questions concerning the relative influence of nucleus and cytoplasm in determining the characters of the offspring arose as soon as it had become probable, in the 1880's, that the nucleus was the vehicle by which the idioplasm, the genetic material, was transmitted. The questions were of importance both for theories of heredity and of development, but these were usually not sharply distinguished. Many proposals attributing the phenomena of heredity and development to living units had been considered in the nineteenth century. Yves Delage, in a comprehensive review (1895), described a large number of such theories. To this division of theoretical biology he gave the name micromerism; it occupied about a third of his extensive treatise of 900 pages. The particles deduced toward the end of the century, such as those of Weismann, tended more and more to be located in the nucleus as cytological knowledge increased. Particle systems invented earlier, like Darwin's gemmules, were usually conceived as extranuclear since they were assumed to arise from cells and to circulate in tissues and body fluids.

The most modern-sounding view was that developed by Hugo de Vries in his "Intracellular Pangenesis" of 1889, for his pangenes were nuclear units, copies of which got into the cytoplasm and thus became responsible for both heredity and development. They did not, however, propagate independently in the cytoplasm and that has become a criterion increasingly used to distinguish nonchromosomal from chromosomal particles.

A. Wilson's Views of Nucleus and Cytoplasm

The status of these questions was also reviewed in 1896 by the cytologist E. B. Wilson in his influential book "The Cell in Development and Inheritance," but Wilson put the emphasis on facts rather than theories. In the second edition (1900), written before the rediscovery of Mendel's principles, Wilson concluded, after a thorough review of the evidence, that the cytoplasm was the seat of the energy-liberating (destructive) metabolism while the nucleus originated the synthetic processes in which "the phosphorous rich substance known as nucleinic acid plays a leading part . . ." (p. 358). Further, "the role of the nucleus in constructive metabolism is intimately related with its role in morphological syntheses and thus in inheritance; for the recurrence of similar morphological characters must in the last analysis be due to the recurrence of corresponding forms of metabolic action of which they are the outward expression" (p. 359). And finally, on the same page, he gives his view which was the one entertained, with some notable exceptions, most generally thereafter: "Considered in all their bearings, however, the facts seem to accord best with the hypothesis that the cytoplasmic organization is itself determined, in the last analysis, by the nucleus; and the principle for which Hertwig and Strasburger contended is thus sustained."

After 1900, questions of nuclear and cytoplasmic inheritance took a more precise form as they concentrated on the transmission mechanism of heredity to which Mendel's rules applied. There was little doubt thereafter that elements transmitted in the nucleus constituted the major system of inheritance, but doubts remained whether hereditary transmission was exclusively vested in the nucleus. The work which threw the most light on this question was that in which specific characters were studied by the methods of experimental breeding.

B. Non-Mendelian Heredity

The study of non-Mendelian heredity was begun in the year following the rediscovery of Mendel's rules by one of the rediscoverers, Carl Correns. A biographical notice of Correns (Stein, 1950) tells us that he began his study of variegation in plants in 1901 but published nothing on it until 1909, when he gave a thorough description of the non-Mendelian inheritance of plastid differences in plants with green and white (or pale yellow) variegation or striping. The inheritance of this condition was strictly maternal, the pollen having no effect on the progeny: flowers on green branches gave green progeny, those on white (or pale) branches gave white progeny, while variegated branches gave green, white, and variegated offspring in irregular ratios. This same non-Mendelian kind of inheritance of green and white spotting (*status albomaculatus*) was found in other plants and was assumed to be due to elements transmitted in the cytoplasm, either the plastids themselves, or other cytoplasmic bodies occurring in two self-replicating forms which determined the development of normal or abnormal plastids.

This clear-cut maternal transmission of determiners of abnormal chlorophyll development in plants was found by other investigators in other plants. Maternal inheritance thus became the clue to transmission outside of the nucleus. But not all cases of maternal effects were found to result from transmission of cell organelles such as plastids in the cytoplasm. Caspari (1948), in a thorough review of cases other than plastid transmission, distinguished two kinds of maternal effects from true cytoplasmic inheritance. These are, first, predetermination of egg or embryo characters by prior action of the genes of the mother; and second, environmentally induced cytoplasmic changes which become transmitted maternally for a number of generations but which gradually disappear. To the latter Jollos (1913), who discovered this effect in protozoa, gave the name Dauer modifications. True cytoplasmic inheritance, illustrated by the plastid differences described above, is characterized by constancy of the maternally transmitted trait in repeated backcrosses to a paternal strain lacking the character or by the exclusion of genic transmission by the substitution through backcrossing of all of the chromosomes of the strain which transmits the trait maternally, thus proving it to be nonchromosomal. At the time of Caspari's review some twenty cases, mostly in plants, met these criteria sufficiently to indicate the existence of a system of inheritance in the cytoplasm to which Wettstein in 1924 had given the name plasmon, meaning the cytoplasm as a hereditary system as distinguished from the genome, a name given by Winkler (1920, p. 165) to the system of genes in the chromosome. Evidences of more or less constant plasmon differences appear chiefly in crosses between local races, subspecies, and varieties of the same species of plants where they are revealed by the influence of a foreign plasmon on genes introduced into it.

In a few cases characters were found to be transmitted constantly only in the cytoplasm. Correns (1908) found such a case in the thistle, the cytoplasm of which allowed only female organs to develop even when combined with genes from a related species in which only hermaphroditic plants exist. Rhoades (1933) proved that a cytoplasmic male sterility in maize was transmitted independently of the genome since it was not changed by substitution, one at a time, of each of the ten marked chromosomes of male-fertile strains of maize. Proof of transmission of a specific property outside of the genome does not of itself constitute proof of cytoplasmic inheritance. L'Héritier (review, 1948) proved that the abnormal sensitivity of a strain of *Drosophila* to the effects of CO_2 was transmitted constantly when each of its four pairs of chromosomes were replaced by marked chromosomes from a normal strain. Subsequently, virus-like particles were shown to be present in the egg cytoplasm of sensitive flies and to be transmitted by transplantation. Similar infective behavior was found to characterize the elements responsible for the transmission of a cytoplasmic sex-ratio factor in *Drosophila prosaltans* and these were shown by Poulson and Sakaguchi (1961) to be in fact spirochetes of the genus *Treponema*. Other examples of particles which are transmitted outside the nucleus are the κ- or killer particles discovered in *Paramecium* by Sonneborn and shown by Preer to contain DNA. These have an infective mode of transmission but can replicate only in cells with a specific genetic constitution. Even closer similarity to a transmitted infection like that of a virus is shown by the so-called milk factor known to be responsible for susceptibility to one form of cancer in mice (Bittner, 1936).

IX. First Principles of Biochemical Genetics

A. Garrod's Hypothesis

The first steps leading toward a chemical interpretation of the actions of genes were taken even before Mendel's rules were rediscovered. In 1899,* the English chemical pathologist, A. E. Garrod, pointed out to the Royal Medical and Chirurgical Society of London that alkaptonuria, the black urine disease detected in infants and continuing throughout life, probably was due to an "abnormal metabolism in the tissues" which was congenital. The abnormality was failure to break down homogentisic acid which in alkaptonurics is excreted in the urine and darkens by oxidation. What connected this with genetics, however, was Garrod's discovery (1901)* that the normal parents of alkaptonuric patients were often blood relatives such as first cousins. Harris (1963), who has reviewed the history and reprinted some of the original documents concerned with such "inborn errors of metabolism," has pointed out that this crucial observation of parental consanguinity came just at the right time. Mendel's laws had recently been rediscovered and William Bateson quickly recognized the significance of Garrod's discovery. Bateson noted (1902, footnote of December 17, 1901)* that "the mating of first cousins gives exactly the conditions most likely to enable a rare and usually recessive character to show itself. . . . First cousins will frequently be

* For literature cf. Harris, 1963.

bearers of similar gametes which may in such unions meet each other, and thus lead to the manifestation of the peculiar recessive characters in the zygote." Garrod's paper of 1902 confirmed and amplified Bateson's interpretation and thus established the first case of recessive inheritance in man. He gave also evidence and reasons for his view that inherited biochemical variations of different kinds may in fact be common not only within a species ("chemical individuality" was his phrase) but could be related to the differentiation of species and to evolution. He subsequently attacked the questions "why alkaptonuric individuals pass the benzene ring of their tyrosine unbroken, and how and where the peculiar chemical change from tyrosine to homogentisic acid is brought about . . . ." His answer to this question in his Croonian Lectures of 1908 was "that the splitting of the benzene ring in normal metabolism is the work of a special enzyme, that in congenital alkaptonuria this enzyme is wanting, whilst in disease its working may be partially or even completely inhibited." He thus introduced a first and basic concept of biochemical genetics: a gene controls a chemical reaction by way of a gene-controlled enzyme.

G. W. Beadle tells us in a succinct account of biochemical genetics (1963) that he, like most other geneticists, was unaware of Garrod's work when in 1935 he, with Boris Ephrussi and later with E. L. Tatum, rediscovered and extended the gene-enzyme concept in experiments with *Drosophila* and *Neurospora.*

In the meantime, other suggestions appeared of enzymes as intermediaries between genes and such characters as coat-color differences in the mouse (Cuénot, 1903) and other mammals (Wright, 1917); the specific chemical reactions associated with mutant genes affecting flower pigments were demonstrated by a group of English investigators which included Onslow, Wheldale, Scott-Moncrieff, Haldane, and others (review in Lawrence and Price, 1940).

B. Gene-Controlled Synthesis of Eye Pigments in Insects*

The mainstream of advance of chemical concepts of gene effects was resumed when mutant genes affecting pigment development in insects were studied, first by Caspari (1933) in the flour-moth *Ephestia* and by Beadle and Ephrussi (1936) in *Drosophila.* By transplanting organs between larvae of different genotypes, diffusible substances (first called hormones) were recognized as gene-controlled. The essential observation in *Drosophila* had been made by Sturtevant (1920, in Caspari, 1960) when he found that the wild-type (normal) allele of a mutant gene which causes

* For review and literature see Caspari (1960).

the eye to be vermillion rather than the normal red produces something which acts like a diffusible substance. This caused a genotypically vermillion eye to develop normal pigment. Beadle and Ephrussi analyzed this phenomenon and discovered a sequence of two successive reactions concerned in the production of normal pigment. The first reaction was blocked by the vermillion mutant, the second by another mutant gene, cinnabar. Two steps in the synthesis of the brown component of the normal pigment were viewed as two different chemical reactions, each controlled by one gene through the mediation of an enzyme specific for that step. In diagrammatic form this can be represented:

Substrate ⟶ substance 1 ⟶ substance 2 ⟶ pigment

A block between 1 and 2 leaves the substance at the vermillion level, between 2 and the end results in cinnabar color. Since other genes investigated did not affect either of these reactions, it was assumed that each gene had one primary function. This was the origin of the "one gene–one enzyme" hypothesis, which served as a useful guide in subsequent work in biochemical genetics. Eventually the substrate was identified chemically by Butenandt and Weidel as tryptophan, substance 1 as kynurenin, and substance 2 as 3-hydroxykynurenin. It is of interest in considering the further development of biochemical and molecular genetics that the working scheme which led to the resolution of this synthesis was based on observations and experimental design which had a purely biological character, viz., use of mutant genes recognized by segregation, and simple tissue transplantation experiments. The reasoning used was of the operational kind employed in modern physics and recommended strongly (and used) by Stadler (1954) in developing key concepts in genetics.

Following the rationale devised for the *Drosophila* experiments, Beadle and his biochemically trained associate E. L. Tatum then turned to a somewhat simpler organism, the breadmold *Neurospora*, an ascomycete. The basic work on the reproduction of this haploid fungus had been done by Bernard O. Dodge (1927) while Carl C. Lindegren (1936) had worked out the genetics of several mutant genes and established the principles of gene location and mapping.

C. Syntheses in *Neurospora*

Tatum and Beadle (1941) worked out the nutritional requirements of *Neurospora crassa* and found that this mold will grow on a synthetic medium containing inorganic salts and biotin with sucrose or a similar source of carbon and of energy. The wild-type mold is able to synthesize all its other requirements such as vitamins or amino acids. They added to the standard or minimal medium as many vitamins and amino acids as could be obtained in a pure state. In this supplemented medium they placed single spores from parents which had been subjected to X rays or ultraviolet light to induce mutations. From each such culture they transferred conidia (asexual spores) to the minimal medium without any supplements. A mutant which could not grow on this medium could grow if the medium were supplemented with some specific vitamin or amino acid. In this way they identified mutants each of which required one specific substance for growth. Each such mutant was regarded as having a specific metabolic deficiency.

Beadle (1959, 1963) has reviewed the work with *Neurospora* and has demonstrated a large number of cases in which the synthesis of a vitamin or an amino acid was controlled by a single mutant gene. This convinced both geneticists and biochemists that the control of metabolism is vested in a network of reactions in which individual steps are controlled by genes which impart specificity to enzymes. Since proteins are made up of amino acids, the proof of genic control of amino acid synthesis seemed like a step toward understanding how genes could control the essential machinery of life.

Tatum (1946, in Beadle, 1959) soon found in the bacteria *Acetobacter* and *Escherichia coli* the same kind of mutants affecting essential syntheses as in *Neurospora*. This was the prelude to the discovery by Lederberg and Tatum (1946) of recombination and a form of sexual process in *E. coli* since Tatum's experiments had provided the essential genetic markers.

D. New Principles from Biochemical Genetics

The development of knowledge of reproduction and heredity in bacteria is discussed in a later section. Here it should be emphasized that the same biochemical mutants were found in almost every species of microorganism studied: bacteria, yeasts, algae, and fungi. Similar biochemical reactions were later identified in cell cultures from higher organisms including man. As Tatum pointed out in his Nobel Lecture (1959), the experiments with *Neurospora* led to four new basic concepts: "(1) That all biochemical processes in all organisms are under genic control, (2) that these overall biochemical processes are resolvable into a series of individual stepwise reactions, (3) that each reaction is controlled in a primary fashion by a single gene . . . , (4) mutation of a single gene results only in an alteration in the ability of the cell to carry out a single primary chemical reaction." The hypothesis underlying all this was that each gene controls the reproduction, function, and specificity of a particular enzyme. The first demonstration of the actual absence of an enzyme in an induced mutant was given by Mitchell and Lein in 1948.

These new principles, reached by about 1950, were far in advance both in breadth of application to biological and evolutionary processes and in the precision of the evidence supporting them of anything foreseen by the early Mendelians. Yet they took off from concepts and methods that had become standard in genetics, to which new dynamic ideas about genes as agents of synthesis were added. They had, of course, been aided in important and essential ways by the newer knowledge and methods of biochemistry. This was the apex of the main line of development of genetics of the classic kind. It also had the character of a watershed on the farther side of which lay the new molecular genetics.

The central idea of the older genetics was that of an element, the gene, which could be conceived and studied operatively, by its behavior in transmission and in controlling steps in a synthesis. The evolution of methods for analyzing the structure and functions of genes in molecular terms brought a new emphasis on the structural aspects of the elements and their activities. It was not that questions about substance were to displace those about behavior, although in the first flush of enthusiasm for the new, some biochemists seemed inclined to renounce or at least to overlook the old way of viewing heredity. This should not occasion surprise since biochemistry and genetics had grown up as parallel streams, largely independent of each other.

The normal orientation and allegiance of biochemistry was to human medicine. It came into contact with genetics in 1902 in the persons of A. E. Garrod and William Bateson but although an essential idea about genic control of enzymes was contributed by Garrod, it was not then followed up either by biochemists or geneticists. Another opportunity for convergence occurred when Gowland Hopkins developed a vigorous school of biochemistry at Cambridge University where Bateson had established the first research center in genetics. Bateson seems to have considered Hopkins as a competitor for talented students rather than as a potential collaborator, and he left Cambridge in 1910 for the John Innes Horticultural Institution. However, the brilliant group assembled around Hopkins had its eventual effect, chiefly by way of J. B. S. Haldane, through whom some of the fructifying effects of genetical ideas and problems later flowed toward students of biochemistry. What biochemistry has done for genetics has since become clearly evident.

E. Genetics and Biochemistry

Less attention has been paid to what genetics has done for biochemistry. One effect of genetics has certainly been to provide additional reasons and incentives for doing biochemistry. Sound quantitative work in biochemistry had, of course, developed without this impetus at a time when biology was almost entirely qualitative and descriptive. Our admiration for the splendid foundations laid both for biochemistry and for such a descriptive field as microscopic anatomy in the last half of the nineteenth and early twentieth century is enhanced rather than diminished by our recognition that in general the work was done without the tempting view of far horizons which became apparent after about 1940. To be sure, both biochemists and morphologists were aware of the bearings of their work on problems of phylogeny and taxonomy but not of the connections between structure at the molecular level and the causes of evolution and of differentiation in the individual organism. These have added new dimensions in time and depth to the outlook of biochemistry. The manner in which biochemical work could be brought to bear on these key problems of life had perforce to flow through a view of the mechanism of continuity and of change which was the prime contribution of genetics. In this sense genetics led the way conceptually as well as in the operational methods by which the means of continuity had been analyzed.

Apart from this general kind of interrelation, there are suggestions of more specific influences passing between genetics and biochemistry, but the direction of flow is difficult to diagnose. I have argued elsewhere (Dunn, 1965, p. 219) that the discovery of gene-controlled reaction sequences in advance of the identification of the reactants was a prime example of a genetic way of thinking to which new biochemical discoveries were a response. Proof that each separate step in a sequence was controlled by a specific enzyme was given at about the same time (mid-1930's) as the statement of the genetic theory. There can be no doubt that the new ideas as described by Tatum earlier in this chapter were basic contributions which led in a new direction in biochemistry. However, changes in this field occurred so swiftly in the years 1935–1941 that the tracing of mutual influences between biochemistry and genetics calls for special historical studies which are still to be made.

X. Tributary Streams Contributing to Modern Genetics

A. Classic and Modern

The motives impelling those who established the main line in genetics were directed toward understanding the gene as the unit of transmission, mutation, recombination, and function. This was their dominant interest, and most of them could be identified primarily as geneticists, usually deriving from botany or zoology or other recognized biological disciplines. But in looking back over the period to about 1860 other streams of advancing knowledge can be discerned which were not at the time directed toward problems of main line genetics, and some of them toward questions which only secondarily concerned biologists.

Many of the tributary streams of development that influenced the great expansion in genetics which began about 1940 can be traced to nineteenth-century origins. Their beginnings were independent of the genetic principles which first came to recognition in 1900. Yet such was the foresight or the breadth of view of a medically trained chemist like Friedrich Miescher, who first isolated a nucleic acid that even in 1871 he supposed it might have a genetic function. When, in 1944, another man of similar training, Oswald Avery, a chemical bacteriologist, proved that a nucleic acid was the hereditary substance, he, too, although an outsider in genetics, recognized the importance of his discovery better than most of his genetically trained contemporaries.

Such instances should warn us against the narrow view to which specialization so easily leads—that the rebirth of a science such as we are now witnessing in the case of genetics will arise solely from the efforts of its professional initiates. It is salutary to recall that Mendel and his first successors who began to build genetics in the first decade of this century were also at the time of their discoveries outsiders.

It is true, nevertheless, that work which began without specific genetic problems in mind provided ideas, stimuli, and especially methods that were essential to the rather unified theory which is at the heart of genetics today. In this theory, the particulate view of inheritance which was the central feature of the theories of Mendel and the Morgan school, has been transformed in order to accommodate the knowledge that the ultimate "particles" of heredity are pairs of nucleotides with only a few dozens of atoms, and that the linear order in which they are arranged in tens of thousands of molecules specifies the functions which they direct in metabolism and in biosynthesis. These functions in turn determine their influence on ontogenetic development and, as acted upon by natural selection, on the course of evolution. The recognition of the basic problems and much of the knowledge which made this transformation possible had been provided by the evolution of genetics in the first half of this century. It seems better not to refer to this as "classic" genetics since that tends to relegate it to a classic period and implies discontinuity with the present "modern" period.

In fact, even the gene as a symbol was not generally conceived as a mere "reckoning unit" (in Johannsen's rather puristic expression). Indeed, concern with the structure and function of the gene was a recurring theme, usually at a speculative rather than factual level. The new knowledge, largely chemical in nature, which was to precipitate the transformation, was simply not available at the earlier period. The chief causes of the transformation, discussed in section X, F, were the ability to manipulate microorganisms, especially bacteria, the attainment of powerful chemical methods which had themselves led to the development of biochemistry based on the work with bacteria and yeasts, and the refinement of cytochemical and cytooptical methods for dealing with the fine structure of chromosomes and other cellular organelles. This may remind us that the older genetics was organismal, dealing with whole multicellular animals and plants, and not with cells as self-sufficient units.

In any case one can discern vigorous growths in other sciences proceeding in advance of, or parallel to, the development of genetics. Of first importance was the growth of cytology directed toward the elucidation of fertilization, mitosis, and meiosis which provided the first rational view of the physical basis of reproduction in higher animals and plants. This began in the early 1870's but a crucial point was Oscar Hertwig's proof in 1875* of the union of one nucleus each from sperm and egg as the essential feature of fertilization. Some of the events of importance for genetics which occurred between that time and the establishment of the rules of meiosis around 1902 have already been referred to in Section V of this article; the period has been well described by E. B. Wilson (1928), by Arthur Hughes (1959) and by William Coleman (1965). Cytology and genetics have been and continue to be so closely interdependent in their developments since 1900 that it hardly seems just to treat either one as an external influence on the other (cf. Schrader, 1948).

Matters stand differently, however, with respect to other streams of development. I shall list a few to show what I have in mind.

1. The development of knowledge of the structure and functions of the nucleic acids from about 1868 when Miescher began his analysis of pus cells to about 1953 when a model of molecular structure and replication of DNA—since confirmed—was attained.

2. The growth of knowledge of protein structure from about the time of completion of the roster of the amino acids found in proteins in the 1920's through the crystallization of the first enzyme in 1926 and the

* See Wilson, 1928.

complete analysis of the first protein in 1955. To do more than point out this sequence and the most important feature (for genetics) of the polypeptide chains of proteins, viz., the linear arrangement of their constituents, would be beyond the scope of this account and the competence of the author.

3. The establishment of the mode of reproduction of those classes of microorganisms which provided experimental materials for investigations of biochemical genetics, the fine structure of the gene, the transforming principle, and the analysis of genic control systems and the genetic code.

4. The development of some methods of quantification and instrumentation which came to play important roles in genetics.

B. The Nucleic Acids

The chief events in the attainment of our present understanding of the structure and biological role of the nucleic acids have been concisely recounted by David Cohen (1965). The older work leading to the view of nucleic acids as polynucleotides was reviewed by Levene and Bass (1931). Study in this field began with the curiosity of a 24-year-old Swiss student, Friedrich Miescher,* concerning the chemical composition of nuclei. His interest in histochemistry had been stimulated by Wilhelm His, the great histologist and Miescher's teacher at Basel. As his first postdoctoral research in 1868 Miescher isolated from the nuclei of human pus cells a substance with high phosphorus content which he called nuclein. He regarded it as representative of a class of phosphorus containing compounds "on an equal footing with the group of protein compounds." Although he had established the elementary composition of this substance early in 1869, Hoppe-Seyler, in whose laboratory at Tübingen the work had been done, refused to sanction publication of the rather startling discovery until he himself and an assistant had verified Miescher's analysis. After Miescher became professor of physiology at Basel in 1872 at the age of 27 he extracted a substance from the sperm of Rhine salmon. Its nitrogen:phosphorus ratio (18.75:5) indicated a compound of a protein (protamin) and nuclein. When separated, nuclein had 9.6% phosphorus shown to be present as phosphoric acid. He soon isolated the same acidic compound from the sperm of other animals and suggested that nuclein was the active genetic material of the sperm.

By the time of Miescher's death in 1895 others had carried his work

* For an account of Miescher's discovery of DNA see Mirsky (1968); for bibliographic references see Cohen (1965) and Levene and Bass (1931).

forward. Piccard, on Miescher's advice, used different extraction procedures on sperm and in 1874 isolated the purine bases guanine and hypoxanthine (as a degradation product of adenine). Altman isolated nucleins from yeast and gave them the name nucleic acids in 1889. By 1900, adenine and xanthine and the pyrimidine bases, thymine and cytosine, had been isolated by Kossell, and uracil by Ascoli. Yeast nucleic acid was shown by Phoebus Levene in 1909 to consist of the four bases adenine, guanine, cytosine, and uracil; phosphoric acid; and a pentose sugar identified as ribose. Thymus nucleic acid was found by Levene in 1930 to contain adenine, guanine, cytosine, and thymine with phosphoric acid and deoxyribose sugar (cf. Levene and Bass, 1931). In the mid-forties the yeast and thymus varieties became known by common usage as ribose and deoxyribose nucleic acids, respectively, or more simply as RNA and DNA.

It was of great importance that at this time the unit of structure of the nucleic acids came to be recognized as the nucleotide consisting of a nitrogenous base (a purine or a pyrimidine) linked to a sugar, which was linked to phosphoric acid. Since two purines and two pyrimidines had been identified in both RNA and DNA, this meant that four kinds of nucleotide were possible. From the first analyses it was assumed that the four bases were present in equal amounts and the four nucleotides could be thought of as joined together in a higher unit, a tetranucleotide. The tetranucleotide hypothesis, a simplistic view of nucleic acid structure, did not survive the proof first provided by Chargaff's analyses (1950) that the bases were not present in equal amounts. Instead he showed that in DNA the amount of thymine always equaled the amount of adenine, and the amount of cytosine equaled the amount of guanine: $T/A = C/G = 1$. It was this set of equivalences which suggested the base-pair idea that formed a keystone in the model proposed by Watson and Crick in 1953.

In the meantime Feulgen and Rossenbeck (1924)* provided one means for the biological identification of the nucleic acids in cells by a reaction specific for DNA. This quickly led to the establishment of the localization of DNA in nuclei since the Feulgen reagent stained chromosomes.

Histochemical studies of tissue sections in the 1940's, led by Brachet's work with the enzyme ribonuclease, established DNA as localized in the chromosomes, RNA as mainly, but not exclusively, cytoplasmic. A discovery pointing to present views of the relation of DNA, RNA, and protein synthesis was made in 1940 by Caspersson and Schultz, and independently by Brachet,* when the presence of RNA in the nucleus

* For references, see Mirsky (1951).

(mainly in the nucleolus) was recognized. Caspersson and his collaborators noted a correlation between the appearance of RNA in the cytoplasm and its appearance in the nucleus. They supposed that RNA passes out of the nucleus into the cytoplasm but, in contrast to our present view, that the first RNA-associated proteins were synthesized in the nucleus. DNA, however, was fixed in the nucleus both in quantity and location, in the chromosomes. The amount of DNA was found to be constant for each haploid set of chromosomes in animal nuclei (Boivin and Vendrely, 1948).* Sperm nuclei and diploid, tetraploid, and octoploid nuclei in the same animal had quantities of DNA in the ratio 0.5: 1:2:4 (Mirsky, 1951). Measurements of DNA in plant cells showed a similar correspondence with chromosome complement over a wide range of ploidy, doubling before mitosis, and reduced by meiosis to half the diploid value in the microspores (Swift, 1950).

C. DNA AS THE TRANSFORMING PRINCIPLE†

The chief advance in understanding the biological role of the nucleic acids came from an unexpected source. It is so well and clearly described in a letter written May 17, 1943, by Oswald Avery to his brother Roy that, through the courtesy of Professor Roy Avery of Vanderbilt University, I reproduce the entire letter except for introductory personal matter. Oswald Avery had just retired as a member of the Rockefeller Institute and had intended to join his brother in Nashville. The men referred to in the first line were Herbert Gasser, director of the Institute, and Thomas Rivers, director of the Institute Hospital. Ernest in the third sentence, was Avery's old friend, the pathologist Ernest Goodpasture, then chairman of the Division of Medicine of the National Research Council. Avery wrote:

Dr. Gasser and Dr. Rivers have been very kind and have insisted on my staying on —providing me an ample budget and technical assistance to carry on the problem that I've been studying. I've not published anything about it—indeed have discussed it only with a few—because I'm not yet convinced that we have as yet sufficient evidence. However, I did talk to Ernest about it in Washington and I hope he has told you first of all—I felt he should know because it bears directly on my coming eventually to Nashville. It is the problem of the transformation of pneumococcal types.

You will recall that Griffith in London some fifteen years ago described a technique whereby he could change one specific type into another specific type through the intermediate R form. For example: Type II → R → Type III. This he accomplished

* For references, see Mirsky (1951).

† For a comprehensive review of transformation as mediated by DNA, cf. Ravin (1961) which contains references to literature cited in Section C.

by injecting mice with a large amount of *heat-killed* Type III cells together with a small inoculum of a *living R* culture derived from Type II. He noted that not infrequently the mice so treated died and from their heart blood he recovered living encapsulated Type III pneumococci. This he could accomplish only by the use of mice. He failed to obtain transformation when the *same* bacterial mixture was incubated in broth. Griffith's original observations were repeated and confirmed both in our lab and abroad by Neufeld and others. Then you remember Dawson with us reproduced the phenomenon *in vitro* by adding a dash of anti-R serum to the broth culture. Later Alloway used *filtered extract* prepared from Type III cells and in the absence of formed elements and cellular debris induced the R culture derived from Type II to become typical encapsulated Type III pneumococcus. This, you may remember, involved several and repeated transfers in serum broth—often as many as 5–6—before the change occurred. But it did occur and once the reaction was induced, thereafter without further addition of the inducing extract, the organisms continued to produce the Type III capsule; that is, the change was hereditary and transmissible in serum in plain broth thereafter. For the past two years, first with MacLeod and now with Dr. McCarty I have been trying to find out what is the chemical nature of the substance in the bacterial extract which induces this specific change. The crude extract (Type III) is full of capsular polysaccharide, C (somatic) carbohydrate, nucleoproteins, free nucleic acids of both the yeast and thymus type, lipids and other cell constituents. Try to find in that complex mixture the active principle! Try to isolate and chemically identify the particular substance that will by itself when brought into contact with the R cell derived from Type II cause it to elaborate Type III capsular polysaccharide, and to acquire all the aristocratic distinctions of the same specific type of cells as that from which the extract was prepared! Some job—full of headaches and heartbreaks. But at last *perhaps* we have it. The active substance is not digested by crystalline trypsin or chymotrypsin. It does not lose activity when treated with crystalline ribonuclease which specifically breaks down yeast nucleic acid. The Type III capsular polysaccharides can be removed by digestion with the specific Type III enzyme without loss of transforming activity of a potent extract. The lipids can be extracted from such extracts by alcohol and ether at $-12°$C. without impairing biological activity. The extract can be de-proteinized by Sevag method—shaking c̄ chloroform and amyl alcohol until protein-free and biuret-negative. When extracted, treated and purified to this extent, but still containing traces of protein, lots of C carbohydrate, and nucleic acids of both the yeast and thymus types are further treated by the dropwise addition of absolute ethyl alcohol, an interesting thing occurs. When alcohol reaches a concentration of about 9/10 volume there separates out a fibrous substance which on stirring the mixture wraps itself about the glass rod-like thread on a spool—and the other impurities stay behind as granular precipitate. The fibrous material is redissolved and the process repeated several times. In short, this substance is highly reactive and on elementary analysis conforms *very* closely to the theoretical values of pure *desoxyribosenucleic* acid (thymus type). Who would have guessed it? This type of nucleic acid has not to my knowledge been recognized in pneumococcus before—though it has been found in other bacteria.

Of a number of crude enzyme preparations from rabbit bone, swine kidney, dog intestinal mucosa, and *pneumococci,* and fresh blood serum of human, dog, and rabbit, only those containing active depolymerase capable of breaking down known authentic samples of desoxyribose nucleic acid have been found to destroy the activity of our substance—indirect evidence but suggestive that the transforming principle as isolated may belong to this class of chemical substance. We have isolated highly purified substance of which as little as 0.02 of a *microgram* is active in inducing transformation—in the reaction mixture (culture medium) this represents a dilution of one part in a hundred million—potent stuff that—and highly specific. This does not leave much room for impurities—but the evidence is not good enough yet. In dilution of 1:1000 the substance is highly viscous as are authentic preparations of desoxyribose nucleic acid derived from fish sperm. Preliminary studies with the ultracentrifuge indicate a molecular weight of approximately 500,000 = a highly polymerized substance.

We are now planning to prepare a new batch and get further evidence of purity and homogeneity by use of ultracentrifuge and electrophoresis. This will keep me here for a while longer. If things go well I hope to go up to Deer Isle, rest awhile—come back refreshed and try to pick up the loose ends in the problem and write up the work. If we are right, and of course that's not yet proven, then it means that nucleic acids are *not* merely structurally important but functionally active substances in determining the biochemical activities and specific characteristics of cells—and that by means of a known chemical substance it is possible to induce *predictable* and *hereditary* changes in cells. This is something that has long been the dream of geneticists. The mutations they induced by X-ray and ultraviolet are always unpredictable, random, and chance changes. If we prove to be right—and of course it is a big if—then it means that both the chemical nature of the *inducing stimulus* is known and the chemical structure of the *substance produced* is also known—the former being thymus nucleic acid—the latter Type III polysaccharides, and both are thereafter reduplicated in the daughter cells—and after innumerable transfers and without further addition of the inducing agent, the same active and specific transforming substance can be recovered far in excess of the amount originally used to induce the reaction—sounds like a virus—may be a gene. But with such mechanisms I am not now concerned—one step at a time—and the first step is, what is the chemical nature of the transforming principle? Someone else can work out the rest. Of course the problem bristles with implications. It touches the biochemistry of thymus type of nucleic acids which are known to constitute the major part of chromosomes but have been thought to be alike regardless of origin and species. It touches genetics, enzyme chemistry, cell metabolism, and carbohydrate synthesis, etc. But today it will take a lot of well-documented evidence to convince anyone that the sodium salt of desoxyribose nucleic acid, protein-free, could possibly be endowed with such biologically active and specific properties, and this evidence we are now trying to get. It's lots of fun to blow bubbles, but it's wiser to prick them yourself before someone else tries to. So there's the story, Roy—right or wrong it's been good fun and lots of work. This supplemented by war work and general supervision of other important problems in the lab has kept me busy as you can well understand. Talk it over with Goodpasture but don't shout it around until we're quite sure or at least as sure as present methods permit. It's hazardous to go off half-cocked, and embarrassing to have to retract later. I'm so tired and sleepy I'm afraid I have not made this very clear. But I want you to know—am sure that you will see that I cannot well leave this problem until we've got convincing evidence. Then I look forward and hope we may all be together—God and the war permitting—and live out our days in peace.

The completed paper of Avery and his co-workers at the Rockefeller Institute, C. M. MacLeod and Maclyn McCarty, appeared in 1944. Although it can now be seen as the decisive initial impetus which launched the new era of molecular genetics and molecular biology, it was not so recognized in the years immediately following its publication. It solved the problem revealed by Griffith's work of 1928 in showing that the transforming principle was DNA which passed out of one variety or type of pneumococcus and into another, but its relation to heredity was not immediately apparent. Convincing evidence of transformations in other microorganisms was lacking until 1951 when Alexander and Leidy proved it for capsular transformations in *Hemophilus influenzae*, and it was soon extended to other species of bacteria.

At the time of Avery's discovery, the function of DNA in the cell was in doubt. That DNA itself was the active principle in the pneumococcal transformations had been made highly probable by McCarty and Avery's demonstration (1946)* that the activity of the transforming agent was removed by treating it with a highly specific enzyme, deoxyribonuclease. Yet it was still possible to think of DNA as a mutagen. Persons familiar only with the genetic systems of diploid organisms found it difficult, however, to equate the differences between the capsular characters of pneumococcal types with gene differences.

Doubts of this sort began to be removed when newly arisen mutant differences in pneumococcus were shown to be transferable by *in vitro* transformation of normal strains with deoxyribonucleates of the mutants (Ephrussi-Taylor, 1951;* Hotchkiss, 1951).* Evidence of transfer and recombination of linked markers by transformation brought the view expressed by Ephrussi-Taylor in 1951 that "crossing-over of molecules" between introduced and host DNA might be the mechanism of transformation. Mirsky, in his 1951 review of chemical aspects of the cell nucleus, considered transformation as "hybridization," foreshadowing the view now generally held.

Mirsky's paper also showed how far the knowledge of localization of DNA in specific hereditary materials—chromosomes—of a variety of organisms had advanced by 1951. DNA was then obviously deemed fit to be the genetic material responsible for transfer of genetic information by a variety of methods. Lederberg's paper (1951) at the same stocktaking meeting as Mirsky's (the fiftieth anniversary of the rediscovery of Mendel's rules) showed what some of these methods are in bacteria. Transformation was classified as "infective" heredity. Lederberg's review of other methods of recombination in bacteria indicated that the ability to use bacterial genetic systems with their high resolving power for

* Cited in Ravin, 1961.

genetical analysis would provide the material for a rapid expansion of knowledge of both the structure and function of the genetic material. A first sign was the discovery in 1952 by Norton Zinder and Joshua Lederberg of the transfer of genes between bacteria by way of transduction through the DNA of bacterial viruses.

D. DNA as Genetic Material in Bacteriophage

In the same year came the proof by Alfred Hershey and Martha Chase (1952) that the DNA of the T2 bacteriophage is competent to transmit the phage genotype, since by differential isotope labeling of phage DNA and protein they were able to show that only the DNA enters the bacterial cell. From this descends the next generation of virus particles with its specific proteins formed from materials in the bacteria, assembled under the direction of the viral DNA.

Later, other workers showed that the specific hereditary characters of tobacco mosaic virus and influenza virus in which RNA is the nucleic acid, are transferred by specific RNA which determines the specific character of the viral protein (cf. H. Fraenkel-Conrat and B. Singer, 1957, in Taylor, 1965).

While these demonstrations that nucleic acids constituted the genetic material were being made, analyses of the genetic systems of microorganisms were producing replicas of the familiar gene maps of diploid animals and plants but with a resolution so much finer that the dimensions of the units of recombination and mutation began to approach those of DNA structure—the nucleotide pair. These developments with bacteria and viruses will be outlined after a brief glance at studies in the field of protein chemistry from which linear structures similar to those in chromosomes and DNA molecules were beginning to take form.

E. Genic Control of Protein Structure

As already stated, no attempt will be made to trace the history of our knowledge of protein structure which occurred parallel to, but, until about 1949, independently of the development of modern genetics. It was generally recognized that genes and proteins, forming the basic living machinery, must be intimately connected. It was a commonly held view until about 1940 that genes and the chromosomes of which they formed a part, were based on protein and owed their specific characters to "side chains" which changed by mutation. There were frequent suggestions that specific genes were in fact specific enzymes, and owed their heterocatalytic properties to this fact. This view changed rather quickly when it was shown that specific genes controlled, by way of identifiable enzymes, individual steps in the synthesis of amino acids.

The roster of individual amino acids in protein had become stabilized at 20 by the 1930's and the primary structure of proteins, as composed of polypeptide chains consisting of amino acids, had assumed definite form during the following decade. A discovery in 1949 which was unexpected, at least by biologists, first brought this amino acid sequence into relation with the genic control of a protein, and led toward the revelation of the significance of amino acid sequence in the structure and functioning of proteins. This discovery came from the application by the physical chemist Linus Pauling and a group of his associates of an electrophoretic method to the separation of proteins from persons with two different alleles of a gene affecting the form of hemoglobin in the red blood cells. A difference in electrophoretic mobility was demonstrated, corresponding to a difference in net electric charge, between hemoglobin formed under the influence of a gene for sickle-cell anemia and that controlled by its normal allele. Pauling, Itano, Singer, and Wells (1949) concluded: "This investigation reveals therefore, a clear case of a change produced in a the first of many such gene-controlled protein differences revealed by the protein molecule by an allelic change involved in synthesis." This was application of this technique. At about the same time, the chemical work of Sanger on the polypeptide chains of insulin independently suggested that the differences found by the Pauling group might be due to differences in amino acid sequence. Sanger in 1955* published the first complete structure of a natural protein, insulin. A biochemist, Vernon Ingram, had already begun a search for the specific structural difference in the two forms of hemoglobin revealed by Pauling's work. By separating certain of the individual peptides of the two forms of the very large hemoglobin molecule and studying them by a combination of paper chromatography and electrophoresis, Ingram was able to show (1957) that the difference due to the sickle gene was localized in a single amino acid substitution. At a single point in one peptide, normal hemoglobin (Hb^A) was found to have a glutamic acid residue with a free carboxyl, while the abnormal hemoglobin (Hb^S) had in its place a neutral residue, valine. Thus the charge difference found by Pauling was accounted for and the change brought about by mutation from Hb^A to Hb^S had been revealed as a single mistake in the synthesis localized in the bonds distinguishing the attachment of a neutral from an acidic amino group. The impression of the simplicity and specificity of this gene effect in a complex molecule with some 600 amino acid residues gave a powerful stimulus to studies of protein structure, in which mutant genes could be made to play a part, as well as to the search for correspondence between the structural mutant change in the protein and the corresponding and controlling mutant change in the DNA. Further evidence of localization of such changes in protein structure came from the subsequent proof by J. A. Hunt and Ingram (1960)* that the gene for another mutant hemoglobin (C), allelic to A and S, originated by another amino acid substitution at the same position, i.e., lysine with a basic effect, replaced the acidifying glutamic acid. Just before this, it had been established that there were four polypeptide chains in the hemoglobin molecule, two identical α-chains and two identical β-chains. The allelic substitutions A to S or to C were all at one point in the β-chain.

The secondary structure of the protein, that is, its pattern of coiling in space (Pauling *et al.*, 1951), and its tertiary structure, that is, its folding due to linkage between reactive groups on the outside of amino acids which are close to each other, were shown to be determined by its primary structure, i.e., the sequence of amino acids in the chain.

It was, however, the study of the tertiary and quaternary structure of proteins, that is, the three-dimensional disposition and cross-linkage of the polypeptide chains which brought closer an understanding of the relation between protein structure and function, as in enzymes.

The first enzymes had been obtained in crystalline form some 40 years before this time (cf. Northrop *et al.*, 1948, for review). There was sometimes a small nonprotein portion, the coenzyme, which in many cases turned out to be a vitamin linked to a nucleotide. The relations of genes to protein structure and to enzyme specificity thereby became one with the problem of genic control of protein synthesis.

Another type of structure in which protein occurs together with a nucleic acid was recognized when W. M. Stanley in 1935 first isolated and crystallized tobacco mosaic virus. The recognition of a class of infectious agents, the filterable viruses, of dimensions much smaller than those of bacteria goes back to Ivanoff and to Beijerinck in the last decade of the nineteenth century. The eventual contributions to genetics arising from the study of viruses will be discussed in Section F, 4.

F. Reproduction in Microorganisms

1. *History of Mycology*†

Knowledge of the mechanism of reproduction has been an essential accompaniment of the development of genetics. In the case of Mendelian

* For comprehensive review and list of literature references, see Ingram (1963).

† References to literature cited in this section are given in Lechevalier and Solotorovsky (1965).

heredity, understanding of the sexual mode of reproduction in flowering plants prepared the way for the discovery, both by providing the technical facility for making controlled crosses and by demonstrating the successful use of experimental methods for studying reproductive functions. As genetics, based on experiments with diploid organisms, developed its analytical methods for resolving the genotype into its elements and arranging these in systems as in the chromosomes, it in turn produced new insights into the biology of reproduction. Genetics in fact grew as a central field of reproductive biology, what Bateson had called "the physiology of descent."

The rapid advances in genetics after 1940, made possible by the use of bacteria, protozoa, viruses, and fungi, especially the filamentous ascomycetes, owed much to the knowledge of life cycles in these forms which had been gained in the eighteenth and nineteenth centuries. Some of the steps in this progress have been pointed out in a recent history of microbiology (Lechevalier and Solotorovsky, 1965) on which the following condensed account is based.

One of the essential first steps was that described by the Italian botanist Micheli (1679–1737). In a work published in 1729 he proved experimentally that the molds *Mucor* and *Aspergillus* (which he had discovered and named) as well as several basidiomycetes gave off spores (Micheli called them seeds) which always reproduced the parental form. He also observed the asci within which the spores were formed. His pupil and successor at Florence, Giovanni Targioni-Tozzetti, confirmed the work of Micheli and extended it to other microorganisms. He seems to have been the first to establish the mode of reproduction by spores of rust and smut fungi which led to discovery of the route of infection of these pathogens. In the same year (1767), Felice Fontana, who had been called from Pisa to Florence, published careful observations on wheat rust which established its parasitic nature and relationship in a mode of reproduction similar to that of the molds described by Micheli.

These good beginnings did not lead to immediate advance in the understanding of fungal reproduction since Bénédict Prévost had to discover for himself the spores of wheat bunt (1807) and incidentally the first practical fungicide (copper sulfate) which stopped the germination of the spores. Later the great German mycologist de Bary worked out the life cycle of wheat rust (1865), divided between the alternate hosts of wheat and barberry. It was he who described sexual processes in some ascomycetes which provided an important preparation for the breeding work with *Neurospora* that contributed so much to biochemical genetics. A further important step was taken by Jules Raulin, a student of Pasteur, whose long paper of 1869 "Chemical Studies on Growth" laid down the methods on which determinations of mineral requirements of molds and thus of chemically defined growth media were based. "One of the most interesting properties of plants," wrote Raulin, "is their ability to grow in artificial media composed exclusively of known chemical components."

The ground work for the cytological study of reproduction in fungi was laid by Pierre Dangeard, who in 1893 observed nuclear fusion in the teleutospores of a rust.

Experimental studies of reproduction in algae were initiated by a young French amateur, Gustave Thuret, who in 1844 discovered the antherozoids of *Fucus*. He established the occurrence of fertilization in this seaweed in 1854 and in 1855 Nathanael Pringsheim observed fertilization in *Vaucheria*.

Pasteur's paper of 1860 laid the foundation for the understanding of alcoholic fermentation and of the physiology and reproduction of yeast, and incidentally it gave rise to biochemistry. "Ever since this milestone," Wildiers wrote in 1901, "there is not a single living organism that we understand better, from the point of view of chemical mechanism, than the yeast." This passage, as cited by Lechevalier and Solotorovsky (1965), occurs in the introduction to an important paper, in which Wildiers (1901) described the discovery of bios, an essential requirement for the growth of yeast which was later fractionated into bios I (inositol) and bios II (biotin and pantothenic acid).

Proof of heterothallism and operative identification of mating types as plus or minus in the bread mold *Mucor* was given by Blakeslee in 1904. It would seem that by this time enough technical knowledge of reproduction and metabolism was available to support breeding analysis in molds, basidiomycetes, algae, and yeasts. However, genetic analysis in thallophytes got under way only much later.

2. *Fungal Genetics*

The bridge leading from the study of mycology for its own sake to fungal genetics owed much to the work of B. O. Dodge (1927), who discovered the heterothallic system in the ascomycete *Neurospora* and of C. C. Lindegren (1932, 1934) who carried out the first genetic analyses in *Neurospora*.* At the same time Zickler (1934) published clear evidence of segregation of spore-color genes in the asci of *Bombardia lunata*. A similar demonstration had already been given by Zattler in

* For an interesting account of these events, see Beadle (1963). A brief history of mycology in relation to genetics is given by Walker (1951).

1924 in a basidiomycete, *Collybia*. Barthelmess (1965) has given a clear and well-illustrated account of these pioneer experiments in fungal genetics (pp. 580–582) with references to the above literature.

The discovery of the sexual cycle in yeast began with the work of Kruis and Satava in 1918, but this did not result in advances in yeast genetics until Winge (1935) worked out the haploid-diploid cycle in brewer's yeast and showed by genetic evidence that this yeast has the same kind of mitotic and meiotic mechanisms as occur in higher plants.

3. *Bacterial Genetics*

There was an even longer gap between the emergence of bacteriology as a science and the first employment of bacteria as material for investigating problems of genetics. Some knowledge of the biology of bacteria, especially of biochemical processes associated with bacterial metabolism, of their relations to fermentations and disease in both animals and plants, and of methods of growing them in pure culture, had been attained by 1900, but bacteria did not contribute importantly to genetics until after 1940.

Discussions of possible reasons for this delay may throw some light on the relation of prior work in bacteriology to the great surge of progress in genetics which occurred after about 1946. The delay in the development of bacterial genetics was discussed by Vernon Bryson in the symposium volume, "Microbiology Yesterday and Today" (1959, pp. 80–99). He suggested that geneticists had not been attracted to the use of microorganisms, which they considered to be mainly bacteria, organisms which lacked interest for students of evolution, mutation, and gene action. For geneticists, prior requirements for experimental material would be those not supplied by microbes: "sexual initiative, phenotypic distinctiveness and genotypic accessibility." This is probably true but does not tell us why these handicaps for geneticists had not seemed to apply for bacteriologists. It should not be surprising to find that scientists follow their interests and that they are most interested in what they know most about. If geneticists had not been interested in bacteria, bacteriologists in general had not been interested in genetic problems as such.

Selman Waksman, in whose honor the symposium cited above was held, gave his views as to the problems that interested microbiologists. Those of yesterday had been fermentation, autotrophy, ecology, symbiosis and parasitism, disease, taxonomy, and especially mechanisms of energy transformation and synthesis which seemed to Waksman to have formed a main bridge leading to today's interests of which metabolism and enzyme mechanisms were paramount, with genetic ones briefly mentioned even though "these have tended to transform completely our understanding of microbial life" (in Bryson, 1959, p. 119). According to Waksman, a major shortcoming of bacteriologists of yesterday was in not having regarded their material as botanists interested in biological problems.

It is true that a dominant interest of bacteriology, at the time of the genetic renaissance in 1900, was in diseases, chiefly human and animal. This was a heritage from the founders of bacteriology, Pasteur and Koch. Also inherited from them was a strong belief in the fixity of bacterial species. To admit that disease-causing bacteria could undergo changes in their hereditary characters seemed to the founders to be a heresy, the defeat of which had permitted the establishment of the "germ theory." Having fought one revolution, it was perhaps too much to expect the classic bacteriologists to take on another one. An explanation of bacterial variations which was less offensive to the traditional attitude was that these were due to environmental changes, such as might occur in culture. But, of course, with no rationale for dealing with hereditary changes, no understanding was possible of the rapid changes in bacterial populations as due to selection. This was only reached after proof of mutation and penetration of ideas from genetics (Braun, 1953).

Salvador Luria in his review of 1947 seems to have put his finger on one of the esesntial causes of delay in the development of bacterial genetics. Luria gives chief place to lack of knowledge and lack of agreement "even on the most elementary facts of reproduction and character transmission in bacteria" (p. 1). Looking through textbooks and general treatises on bacteriology before about 1950 seldom reveals a rubric "reproduction." It was apparently not an active or important field of bacteriological research. Bacteria seem to have lacked those very qualifications that caused higher plants to supply the initial impetus to the development of genetics.

Although they were thus precluded from leading the way in analytical genetics, work with bacteria provided biochemical insights, methods of thinking and techniques which, in the tradition of Pasteur, Duclaux, and Buchner, led to the foundation of biochemistry largely on the basis of research with microorganisms. The great role subsequently played by bacteria in the study of molecular genetics stemmed from this knowledge and not, as in the case of most other materials for genetic research, a prior knowledge of reproductive mechanisms.

A second cause of delay in bacterial genetics pointed out by Luria was "ignoring delicate population problems involved in distinguishing between cell characters and culture characters" (1947, p. 2). The most persuasive argument that these had been essential deterrents to progress in bacterial genetics is found in the fact that even the first steps taken to remove them resulted in immediate and dramatic progress.

Luria's review, cited above, was written only three years after the application of the methods of thought characteristic of population genetics had shown how the spontaneous origin of rare mutations in bacteria could be proved. This was the problem worked out by Luria and Delbrück (1943); it can be viewed as marking a transition point after which bacterial populations could be studied as genetic systems like other organisms. The fact of special importance was that Luria and Delbrück created a general model together with methods of testing it, and gave preliminary evidence that resistance to a bacteriophage may arise as a spontaneous mutation in populations of colon bacilli.

The other landmark publication that helped to bring bacteria within the ambit not only of genetics but of a more general biology, had appeared only a few months before Luria's review was written. This was the paper of Lederberg and Tatum (1946), which by its proof of the occurrence of gene recombination in a special strain of *E. coli*, brought the first convincing evidence of a sexual process in bacteria. This was a step in the realization, still to come, that a Mendelian system of heredity would prove to be as characteristic of bacteria as of most other organisms. It thus gave more breadth and substance to a view of living forms as having a unity based on a particulate mechanism of heredity.

In 1945 had appeared Tatum's paper applying to bacteria the methods he and Beadle had worked out in initiating the study of biochemical genetics in *Neurospora* (Tatum and Beadle, 1941). In the short space of about three years (1943–1946), all of these basic ideas were confirmed and extended and bacterial genetics was firmly launched. It is interesting that the study of mutation played a key role here as it had in the beginning of genetics and in changing its course in 1927.

In bacteria, a sexual process was first revealed by the occurrence of genetic recombination. But since sexual recombination in bacteria involved whole cells, knowledge of what was transferred was not required in proving the fact of transfer. Subsequently the mechanism of transfer of the bacterial chromosome during conjugation was worked out (Jacob and Wollman, 1961). By the time a second means of transfer of genes between certain clones of bacteria, namely, transduction, had been discovered (Zinder and Lederberg, 1952), it was possible to see what the transducing phage and the transforming principle had in common. This was DNA; thereafter it was evident that transformation, conjugation, and transduction—discovered in that order—represented various modes of transfer of DNA. These must have evolved in bacteria in response to the selective pressures which in higher plants and animals had made sexual communication the prevailing mode of recombination.

After 1952, there were thus three ways, instead of none, as formerly, of analyzing the genetic systems of bacteria. Even though these were applicable to only a few species, and sometimes only to special clones within a species, they made possible a rapid localization of thousands of mutational sites within the bacterial genome. Although the techniques were different from those used with diploid organisms, the principles established by plant geneticists proved to be adequate guides to the new materials. Sites of mutation could be separated by recombination, by transduction, and by transformation of linked markers, and with a resolving power incomparably greater than ever attained with other organisms. The technical advantages of bacteria, detection of events of great rarity in the enormous populations that could be observed, methods for selecting and preserving rare mutants or new forms arising by recombination, soon put them at the apex of the advancing front of genetics.

Within a few years, in the mid-1950's, maps of the linkage groups of *Escherichia coli* and of *Salmonella typhimurium* took form. The elements separable by recombination were found to be arranged linearly in a circular ring-shaped map. One of the firstfruits of the new mapping work was the discovery by Demerec and his associates (Demerec and Hartman, 1956; Hartman, 1956) of clusters of genes in the enteric bacteria, *Salmonella,* each cluster associated with the synthesis of a single amino acid. Moreover, the spatial order of genes within the cluster was found to follow the temporal sequence of the steps in the reaction sequence. A similar clustering and order of the sites controlling the synthesis of the enzyme tryptophan synthetase in the colon bacillus, *Escherichia coli,* was revealed by the work of Charles Yanofsky and his associates. This work with bacteria made it possible to test ideas about the connection between physical association or contiguity of genes and their integration in a common biochemical function. The problem had been perceived through cases first observed in higher organisms but it first became susceptible of proof in bacteria.

This experience was to repeat itself frequently in the brief period since the development of bacterial genetics. The rapidity of genetic analysis and the ability to identify both individual sites of mutation and the elementary or primary chemical substances controlled by them brought about a rapid expansion of the horizons of genetics to include the control of biosynthetic functions.

4. *Viruses*

The relation between the study of viruses and the growth of genetics was in one sense the reverse of that between the study of reproduction in plants and the discovery of Mendelian heredity. In the latter case it was hybridization, made possible by sexual reproduction, which guided the formulation of questions about inheritance. In the case of viruses, on the other hand, the prior knowledge of particulate heredity led to interpretations of the facts of mixed infections of bacterial viruses which revealed the mode of reproduction of virus particles. Once it had been shown that a virus particle was essentially a string of genes enclosed in a protein coat, viruses suddenly changed their status in biology. From being the most mysterious form of living matter they quickly became that form of which the genetic behavior, even at the molecular level, was best understood.

Infective agents which could pass through the pores of fine porcelain filters had been known since Ivanowski's observations of 1892 on the mosaic disease of tobacco plants.* A filter-passing agent was found by Löffler and Frosch in 1898 in cattle with hoof-and-mouth disease. The recognition of the tobacco mosaic virus as a filterable living agent which multiplies in the cells of growing tissues dates from Beijerinck's paper of 1900. He called it *contagium vivum fluidum.* His proof that it reproduced itself identified it as living substance. Other viruses with this property identified in the first decade of this century were: yellow fever, 1901; fowl leukemia, 1908; Rous sarcoma in fowls, 1911.

Proof of the particulate nature of viral infections came first from counting the number of separate vesicles formed on the skin of a rabbit when rubbed with vaccinia virus. This method, invented by Calmette and Guerin, was used by Steinhardt, Israel, and Lambert in their proof in 1913 of the multiplication of vaccinia viral particles in cell cultures. The view that viruses consisted of bodies generally invisible in the light microscope had been attained by experimental methods before they were revealed by the electron microscope about 1940. In fact, the electron micrograph of tobacco mosaic virus particles by Kausche in 1938 was one of the first contributions of the newly developed instrument to biology.

a. *Discovery of Bacteriophage, 1915–1917.* Of greatest importance for genetics was the discovery of viruses parasitizing bacteria, the bacteriophages. The first observation was that of the English bacteriologist

* References to literature in this and the following three paragraphs are in Lechevalier and Solotorovsky (1965).

F. W. Twort, who in 1915 reported the discovery of a filter-passing agent which caused certain bacteria to lyse and cultures thus to become transparent. Twort's work attracted little notice until the lytic principle was rediscovered and named bacteriophage by the French-Canadian bacteriologist Félix d'Herelle about 1917. Different bacteriophages proved to be obligate parasites of specific strains of bacteria. The presence of the parasite was revealed by the presence of a plaque or clear area in which the bacteria growing on a solid medium had undergone lysis.

In 1935 the American virologist Wendell M. Stanley identified from leaves infected with tobacco mosaic disease a crystalline protein and regarded the virus as "an autocatalytic protein." In 1937, however, the English biochemists F. C. Bawden and N. W. Pirie showed that the virus was in fact a nucleoprotein, later shown to be ribosenucleic acid in association with a protein.

b. *Recombination, 1948.* Modern virus research leading to our present understanding of the structure and genetics of bacterial viruses was opened by Max Delbrück and Emory Ellis in 1940 (cf. Beadle, 1963; Ravin, 1965, Chapter 2). They developed methods for studying phage infection in individual bacterial cells. This made possible the proof that different phage lines could be crossed (Delbrück and Bailey, 1946, in Taylor, 1965) and that they undergo genetic recombination (Hershey and Rotman, 1948, in Taylor, 1965). The proof in 1952 that the genetic system of a virus is in its core of nucleic acid (DNA in the case of bacteriophage) has already been recounted (p. 65). Thereafter the phage genetic system could be analyzed and controlled like that of other organisms.

c. *Episomes, 1958.* Studies of bacteriophage also revealed a new form of relationship between genetic systems of different organisms. Lwoff discovered (1952) (cf. Ravin, 1965, Chapter 2) that bacterial virus particles may exist in a nonvirulent or temperate form. The latter may occur in alternating forms which Lwoff called phage and prophage. In the prophage state the genetic material DNA of the phage is an integral part of the bacterial DNA and replicates with it. When the virus DNA leaves the bacterial chromosome, probably by crossing-over, it can again multiply freely and independently, and may become virulent and cause lysis of the bacterium. A genetic particle which like phage can exist either free or as part of a chromosome was called an episome by Jacob and Wollman (1958 recorded in 1961), who applied the term first to the fertility factor in *Escherichia coli.* The idea of a genetic particle alternating between freedom from an organized physical system and subordination to it was novel and opened new possibilities for investigating the manner in which control mechanisms operate in cell differentiation and in development.

G. Development of Methods and Instrumentation

There is little doubt that the most important methodological contributions to and from genetics have been logical and statistical ones by which theoretical models have been prepared for testing. The development of such methods was especially active in the last half of the nineteenth century. J. T. Merz in the second volume of his "History of European Thought in the 19th Century" characterized the new views then attained as "the statistical view of nature." This was a result of the discovery, itself due to statistical thinking, of the atomic and molecular organization of matter. The behavior of matter had thenceforth to be dealt with in terms of units which could be manipulated by probability theory and statistical methods. Although its roots derive from an earlier time, Willard Gibbs' "Statistical Mechanics" of 1902 signalized the change which was to transform chemistry and physics. In that same year, Rutherford and Soddy gave the first proof of the transmutation of elements, the radioactive decay of thorium by emission of subatomic particles.

Of special importance for the later development of genetics had been the recognition by Quetelet, Pearson, Yule, and others who laid the foundations of mathematical statistics, of the need for methods of dealing with populations of units, whether of individual men or animals or of molecules as in a gas.

1. *Statistical Methods*

The introduction of statistical methods for dealing with biological variation is usually attributed to Quetelet (1846)* and those for studying inheritance are said to stem from Galton's "Natural Inheritance" of 1889. The latter book provided the origins of biometry. It must not be forgotten, however, that the essential principle of segregation, a statistical conception, was discovered without knowledge of, or recourse to, the methods of Quetelet or of Galton. Nevertheless, it is true that the growth of genetics after 1900 occurred at a time of changing modes of thinking in science. These were spreading toward biologists but not all of them recognized or welcomed the change. Johannsen, Pearson, and Yule certainly did. T. H. Morgan, whose influence on genetics far exceeded theirs,

* For an interesting account of Quetelet's work and its relation to probability analysis see Gillispie (1963).

did not. There is a story that when Morgan began to teach genetics* at Columbia about 1908 he was asked by a statistically minded colleague what mathematical preparation he would suggest for a student intending to study genetics. "Addition and subtraction, certainly," said Morgan, "multiplication would be useful, and probably long division, too." This is probably apocryphal, but it emphasized a common opinion that the essential notions of genetics could be obtained without sophisticated mathematical or statistical training. While simple, the statistical theory by which Mendel explained his results (reached independently by Correns just before 1900) was rigorous and brought the phenomena of inheritance, and later those of biological evolution, into an order which was both obvious and general since shared with other natural phenomena.

The twentieth century, dominated by counting in all fields of science, would have little room for nonmechanistic thinking. This was one of the lessons that Karl Pearson's "Grammar of Science" impressed upon the generations of students to whom its successive editions beginning in 1892 became a kind of bible. This was a climate fit for statistical genetics, and the methods developed by Pearson and the other founders of the journal *Biometrika* in 1902 were on hand when needed. There was an unfortunate lag in application of these useful statistical ideas in the early 1900's because of the misunderstandings followed by enmity between the biometricians led by Weldon and Pearson and the Mendelians led by Bateson. But in 1918 R. A. Fisher, who was both a biometrician and a Mendelian, showed the opposition to be unfounded and followed this up by developing a body of methods which formed a firm foundation for the mathematical statistics underlying modern genetics.

2. *Instrumentation*

During the period 1910–1940, and especially toward the end of it, there were being prepared, largely by physical chemists, many of the instruments and methods which in the 1940's and 1950's produced a veritable eruption of new knowledge about the structure of proteins and nucleic acids. This in turn stimulated the rapid development of genetics at the molecular level. An important element responsible for the speed of transformation of understanding about these macromolecules was that much of the development of the instrumentation was carried out by the scientists who required it for the solution of their own theoretical problems.

This was clearly true of X-ray crystallography. The Braggs, Sir William

* Morgan's course, however, was always called "Experimental Zoology."

and Sir Lawrence, father and son, had taken immediate advantage of von Laue's discovery that X rays could be diffracted by crystals; in 1912 they worked out the structure of the crystal of common salt. Subsequently they and Linus Pauling applied this powerful method to inorganic silicates. J. D. Bernal's laboratory led the way in getting diffraction patterns of wet protein crystals (1935), and aided by Pauling's happy supposition that peptide bonds in proteins should lead to helical arrangements, Perutz and Kendrew,* by introducing heavy marker atoms for X-ray reflections, calculated the atomic positions in a protein. But the big pay-off of crystallography for genetics came when J. D. Watson and Francis Crick in 1953 worked out a double helical structure for DNA based on the X-ray diffraction patterns obtained by M. H. F. Wilkins and R. E. Franklin.†

The high speed ultracentrifuge for the rapid sedimentation of large molecules such as proteins and nucleic acids was developed by Svedberg in the 1920's. When fitted with optical systems to record the rate of sedimentation, it gave essential data on molecular weights and aided the classification of macromolecules by relative mass and shape. Separation and size classification of cellular organelles, such as ribosomes as well as ribosomal RNA molecules, were made possible by this instrument which thus contributed directly to the solution of the mechanisms of transcription and translation of genetic information. The importance of the development of this instrument was recognized by the award of a Nobel prize to Svedberg in 1926 (cf. Svedberg and Pedersen, 1940).

Chromatography by which peptides and amino acids are separated by relative solubilities in different solvents originated in independent observations of an American oil engineer D. D. Day (1889–1925) and a Russian botanist T. S. Tswett (1872–1919).‡ Day noted differential penetration of porous rocks by hydrocarbons with specific chemical differences, while Tswett as early as 1903 had used absorbent columns to separate, by different rates of migration, leaf pigments differing in chemical constitution. Paper partition chromatography as applied to the separation of parts of macromolecules took on new forms and usefulness when developed by A. S. P. Martin and R. L. M. Synge. In 1942 they adapted paper chromotography to the separation of some amino acids which differed in their relative solubilities in different solvents. The mi-

* For a spirited account of this see Watson (1968).

† For X-ray analysis of protein structure see Crick and Kendrew (1957 with bibliography).

‡ For a history of chromatography with citations of works in this paragraph, see Heftmann (1961).

gration of such residues could be followed first in one direction in one solvent and then in a direction at right angles to this by turning the paper strips through 90 degrees and allowing migration in another solvent. Chargaff used this method in proving in 1950 that in DNA the members of each purine–pyrimidine pair occurred in equimolar quantities. Frederic Sanger used it in deducing the full structure of the insulin molecule (1955) and his important proof that each kind of protein has its own specific arrangement of amino acids. Paper partition chromatography led to the development of electrophoretic methods of separation of proteins on paper (e.g., Pauling *et al.*, 1949), and later on starch and other gels (Smithies, 1956). When partition chromatography and electrophoresis were combined, as in Ingram's work (1957), they permitted reliable separation of individual peptides from a protein digest and made it possible to localize a change brought about by mutation. Martin and Synge were awarded a Nobel prize in 1952 for the development of chromatographic methods.

Among other methods of more general applicability should be cited the application of ionizing radiations, such as X rays and γ rays, to the experimental analysis of the mutation process and the structure of the hereditary material. This began soon after the discovery of X rays by Roentgen in 1895 and of radioactivity by Becquerel a few months later. The earlier attempts to produce changes in bacteria (e.g., Aschkenass and Caspari in 1901) and in higher plants (e.g., Gager, 1908) were unable to utilize the rationale provided by development of the concept of the gene system. Radiation genetics therefore dates from Muller's proof of mutation induction by X rays in 1927. The effect of the discovery and use of radiation was as important for biology and genetics as it had been earlier for physics, but the subject is too vast to be treated here.

As the earlier development of the microscope* had been an essential preparation for cytology and cytogenetics in the 1930's, the use of magnetic fields as lenses for electrons made it possible to visualize objects with dimensions far smaller than the wavelength of light. The use of magnifications of hundreds of thousands of diameters, facilitated by the later refinements of the electron microscope, brought rapid confirmation of the presence in organisms of structures which had been inferred from experimental observations. The bacterial genonema, the structure of viruses, and many other aspects of fine structure could be rapidly and

* Cf. Freund and Berg (1963) for a historical account of the development of microscopy in biology.

unambiguously established. The history of the application of the light microscope to genetic problems tends to be repeated with the electron microscope. The confidence engendered by rapid testing of theories concerning structure leads to an increase in the speed of discovery comparable to the increase in the degree of resolution. However, the historical development of both light and electron microscopy are beyond the scope of this chapter.

Of special importance for the understanding of turnover and replication processes of large molecules like DNA and protein was the use of isotopes of elements such as carbon, nitrogen, sulfur, and phosphorus differing in weight or in radioactivity from the forms usually found in living organisms. A case in point was Rudolf Schoenheimer's use of the stable ^{15}N isotope of nitrogen in establishing what he referred to as "The Dynamic State of Body Constituents" (1942). These lectures, delivered in 1940, caused a fundamental change in attitude toward such constituents as proteins, which were shown to undergo rapid molecular regeneration; they also paved the way for the study of the synthesis and mode of replication of large molecules. Later, unstable isotopes were used in the process of autoradiography by which compounds such as nucleic acids could be followed by the disintegrations registered on photographic plates of radioactive labels such as ^{3}H (tritium) which had been taken up by the compound under study. An example of this technique was Taylor's proof (1965, p. 296) of the semiconservative character of the replication of chromosomes.

In much of the recent work in genetics and molecular biology, high-speed computers have played an increasingly important role, not only in dealing with intricate computations, as in estimating angles in X-ray diffraction patterns, but in testing theoretical models. These tests have sometimes led to rejection of hypotheses even before an experimental test was made.

XI. The Transformation of Biology

It is certainly a risky operation to try to view recent events in perspective. Perspective implies distance whether optical or historical. Recent or near events should thus have nearer events to provide criteria for comparison, and since the latter will usually lie in the future, the judgment of recent events will necessarily partake of the nature of prediction. On rare occasions this is a risk worth taking. The period since 1953 in the development of genetics provides such an occasion.

A. "Particle Biology"

The history of biology before 1900 was replete with proposals for solving the problems of the activities of living material in terms of living units or particles. Contrary to prevailing opinions, in the older biology organisms were not always viewed as wholes but often as aggregates or systems of interacting parts. Atoms, as indivisible parts composing complex substances, were conceived in the age of Pericles. Lucretius' great poem about them was a restatement of Democritus. Living and nonliving matter showed no discontinuity in this respect: all substance was divisible into indivisible subunits, and by this flexibility it was provided with a means of operating and changing with time. The problem of explaining change was obviously more open and more pressing with living beings, which were in a state of change from conception to death, than with inorganic nature in which change was slow and often imperceptible.

In the nineteenth century changes occurring in organisms, both in ontogeny and in phylogeny, were often thought of in terms of the distribution and activities of vital units. These hypothetical elements, however they were called (gemmules, pangenes, ids, determinants, micellae), performed no really constructive service in the development of biology until after 1900. Then the proof of the existence and behavior of segregating units as the mechanism of inheritance led to what has been called "particle biology." The central idea was that the system of heredity was particulate. Now, with the hindsight provided by discoveries of the last decade, we can see that "quantum biology" might have been a better designation than particle biology since the essential feature is the discrete nature of the elements first recognized by segregation and recombination.

This had been expressed most clearly by the physicist Erwin Schrödinger in his Dublin lectures of 1943. When published in 1945, under the title "What is Life?" they brought to the attention of biologists and especially of geneticists those views of the physical nature of the gene and of mutation which had been formulated by another physicist, Max Delbrück,* whose interest in turn had been aroused by the mutation-induction studies of H. J. Muller and N. Timoféeff-Ressovsky. In his chapter on "Delbrück's Model Discussed and Tested" Schrödinger wrote (p. 57): "Consequently we may safely assert that there is no alternative to the molecular explanation of the hereditary substance." This, with previous discussions of Muller and of Delbrück, provided the intellectual preparation for what was to develop in the next decade. Schrödinger's con-

* For a discussion of Delbrück's relation to the origins of molecular genetics see Cairns, Stent and Watson (1966) and Stent (1968).

ception of the hereditary substance as transmitting instructions in the form of a code was also prophetic.

With the refinements made possible by the ability to detect very low frequencies of recombination, as in bacteria and viruses, the ultimate elements have proved to be pairs of nucleotides with a few dozens of atoms. As parts of a continuous chain in DNA they are hardly to be designated as particles. The first proofs of discrete elements—genes—came from statistical evidence and the units so recognized were abstractions. A corollary of the statistical theory of the gene was the existence of a steady state of frequencies of such elements in random mating populations, and evolutionary changes could then be thought of as due to disturbances in the steady state.

The elements composing the genetic system were, however, nearly always conceived as parts of living matter. They were never pure abstractions except in the minds of such purists as Johannsen, who denied them a physical existence. For most biologists their material basis became evident when they were shown to be parts of chromosomes and to be alterable by physical means such as quanta of radiation.

It is an interesting reflection that the essential feature of the transmission system of heredity has retained its particulate character even though there is no corresponding system of "particles" within the continuous sequence of nucleotides composing the nucleic acid molecule. The durable and useful aspect of gene and "particle" alike is thus the abstract one, as it was when the particulate view of heredity was first conceived by Mendel.

B. Expansion of Genetics

It was at the end of a period of refined study of the hereditary material by methods of genetic analysis that ideas basic to genetics underwent a sudden expansion and were seen to underlie all biological processes and properties. One crude way of explaining how in the ten years from 1953 to 1963 genetics broke out of its traditional boundaries and saw its basic theories assume a universal character is to say that genes operate not only as the means of transmission of life, as genetics had already shown, but also as the means guiding the synthesis and functioning of all living material. Geneticists used to be properly skeptical of statements about the "primary" effect or product of a gene, since, when materials for study of such questions consisted mainly of higher plants and animals, there was little likelihood that a primary action could be isolated and identified. Now, however, thanks to research on microorganisms and cell-free systems derived from them, we know without much doubt what the primary product of a gene is. It is a complementary replica of that segment of DNA, its mirror image in RNA, which will transfer and translate the gene image, through one or two intermediate steps, into a sequence of amino acids forming a polypeptide chain of a protein. That, in essence, is what the genetic code means.

The coding problem seems to have been first formulated by Dounce (1952),* who suggested that a sequence of three nucleotide bases (a triplet) might determine a "code" for each of the 20 kinds of amino acids in a protein. George Gamov effectively stated a coding hypothesis in 1954 and 1955 which connected a specific nucleotide triplet with a specific amino acid for incorporation in protein. Elements essential for the final solution were Crick's statement of the adaptor or transfer RNA (tRNA) hypothesis in 1958; and Jacob and Monod's proposal of the messenger RNA (mRNA) hypothesis in 1961.*

The proof of operation of a code was provided in 1961 by Nirenberg and Matthaei's demonstration that, in a cell-free extract from *Escherichia coli*, protein synthesis was dependent upon template or mRNA. This made it possible to test the messages carried by natural mRNA as well as those of synthesized polynucleotides. The first proof was that poluridylic acid specifically caused the incorporation of phenylalanine into protein. The code in mRNA for this amino acid was therefore UUU.*

The deciphering of the code does not merely demonstrate the ability to represent the message in the genetic material by a linear sequence of three-letter code words each representing an amino acid. A view of that kind of pattern on a cruder scale had been attained by the methods of formal genetics which portrayed the genetic material as a linear succession of elements, each of which could assume specifically different functional forms. What gives the new view its universal character is the ability, attained by chemical means, to accomplish *in vitro* the incorporation of identifiable building blocks—amino acids—into protein in the same way in which the gene does it *in vivo*. If, as is now evident, genes are responsible for synthesis of proteins which constitute the vital machinery in all forms of life, they are then at the center, at the controls, so to speak, of living activity and they thereby become the biological rather than only genetic units.

Such a view of the primacy of the gene material not only in the present functioning of protoplasm but in the origin of organized living material had been expressed by H. J. Muller long before its proof by physico-

* Good bibliography of publications on the code is in the recently published paper of Matthaei *et al.* "An Experimental Analysis of the Genetic Code," in Brink, (1967, pp. 105–145); other literature is in Lanni (1964).

chemical methods during the last decade. Muller's retrospective discussion of "The Gene as the Basis of Life" (the title of his 1929 paper) formed a fitting climax to the proceedings of the Mendel Centennial Symposium of 1965 (Brink, 1967). This paper should be consulted for a view as to how molecular biology developed by means of the methods of physical and biological chemistry.

Dedication

Dedicated to Theodosius Dobzhansky in honor of his 70th birthday, January 25th, 1970, in friendship and affection.

REFERENCES

Allen, C. E. (1917). *Science* **46**, 466–467.

Aschkenass, E., and Caspari, W. (1901). *Arch. ges Physiol.* **86**, 603–618.

Auerbach, C. (1945). Cf. Auerbach in Brink (1967, pp. 67–80).

Avery, O., MacLeod, C., and McCarty, M. (1944). *J. Exptl. Med.* **79**, 137–158 (reprinted in Taylor, 1965).

Babcock, E. B. (1949). *Port. Acta Biol., Ser. A*, 1–46.

Barthelmess, A. (1952). "Vererbungswissenschaft." Orbis Academicus, Verlag Karl Albert, Freiburg München.

Barthelmess, A. (1965). "Grundlagen der Vererbung." Akad. Verlagsges. Athenaion, Konstanz.

Bateson, W. (1894). "Materials for the Study of Variation." Macmillan, New York.

Bateson, W. (1902). Reports to the Evolution Committee of the Royal Soc. Report I, Part II, pp. 142–160. Harrison and Sons, London.

Bateson, W., Saunders, E. R., and Punnett, R. C. (1905). Reports to the Evolution Committee of the Royal Soc. Report ii, p. 89. Harrison and Sons, London.

Bateson, W. (1928). *In* "The Scientific Papers of William Bateson" (R. C. Punnett, ed.), 2 vols. Cambridge Univ. Press, London and New York.

Baur, E. (1911). "Einfuhrung in die experimentelle Vererbungslehre," Borntraeger, Berlin (2nd ed., 1914, 3rd ed., 1919).

Beadle, G. W. (1959). *Science* **129**, 1715–1726.

Beadle, G. W. (1963). "Genetics and Modern Biology." Mem. Am. Phil. Soc., Philadelphia, Pennsylvania.

Belling, J. (1926). *Biol. Bull.* **50**, 355–363.

Bennett, J. H., ed. (1965). "Experiments in Plant Hybridization by Gregor Mendel" (with commentary by R. A. Fisher). Oliver & Boyd, Edinburgh and London.

Bernal, J. D. (1965). "Science in History." 3rd ed. Watts, London.

Bernstein, F. (1924). *Klin. Wochschr.* **3**, 1495–1497.

Bittner, J. J. (1936). *Science* **84**, 162.

Blakeslee, A. F. (1904). *Proc. Am. Acad. Arts Sci.* **40**, 205–319.

Blakeslee, A. F. (1936). *Brooklyn Botan. Garden Mem.* **4**, 29–40.

Boyer, S. H., IV (1963). "Papers on Human Genetics." Prentice-Hall, Englewood Cliffs, New Jersey.

Braun, W. (1953). "Bacterial Genetics." Saunders, Philadelphia, Pennsylvania.
Bresch, C. (1964). "Klassische und molekulare Genetik." Springer, Berlin.
Bridges, C. B. (1916). *Genetics* **1**, 1–52 and 107–163.
Bridges, C. B. (1919). *Anat. Record* **15**, 351–358.
Bridges, C. B. (1935). *J. Heredity* **26**, 60–64.
Bridges, C. B. (1936). *Science* **83**, 210–211.
Brink, R. A., ed. (1967). "Heritage from Mendel." Univ. of Wisconsin Press, Madison, Wisconsin.
Bryson, V., ed. (1959). "Microbiology Yesterday and Today." Inst. Microbiol., Rutgers Univ. Press, New Brunswick, New Jersey.
Cairns, J., Stent, G. S., and Watson, J. D., eds. (1966). "Phage and the Origins of Molecular Biology." Cold Spring Harbor Lab. Quant. Biol. Cold Spring Harbor, New York.
Carlson, E. A. (1966). "The Gene: A Critical History." Saunders, Philadelphia, Pennsylvania.
Caspari, E. (1948). *Advan. Genet.* **2**, 1–66.
Caspari, E. (1960). *Perspectives Biol. Med.* **4**, 26–39.
Caspersson, T., and Schultz, J. (1938). *Nature* **142**, 294.
Castle, W. E. (1903). *Proc. Am. Acad. Arts Sci.* **39**, 223–242 (reprinted in Křiženecký, 1965b).
Chargaff, E. (1950). *Experientia* **6**, 201–209 (reprinted in Taylor, 1965).
Chetverikov, S. S. (1927). *Verhandl. 5th Kong. Vererb., Berlin,* Vol. 2, pp. 1499–1500; vide Dobzhansky, T. (1967). *Genetics* **55**, 1–3.
Cleland, R. E. (1962). *Advan. Genet.* **11**, 147–237.
Cohen, D. (1965). The Biological Role of the Nucleic Acids." American Elsevier, New York.
Coleman, W. (1965). *Proc. Am. Phil. Soc.* **109**, 124–158.
Correns, C. (1900). Cited in Stern and Sherwood (1966).
Correns, C. (1902). *Ber. Deut. Botan. Ges.* **20**, 161–172.
Correns, C. (1908). *Ber. Deut. Botan. Ges.* **36**, 686–701.
Correns, Carl, (1909). *Zeits. f. ind. Abst.-u-Vererb.-Lehre* **1**, 291–329.
Creighton, H. B., and McClintock, B. (1931). *Proc. Natl. Acad. Sci. U.S.* **17**, 492–497.
Crew, F. A. E. (1966). "The Foundations of Genetics." Pergamon Press, Oxford.
Crick, F. H. C., and Kendrew, J. (1957). *Adv. Prot. Chem.* **12**, 133–214.
Cuénot, L. (1903). *Arch. Zool. Exptl. et Gén. 4e ser. Notes et revue* **xxxiii**.
Darlington, C. D. (1939). "Evolution of Genetic Systems." Oliver & Boyd, Edinburgh and London.
Darwin, C. (1876). "The Variation of Animals and Plants under Domestication." 2nd ed. rev. (1st English ed., 1868). Appleton, New York.
Darwin, C. (1877). "Effects of Cross and Self-Fertilization." Appleton, New York.
Darwin, F. (1887). "Life and Letters of Charles Darwin," Vol. 1. Appleton, New York.
Delage, Y. (1895). "La structure du protoplasme et les théories rus l'hérédité." Gallimard, Paris.
Demerec, M., and Hartman, Z. (1956). *Carnegie Inst. Wash. Publ.* **612**, 5–33.
de Vries, H. (1889). Intracellulare Pangenesis." Fischer, Jena (English translation by C. S. Gager, Open Court, Chicago, Illinois).

de Vries, H. (1901–1903). "Die Mutationstheorie," Vols. I and II.
Veit. u. Co., Leipzig (English translation 1909–1910 by J. B. Farmer and A. D. Darbishire, "The Mutation Theory." Open Court, Chicago, Illinois).
Dobzhansky, T. (1951). "Genetics and the Origin of Species," 2nd ed. (1st ed., 1937). Columbia Univ. Press, New York.
Dobzhansky, T. (1962). "Mankind Evolving." Yale Univ. Press, New Haven, Connecticut.
Dodge, B. O. (1927). *J. Agr. Res.* **35**, 289–305.
Dunn, L. C., ed. (1951). "Genetics in the 20th Century." Macmillan, New York.
Dunn, L. C. (1965). "A Short History of Genetics." McGraw-Hill, New York.
Emerson, R. A. (1914). *Nebraska, Univ., Agr. Expt. Sta., Res. Bull.* **4**, 1–35.
Emerson, R. A., Beadle, G. W., and Fraser, A. C. (1935). *Cornell Univ., Agr. Expt. Sta. Mem.* **80**, 1–83.
Fisher, R. A. (1930). "The Genetical Theory of Natural Selection." Oxford Univ. Press (Clarendon, London and New York.
Fisher, R. A. (1936). *Ann. Sci.* **1**, 115–137 (reprinted with comment in Bennett, 1965, and C. Stern and Sherwood, 1966).
Ford, E. B. (1953). *Advan. Genet.* **5**, 43–87.
Ford, E. B. (1964). "Ecological Genetics." Wiley, New York.
Freund, H., and Berg, A. (1963). "Geschichte der Mikroskopie; Leben und Werk grosser Forscher," Vol. I. Umschau-Verlag, Frankfurt/Main.
Galton, F. (1889). "Natural Inheritance." Macmillan, New York.
Garrod, A. E. (1902). *Lancet* **II**, 1616 (reprinted in Harris, 1963).
Garrod, A. E. (1908). *Lancet* **II**, 1, 142, 173, and 214.
Garrod, A. E. (1909). "Inborn Errors of Metabolism" (a revision of Garrod, 1908). Oxford Univ. Press London and New York.
Gillispie, C. C. (1963). *In* "Scientific Change" (A. C. Crombie, ed.) 431–453. Basic Books, New York.
Glass, B. (1947). *Quart. Rev. Biol.* **22**, 196–210.
Glass, B. (1953). *In* "Studies in Intellectual History" (G. Boas, ed.), pp. 148–160. Johns Hopkins Press, Baltimore, Maryland.
Glass, B. (1963). *In* "Scientific Change" (A. C. Crombie, ed.), pp. 521–541. Basic Books, New York.
Grant, V. (1956). *Am. Sci.* **44**, 158–178.
Gustafsson, A. (1963). *In* "Recent Plant Breeding Research," pp. 89–104. Wiley, New York.
Haldane, J. B. S. (1932). "The Causes of Evolution." Longmans, Green, New York.
Hardy, G. H. (1908). *Science* **28**, 49–50 (reprinted in Peters, 1959).
Harris, H. (1963). "Garrod's Inborn Errors of Metabolism." Oxford Univ. Press, London and New York.
Hartman, P. E. (1956). *Carnegie Inst. Wash. Publ.* **612**, 35–61.
Heftmann, E., ed. (1961). "Chromatography." Reinhold, New York.
Hershey, A. D., and Chase, M. (1952). *J. Gen. Physiol.* **36**, 39–56 (reprinted in Taylor, 1965).
Hughes, A. F. W. (1959). "A History of Cytology." Abelard-Schuman, New York.
Iltis, H. (1924). "Gregor Mendel: Leben, Werk und Wirkung." Springer, Berlin.
Ingram, V. M. (1957). *Nature* **180**, 326–328.
Ingram, V. M. (1963). "The Hemoglobins in Genetics and Evolution." Columbia Univ. Press, New York.

Jacob, F., and Wollman, E. L. (1961). "Sexuality and the Genetics of Bacteria." Academic Press, New York.

Jakubiček, M., and Kubíček, J. (1965). "Bibliographia Mendeliana." Universitni Knikovna v Brně (361 citations of works by and relating to Mendel from 1853 to Dec. 1964. Prepared for the 100th anniversary celebration in Brno, 1965).

Johannsen, W. (1903). Über Erblichkeit in Populationen und in reinen Linien." Fischer, Jena (English translation in Peters, 1959).

Jollos, V. (1913). *Biol. Zentr.* **33**, 222–236.

Koltzoff, N. (1939). "Les molécules héréditaires." Hermann, Paris.

Krius, K., and Satara, J. (1918). Cited in Winge (1935).

Křiženecký, J. (1965a). "Gregor Mendel 1822–1884. Texte und Quellen zu seinem Wirken und Leben." Barth, Leipzig.

Křiženecký, J. (1965b). "Fundamenta Genetica" (the revised edition of Mendel's classic with a collection of original papers published during the rediscovery era). Publ. House Czecho. Acad. Sci., Prague.

Lanni, F. (1964). *Advan. Genet.* **12**, 1–141.

Lawrence, W. J. C., and Price, J. R. (1940). *Biol. Rev. Cambridge Phil. Soc.* **15**, 35–58.

Lechevalier, H. A., and Solotorovsky, M. (1965). "Three Centuries of Microbiology." McGraw-Hill, New York.

Lederberg, J. (1951). *In* "Genetics in the 20th Century" (L. C. Dunn, ed.), pp. 263–289. Macmillan, New York.

Lederberg, J., and Tatum, E. L. (1946). *Nature* **158**, 558 (reprinted in Peters, 1959).

Levene, P. A., and Bass, L. W. (1931). "Nucleic Acids." Chem. Catalog Co., New York.

L'Héritier, P. (1948). *Heredity* **2**, 325–348.

Li, C. C. (1967). *Am. J. Human Genet.* **19**, 70–74.

Lindegren, C. C. (1932). *Bull. Torrey Botan. Club* **59**, 119–138.

Lindegren, C. C. (1934). *Am. J. Botany* **21**, 55–65.

Lindegren, C. C. (1936). *J. Genet.* **32**, 243–256 (reprinted in Taylor, 1965).

Lock, R. H. (1906). "Recent Progress in the Study of Variation, Heredity and Evolution." Murray, London.

Luria, S. (1947). *Bacteriol. Rev.* **11**, 1–40.

Luria, S., and Delbrück, M. (1943). *Genetics* **28**, 491–511.

Matthaei, H., Kleinkauf, H., Heller, G., Voigt, H. P., and Matthei, H. (1967). *In* "Heritage from Mendel" (R. A. Brink, ed.), 105–145. Univ. of Wisconsin Press, Madison, Wisconsin.

Maupertuis, P. L. Mde (1751). Vide Glass (1947).

Mayr, E. (1963). "Animal Species and Evolution." Harvard Univ. Press, Cambridge, Massachusetts.

McClung, C. E. (1901). *Anat. Anz.* **20**, 220–226, reprinted in Křiženecký (1965b).

Mendel, G. (1866). *Verhandl. Naturforsch. Ver. Brunn* **4**, 3–47; reprinted *ibid.* **49**, (1911) (a new and corrected translation into English is in C. Stern and Sherwood, 1966).

Mendel, G. (1965). "Autobiographia Juvenilis," Ad centesimum doctrinae geneticae anniversarium 1965. Universitas Purkyniana Brunensis, eds. Vanýsek, J., and Rozehnal, B.

Merz, J. T. (1907). "The History of European Thought in the 19th Century." Vol. II Blackwood, London.
Miescher, F., (1871). Hoppe-Seyler's med.-chem. Unters. **4**, 441.
Mirsky, A. E. (1951). *In* "Genetics in the 20th Century" (L. C. Dunn, ed.), pp. 127–153. Macmillan, New York.
Mirsky, A. E. (1968). *Sci. Am.* **218**, 78–88.
Mitchell, H. K., and Lein, J. (1948). *J. Biol. Chem.* **175**, 481.
Morgan, T. H. (1914). "Heredity and Sex," 2nd rev. ed. Columbia Univ. Press, New York.
Morgan, T. H. (1926). "The Theory of the Gene." Yale Univ. Press, New Haven, Connecticut.
Morgan, T. H., Sturtevant, A. H., Muller H. J., and Bridges C. B. (1915). "The Mechanism of Mendelian Heredity." Holt, New York.
Mourant, A. E. (1954). "The Distribution of the Human Blood Groups." Blackwell, Oxford.
Muller, H. J. (1918). *Genetics* **3**, 422–499.
Muller, H. J. (1927). *Science* **66**, 84–87.
Muller, H. J. (1929). *Proc. Int. Congr. Plant Sci.,* Ithaca, New York **1**, 897–921.
Muller, H. J. (1947). *J. Heredity* **38**, 259–270 (Nobel Lecture).
Muller, H. J. (1951). *In* "Genetics in the 20th Century" (L. C. Dunn, ed.), pp. 77–99. Macmillan, New York.
Němec, B. (1965). See Křiženecký (1965b, pp. 7–13).
Northrop, J. H., Kunitz, M., and Herriott, R. M. (1948). "Crystalline Enzymes," 2nd ed. Columbia Univ. Press, New York.
Olby, R. C. (1965). *Heredity* **20**, 636–638.
Olby, R. C. (1966). "Origins of Mendelism." Constable, London.
Painter, T. S. (1933). *Science* **78**, 585–586.
Pauling, L., Itano, H. S., Singer S. J., and Wells, I. B. (1949). *Science* **110**, 543–548.
Pauling, L., Corey, R. B., and Branson, H. R. (1951). *Proc. Natl. Acad. Sci. U.S.* **37**, 205–207.
Pearson, K. (1904). *Phil. Trans. Roy. Soc. London* **A203**, 53–86.
Peters, J. A. (1959). "Classic Papers in Genetics." Prentice-Hall, Englewood Cliffs, New Jersey.
Poulson, D. F., and Sakaguchi, B., *Science* 133, 1489–90 (1961).
Ravin, A. W. (1961). *Advan. Genet.* 10, 61–163.
Ravin, A. W. (1965). "The Evolution of Genetics." Academic Press, New York.
Stanley, W. M. (1935). *Science* 81, 644–645.
Rhoades, M. M. (1933). *J. Genet.* **27**, 71–93.
Rhoades, M. M., and McClintock, B. (1935), *Botan. Rev.* **1**, 292–325.
Richter, O. (1940). *Verhandl. Naturforsch. Ver. Brunn* **72**, 110–173.
Roberts, H. F. (1929). "Plant Hybridization Before Mendel." Princeton Univ. Press, Princeton, New Jersey.
Romanes, G. J. (1895). "Darwin and After Darwin," Vol. II. Longmans, Green, New York.
Schoenheimer, R. (1942). "The Dynamic State of the Body Constituents." Harvard Univ. Press, Cambridge, Massachusetts.
Schrader, F. (1948). *Science* **107**, 155–159.
Schrödinger, E. (1945). "What is Life? The Physical Aspect of the Living Cell." Macmillan, New York.

Sheppard, P. M. (1958). "Natural Selection and Heredity." Hutchinson, London.
Stadler, L. J. (1928). *Science* **68**, 186–187.
Stadler, L. J. (1954). *Science* **120**, 811–819.
Stadler, L. J., and Sprague, G. F. (1936), *Proc. Natl. Acad. Sci. U.S.* 572–591.
Stadler, L. J., and Uber, F. M. (1942), *Genetics* **27**, 84–118.
Stein, E. (1950). *Naturwissenschaften* **37**, 457–463.
Stent, G. S. (1968). *Science* **160**, 390–395.
Stern, C. (1931). *Biol. Zentr.* **51**, 547–587.
Stern, C., Sherwood, E., ed. (1966). "The Origin of Genetics, a Mendel Source Book." Freeman, San Francisco, California [a new translation into English of Mendel's paper of 1866 and of de Vries's "rediscovery" paper of 1900 and that of Correns (1900) with comments by C. Stern, R. A. Fisher, and S. Wright].
Stubbe, H. (1938). "Genmutation." Borntraeger, Berlin.
Stubbe H. (1965). "Kurze Geschichte der Genetik bis zur Wiederentdeckung der Vererbungsregeln Gregor Mendels," 2nd ed. Fischer, Jena. (1st ed. 1963).
Sturtevant, A. H. (1920). *Proc. Soc. Exptl. Biol. and Med.* **27**, 70–71.
Sturtevant, A. H. (1965). "A History of Genetics." Harper, New York.
Svedberg, T., and Pedersen, K. O. (1940). "The Ultracentrifuge." Oxford Univ. Press (Clarendon), London and New York.
Swift, H. (1950). *Proc. Natl. Acad. Sci. U.S.* **36**, 643–654 (reprinted in Taylor, 1965.)
Tatum, E. L. (1945). *Proc. Natl. Acad. Sci.* **31**, 215–219.
Tatum, E. L. (1946). Cold Spring Harbor Sympos. Quant. Biol. 2.
Tatum, E. L. (1959). *Science* **129**, 1711–1715.
Tatum, E. L., and Beadle, G. W. (1941). *Proc. Natl. Acad. Sci. U.S.* **27**, 499–506.
Taylor, J. H., ed. (1965). "Selected Papers on Molecular Genetics." Academic Press, New York.
Timoféeff-Ressovsky, N. (1937). "Experimentelle Mutationsforschung in der Vererbungslehre." Steinkopff, Dresden and Leipzig
Timoféeff-Ressovsky, N., Zimmer, K. G., and Delbrück, M. (1935). *Nachr. Ges. Wiss. Goettingen, Math.-Physik. Kl. VI*, 189–245.
Waksman, S. A. (1959). *In* Bryson, 1959.
Walker, J. C. (1951). *In* "Genetics in the 20th Century" (L. C. Dunn, ed.), pp. 527–554. Macmillan, New York.
Wallace, H. A., and Brown, W. L. (1956). "Corn and its Early Fathers." Michigan State Univ. Press, East Lansing, Michigan.
Watson, J. D. (1965). "Molecular Biology of the Gene." Benjamin, New York.
Watson, J. D., and Crick, F. H. C. (1953). *Nature* **171**, 737–738 (reprinted in Taylor, 1965).
Watson, J. D. (1968). "The Double Helix." Athenaeum, New York.
Weinberg, W. (1908). *Jahresber. Ver. Vaterländische Naturk. Wurttemberg* **64**, 369–382 (reprinted in Boyer, 1963).
White, M. J. D. (1945). "Animal Cytology and Evolution." Cambridge Univ. Press, London and New York.
Wilson, E. B. (1896). "The Cell in Development and Inheritance." Columbia Univ. Press, New York (reprinted, 1966, with introduction by H. J. Muller, Johnson Reprint Corp., New York).

Wilson, E. B. (1900). "The Cell in Development and Inheritance," 2nd ed. Macmillan, New York.
Wilson, E. B. (1902). *Science* **14**, 99 (reprinted in Křiženecký, 1965b).
Wilson, E. B. (1928). "The Cell in Development and Heredity," 3rd ed. with corrections. Macmillan, New York (the last edition of this classic of cytology).
Winge, Ö. (1935). Compt. Rend. Trav. Lab. Carlsberg **21**, 77.
Winkler, H. (1920). "Verbreitung und Ursache der Parthenogenesis im Pflanzen-und-Tierreiche." Fischer, Jena.
Wolff, S., ed. (1963). "Radiation Induced Chromosome Aberrations." Columbia Univ. Press, New York.
Wright, S. (1917). *J. Heredity* **9**, 521–527.
Wright, S. (1931). *Genetics* **16**, 97–159.
Wright, S. (1967). *In* "The Origin of Genetics, a Mendel Source Book" (C. Stern and E. Sherwood, eds.), pp. 173–175. Freeman, San Francisco, California.
Yule, G. U. (1902). *New Phytologist* **1**, 193–207 and 223–238.
Zinder, N. D., and Lederberg, J. (1952). *J. Bacteriol.* **64**, 679–699 (reprinted in Peters, 1959).
Zirkle, C. (1935). "The Beginnings of Plant Hybridization." Univ. of Pennsylvania Press, Philadelphia, Pennsylvania.
Zirkle, C. (1946). *Trans. Am. Philo. Soc.* [N. S.] **35**, 91–151.

II THE STRUCTURE OF THE NUCLEIC ACIDS

HENRY M. SOBELL

I. Introduction

The nucleic acids form a class of biological polymers which are uniquely specialized in their ability to store and to transmit genetic information in living systems. The information transfer properties of these macromolecules are known primarily to reflect hydrogen-bonding specificity between the complementary purines and pyrimidines, adenine and thymine (or uracil), guanine and cytosine. The base-pairing scheme proposed by Watson and Crick (1953) in their structural hypothesis for the DNA molecule has gained wide acceptance and provides a natural explanation for information transfer during the replication of DNA, and during the transcription of DNA into messenger RNA. However, other types of purine-pyrimidine interactions may also be used in biological systems. On the level of messenger RNA translation, for example, it seems likely that the degeneracy of the genetic code partly reflects additional base-pairing combinations besides the usual Watson-Crick combinations (Crick, 1966). Similarly, macromolecular entities such as the ribosome or transfer RNA could use additional interactions to stabilize their structures or to perform their biological functions (Spirin, 1963; Holley *et al.,* 1965). Base analog mutagens, such as 2-aminopurine and 5-bromouracil, are known to have chemical and physical properties which differ from the naturally occurring bases and these may alter their hydrogen-bonding properties (Freese, 1959; Lawley and Brookes, 1962).

This chapter begins by summarizing recent advances in understanding the hydrogen-bonding properties of purine and pyrimidine monomer derivatives in solution and in the crystalline state. These studies have given additional insight into the nature of intermolecular forces involved in purine-pyrimidine base pairing, and are therefore directly related to the following section which describes the structure and physical chemistry of synthetic and naturally occurring nucleic acid complexes. Further sections discuss recent advances in understanding the three-dimensional structure of DNA and RNA, and include a summary of data now available on the primary and secondary structures of transfer RNA, and of 5 S ribosomal RNA. It is anticipated that a great deal of additional nucleotide sequence data will become available on RNA, as well as on DNA, in the years ahead.

II. Hydrogen-Bonding Specificity between Monomer Purines and Pyrimidines in Solution

A. Infrared Spectroscopy and Nuclear Magnetic Resonance Studies

The techniques of infrared spectroscopy and nuclear magnetic resonance have shown that considerable hydrogen-bonding specificity exists between monomer purine and pyrimidine derivatives in nonpolar solvents (Hamlin *et al.,* 1965; Küchler and Derkosch, 1966; Pitha *et al.,* 1966;

Kyogoku *et al.*, 1966, 1967a,b; Katz and Penman, 1966; Shoup *et al.*, 1966; Miller and Sobell, 1967). These studies have shown that adenine and uracil derivatives, as well as guanine and cytosine derivatives, interact to form a hydrogen-bonded dimer much more strongly with each other than with themselves. In these studies, derivatives of the monomer base pairs have been synthesized which are soluble in nonpolar solvents. Nonpolar solvents have been used to maximize hydrogen bonding between these compounds and to minimize base stacking which is found in aqueous solution (Ts'o *et al.*, 1963; Jardetzky, 1964; Solie and Schellman, 1968). In addition, these solvents allow accurate infrared measurement of the amino and N–H stretch frequencies, and permit observation of the proton signals from these groups by nuclear magnetic resonance.

Figure 1 shows the infrared spectrum of dilute and concentrated solu-

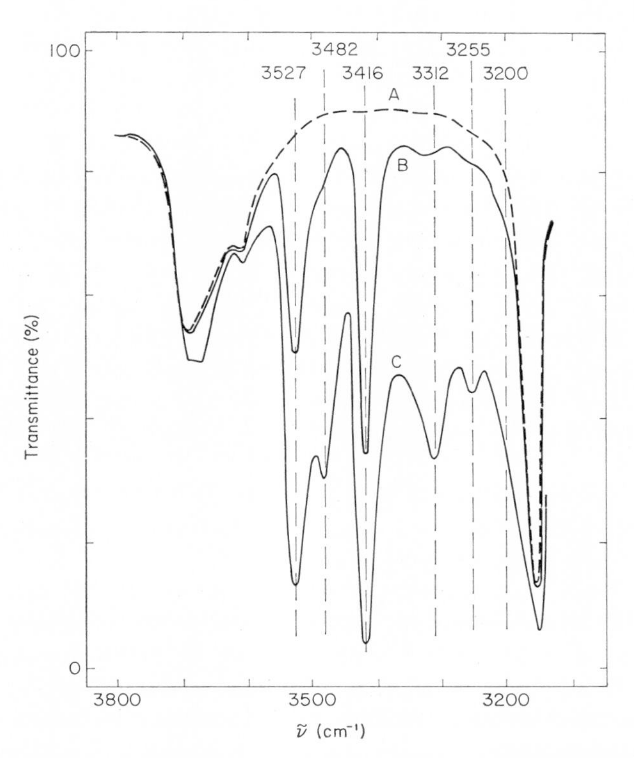

FIG. 1. Infrared spectra of 9-ethyladenine in deuterochloroform from 3100 to 3800 cm^{-1}. Silica cells having a path length of 1 mm were used. A. Dashed line is the solvent spectrum. B. 9-Ethyladenine at a concentration of 0.022 *M*. C. 9-Ethyladenine saturated at 21°C (about 0.1 *M*). (Redrawn from Hamlin *et al.*, 1965.)

tions of 9-ethyladenine. The spectrum of the dilute solution (0.022 M) shows two prominent peaks at 3416 and 3527 cm^{-1}, corresponding to the symmetric and antisymmetric stretch frequencies of the amino group of adenine.

These stretch frequencies correspond to the two types of fundamental vibrations of the amino group. In the symmetric stretching mode, both protons of the amino group simultaneously vibrate toward or away from the amino nitrogen atom

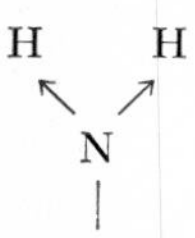

In the antisymmetric stretching mode, the protons alternately vibrate toward and away from the amino nitrogen

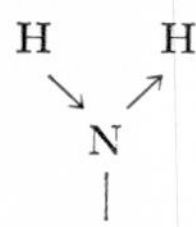

The effect of hydrogen bonding is to decrease the force constant, k, of the N—H bond. The absorption band is therefore shifted to lower frequency, i.e., as a first approximation, the frequency of vibration is given by the equation for a harmonic oscillator, $\nu = (1/2c)\sqrt{k/m}$, where ν is the frequency in cm^{-1}, c is the velocity of light in cm/sec, m is the reduced mass in grams, and k the force constant for the bond in dynes/cm (for descriptions of the techniques of infrared spectroscopy, see Colthup *et al.*, 1964; for the use of infrared spectroscopy and nuclear magnetic resonance in studying the hydrogen bond, see Pimental and McClellan, 1960).

The more concentrated solution (approximately 0.1 M) shows additional peaks at 3482, 3312, 3255, and a shoulder at 3200 cm^{-1}, indicating hydrogen bonding between adenine residues.

Figure 2 shows the infrared spectra of dilute (0.019 M) and concentrated (0.077 M) solutions of cyclohexyluracil. Both spectra show a prominent band at 3392 cm^{-1}, corresponding to the nonbonded N—H stretching vibration on uracil. An additional shoulder appears in the more concentrated solution at 3210 cm^{-1}, indicating hydrogen bonding between uracil residues.

Figure 3 shows the infrared spectra of varying molar ratios of 9-ethyladenine and cyclohexyluracil. Although dilute solutions of these compounds show no evidence of hydrogen bonding, upon mixing these solutions new peaks appear at 3490, 3330, 3260, and 3210 cm^{-1}. This indicates preferential hydrogen bonding between adenine and uracil rings. The stoichiometry of the interaction can be determined by plotting optical density of a particular association band as a function of the mole percent adenine and uracil (Fig. 4). The optical density of the dimer band is seen to be maximal at a 1:1 mixture of these compounds.

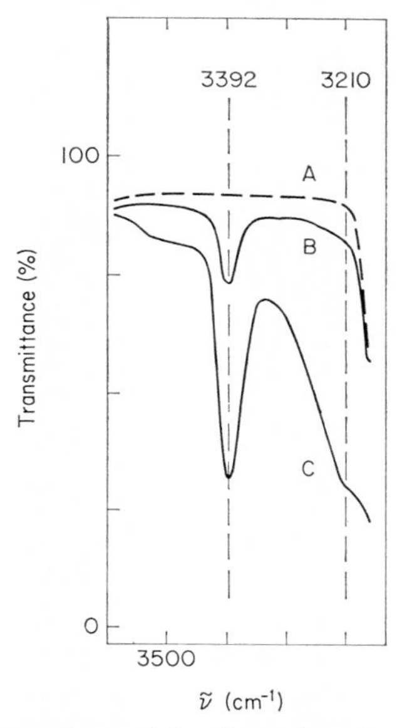

FIG. 2. Infrared spectra of 1-cyclohexyluracil in deuterochloroform from 3200 to 3500 cm^{-1}. The path length of the solutions was 0.2 mm and NaCl windows were used in the cells. A. The dashed line is the solvent spectrum. B. 1-Cyclohexyluracil at a concentration of 0.019 *M*. C. 1-Cyclohexyluracil saturated at 21°C (0.077 *M*). (Redrawn from Hamlin *et al.*, 1965.)

It is possible to determine the equilibrium constants for this interaction, as well as for the self-interactions, from infrared data such as these. Kyogoku *et al.* (1967a) have determined the equilibrium constants for the uracil-uracil, adenine-adenine, and adenine-uracil interactions at various temperatures, and have calculated the heats of formation and the entropy changes associated with complex formation (see Table I). In all cases, the $\Delta S°$ value is about 12 entropy units, not unexpected considering the similarity of these molecules. However, it is seen that the adenine-uracil interaction involves a $\Delta H°$ value of 6.2 ± 0.6 kcal/mole, compared with the $\Delta H°$ values of 4.3 ± 0.4 kcal/mole and 4.0 ± 0.8 kcal/mole observed for the uracil-uracil and adenine-adenine interac-

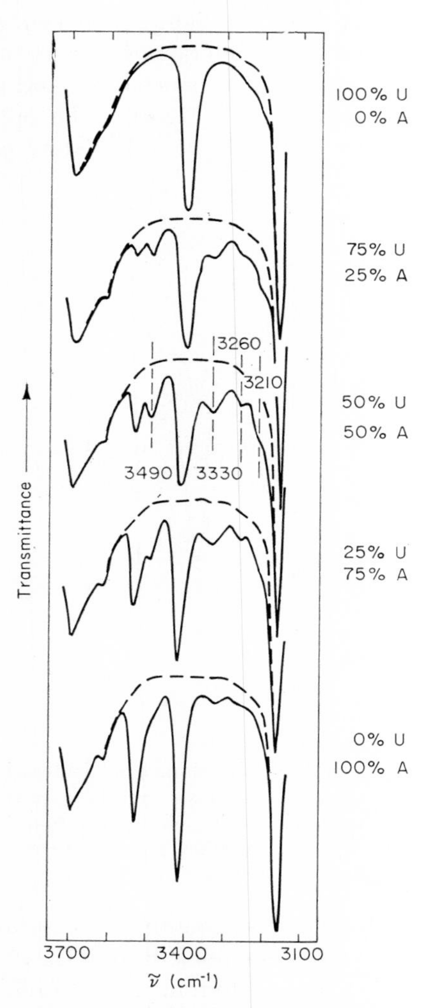

FIG. 3. A continuous mixing study demonstrating hydrogen bonding between 9-ethyladenine and 1-cyclohexyluracil in deuterochloroform solution from 3100 to 3700 cm^{-1}. The path length was 1 mm in these experiments, and silica cells were used. Dashed lines represent the solvent and cell background, while solid lines represent the actual spectra. The total concentration in all cases was 0.022 *M* and spectra from varying molar ratios of adenine and uracil derivatives were recorded. (Redrawn from Hamlin *et al.*, 1965.)

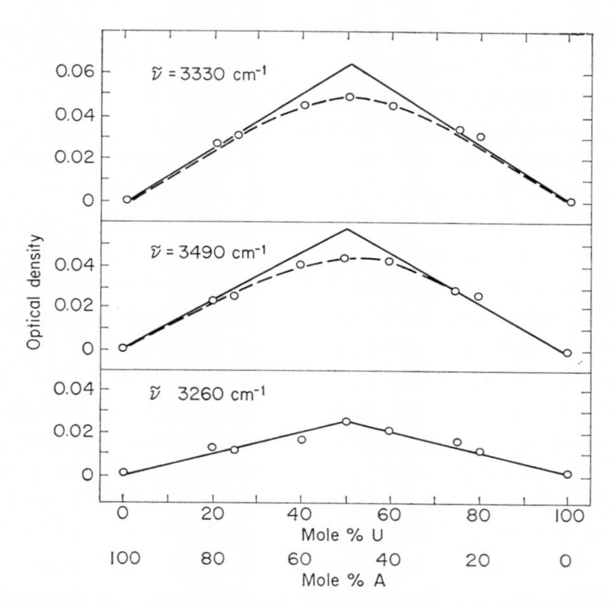

FIG. 4. The change in optical density of three bands due to hydrogen-bonded species present in solutions of ethyladenine and cyclohexyluracil. The abscissa represents the molar ratio of uracil and adenine derivatives. The change in optical density in the lowest curve is too small to show the difference between dashed and solid lines. (Redrawn from Hamlin *et al.*, 1965.)

tions, respectively. The difference of 2.0 kcal between ΔH°_{AU} and ΔH°_{AA} (or ΔH°_{UU}) is small, but has an important effect on the association constants. The association constant between adenine and uracil, K_{AU}, is 15 times larger than K_{UU}, which is about twice as large as K_{AA} at 25°C. These values mean that, if equal volumes of 0.01 M solutions of adenine and uracil are mixed at 25°C, 25.8% of each dissolved substance forms

TABLE I
THERMODYNAMIC FUNCTIONS[a]

Form of dimer Wavenumber (cm⁻¹) of measurements	Uracil-uracil 3395	Adenine-adenine 3527	 3416	 Mean	Adenine-uracil 3527	 3404	 Mean
$-\Delta H^\circ$ (kcal/mole of dimer)	4.3 ± 0.4	4.2	3.8	4.0 ± 0.8	6.3	6.1	6.2 ± 0.6
$-\Delta S^\circ$ (entropy units)	11.0 ± 1.0	12.1	10.7	11.4 ± 2.0	12.8	10.8	11.8 ± 1.2

[a] From Kyogoku *et al.* (1967a).

the complex dimer, 3.0% of uracil forms the uracil-uracil dimer, 1.6% of adenine forms the adenine-adenine dimer, and 71.2% of uracil and 72.6% of adenine are left as monomer.

The hydrogen-bonding interaction between 9-ethyladenine and 1-cyclohexyluracil has also been studied by the technique of nuclear magnetic resonance (NMR) (Katz and Penman, 1966) (for a summary of the principles of NMR, see Pople *et al.*, 1959; for the application of NMR to the study of the hydrogen bond, see Pimental and McClellan, 1960). Hydrogen bond formation is known to decrease the magnetic shielding of the protons involved, resulting in a downfield chemical shift. Figure 5 shows the behavior of the hydrogen-bonding proton resonances of adenine and uracil derivatives in deuterochloroform in the presence and absence of the complementary base. The very large downfield shift of the uracil N–3 proton resonance with increasing concentrations of adenine indicates a large degree of association between these compounds. The adenine amino resonance shows a smaller downfield shift which indicates that these proton resonances are less sensitive than the N—H protons to association. It is observed, however, that there is a large deviation from linearity of the adenine amino-proton resonance at high uracil to adenine ratios. This suggests that the association between adenine and uracil at this elevated concentration is not simply 1:1. The data are consistent with the formation of a sizable amount of 2:1 trimer species formed between the uracil and adenine derivatives. This finding is of particular interest with respect to the crystal studies to be discussed in the next section.

The preferential interaction between guanine and cytosine derivatives has also been studied by infrared and nuclear magnetic resonance methods. Kyogoku *et al.* (1966) have demonstrated that 2′,3′-benzylidine-5′-trityl guanosine and the correspondingly substituted cytidine form strong association bands in a 1:1 mixture of these compounds in deuterochloroform, indicating complex formation. Similar findings for different guanine and cytosine derivatives have also been reported (Pitha *et al.*, 1966).

Proton magnetic resonance has also been used to study the guanine-cytosine interaction (Katz and Penman, 1966; Shoup *et al.*, 1966). Figure 6 shows the proton magnetic resonance spectrum of 9-ethylguanine and 1-methylcytosine individually, and as a 1:1 mixture, in dimethylsulfoxide. Although dimethylsulfoxide forms hydrogen bonds with the purine and pyrimidine derivatives, as shown by a temperature dependence of the chemical shifts of the N—H and NH_2 protons of individual solutions of these compounds, further downfield shifts are observed when complementary purines and pyrimidines are mixed, indicating preferential interaction. The interaction appears to be extremely specific, for little or no guanine-guanine, guanine-adenine, guanine-thymine, or adenine-cytosine interactions are observed at these concentrations (see Table II). This is remarkable, since these compounds potentially are capable of

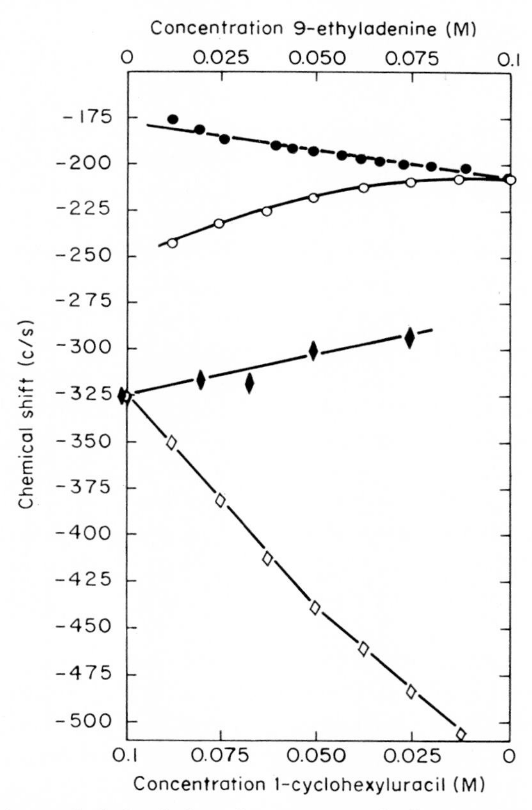

FIG. 5. Chemical shift of the adenine amino hydrogens (○, ●) and uracil N_3 protons (□, ■) in the presence and absence of the complementary species. The measurements were carried out at 25°C in deuterochloroform, to which a small amount of dimethyl sulfoxide was added to serve as an internal marker. (●, ■) Dilution curves for the individual species; (○, □) results obtained when sufficient complementary nucleoside derivative was added to bring the total nucleoside concentration to 0.1 *M*. (Redrawn from Katz and Penman, 1966.)

forming hydrogen-bonded complexes with each other. There is no obvious reason for this high degree of specificity. Additional factors appear to be involved in the hydrogen bonding specificity between these compounds and these are not completely understood at the present time.

In addition to demonstrating interactions between the naturally occurring purine and pyrimidine base derivatives, infrared studies have been carried out which demonstrate hydrogen bonding between synthetic purine and pyrimidine analogs. Miller and Sobell (1967) have shown

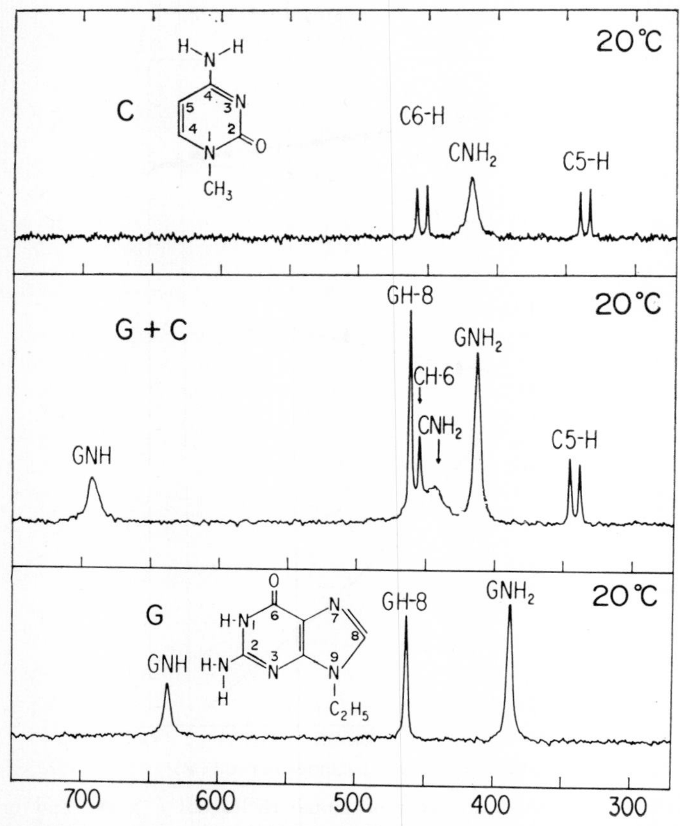

FIG. 6. NMR spectra of 9-ethylguanine (G) and 1-methylcytosine (C) in DMSO-d_6. The spectra were measured with a Varian A-60 spectrometer. Each spectrum was calibrated by means of audio frequency side bands. The abscissa is given in cps as downfield shifts from an internal TMS standard. (Redrawn from Shoup *et al.*, 1966.)

that alkylated derivatives of 2-aminopurine and 2,6-diaminopurine preferentially hydrogen bond with uracil and bromouracil derivatives in deuterochloroform. Similar, although more extensive, studies have been reported by Kyogoku *et al.* (1967b) for a larger series of synthetic base analogs. The results of these studies are summarized in Fig. 7.

TABLE II
SHIFTS IN THE RESONANT FREQUENCY OF THE N_1 PROTON OF GUANOSINE AND THE N_3 PROTON OF URIDINE UPON THE ADDITION OF OTHER NUCLEOSIDES[a,b]

Guanosine	
Nucleoside added	Shift in cycles/sec of N_1 proton
Adenosine	−0.1
Guanosine	−7.1
Cytidine	−134.7
Uridine	−1.2
Uridine	
Nucleoside added	Shift in cycles/sec of N_3 proton
Adenosine	−8.2
Guanosine	0
Cytidine	−0.6
Uridine	0

[a] From Katz and Penman (1966).

[b] The solvent was an equal volume mixture of dimethylsulfoxide and benzene. The chemical shifts were measured at 60 *M*c/sec and −4°C. The dimethylsulfoxide resonance was used as an internal standard. The N_1 proton of guanosine and the N_3 proton of uridine were measured at a base concentration of 0.05 *M*. Then additional nucleosides were added to bring the total molarity to 0.2 *M* and the proton frequency measured again. The table gives the added compound and the net shift between the two measurements. Because of broadening due to the ^{14}N quadripole moment, the accuracy of the relative shift measurement was limited to about ± 1.0 cycles/sec.

B. Ultraviolet Hypochromism

Along with these infrared and nuclear magnetic resonance studies, ultraviolet absorption studies have been carried out on solutions containing similar nucleoside derivatives in chloroform (Thomas and Kyogoku, 1967). These studies demonstrate that significant hypochromism is associated with hydrogen bonding between complementary base pairs, indicating that stacking of bases in polynucleotide complexes is not the sole factor promoting hypochromism of electronic transitions in purine and pyrimidine heterocyclic rings.

Figure 8a shows the ultraviolet absorption spectra of chloroform solutions containing adenine-uracil in varying molar ratios. If one plots the absorbance at 265 mμ versus mole fraction adenine, it is observed that there is a significant negative deviation from Beer's law at equimolar mixtures of these compounds (Fig. 8b). In view of the evidence presented earlier, this hypochromism appears to be correlated with the formation of a 1:1 hydrogen-bonded dimer. The equilibrium constant for dimer formation at 25°C, as determined by infrared measurements, is 1.0×10^2 liters/mole. In the equimolar mixture, with $[C_T] = 0.0075$, for example,

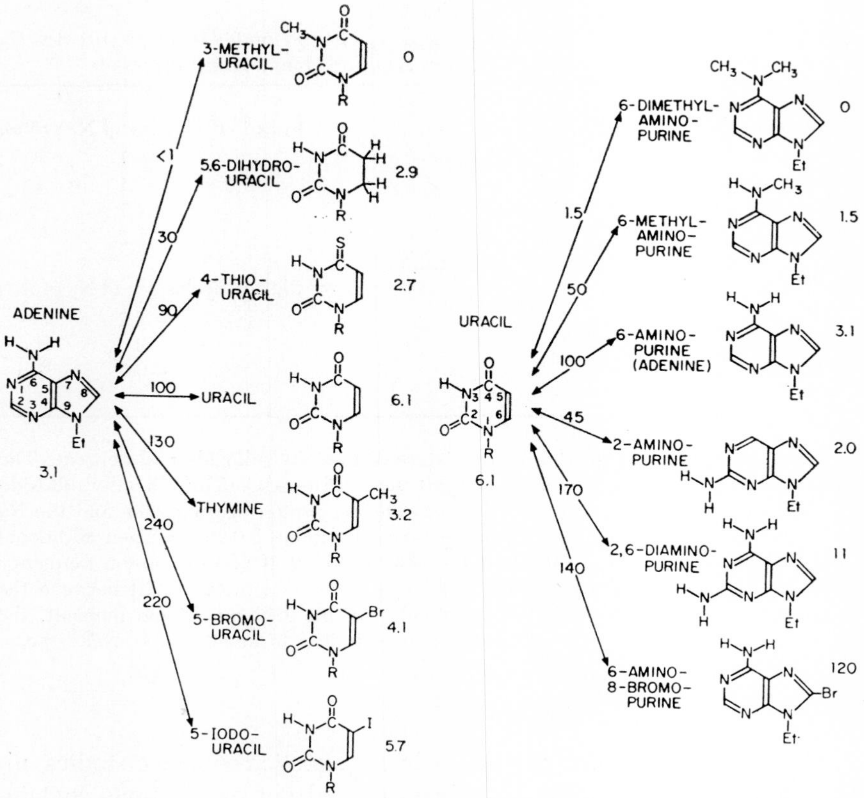

FIG. 7. Association constants (liter/mole) between various substituted 1-cyclohexyluracil derivatives and 9-ethyladenine derivatives in deuterochloroform solution at 25°C. Figures near the double-headed arrows are the association constants between adenine and uracil derivatives, while the numbers shown to the right of the structural formulae are the self-association constants (A) Association of 9-ethyladenine with uracil derivatives. (B) Association of 1-cyclohexyluracil with adenine derivatives. (Redrawn from Kyogoku *et al.*, 1967b.)

FIG. 8. Ultraviolet absorption spectra of chloroform solutions of A and U mixtures as indicated. Total solute concentration is 0.0075 *M* and sample-cell length is 0.00906 cm. In this and in subsequent figures, the lowest curve shows the residual absorption when chloroform is placed in both sample and reference beam cells. (Redrawn from Thomas and Kyogoku, 1967.)

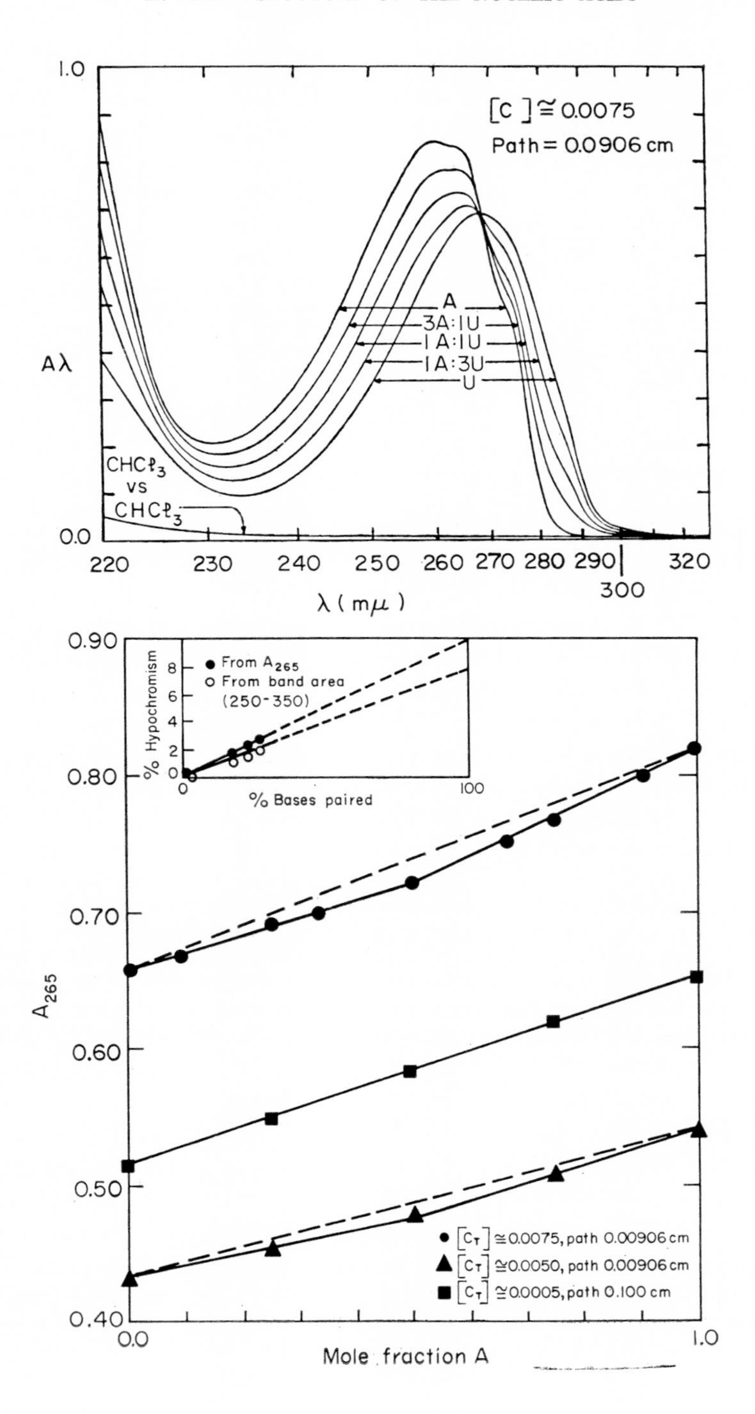
[C] ≅ 0.0075
Path = 0.0906 cm
A
3A:1U
1A:1U
1A:3U
U
Aλ
CHCl3 vs CHCl3
λ (mμ)
From A265
From band area (250-350)
% Hypochromism
% Bases paired
A265
[CT] ≅ 0.0075, path 0.00906 cm
[CT] ≅ 0.0050, path 0.00906 cm
[CT] ≅ 0.0005, path 0.100 cm
Mole fraction A

23% of each base is present in the dimer form, and the relative hypochromism observed is between 2.0 and 2.5% of the expected band absorbance. At $[C_T]$ less than 0.0005, fewer than 2% of the bases remain paired and no hypochromism can be detected. The results for several values of $[C_T]$ are summarized in the Fig. 8b inset, where it is observed that an extrapolated value of 8–10% relative hypochromism would be expected if all the bases exist as hydrogen bonded dimers. The percent relative hypochromism at wavelength λ for an equimolar mixture of interacting compounds X and Y is defined as $100\{1 - [2A_\lambda^{X+Y}/(A_\lambda^X + A_\lambda^Y)]\}$, where A_λ^X, A_λ^Y, and A_λ^{X+Y} are absorbances at λ for solutions of X, Y, and the mixture of X and Y, respectively, at the same solute concentration. To obtain the overall band hypochromism, the respective integrated band intensities are substituted for absorbance. Unfortunately, overlapping of bands makes the quantitative evaluation of band areas difficult.

Since the adenine and uracil absorption bands are closely spaced, it is difficult to compare relative integrated band intensities. This difficulty is obviated by replacing uracil with bromouracil in similar studies, as shown in Fig. 9. It is found that the absorption band of bromouracil occurs at 287 mμ, compared to 268 mμ for uracil. Hypochromism can be seen to be localized almost entirely in the 287 mμ band of bromouracil, the 260 and 266 mμ bands of adenine being unperturbed by base pairing. Since bromine substitution at the C-5 position of uracil enhances its interaction with adenine, the dimerization constant being 2.4×10^2 liters/mole, one should expect to find a somewhat larger hypochromic effect for adenine-bromouracil than for adenine-uracil mixtures at comparable values of $[C_T]$. It is found that there is 6–7% hypochromism where $[C_T] = 0.005$ (30% bases paired), 1.5% at $[C_T] = 0.0005$ (5.4% bases paired), and not detectable at $[C_T] = 0.00005$ (less than 2% bases paired). Thus, about 20% relative hypochromism is expected for this interaction if all monomers exist as hydrogen-bonded dimers. This is about twice the hypochromism observed in the adenine-uracil interaction and this may reflect the more precise measurement of integrated band intensities in this study.

The interaction between hypoxanthine and cytosine as well as guanine and cytosine derivatives has been studied in a similar manner. In these studies one sees both a positive deviation (hyperchromism), as well as a negative deviation (hypochromism) from Beer's law. For instance, for guanine and cytosine mixtures, the observed absorbances show a negative Beer's law deviation at wavelengths higher than 267 mμ and a positive deviation at shorter wavelengths. Interpretation of hypochromism in the

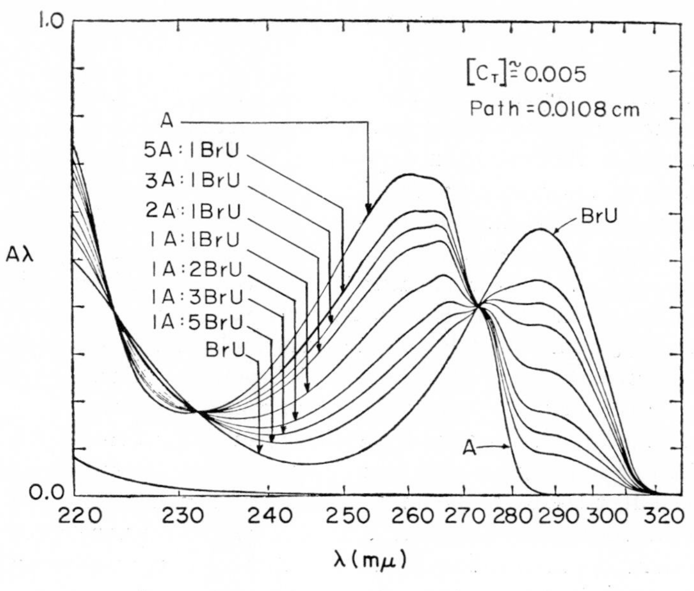

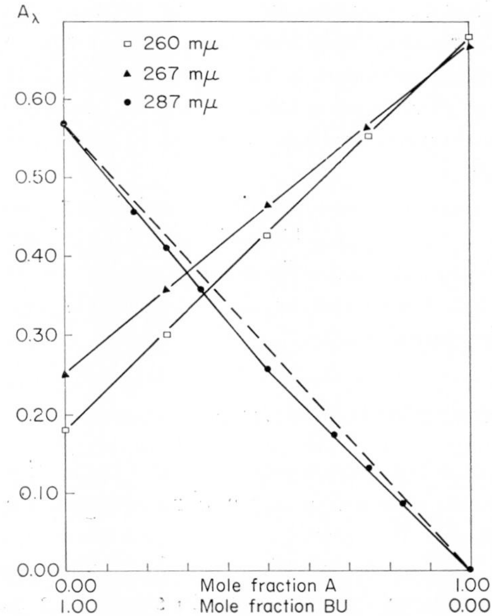

FIG. 9. (a) Ultraviolet absorption spectra of chloroform solutions of A and BrU mixtures as indicated. Total solute concentration is 0.005 *M* and sample-cell length is 0.0108 cm. (Redrawn from Thomas and Kyogoku, 1967.)

(b) Comparison of observed and Beer's law absorbances (dashed lines) at various wave lengths for solutions containing different relative amounts of adenine and bromouracil derivatives. The total concentration, $[C_T]$, of adenine and bromouracil derivatives is 0.005 *M*.

high wavelength region (277 and 320 mμ) localized predominantly in the cytosine band is straightforward. The effect is greatest in equimolar mixtures and is concentration dependent, so that the hypochromism is attributable to the formation of a 1:1 guanine-cytosine dimer. The absorption spectra of guanine and cytosine solutions alone follow Beer's law closely at these higher wavelengths, indicating that the hypochromism in the guanine and cytosine mixtures is not complicated by self-association. Finally, the assumption of reasonable values for the dimerization constants (K_{GC} approximately 10^4–15^5 liters/mole and K_{GG} approximately 5×10^3 liters/mole) show that the hypochromism is linearly related to the fraction of guanine-cytosine dimers in mixed solutions.

On the other hand, apparent hyperchromism in the 240–277 mμ region in solutions containing guanine and cytosine is not as readily explained. First, the effect seems to be independent of concentration and is not maximal in equimolar mixtures. Second, the absorption spectrum of guanine alone does not follow Beer's law at these lower wavelengths, the extinction coefficient of guanine decreasing significantly (approximately 10–20%) with increasing concentration. The large hypochromism in concentrated guanine solutions is undoubtedly due to self-association by hydrogen bonding. For these reasons, it is not possible to give a simple explanation of the hyperchromism observed in guanine-cytosine mixtures at these wavelengths.

In addition to these studies, solutions containing pairs of noncomplementary bases A + C, A + G, U + G, and U + C have also been examined but show no unusual effects attributable to heterologous base-pairing interactions. These results are in complete agreement with the infrared results described earlier.

C. Column Chromatographic Evidence

Specific interaction between complementary nucleosides has been demonstrated in aqueous solution using a chromatographic system (Tuppy and Küchler, 1964a,b). It is possible to prepare chromatographic columns containing single nucleosides covalently linked to an Amberlite cellulose resin. Such a column can then be used to separate a sample containing different nucleoside bases. For example, a guanosine-Amberlite column can be used to separate a mixture of cytidine and uridine (Fig. 10). Cytidine is selectively retarded by the column, i.e., uridine is eluted first from the column, then cytidine. If, however, these same nucleosides are passed down an adenosine-Amberlite column, the nucleosides come off in reverse order. These effects are abolished in 7 *M* urea, conditions known to abolish hydrogen bonding in proteins and nucleic acids. A detailed study has shown that methylating the ring nitrogens normally involved in hydrogen bonding prevents separation (Tuppy and Küchler, 1964b). Thus, compounds such as N–3-methyluridine and dimethylcytidine are not retarded by this chromatographic system. The data strongly suggest that these interactions involve hydrogen bonding between complementary base pairs, although they do not exclude additional non-

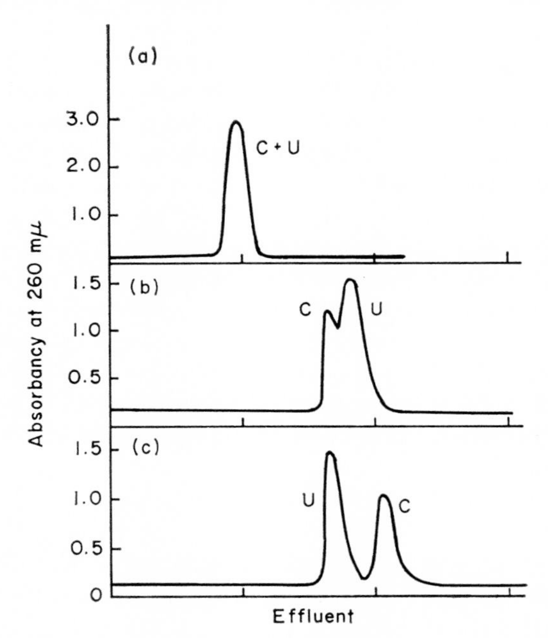

FIG. 10. Elution diagram uridine (U) and cytidine (C) on columns of (a) Amberlite hydrazide, (b) adenosine-Amberlite and, (c) guanosine-Amberlite. (Redrawn from Tuppy and Küchler, 1964a.)

specific stacking interactions which are known to occur in aqueous solution.

This chromatographic system has proven useful in studying the hydrogen-bonding properties of the synthetic base analog fluorodeoxyuridine (Labana and Sobell, 1967). It has been suggested that base analog mutagens such as this are more likely to undergo tautomeric change, and that their mutagenic properties are related to this (Freese, 1959). However, another possibility involves ionization. Fluorouracil is more acidic than thymine (the pK_a of fluorouracil is 7.6; the pK_a of thymine is 9.8), and one would therefore expect that, at neutral pH, approximately 20%

of the fluorouracil molecules would exist in the anionic form. In this ionized state, fluorouracil could interact with guanine (Lawley and Brookes, 1962; Litman, 1957; Meselson and Stahl, 1966).

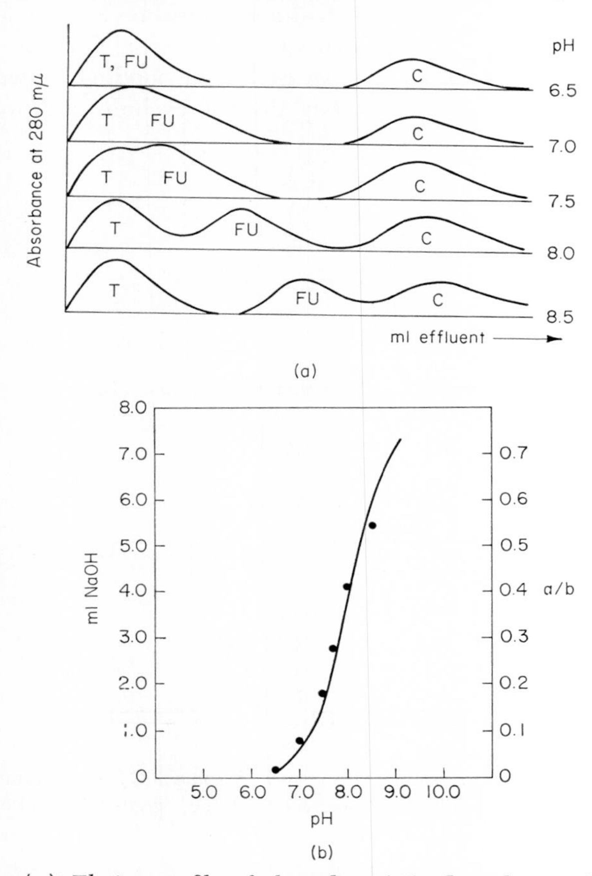

FIG. 11. (a) Elution profile of thymidine (T), fluorodeoxyuridine (FU), and deoxycytidine (C) from a guanosine-Amberlite column at various pH values. (b) A plot of the relative retardation parameter, a/b, as a function of pH. The titration curve of fluorodeoxyuridine (at 5°C) has been superimposed. See text for explanation.

Recently, it has been possible to demonstrate a pH-dependent interaction between fluorodeoxyuridine and guanosine, as well as with adenosine, using the Tuppy-Küchler chromatographic system. A nucleoside mixture containing deoxycytidine, thymidine, and fluorodeoxyuridine is passed down a guanosine-Amberlite column maintained at a constant pH and ionic strength. At acidic pH, pH 6.5, it is observed that thymidine and fluorodeoxyuridine travel as a single peak, followed by deoxycytidine (see Fig. 11a). A slight broadening of this peak is observed when these nucleosides are separated at pH 7.0. However, at pH 7.5, 8.0, and 8.5, the fluorodeoxyuridine peak becomes progressively better separated from the thymidine peak, being more and more retarded by the guanosine column as the pH becomes more alkaline. A convenient measure of the relative retardation of fluorodeoxyuridine compared with thymidine and deoxycytidine is the ratio of the distance between the thymidine and

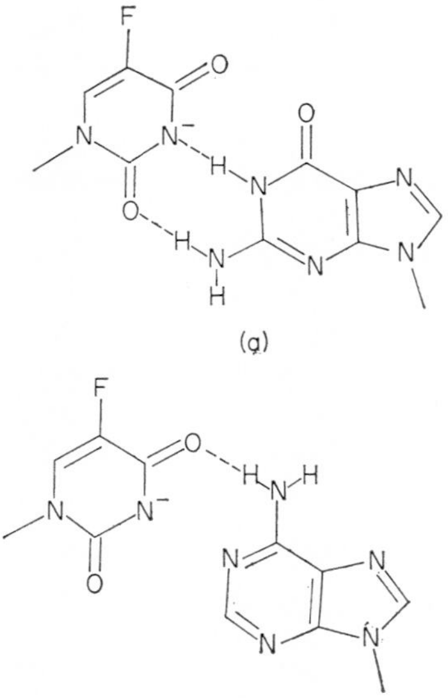

FIG. 12. A schematic diagram showing the interaction between (a) guanosine and fluorodeoxyuridine (ionized) (b) adenosine and fluorodeoxyuridine (ionized).

fluorodeoxyuridine peaks (here, called *a*) to the distance between thymidine and deoxycytidine peaks (here, called *b*). Figure 11b shows a plot of this *a/b* ratio versus pH. For convenience, the titration curve of fluorodeoxyuridine is also shown. It is seen that the maximal rate of change in the *a/b* retardation parameter occurs at pH 8.2, the pK_a value of fluorodeoxyuridine at 5°C. This indicates that the retardation of fluorodeoxyuridine by the column is correlated with the ionization of the N–3 proton on fluorodeoxyuridine. One possible interaction is that shown schematically in Fig. 12a. However, one can imagine other more nonspecific interactions responsible for the retardation effect. Forces such as base-stacking interactions or ionic interactions with the resin may be important. Evidence that such forces are not solely responsible for the retardation is provided by a control study in which thymidine, deoxycytidine, and fluorodeoxyuridine are passed down an adenosine-Amberlite column, the interaction again studied as a function of pH. This has demonstrated that fluorodeoxyuridine loses its affinity for the adenosine column, migrating progressively faster than thymidine, although slower than cytidine, as the pH becomes more alkaline. The rate of change of this loss of affinity for the adenosine-Amberlite resin again is maximal at the pK_a value of fluorodeoxyuridine. A reasonable explanation for this pH-dependent chromatographic behavior of fluorodeoxyuridine involves interactions which are schematically summarized in Figs. 12a and b.

III. X-Ray Crystallography

Of particular interest in relation to these solution studies has been the isolation of crystalline complexes containing monomer purine and pyrimidine derivatives and the determination of their hydrogen-bonded, base-paired configurations by X-ray crystallography. These studies have revealed rather surprising information concerning the manner in which different purines and pyrimidines associate in the solid state.

A. Purine–Pyrimidine Crystalline Complexes

1. *Guanine–Cytosine Complexes*

Four crystal structures containing different guanine-cytosine derivatives have been determined by X-ray analysis (O'Brien, 1963, 1967; Sobell *et al.*, 1963; Haschemeyer and Sobell, 1964, 1965b). One such structure, a crystalline complex containing the nucleosides, deoxyguanosine and bromodeoxycytidine, is shown in Fig. 13. The halogenated cytidine derivative was chosen, since this greatly simplifies the X-ray analysis. It is seen that the guanine and cytosine rings form a planar complex, connected together by three hydrogen bonds. Adjacent base pairs are stacked one on another to form an infinite column along the short *a* axis (the *a* axis is perpendicular to the plane of the figure and projects up toward the reader). In addition to the hydrogen bonding between purine and pyrimidine rings, there is additional hydrogen bonding in the structure. Of particular interest is a spiral arrangement of hydrogen bonds along the *a* axis involving the carbonyl oxygen O-6 of guanine bonded to a neighboring cytidine sugar hydroxyl group O-3′, which, in turn, is hydrogen bonded to the hydroxyl group O-3′ of a guanosine sugar residue. The latter bonds to the N-4 amino group on cytidine in the cell below. One wonders what effect additional hydrogen bonding, such as this, has in determining the base-pairing configuration which is observed in this study.

Evidence that the Watson-Crick type pairing configuration is a particularly stable type of association between guanine and cytosine is

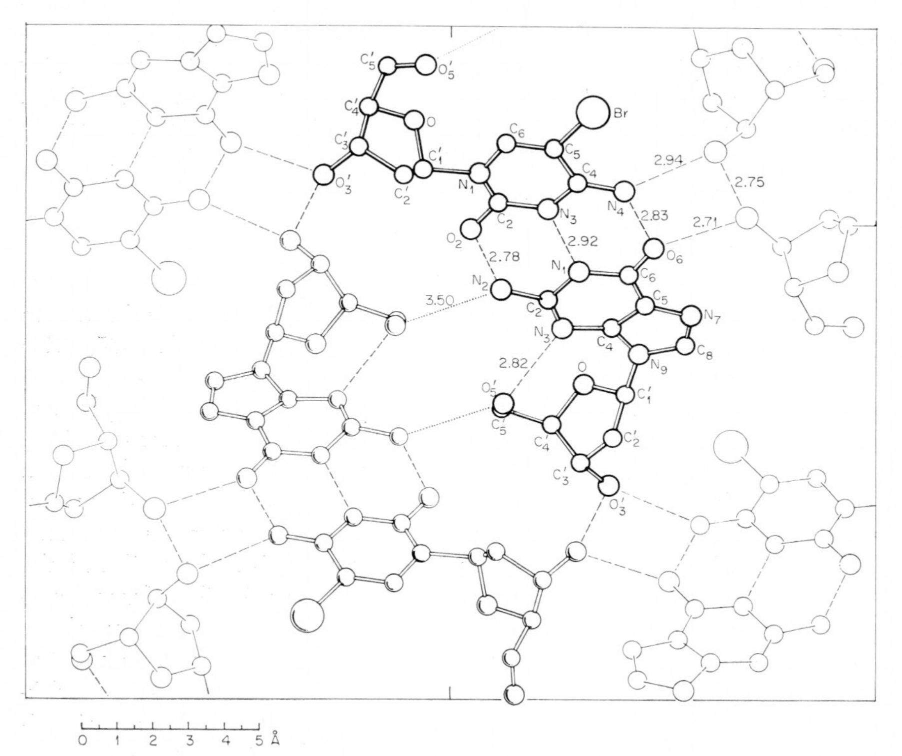

Fig. 13. A projection of the deoxyguanosine-bromodeoxycytidine crystal structure viewed down the short *a* axis. Dashed lines indicate hydrogen bonds. (Redrawn from Haschemeyer and Sobell, 1965b.)

provided by studies of other crystalline complexes containing different guanine and cytosine derivatives (O'Brien, 1963, 1967; Sobell *et al.*, 1963). Complexes containing 9-ethylguanine-1-methylcytosine (or -1-methyl-5-fluorocytosine) and 9-ethylguanine-1-methyl-5-bromocytosine exhibit the Watson-Crick base pairing configuration, although their crystal structures bear little similarity. (The alkylated derivatives of these compounds have been used to simulate more closely the glycosidic linkages in polynucleotides. This prevents hydrogen bonding from occurring at this site and also helps to maintain the normal tautomeric forms found in the nucleosides.) Thus, the former structures consist of sheets of base pairs held together by hydrogen bonds between neighboring guanine residues, hydrogen bonding involving the N-2 amino group on one guanine residue interacting with the N-3 ring nitrogen on an adjacent guanine residue. The latter structure, however, is an angulated structure in which adjacent base pairs are hydrogen bonded together in a completely different manner (the reader is referred to the original papers in order to compare these structures in greater detail). These findings suggest that lattice forces play an insignificant role in determining the base-pairing configuration which is observed in these crystal structures; however, this may not necessarily be the case for *all* crystals of this type which have been studied.

2. *Adenine–Thymine (or Uracil) Complexes*

Complexes between adenine and uracil (or thymine) derivatives have also been found in the solid state (Hoogsteen, 1959, 1963; Mathews and Rich, 1964; Haschemeyer and Sobell, 1963, 1965a; Katz *et al.*, 1965, 1967; Sakore *et al.*, 1969a; Sakore and Sobell, 1969a). Figure 14 shows the structure of a crystalline complex between 9-ethyladenine and 1-methyluracil. In this structure, the adenine and uracil rings are not oriented in a Watson-Crick fashion. Instead, the ring nitrogen N-3 of uracil hydrogen bonds with the imidazole nitrogen N-7 of adenine, not with N-1, as predicted in the Watson-Crick scheme. Neighboring base pairs are connected by hydrogen bonds between the carbonyl oxygen O-2 of uracil and the amino group of adenine, forming a sheetlike structure. A still different base-pairing configuration has been found to occur in crystalline complexes containing adenine and bromouracil derivatives. This is demonstrated in two structures: a complex between 9-ethyladenine-1-methyl-5-bromouracil (shown in Fig. 15) and a complex between adenosine-5-bromouridine (shown in Fig. 16). The first of these structures has almost identical unit cell constants, molecular packing, and overall hydrogen bonding as the 9-ethyladenine-1-methyluracil structure (compare with Fig. 14), and differs only in the orientation of the purine and pyrimidine rings. The adenine and bromouracil rings are oriented such that the carbonyl oxygen, O-2, rather than O-4, of bromouracil interacts with the amino group of adenine. This isomorphism between structures is possible, since the 1-methyl-5-bromouracil molecule is very similar in shape to the 1-methyluracil molecule, and, in addition has a pseudo-twofold axis of symmetry connecting the C-6 and N-3 ring atoms. Thus, a rotation of 180° about this axis would interchange the positions of the bromine atom and the methyl group (van der Waal's radii, 1.95 and 2.0

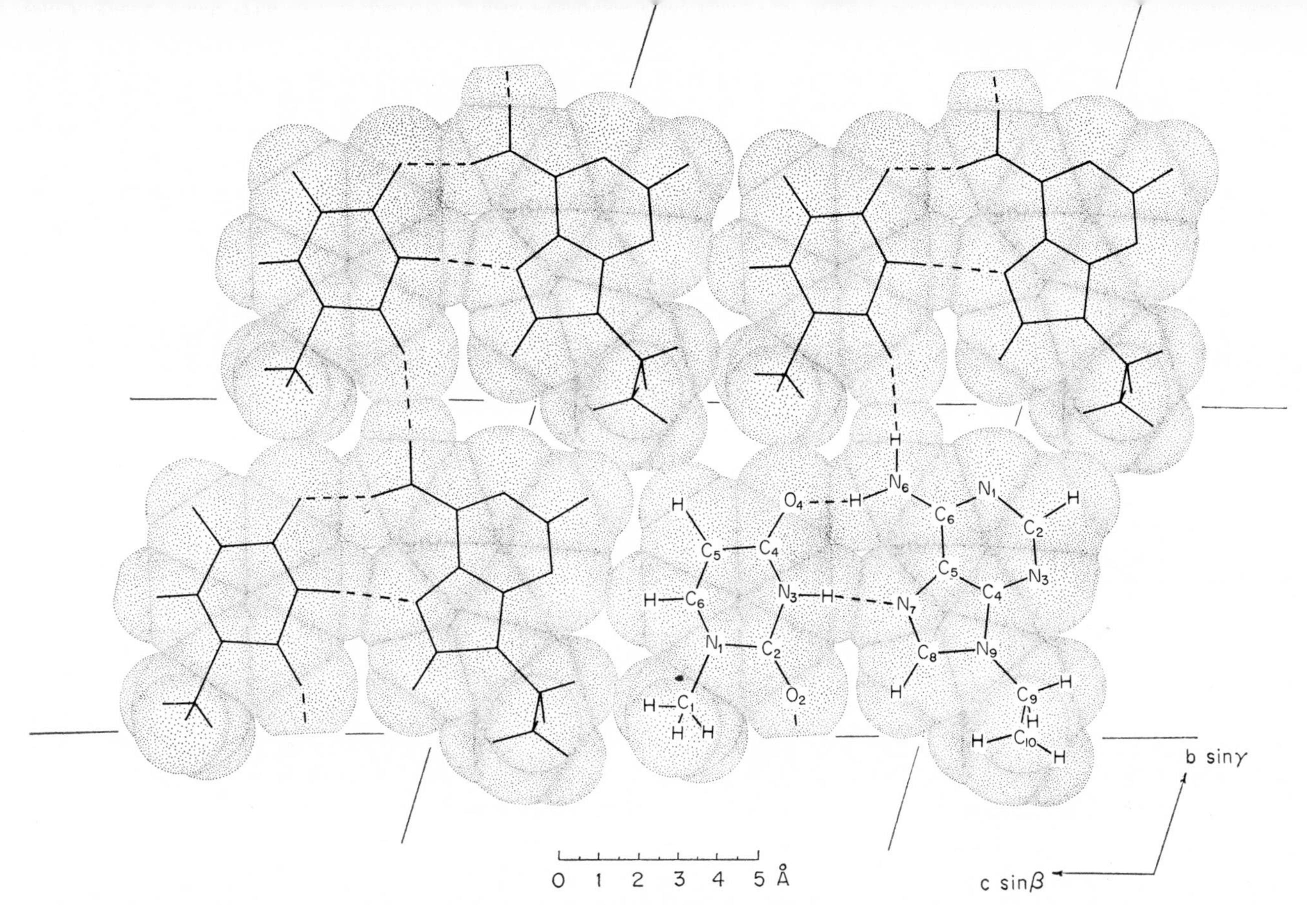

Fig. 14. A van der Waals packing diagram of the 9-ethyladenine-1-methyluracil crystal structure. Dashed lines represent hydrogen bonding. The view is down the *a* axis. (Redrawn from Mathews and Rich, 1964).

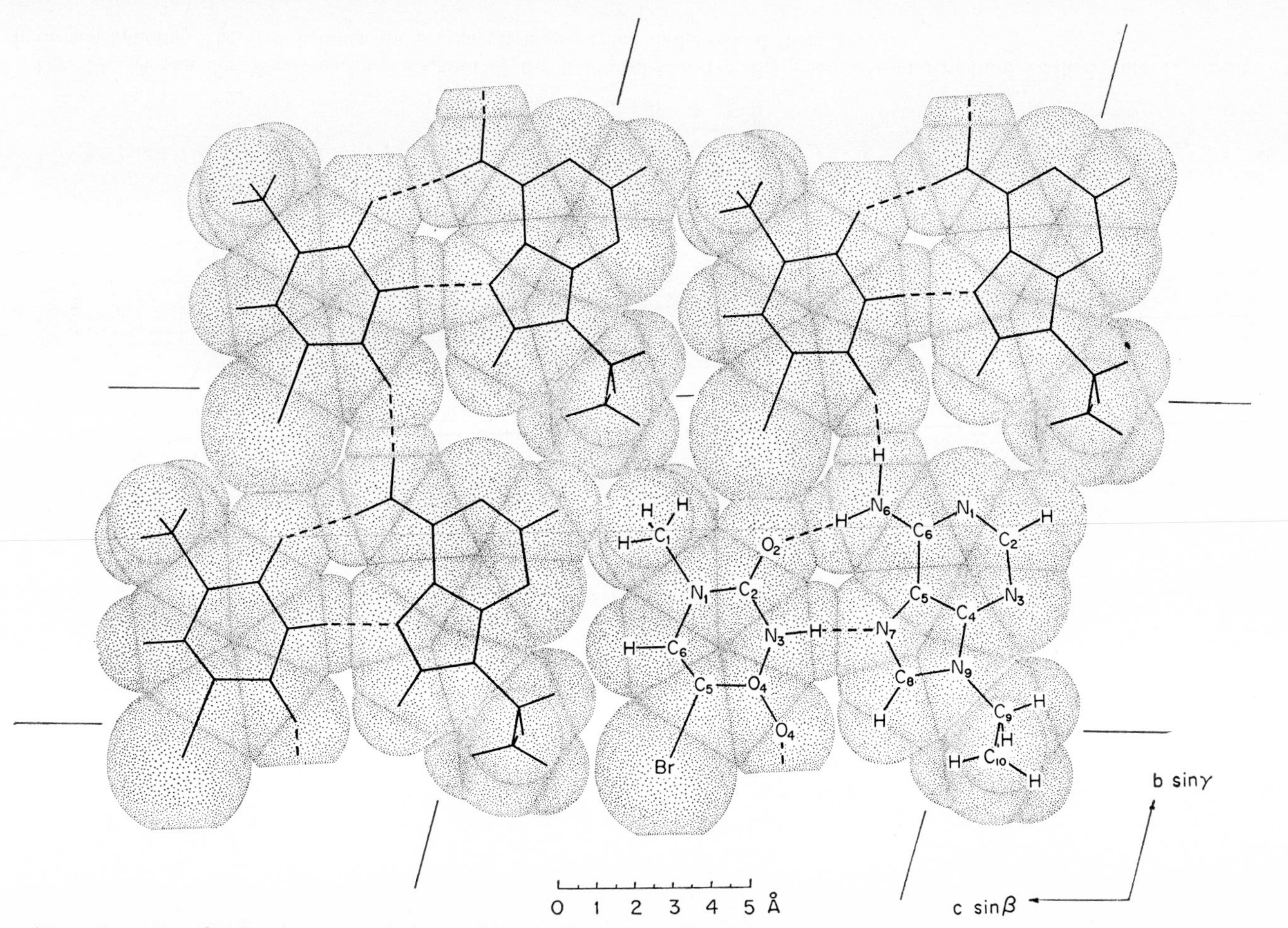

Fig. 15. A van der Waals packing diagram of the 9-ethyladenine-1-methyl-5-bromouracil crystal structure. Dashed lines represent hydrogen bonds. The view is down the *a* axis. (Redrawn from Katz *et al.*, 1965.)

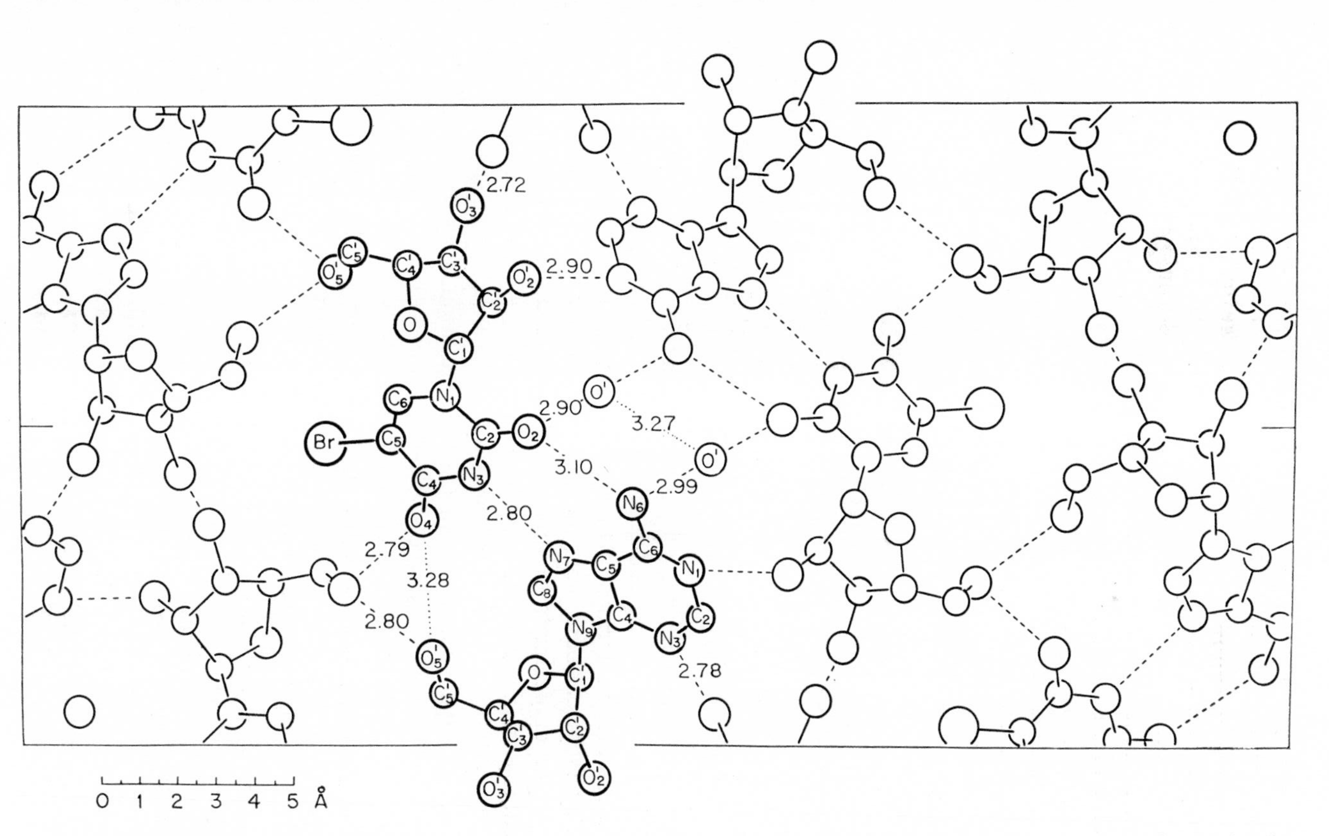

FIG. 16. Schematic diagram showing the adenosine-bromouridine crystal structure. Dashed lines represent hydrogen bonds. Dotted lines show other contacts of interest. (Redrawn from Haschemeyer and Sobell, 1963.)

Å, respectively), as well as the positions of the two carbonyl oxygens, O-2 and O-4, and the ring atoms, N-1 and C-5. In fact, it is sterically possible to have a mixture of different base-pairing configurations in the same crystal lattice because of this. Such a pairing disorder has been found to occur in the 9-ethyladenine-1-methyl-5-bromouracil complex. Both configurations (shown in Figs. 14 and 15) exist between the adenine and bromouracil rings in the same crystal structure. The Hoogsteen configuration occurs 6% of the time while the "reversed" Hoogsteen configuration occurs 94% of the time. If one considers the Hoogsteen configuration and the reversed Hoogsteen configuration as two states populated according to a Boltzmann distribution

$$N_1/N_2 = e^{-(E_1-E_2)/kT}$$

where N_1 corresponds to the number of molecules in state 1, N_2 corresponds to the number of molecules in state 2, E_1 and E_2 are the energies associated with these states, k is Boltzmann's constant, and T is the absolute temperature, one can calculate the difference in energy between the Hoogsteen and reversed Hoogsteen configurations at 25°C. The energy difference between the two states, calculated in this fashion, is 1.6 kcal/mole.

This same base-pairing configuration occurs in a nucleoside complex containing adenosine and bromouridine (shown in Fig. 16). These compounds interact to form a planar complex with a reversed Hoogsteen geometry. Adjacent base pairs stack one on another along the *a* axis (the *a* axis is perpendicular to the plane of the figure and projects up toward the reader) in a manner resembling the deoxyguanosine-bromodeoxycytidine crystal structure. Additional hydrogen bonding occurs between neighboring sugar hydroxyl groups, and, in addition, a water molecule is incorporated into the structure (shown as circles containing O′). These findings suggest that the reversed Hoogsteen base-pairing configuration may be specifically related to the presence of the bromine atom on the uracil ring, which may act to perturb the intermolecular forces which are involved in complex formation.

3. *Complexes Involving Synthetic Base Analogs*

We have now investigated the structures of a large number of complexes in which derivatives of adenine and the adenine base analogs, 2,6-diaminopurine and 2-aminopurine, are cocrystallized with a variety of halogenated uracil derivatives in an attempt to understand the general effect halogen substitution has on purine-pyrimidine base pairing. These studies, summarized in Table III and depicted schematically in Fig. 17,

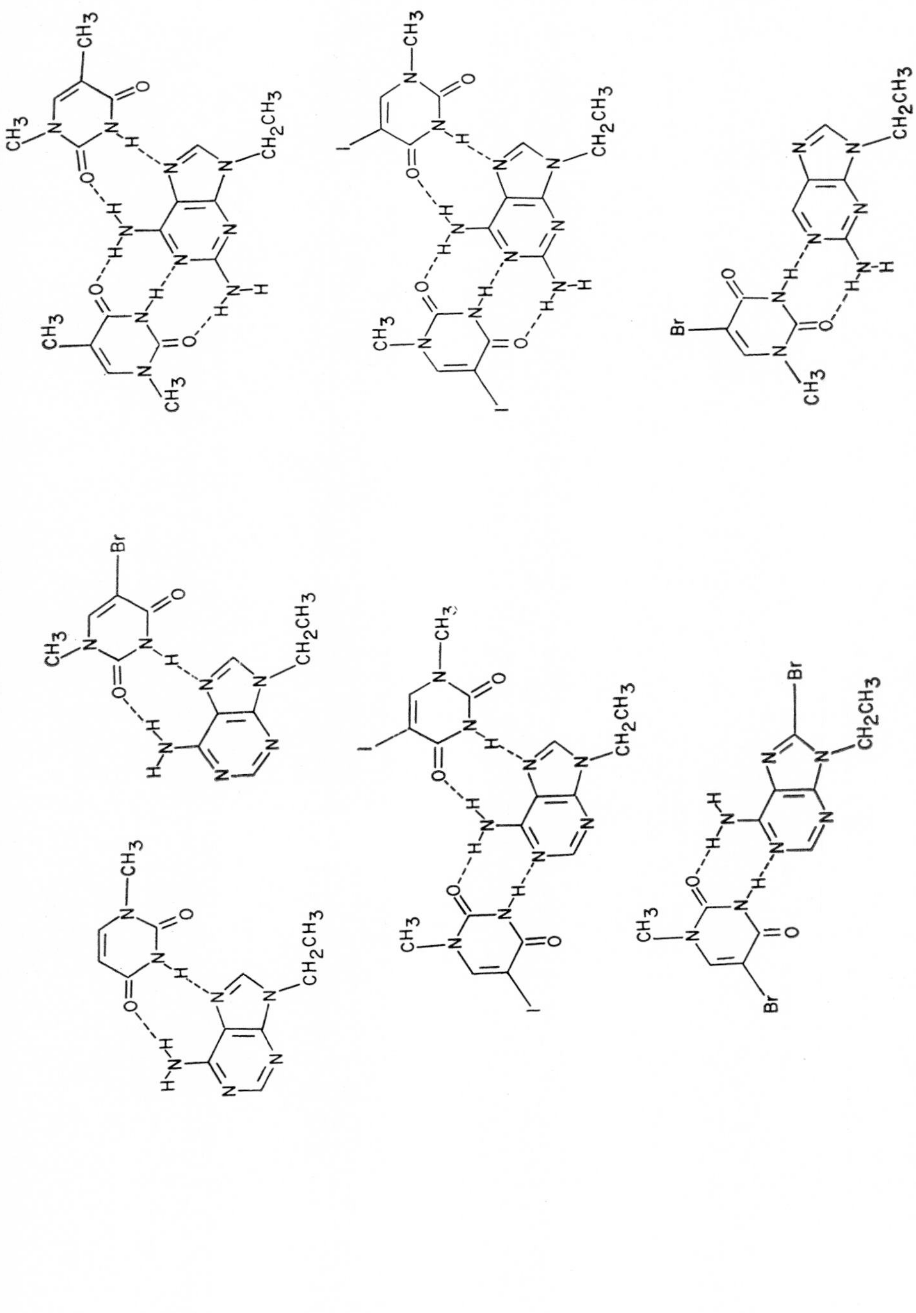

FIG. 17. Schematic drawing of base pairing configurations found in purine-pyrimidine crystalline complexes.

TABLE III

SUMMARY OF COCRYSTALLIZATION STUDIES BETWEEN ADENINE AND URACIL DERIVATIVES

Compounds	Stoichiometry	Crystal data	Base-pairing configuration	Reference
9-Methyladenine: 1-methylthymine	1:1	Monoclinic, P2/m	Hoogsteen	Hoogsteen (1959, 1963)
9-Methyladenine: 1-methyl-5-bromo-uracil	1:1	Triclinic, $P\bar{1}$	Hoogsteen	Baklagine *et al.* (1966)
9-Ethyladenine: 1-methyluracil	1:1	Triclinic, $P\bar{1}$	Hoogsteen	Mathews and Rich (1964)
9-Ethyladenine: 1-methyl-5-fluorouracil	1:1	Triclinic, $P\bar{1}$	Hoogsteen	Tomita *et al.* (1967)
9-Ethyladenine: 1-methyl-5-bromo-uracil	1:1	Triclinic, $P\bar{1}$	Reversed Hoogsteen	Katz *et al.* (1965, 1967)
9-Ethyladenine: 1-methyl-5-iodo-uracil	1:2	Monoclinic, $P2_1/c$	Hoogsteen Reversed Watson-Crick	Sakore *et al.* (1969a)
9-Ethyl-8-bromo-adenine: 1-methyl-5-bromo-uracil	1:1	Monoclinic, $P2_1/c$	Reversed Watson-Crick	Tavale *et al.* (1969)
Adenosine: 5-bromo-uridine	1:1	Orthorhombic, $P22_12_1$	Reversed Hoogsteen	Haschemeyer and Sobell (1963, 1965a)

Adenosine: 5-iodo-uridine	1:1	Orthorhombic, $P22_12_1$	Reversed Hoogsteen	Sobell (1969)
Adenosine: thymine riboside	1:1	Orthorhombic, $P22_12_1$	Reversed Hoogsteen	Sakore and Sobell (1969a)
9-Ethyl-2,6-di-aminopurine: 1-methylthymine	1:2	Triclinic, $P\bar{1}$	Watson-Crick Reversed Hoogsteen	Sakore *et al.* (1969b)
9-Ethyl-2,6-di-aminopurine: 1-methyl-5-iodo-uracil	1:2	Monoclinic, $P2_1/c$	Reversed Watson-Crick Hoogsteen	Labana and Sobell (1967); Sakore *et al.* (1969b)
9-Ethyl-2-amino-purine: 1-methyl-5-fluorouracil	1:1	Monoclinic, $P2_1/c$	Watson-Crick	Sobell (1966); Mazza *et al.* (1969)
9-Ethyl-2-amino-purine: 1-methyl-5-bromouracil	1:1	Monoclinic, $P2_1/c$	Watson-Crick	Mazza *et al.* (1969)

have shown that adenine and uracil derivatives cocrystallize primarily as 1:1 complexes and, in all but one case (9-ethyl-8-bromoadenine:1-methyl-5-bromouracil), involve Hoogsteen or reversed Hoogsteen base-pairing configurations. There seems to be no clear correlation between the reversed Hoogsteen base-pairing configuration and halogen substitution at C-5 on the uracil ring. If the halogen acts to perturb the base-pairing interaction, its effect must be very small and can easily be masked by crystal lattice forces. However, this may not be the case with complexes between diaminopurine and uracil derivatives. 9-Ethyl-2,6-diaminopurine forms 1:2 complexes with 1-methylthymine and with 1-methyl-5-iodouracil: the first involving Watson-Crick and reversed Hoogsteen base-pairing configurations, the second involving reversed Watson-Crick and Hoogsteen base-pairing configurations (Sakore *et al.*, 1969b; Labana and Sobell, 1967). Although more data are necessary before firm conclusions can be drawn, these findings suggest that halogen substitution on uracil can be an important factor determining the base-pairing interaction between these compounds.

The 9-ethyladenine:1-methyl-5-iodouracil structure is the first adenine-uracil complex containing a 1:2 stoichiometric ratio of these compounds. In addition to Hoogsteen base pairing, a reversed Watson-Crick type base-pairing configuration is seen and it is of interest to understand whether this second configuration reflects some altered interaction between the adenine and iodouracil derivatives. With this in mind, we have cocrystallized 9-ethyl-8-bromoadenine with 1-methyl-5-bromouracil (unfortunately, attempts to cocrystallize this adenine derivative with 1-methyl-5-iodouracil have not been successful thus far). Earlier model building studies had suggested that the bulky bromine atom at the 8 position on the adenine ring might sterically prevent base pairing from occurring on the N-7 imidazole side of adenine, leaving only Watson-Crick or reversed Watson-Crick base pairing possible. Of the two, we have found the reversed Watson-Crick base-pairing configuration (Tavale *et al.*, 1969). This, along with the data from the cocrystallization studies with diaminopurine, strongly suggests that the reversed Watson-Crick configuration reflects an altered base-pairing interaction with these halogenated uracil derivatives; the possible biological significance of this is discussed in Section III, D.

2-Aminopurine forms 1:1 complexes with two halogenated uracil derivatives. In both cases, the configuration used is a Watson-Crick type base-pairing configuration (Sobell, 1966; Mazza *et al.*, 1969). Further data are needed before any statement can be made concerning the effect of halogen substitution on base pairing between these compounds.

4. *A Hypoxanthine–Uracil Complex*

Of interest with regard to Crick's "wobble" hypothesis (discussed in detail in Section V, B), has been the isolation of a crystalline complex between 9-ethylhypoxanthine and 5-fluorouracil (Kim and Rich, 1967). Unfortunately, in this study the uracil molecule is unsubstituted on the glycosidic ring nitrogen, and base pairing has been found to involve the proton on this nitrogen and the N-7 imidazole nitrogen on hypoxanthine. This, therefore, makes the biological meaning of this structure questionable. An additional hydrogen bond is formed between the O-2 fluorouracil carbonyl oxygen and the N-1 proton on a neighboring hypoxanthine residue. Adjacent fluorouracil molecules form a hydrogen-bonded dimer involving two N—H. . .O hydrogen bonds. This purine-pyrimidine complex is unique in that only one hydrogen bond connects adjacent bases, and it may be relevant that infrared solution studies have failed to demonstrate significant interaction between similar compounds in nonpolar solvents (Kyogoku *et al.*, 1969).

B. Mixed Purine–Purine Crystalline Complexes

1. *An Adenine–Hypoxanthine Complex*

Another structure of interest with respect to Crick's wobble hypothesis is a complex, recently solved, between 9-ethyl-8-bromohypoxanthine and 9-ethyl-8-bromoadenine; this is shown in the Fourier electron density map in Fig. 18 (Sakore and Sobell, 1969b). In this structure, the adenine and hypoxanthine rings are held together by two hydrogen bonds, an N—H. . .N hydrogen bond between the N-1 ring nitrogen of hypoxanthine and the N-7 imidazole nitrogen of adenine, and an N—H. . .O hydrogen bond between the amino group of adenine and the carbonyl oxygen of hypoxanthine. This hydrogen-bonding arrangement is one of two arrangements possible which involves two hydrogen bonds between these bases. The other arrangement, suggested by Crick as a "wobble" pair, involves hydrogen bonding on the N-1 side of adenine, and will be discussed in Section V, B. The present pairing configuration almost certainly is used in a synthetic polynucleotide complex between polyinosinic acid and polyadenylic acid, as discussed in Section IV, 6.

C. Mixed Pyrimidine–Pyrimidine Crystalline Complexes

1. *Cytosine–Uracil Complexes*

Voet and Rich (1969) have recently reported a structure containing cytosine and 5-fluorouracil hydrogen bonded together in a sheetlike structure. A similar structure containing 1-methylcytosine and 5-fluorouracil has also been reported (Kim and Rich, 1969). Hydrogen bonding in both structures involves an N—H . . . N hydrogen bond between N-1 of fluorouracil and N-3 of cytosine, and an N—H . . . O hydrogen bond between the amino group of cytosine and the carbonyl oxygen, O-2, of fluorouracil. Adjacent fluorouracil molecules are related across a center of symmetry and are hydrogen bonded together, hydrogen bonding in-

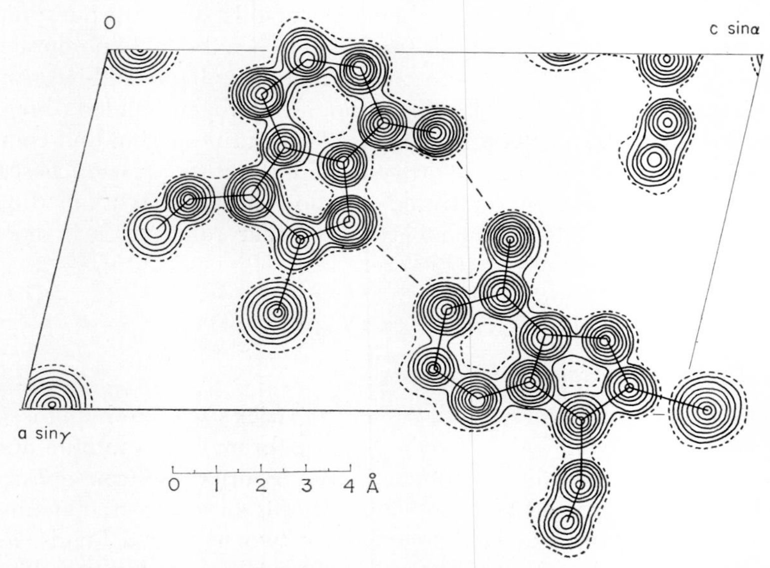

FIG. 18. Fourier electron density map of the 9-ethyl-8-bromohypoxanthine-9-ethyl-8-bromoadenine crystalline complex. Contours are drawn at approximately 1 electron/Å^3, except for the bromine atoms, where each contour represents approximately 5 electrons/Å^3. Hydrogen bonds are shown by dashed lines. (Redrawn from Sakore and Sobell, 1969b.)

volving the carbonyl oxygen, O-4, of one molecule interacting with the N-3 ring nitrogen proton on the other symmetry-related molecule. Again, since the mixed pyrimidine-pyrimidine hydrogen bonding involves the glycosidic nitrogen position, the biological meaning of these structures is uncertain.

D. CONCLUDING REMARKS

A major incentive for studying the base-pairing properties of purines and pyrimidines in the solid state has been to understand whether halogen substitution on the uracil ring can alter the relative stabilities of different types of purine-pyrimidine hydrogen-bonded configurations in solution which exist prior to cocrystallization. Although it is not possible to correlate directly configurations found in the solid state with configurations prevalent in solution, there is now a large body of information concerning the interactions of purines and pyrimidines which strongly suggests this (for other recent reviews in this area, see Hoogsteen, 1968; Voet and Rich, 1969). For these reasons, the altered adenine-bromouracil and adenine-iodouracil base-pairing configuration is of particular interest since it may reflect a perturbation in intermolecular forces not present in the adenine-thymine interaction. An altered hydrogen-bonding interaction between adenine and bromouracil could play an important role in bromouracil-induced mutagenesis, and may also alter the rate of DNA replication in cells grown in the presence of bromouracil. Hanawalt and Ray (1964) have shown that *Escherichia coli* grown in the presence of bromouracil replicate their DNA at a significantly slower rate than when grown in the presence of thymine. When confronted with both thymine and bromouracil in the media, bacteria selectively utilize thymine to bromouracil for normal and repair DNA synthesis (Hackett and Hanawalt, 1966; Kanner and Hanawalt, 1968). These observations could reflect the selective permeability of thymine over bromouracil at the cell membrane level, or could reflect different rates at which thymine and bromouracil are converted to their deoxyribonucleotide triphosphate derivative. Alternatively, the observations can be explained at the level of DNA polymerization by the DNA polymerase enzyme. Since base pairing between the template DNA strand and the incoming monomer deoxyribonucleotide triphosphate is an important prerequisite in ensuring the complementarity of base sequence during DNA replication, an altered adenine-bromouracil base-pairing interaction could interfere with the polymerization reaction and could also increase the probability for base-pairing errors during DNA replication. Thus, a reversed Watson-Crick type base-pairing configuration between, say, bromodeoxyuridine triphosphate and adenine (on the template strand) may have to be "rejected" by the polymerase enzyme for steric reasons. Only when these compounds interact with the proper geometry (i.e., the Watson-Crick geometry), can polymerization proceed. Events such as these would slow down the rate at which DNA is synthesized and would result in the preferential utilization of thymine to bromouracil by cells during normal and repair DNA synthesis. Once bromouracil is incorporated into the growing strand, subsequent replication, i.e., involving deoxyadenosine triphosphate pairing with bromouracil (on the template strand), would also be inhibited. Moreover, errors in replication may occur. Thus, if large numbers of monomer deoxyadenosine triphosphate-bromouracil pairings were not of the Watson-Crick type and therefore had to be rejected by the enzyme, this would increase the probability that deoxyguanosine triphosphate could "slip in" and pair with bromouracil. The latter interaction might be associated with an altered tautomeric state of bromouracil (Freese, 1959), or with an ionized state, as discussed earlier. Such a mechanism would predict that errors in replication should be more prevalent than errors in incorporation, and there is some evidence which suggests this (Strelzoff, 1961; Howard and Tessman, 1964).

At the present time it is not possible to state in detail what forces are responsible for the purine-pyrimidine hydrogen-bonding specificity which has been observed in solution and in the solid state. Detailed quantum mechanical calculations have been carried out to calculate the energies associated with different purine-pyrimidine combinations and configurations (see, for example, Pullman *et al.*, 1966, 1967; Nash and Bradley, 1966; Pollak and Rein, 1966). The results of these studies have shown that the major forces responsible for the specificity of purine-pyrimidine interaction are electrostatic in nature, primarily of the monopole-monopole type. Other forces include monopole-induced dipole and dispersion forces, these forces apparently being only of secondary importance. An interesting conclusion that has resulted from these studies has been that the Hoogsteen and reversed Hoogsteen structures are significantly lower in energy than the adenine-thymine Watson-Crick structure (for reviews in this area, see Pullman and Pullman, 1968); and this has suggested that the solid state configurations reflect particularly stable solution associations. However, calculations such as these use semi-empirical molecular orbital methods which involve significant approximation. Small changes in pairing energies may not be readily detected and this could explain discrepancies between theory and experiment. For example, Pullman and Caillet (1967) have calculated the most stable configurations between 2, 6-diaminopurine and thymine to be reversed Watson-Crick and Hoogsteen configurations, while the crystal structure, 9-ethyl-2,6-diaminopurine:1-methylthymine, demonstrates Watson-Crick and reversed Hoogsteen configurations.

Although the hydrogen bond is thought to be primarily electrostatic in nature, there may be a large covalent component in hydrogen-bonding between purines and pyrimidines. Thus, the consistently short N—H...N interactions observed in these studies between the purine and pyrimidine rings (2.80 Å), and the observation that significant ultraviolet hypochromism accompanies complex formation (Thomas and Kyogoku, 1967), may indicate that a sizable amount of quantum mechanical exchange occurs between the purine and pyrimidine rings. In addition, factors such as molecular polarizability and group polarizability are important parameters determining the magnitude of London forces between molecules, and the presence or the absence of heavy halogens on the uracil ring almost certainly alters these parameters. Careful crystallographic data and more complete quantum mechanical calculations are required, and these almost certainly will lead to a deeper understanding of hydrogen bonding in nucleic acid structures.

IV. Structure of Synthetic and Naturally Occurring Polynucleotides

There is now a large amount of information on the structure and physical chemistry of synthetic and naturally occurring polynucleotide complexes. Recent reviews emphasizing the physical chemistry of polynucleotides have been published (Felsenfeld and Miles, 1967; Michelson *et al.*, 1967), and this section will therefore concentrate primarily on structural studies of polynucleotides.

A. The Structure of Synthetic Polynucleotide Complexes

1. *Polyadenylic Acid*

Polyadenylic acid (poly A) exists in two alternate structural forms at different pH values. At neutral pH, poly A is primarily a random coil with short, single-stranded helical regions stabilized largely by stacking interactions between neighboring bases (see, for example, Leng and Felsenfeld, 1966; Brahms *et al.*, 1966). At acidic pH, however, poly A is known to assume a rigid double helical structure; this has been studied in considerable detail by fiber X-ray diffraction (Rich *et al.*, 1961).

Acidic solutions of polyadenylic acid become extremely viscous if allowed to dry, and it is possible to pull fibers of poly A which show strong negative birefringence. Negative birefringence is characteristic of highly ordered nucleic acid complexes and indicates that the purine and pyrimidine rings are oriented approximately at right angles to the fiber axis. The diffraction pattern of polyadenylic acid is shown in Fig. 19. There is a very intense meridional reflection at a spacing of 3.8 Å, corresponding to the distance between adjacent adenine rings along the fiber axis direction. Since the thickness of a purine ring is known to be about 3.4 Å, this indicates that adenine rings are not exactly perpendicular to the fiber axis. Furthermore, the absence of meridional reflections up to the 3.8 Å spacing is characteristic of helical diffraction patterns. Careful analysis of this fiber pattern has shown that there are four layer lines. The first layer line has a spacing of 15.2 Å, and the fourth layer line contains the meridional 3.8 Å spacing. The diffraction pattern does not allow one to determine the structure of the complex unambiguously. For instance, a possible structure compatible with this diffraction pattern consists of a

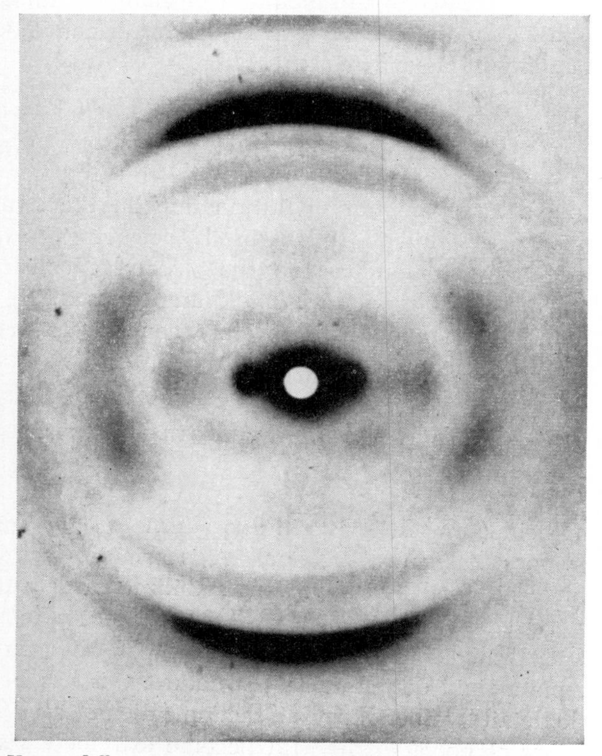

FIG. 19. X-ray diffraction photograph of polyadenylic acid. The fiber axis is vertical, relative humidity 80 per cent. The strong reflection on the meridian is at 3.8 Å. (From Rich *et al.*, 1961.)

single-stranded poly A helix, with four adenylic acid residues per complete turn of the helix. This would then mean that adjacent adenylic acid residues would be related by a translation of 3.8 Å and a twist of 90°. However, such a structure is not possible for steric reasons, since the ribose-phosphate backbone is not long enough to connect neighboring adenylic acid residues. The diffraction pattern, on the other hand, is compatible with a double-stranded helical structure consisting of two poly A strands wrapped around each other with the sugar-phosphate backbones running in the same direction. This is shown schematically in Fig. 20. Adjacent adenine residues on different strands are held together by hydrogen bonds, as shown in Fig. 21. This type of hydrogen-bonded configuration is seen in the crystal structure of adenine hydrochloride hemihydrate (Cochran, 1951). It is of interest that the double helical poly A structure is stable only at acidic pH; at pH values higher than

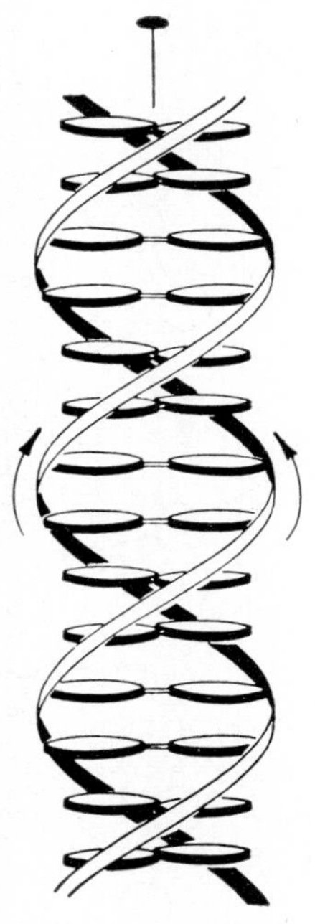

FIG. 20. Schematic representation of polyadenylic acid. Each band represents the ribose-phosphate backbone, while the bars stand for pairs of adenine bases. (From Rich *et al.*, 1961.)

6.5 only the single-stranded form is found. An important additional factor which stabilizes the double helical structure is an internal salt linkage between the negatively charged phosphate group on one strand and the protonated N-1 ring nitrogen on adenine (see Fig. 21). In addition, a hydrogen bond is formed between this phosphate group and the neighboring amino group on adenine which provides additional stability to the complex.

It should be noted that this type of helix is fundamentally very different from the DNA helix. The DNA structure involves purine-pyrimidine hydrogen bonding and the sugar-phosphate chains are intertwined in opposite directions. Polyadenylic acid, however, utilizes purine–purine hydrogen bonding and the sugar-phosphate chains are oriented parallel to each other. This reflects its twofold axis of symmetry along the helix axis.

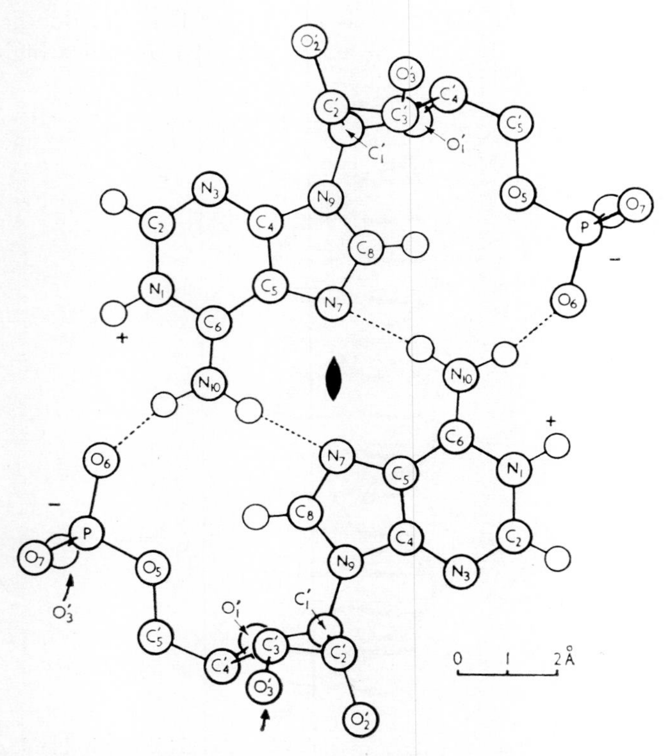

FIG. 21. The two-stranded molecule of polyadenylic acid as viewed down the fiber axis. Each adenine residue forms three hydrogen bonds. An extra O_3-atom is shown to indicate the backbone connections. For clarity the hydrogen atoms on the sugar have been omitted. (Redrawn from Rich *et al.*, 1961.)

2. *Polycytidylic Acid*

Like polyadenylic acid, polycytidylic acid (poly C) is known to exist in two alternate forms at different pH values. At neutral pH, poly C contains small regions of single-stranded helical configuration stabilized primarily by interplanar stacking forces between neighboring pyrimidine rings (Fasman *et al.*, 1964). At acidic pH, however, a stable complex is formed which is highly ordered and which gives an extremely detailed fiber diffraction pattern (Langridge and Rich, 1963). Infrared and titrometric studies have shown that the complex contains half the cytosine residues protonated on the N-3 ring nitrogen (Hartman and Rich, 1965; Akinrimisi *et al.*, 1963). The diffraction pattern contains a strong meridional reflection at 3.11 Å, indicating that the cytosine bases are

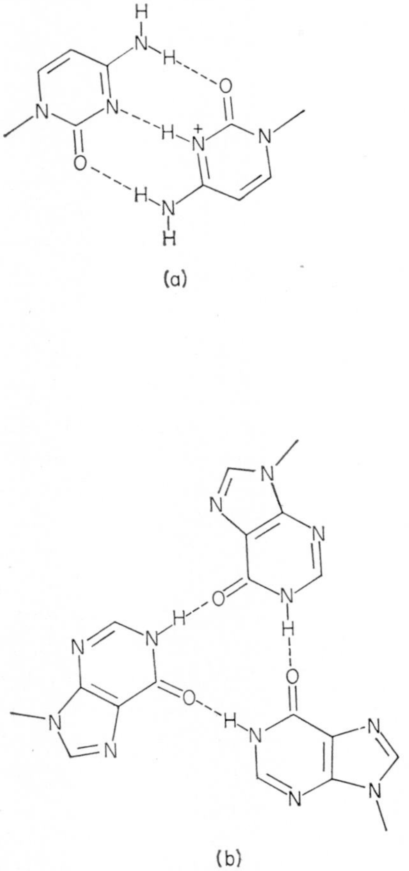

(a)

(b)

FIG. 22. Schematic diagram showing the hydrogen bonded configuration in (a) polycytidylic acid (acidic form) (b) polyinosinic acid (neutral form).

oriented at approximately right angles to the fiber axis. There are six layer lines, the first layer line occurring at 18.65 Å and the sixth layer line occurring at 3.11 Å. The diffraction data and physicochemical evidence concerning complex formation are explained by a double helical structure in which cytosine residues are held together by hydrogen bonds, as shown in Fig. 22a. This rather unusual type of hydrogen-bonding configuration involves half the cytosine residues protonated on its ring nitrogen and can be seen in the crystal structure of cytosine-5-acetic acid (Marsh *et al.*, 1962). The poly C helix resembles the poly A helix in that the sugar-phosphate backbones of both chains run parallel to each other. This is a direct consequence of having a twofold axis of symmetry along the helix axis in both these structures.

3. *Polyinosinic acid*

Polyinosinic acid (poly I) is capable of forming a triple-stranded helical molecule at neutral pH (Rich, 1958). The structure probably consists of three poly I strands, held together by hydrogen bonds between hypoxanthine residues (shown in Fig. 22b). The structure is similar to the previous ones in that all three helical chains are oriented parallel to each other and are on the outside of the molecule where they can interact with solvent and cations.

4. *Polyadenylic Acid–Polyuridylic Acid*

In addition to forming a complex with itself, poly A can interact with polyuridylic acid (poly U) to form double- and triple-stranded helical structures (Warner, 1957; Felsenfeld *et al.*, 1957; Stevens and Felsenfeld, 1964). The stoichiometry of these interactions can be followed rather easily using the method of continuous variation, as described earlier with the infrared and nuclear magnetic resonance monomer studies. One plots the optical density at 260 mμ of various mixtures of poly A and poly U in which the total number of molecules is kept constant (see Fig. 23). A decrease in optical density is observed with the formation of an ordered helical complex, i.e., the "hypochromic effect." In the absence of Mg^{++}, it can be seen that the stoichiometry of the poly A–poly U complex is 1:1. However, in the presence of Mg^{++}, a 2:1 complex forms between poly U and poly A. The kinetics and thermodynamics of complex formation have been studied in detail (Stevens and Felsenfeld, 1964; Blake *et al.*, 1967); the reaction $2\ (A + U) \rightleftharpoons A + (A + 2\ U)$ is second order in poly U and strand disproportionation is favored by elevated Mg^{++} concentration and the addition of heat.

The structure of the 1:1 complex has been studied by fiber X-ray diffraction (Sasisekharan and Sigler, 1965). The layer line spacing is 30.9 Å, with a meridional reflection on the tenth layer line at 3.09 Å. This indicates 10 base pairs per turn of the helix. It is possible, however, that, like reovirus RNA, the helix may be 11-fold with adjacent helices related by 3-fold symmetry (see Section IV, C). The overall intensity distribution closely resembles the pattern obtained from reovirus RNA, and therefore almost certainly involves a Watson-Crick, rather than a Hoogsteen, base pairing arrangement.

Unfortunately, little structural information is currently available on the 2 poly U–1 poly A complex. A reasonable model involves the second poly U strand being accommodated in the deep groove of the 1:1 adenine-uracil Watson-Crick helix, the pairing configuration being the same as that found in the monomer base-pairing studies in the solid state, i.e., the Hoogsteen configuration. However, Miles (1964) has presented infrared evidence which favors the second poly U strand interacting in

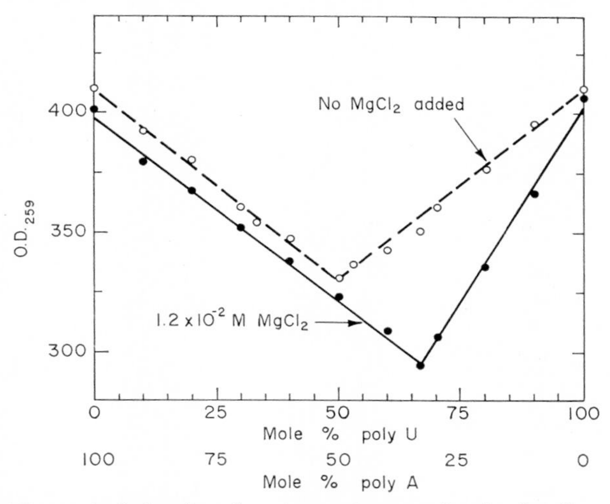

FIG. 23. The optical density of various mixtures of polyadenylic acid and polyuridylic acid. Optical densities were measured 2 hours after mixing. All solutions are in 0.1 M NaCl, 0.01 M glycylglycine, pH 7.4, $T = 25°C$. (From Felsenfeld *et al.*, 1957.)

a reversed Hoogsteen fashion. Although this latter model seems less likely for steric reasons, it cannot be ruled out until a careful X-ray analysis has been completed.

5. *Polyinosinic Acid–Polycytidylic Acid*

Because of initial difficulty synthesizing polyguanylic acid with the polynucleotide phosphorylase enzyme, structural studies have primarily centered on the related polymer, polyinosinic acid (poly I). These studies have demonstrated that, in addition to being able to form a triple helix with itself, poly I can form numerous complexes with other homopolymers.

One such complex, a 1:1 complex between poly I and poly C, forms at neutral pH (Davies, 1960). The diffraction pattern differs distinctly from that obtained with 1:1 poly A–poly U mixtures and from double helical viral RNA; the layer line spacing is 35 Å and a meridional reflection occurs at 3.0 Å. This means that there are 11.6 base pairs per turn of the helix, each base pair related to the next by a twist of 31° and a translation of 3.0 Å (this is to be compared with the poly A–poly U structure, which has a twist of 36° and a translation of 3.09 Å). The overall intensity distribution, however, is similar to the A form of DNA and

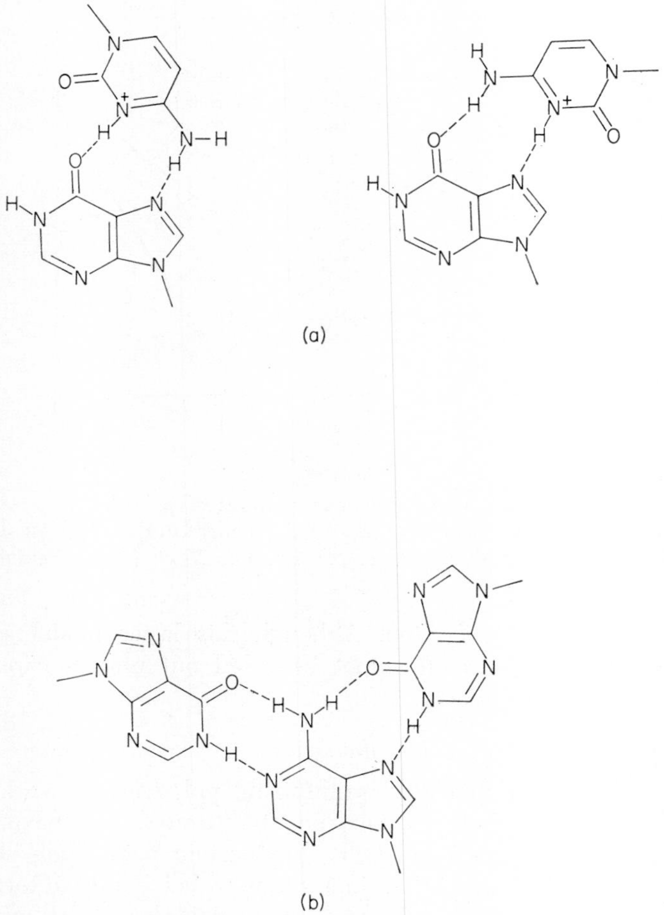

(a)

(b)

FIG. 24. Schematic diagram showing the hydrogen bonded configurations in (a) polyinosinic acid–polycytidylic acid (acid form) (b) 2 polyinosinic acid–1 polyadenylic acid.

therefore the structure almost certainly consists of hypoxanthine and cytosine residues hydrogen-bonded together in a Watson-Crick fashion.

Poly I and poly C form a different complex near pH 3 in which polycytidylic acid is protonated on the ring nitrogen, N-3 (Giannoni and Rich, 1964). Sedimentation studies have shown that the protonated complex has a 1:1 stoichiometry. The poly I–poly C^+ complex has been found to be more stable than the neutral complex, as judged by its thermal stability. The structure of the complex is not yet known in detail; however, X-ray studies have shown it to be different from the neutral complex (Tomita and Rich, 1965). The kinetics of hydroxylmethylation of the amino group on cytidylic acid by formaldehyde suggest that the amino group of cytosine is involved in hydrogen bonding. Likely base pairing configurations for this complex are shown schematically in Fig. 24a.

6. *Polyinosinic Acid–Polyadenylic Acid*

Poly I has been shown to form both 1:1 and 2:1 helical complexes with poly A (Rich, 1959). The diffraction pattern of a 1:1 mixture of poly A and poly I is shown in Fig. 25. The diffraction photograph is somewhat

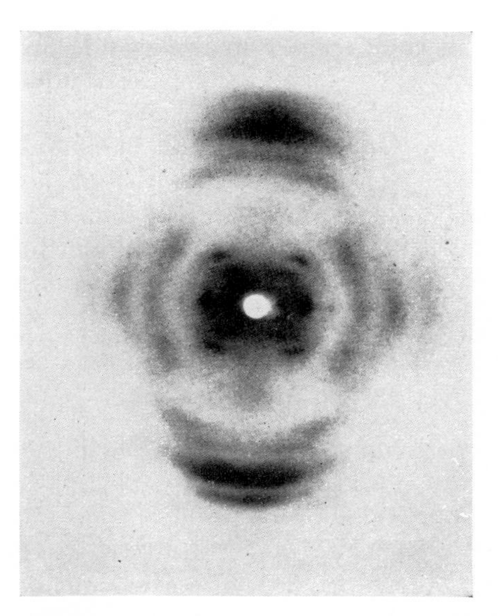

FIG. 25. X-ray diffraction photograph of a fiber of polyadenylic acid plus polyinosinic acid (1:1). The fiber is at a relative humidity of 66% in a helium-filled camera. The fiber is tilted slightly from the vertical. (From Rich, 1959).

similar to that obtained from the B form of DNA. The helix has a pitch of 38.8 Å and has about 11 1/2 residues per turn. It forms a hexagonal lattice, with $a = 24.4$ Å. The strong reflections in the 3–4 Å region are again due to stacking of planar bases at right angles to the fiber axis.

Probable structures for the 1:1 and 1:2 complexes are shown schematically in Fig. 24b. The 1:1 complex involves adenine and hypoxanthine residues interacting either on the N-1 side of adenine or on its N-7 side (it is not currently known which of these is preferred). The 1:2 complex uses both pairing configurations. There may be a formal similarity between this triple helical complex and the 2 poly U–1 poly A complex (assuming the latter uses the Hoogsteen geometry) in that both complexes utilize similar pairing configurations and have the same ribose-phosphate backbone polarity. It is of interest that one of these pairing configurations has been seen earlier in the cocrystallization study between adenine and hypoxanthine derivatives (see Section III, B).

7. *Polyinosinic Acid–Polyadenylic Acid–Polythymidylic Acid*

Rich (1960) has described the formation of the triple stranded helix, poly I–poly A–poly T. The complex can be formed by adding poly T to a poly I–poly A double helix, or by adding poly I to a poly A–poly T double helix. It is not known whether the two resulting triple helical structures are identical. For example, poly A–poly I may utilize hydrogen bonding on the N-1 adenine side (see Section IV, 6). The additional poly T strand may hydrogen bond on the adenine N-7 side, forming a Hoogsteen-type base pair. Poly A–Poly T, on the other hand, may be a Watson-Crick structure and involve hydrogen bonding on the N-1 adenine side. The addition of poly I to this helix may then involve hydrogen bonding on the adenine N-7 side. It is possible, therefore, that this complex has two different configurations depending on the manner in which complex formation is allowed to occur. There is currently no X-ray diffraction data available on this complex.

8. *Oligoguanylic Acid–Polycytidylic Acid*

Because of the tendency of polyguanylic acid to form stable macromolecular complexes with itself (Ralph *et al.*, 1962), it has been difficult to investigate the complexing properties of this polymer with other homopolynucleotides. If guanine oligonucleotides are used in place of poly G, however, the interference due to self-bonded guanine structures is less pronounced. Lipsett (1964) has investigated the properties of guanine oligonucleotides and has shown that oligo G and poly C interact readily to form a 1:1 double helical complex. Moreover, at elevated salt concentrations, low temperature, and slightly acidic pH, one finds evidence for 2:1 complex formation between poly C and oligo G. The formation of this complex has been studied carefully by Howard *et al.* (1964) using infrared spectroscopy, and it has been shown that half the cytosine residues are protonated in complex formation. Although the pK_a of cytosine is 4.3, one would expect a large cooperative effect to accompany complex formation altering the effective pK_a value of cytosine (this was observed previously in the half-protonated poly C structure). The structure of this triple-stranded complex is not known. However, one structure (shown in Fig. 26b) is of particular interest in view of recent evidence demonstrating the ability of double helical DNA-like polymers of defined sequence to form triple-stranded complexes with single-stranded RNA polymers of homologous sequence. This will be discussed in Section IV, A, 10.

9. *Monomer–Polymer Interactions*

Of interest in relation to the above discussion of triple-stranded helical polynucleotide complex formation is the observation that analogous complexes exist between monomer purine nucleosides and nucleotides and their complementary pyrimidine-containing polymer in aqueous solution (Howard *et al.*, 1966; Huang and Ts'o, 1966). Thus, adenosine, deoxyadenosine, 2,6-diaminopurine riboside, and related compounds form triple-stranded complexes with poly U, while guanosine and its monophosphate form triple-stranded complexes with poly C. Purine ribosides, such as tubercidin (7-deazaadenosine) and inosine, are unable to form complexes with their complementary polymer; this implies that an important factor stabilizing these interactions is triplex formation, in which the monomer is hydrogen bonded to two polypyrimidine-containing strands. Likely base-pairing configurations for these interactions are shown in Fig. 26.

Complex formation such as this has recently been shown to impart specificity to spontaneous oligonucleotide synthesis directed by a polyribonucleotide template (Sulston *et al.*, 1968). Thus, polyuridylic acid has been found to facilitate the formation of nucleotide bonds between adenylic acid and adenosine in the presence of the water-soluble condensing agent, carbodiimide, but not between adenylic acid and guanosine, cytidine, or uridine. Similarly, polycytidylic acid promotes dimerization of guanosine and guanine monophosphate, but not with other nucleoside and nucleotide combinations. The Watson-Crick base-pairing rules therefore apply to these nonenzymatic syntheses experiments, and observations such as these may well be relevant to prebiotic chemical synthesis of polynucleotides. The sugar-phosphate linkages, however, have been found to involve 5′—5′, 2′—5′, as well as 3′—5′ phosphodiester bonds, and therefore polymerization lacks this element of specificity.

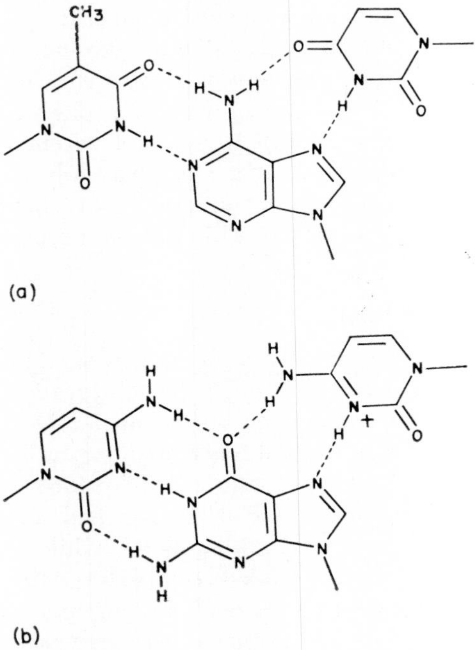

(a)

(b)

FIG. 26. Schematic diagram showing hydrogen bonded configurations postulated in triple helical polynucleotide complexes. (a) adenine: thymine: uracil (b) guanine: cytosine: cytosine+ (Miller and Sobell, 1966).

10. *The dAT Copolymer and Other Synthetic DNA-Like Polymer Complexes*

The alternating dAT copolymer was one of the first DNA-like polymers synthesized (Schachman *et al.*, 1960), and its structure has been studied in considerable detail by fiber X-ray diffraction (Davies and Baldwin, 1963). As with DNA, the type of diffraction pattern obtained from oriented fibers of this polymer depends on the relative humidity and the type of cation present. The diffraction data indicate that the dAT copolymer is very similar in structure to naturally occurring DNA, and is therefore a Watson-Crick helix. Similar studies have been carried out on the related polymer, dABU, which contains perfectly alternating adenine-bromouracil sequences. The structure of this polymer has been found to be very similar to the dAT copolymer.

There are now a large number of DNA-like polymers available for structural study. These polymers have been made available by Dr. H. G. Khorana and co-workers, who have synthesized short-chain oligonucleotides of defined sequence by chemical methods, and have used these oligonucleotides to prime the DNA polymerase system to make long-chain polymers with defined sequence. Unfortunately, structural studies on these polymers have not been completed, and a detailed description of their structure cannot be given at the present time.

Recently, however, some interesting observations have been made concerning the ability of several DNA-like polymers to complex with polyribonucleotides which have complementary base sequences (Riley *et al.*, 1966; Morgan and Wells, 1968). Ultracentrifugation studies have demonstrated that the homopolymer poly dA:dT readily complexes with poly rU, forming a triple stranded dA:dT:rU structure. The alternating dAT copolymer, however, cannot form the analogous complex with poly rAU. Poly dAG:dTC (a copolymer containing a perfectly alternating sequence of guanine and adenine on one strand, and thymine and cytosine on the other) complexes readily with poly rUC, complex formation being favored by high salt and acidic pH. One cannot, however, demonstrate interaction between this polymer and poly rAG. It is of interest that the related alternating polymer, poly dTG:dAC, cannot complex with either poly rUG or poly rAC.

These observations are readily understandable if the structure of these triple-stranded polymer complexes involves hydrogen-bonded configurations shown in Fig. 26. Polymers which contain purines on one strand and pyrimidines on the other, such as poly dA:dT and poly dAG:dTC, can accommodate an additional pyrimidine-containing strand in the deep groove of the Watson-Crick helix. This additional interaction uses the Watson-Crick type base-pairing rules, i.e., adenine = uracil, guanine = cytosine, but involves hydrogen bonding to the N-7 imidazole nitrogen of adenine and guanine (Miller and Sobell, 1966). The configuration was seen earlier in the adenine-thymine cocrystallization studies, and probably exists in the 2 poly U:1 poly A and 2 poly C:oligo G polyribonucleotide complexes, as discussed above.

11. *A Structural Hypothesis to Explain the Origin of the dAT Alternating Copolymer*

Of particular interest to the above discussion of polynucleotide structure is a possible explanation for the formation of the perfectly alternating dAT copolymer. It is not understood why the alternating adenine-thymine sequence is synthesized *de novo* by the *E. coli* and *Bacillus subtilis* DNA polymerase enzymes, while homopolymers of dG and dC, as well as dI and dC, are formed under similar conditions (Schachman *et al.*, 1960; Radding and Kornberg, 1962). It is possible that the formation of these polymers reflects a peculiarity of the enzyme surface which plays an important role in the initial stages of polymerization. On the other hand, this phenomenon may at least partly reflect a difference in secondary structure of the dinucleotide complexes, pApA-pTpT, pApT-pApT, and pTpA-pTpA. Structural considerations (to be described below) suggest that the pApA-pTpT complex may be either a Hoogsteen structure or a triple-stranded structure, while the pApT-pApT and pTpA-pTpA complexes are Watson-Crick structures.

The *de novo* synthesis may proceed initially by a random polymerization of dATP and dTTP to form dinucleotides of the type pApA, pTpT, pApT, and pTpA (more specifically, dinucleoside triphosphates of this type). Due to selective interaction between these species, complexes of pApA-pTpT, pApT-pApT, and pTpA-pTpA would result, the first having either a Hoogsteen structure or a triple-stranded structure of the type 2 pTpT:1 pApA, the latter two having a Watson-Crick configuration. These configurations need not be stable associations in solution, but may be transient associations having lifetimes of, perhaps, 10^{-4} or 10^{-5} seconds. The polymerase enzyme may selectively require a Watson-Crick structure to bind to its surface to initiate polymerization. It would therefore interact only with the dinucleotide complex pApT-pApT (or with pTpA-pTpA). Synthesis of the dAT copolymer may then require a juxtaposition of two dinucleotide triphosphate complexes to form a tetranucleotide complex, this step probably facilitated by the enzyme. Further synthesis could then proceed by a template-slippage mechanism (Kornberg *et al.*, 1964) until several long dAT molecules were made. These molecules would then act as templates to autocatalyze further polymer synthesis.

Structural considerations are as follows. First, consider the possible secondary structures for the dinucleotide complex, pApA-pTpT. This complex can assume a Watson-Crick pairing configuration,

$$\begin{array}{l}\xrightarrow{\hspace{3em}}\\ 5'\,\mathrm{pApA}\\ \quad\;\;\mathrm{TpTp}\,5'\\ \quad\;\;\xleftarrow{\hspace{3em}}\end{array}$$

or a Hoogsteen configuration,

$$\begin{array}{l}\xrightarrow{\hspace{3em}}\\ 5'\,\mathrm{pApA}\\ 5'\,\mathrm{pTpT}\\ \xrightarrow{\hspace{3em}}\end{array}$$

These structures differ not only in their base-pairing configurations, but also in the polarity of their sugar-phosphate backbones. In the Watson-Crick structure, the chains have opposite polarity, while in the Hoogsteen structure, the chains have the same polarity. The complex may use both base-pairing configurations simultaneously to form a triple stranded complex containing 2 pTpT dinucleotides to 1 pApA dinucleotide. This would be analogous to the 2:1 poly U–poly A complex discussed previously. Evidence that significant trimerization can occur between monomer adenine and uracil derivatives (although, in nonpolar solvents) has been cited earlier (see Section II, A).

Next, consider possible secondary structures for the dinucleotide complexes, pApT-pApT and pTpA-pTpA. These complexes can easily assume a Watson-Crick configuration,

$$\begin{array}{ll} \overrightarrow{5'\ \mathrm{pApT}} & \text{and } \overrightarrow{5'\ \mathrm{pTpA}} \\ \quad \underleftarrow{\mathrm{TpAp}\ 5'} & \qquad \underleftarrow{\mathrm{ApTp}\ 5'} \end{array}$$

However, a structural difficulty immediately arises when one attempts to construct a molecular model of these dinucleotides using the Hoogsteen base pairing configuration. It is found that the normally occurring *anti* conformation of the purine and pyrimidine deoxyribonucleotide must be converted to the *syn* conformation, in order to build a sterically acceptable model (the terms *anti* and *syn* refer to the relative orientation of the sugar residue to the base and are discussed in detail by Donohue and Trueblood, 1960). Present evidence, based on structural studies of naturally occurring and synthetic polynucleotides, as well as from crystal structure studies of nucleic acid components, strongly suggest that the *anti* conformation is the preferred rotational isomer for these compounds, particularly for pyrimidine nucleotides. This, therefore, combined with current knowledge of the structure of the dAT copolymer, makes the Hoogsteen structure unlikely for these complexes.

In summary then, the dinucleotide with alternating sequence almost certainly tends to form a Watson-Crick structure, while the dinucleotides with homo sequence could preferentially form Hoogsteen or triple-stranded structures. These latter structures may not readily bind to the enzyme surface, and even if binding occurs, chain growth may somehow be impaired early in synthesis due to triple-stranded complex formation.

Remaining to be explained, however, is the observation that homopolymers of dG:dC and dI:dC (as well as the closely related halogenated series, dG:dBC and dI:dBC) are formed with no detectable alternating sequence. The formation of homopolymers is not entirely unexpected, since there is considerable evidence that purines and pyrimidines tend to stack in solution (Ts'o *et al.*, 1963), and this could influence the probability for spontaneous dimerization (and, perhaps, for longer chain growth) of homopolymer sequences early in *de novo* synthesis. However, even when primed with oligonucleotides of the type $(dGC{:}dCG)_n$, the DNA polymerase is unable to achieve long-chain synthesis. This may reflect the stability of sequences such as these in inhibiting the template-slippage mechanism proposed for DNA polymerase reiteration (Kornberg *et al.*, 1964). The formation of dG:dC involves slippage of only one nucleotide base pair held by three hydrogen bonds, while the formation of dGC:dCG would involve slippage of two base pairs held by six hydrogen bonds. The dI:dC polymer, however, would involve slippage of only one nucleotide base pair held by two hydrogen bonds, compared with the dIC:dCI polymer, which would require slippage of two base pairs held by four hydrogen bonds. In analogy with the dAT copolymer, therefore, this alternating copolymer should be able to prime the DNA polymerase to form long-chain alternating dIC copolymers. For *de novo* synthesis, however, homopolymer formation of the type dI:dC and dG:dC is not entirely unexpected from the considerations above. The lack of *de novo* formation of other mixed polymers, such as dTG:dAC, can be explained with similar arguments.

Even considering the above arguments, however, the ease with which the DNA polymerase makes the dAT copolymer is highly suggestive that enzyme specificity plays an important additional role in *de novo* synthesis. Perhaps this reflects the molecular evolution of the enzyme, whose structure may have evolved in response to structural problems posed by dA:dT oligonucleotides in the primitive seas billions of years ago.

B. A Brief Review of DNA Structure

This section briefly reviews current knowledge concerning the different structural conformations of the DNA molecule as evidenced by their fiber X-ray diffraction patterns. A lengthy review of work in this area has not been attempted here, and the reader is referred to the original papers for a more complete discussion.

DNA is known to exist in three main structural conformations, denoted A, B, and C. In all cases, the diffraction data are explained by the same basic double helical Watson-Crick structure, modified slightly as regards the inclination of the base pairs to the helix axis and the angular twist and axial translation relating adjacent base pairs.

Figure 27a and d shows diffraction photographs of Li DNA in the B conformation, taken at 92 and 66% relative humidity. The pattern taken at lower relative humidity indicates a high degree of crystallinity, as evidenced by its discontinuity. Both patterns reflect a common structure having ten nucleotide base pairs per helix turn in which each base pair lies perpendicular to the helix axis (see Fig. 28a). Thus, adjacent base pairs are related by a twist of 36° and a translation of 3.4 Å along the helix axis. Space-filling van der Waals' models (not shown here) indicate

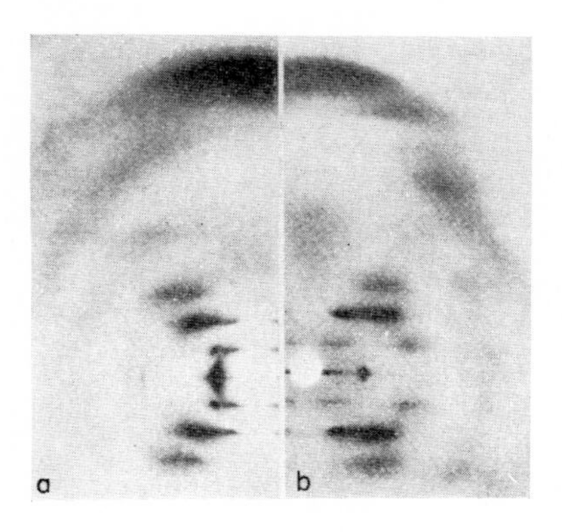

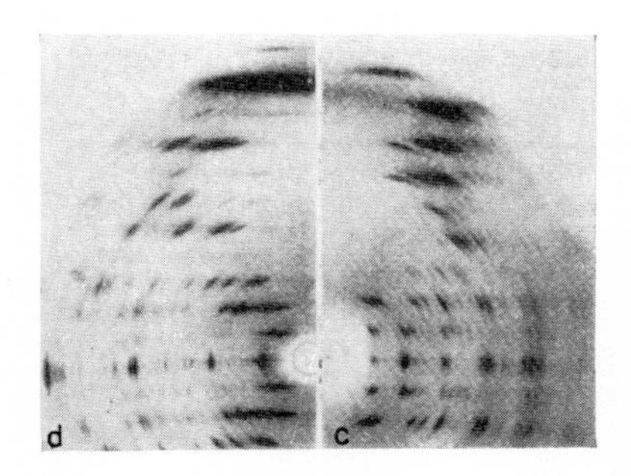

FIG. 27. X-ray diffraction patterns of DNA in the B conformation (a) and C conformation (b). X-ray diffraction patterns of crystalline DNA in the A conformation (c) and in the B conformation [Li salt of DNA] (d). (From Hamilton, 1968.)

that there are two grooves in the molecule, a narrow groove and a deep groove. The latter groove lies on the N-7 imidazole side of guanine and adenine residues. The B conformation exists in intact sperm heads (Wilkins and Randall, 1953), and most probably is the conformation found *in vivo*. Complete details of the structure analysis are described by Langridge *et al.* (1960).

Figure 27c reproduces the diffraction pattern of Na DNA in the A conformation, obtained at 75% relative humidity. Under these conditions, water is lost from the structure and this alters the conformation of the molecule, there now being eleven nucleotide base pairs per turn of the helix with a separation of 2.56 Å along the helix axis. The shortened interbase separation reflects a twenty degree inclination each base now makes with a plane perpendicular to the helix axis (see Fig. 28b). Although this alternative conformation is not thought to be the *in vivo* conformation of DNA, it is of interest since double helical RNA and an RNA-DNA hybrid helix have been found to assume this conformation over a wide humidity range (see sections below). The crystallinity of

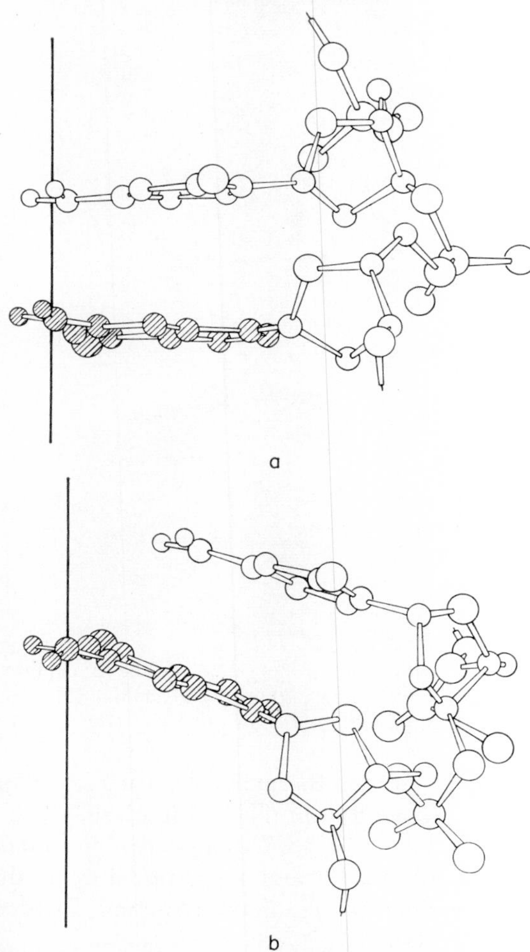

FIG. 28. (a) Projection of two nucleotides in the B conformation of DNA showing bases horizontal and at right angles to the helix axis. (b) Projection of two nucleotides in the A conformation of DNA showing bases inclined at 20°. (From Hamilton, 1968.)

the diffraction pattern and its wealth of data has allowed a detailed structure analysis to be done (Fuller *et al.*, 1965), and this has given strong additional evidence supporting the correctness of the basic Watson-Crick model.

Figure 27b shows the diffraction pattern from the C form of DNA, obtained as a lithium salt at 44% relative humidity. The pattern closely resembles that obtained from the B form of DNA, except that it indicates 9.3 residues per helix turn, with an axial separation of 3.32 Å. A detailed study (Marvin *et al.*, 1961) has shown that the base pairs are inclined —6° to a plane perpendicular to the helix axis, and each base pair, in addition, is hydrogen bonded together with a propeller-like twist of 5°. This results in moving the base pairs 1.5 Å further away from the position of the helix axis in the B conformation. Although the narrow groove of the molecule is slightly larger, the C conformation resembles the B conformation in its general aspects.

These studies, taken together, lend strong verification to the correctness of the basic Watson-Crick structure. However, the possibility remained that DNA could have a hydrogen-bonded base-pair arrangement similar to that found in the adenine-thymine cocrystallization studies, i.e., a Hoogsteen base-pairing arrangement. This possibility has been explored (Langridge and Rich, 1960), and largely eliminated by use of difference Fourier techniques (Arnott *et al.*, 1965). These studies, however, cannot rigorously exclude the possibility that Hoogsteen or other non-Watson-Crick–type base-pairing arrangements are present in small domains of the DNA molecule, although there is no evidence at the present time which indicates this.

C. Recent Advances

Since the early studies of Rich and Watson (1954), a great deal of detailed information has emerged with regard to the three-dimensional structure of RNA. These studies have demonstrated that double helical RNA from different sources has a unique configuration, which, although resembling the DNA structure, has significant differences.

1. *Double Helical Viral RNA*

Langridge and Gomatos (1963) were the first to obtain crystalline diffraction patterns from RNA obtained from the double-stranded RNA plant virus, reovirus. Subsequently, similar diffraction patterns were obtained by Tomita and Rich (1964) with wound tumor virus RNA and by Langridge *et al.* (1964) with MS-2 viral replicative form RNA.

Reovirus RNA has been found to exist in two crystalline forms, called α and β. Although there are no appreciable differences in the overall diffracted intensities for both forms, the β-type gives the more crystalline diffraction pattern, as evidenced by its discontinuity (Arnott *et al.*, 1967a). Both patterns have been indexed on a hexagonal unit cell, with a close to 40 Å and c (along the fiber axis) close to 30 Å. Packing con-

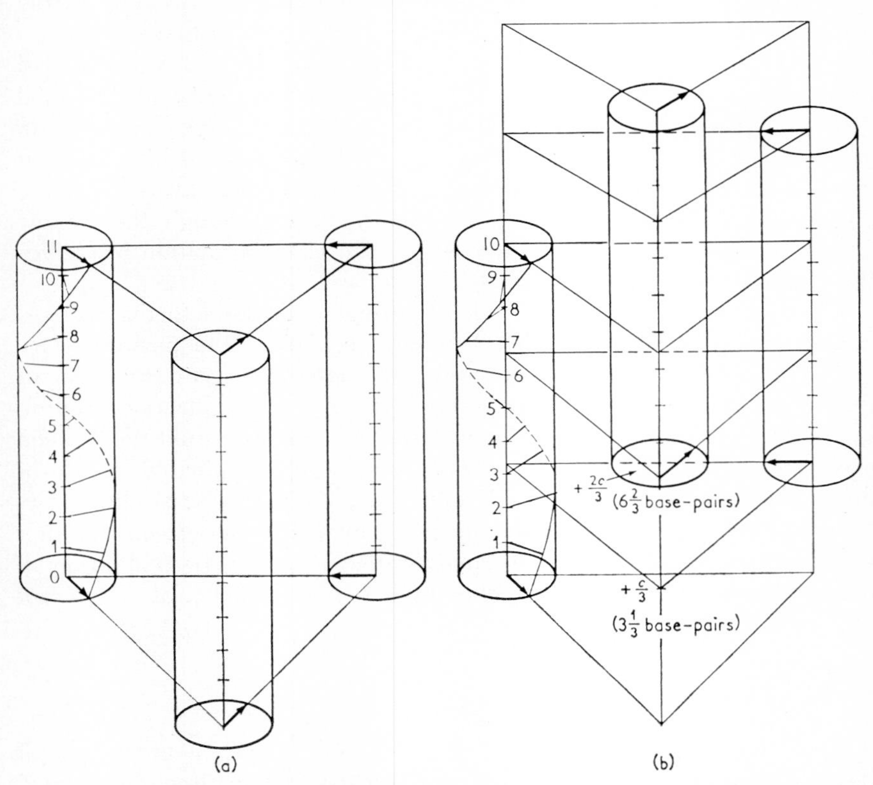

FIG. 29. (a) Three 11-fold helices related by a rotation triad. (b) Three 10-fold helices related by a left-handed screw triad. (Figure kindly supplied by Dr. M. Spencer.) (From Arnott *et al.*, 1967a.)

siderations for a Watson-Crick–type double helix, as well as a detailed study of the diffracted intensity on higher order layer lines, strongly suggest the existence of threefold symmetry in packing adjacent RNA molecules (Arnott *et al.*, 1967a, b). These studies have shown that if the helix has eleven base pairs per turn of the helix, neighboring RNA molecules must be related by a twist of 120°, i.e., a threefold rotation axis. If, however, the helix has ten base pairs per turn, then this necessitates that three RNA molecules be related in position by a twist of

—120° and a translation of 10 Å, i.e., a left-handed threefold screw operation (see Fig. 29). Both models give the expected 30 Å periodicity along the fiber axis. In the tenfold model, the base pairs are inclined 11° to a plane perpendicular to the helix axis, while the elevenfold model predicts a 14° inclination.

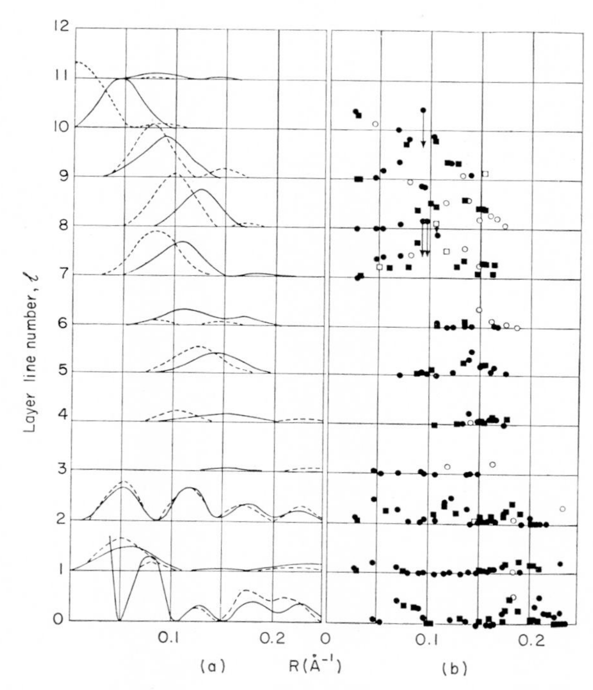

FIG. 30. (a) The calculated cylindrically averaged squared Fourier transform of RNA 10 (— — —) and RNA 11 (———). (b) The corresponding observed intensities derived from α RNA (□, □) and β RNA (○, ○) data; the less reliable values are shown □, ○. Where necessary, arrows indicate the layer line on which the observed intensities occur. (From Arnott *et al.*, 1967b.)

Molecular models such as these must satisfy stringent stereochemical criteria. Moreover, the diffraction pattern of both models can be calculated and compared with the observed diffraction pattern to establish the unique structure. The results of such a comparison are shown in Fig. 30.

The diffraction pattern from a helical molecule can be calculated from the following expression (Cochran *et al.*, 1952):

$$F(R,\psi,l/P) = \sum_j \sum_n f_j J_n(2\pi R r_j) \exp i[n(\psi + \pi/2 - \phi_j) + 2\pi l z_j/c]$$

where $F(R,\psi,l/P)$ is the structure factor amplitude and phase of the reflection given by the reciprocal space cylindrical coordinates $(R, \psi, l/P)$. The summation extends over j atoms in the helix repeat, and over the Bessel function terms of order n, the allowed values for n on a given layer line, l, given by $m/p + n/P = l/c$, where m is an integer assuming arbitrary values, p is the distance between adjacent residues projected along the helix axis, P is the pitch of the helix, and c is the crystallographic repeat along the helix axis. This expression is modified to include contributions due to symmetry-related RNA molecules (see, for example, Fuller *et al.*, 1967), and is cylindrically averaged about ψ (Davies and Rich, 1959).

In contrast with the distinctive differences of the ten- and elevenfold helices of DNA (the B and A conformation, respectively), both RNA models give very similar calculated diffraction patterns. This is particularly evident on the equatorial layer line ($l = 0$) and on the first and second layer lines ($l = 1, 2$), where the calculated intensities are almost identical. In both cases, intensities on the third, fourth, fifth, and sixth layer lines are relatively weak, while on the seventh, eighth, and ninth they are very strong. The eleventh layer line is effectively absent for both models. The main differences in the calculated diffraction patterns for both models lie, not in the relative heights of the intensity maxima, but in their position. In general, the positions of the observed maxima on $l = 4, 5, 7$, and 9 are in better agreement with the elevenfold model than for the tenfold model, although the evidence for this is not conclusive. For this reason, Fourier synthesis methods have been used as additional evidence to decide between these models (Arnott *et al.*, 1967a). These methods involve calculation of not only the amplitude of the diffracted wave, but also its relative phase. A Fourier synthesis is then calculated which uses the observed diffracted amplitudes and the calculated phase angles. The results of these calculations (the details of which will not be discussed here) indicate that the elevenfold model best accounts for the diffraction data.

The molecular packing of the elevenfold model has been studied in considerable detail. This has shown that hydrogen bonding probably occurs between the 2′-ribose hydroxyl of one molecule and an oxygen of a sugar-phosphate chain on a neighboring molecule. It is not clear as yet whether this has any biological significance.

2. *Fragmented Ribosomal RNA*

Detailed diffraction patterns have been obtained from fragmented yeast RNA. This RNA was first thought to be intact transfer RNA (Spencer, 1963), but subsequent study has shown this to be fragmented

RNA of ribosomal origin (Spencer and Poole, 1965). The diffraction pattern resembles that given by reovirus RNA and strongly resembles the DNA A-type pattern. The similarity shows that these ribosomal fragments are predominantly Watson-Crick–type structures, although large regions of ribosomal RNA may assume more varied configurations in its unfragmented form. Unlike the double helical RNA of viral origin, it appears that ribosomal RNA may be a tenfold helix, with a pitch of 29 Å, each base pair making an angle of 20° with a plane perpendicular to the helix axis (Fuller *et al.*, 1967). However, the X-ray data are not adequate to rule out the elevenfold model or some nonintegral helix. The packing of adjacent ribosomal RNA helices again appears to be in groups of three, being related by a left-handed screw triad. However, the lack of definition in these noncrystalline ribosomal RNA patterns does not allow an unambiguous structure determination.

3. *An RNA-DNA Hybrid Helix*

Of interest with regards RNA-DNA interaction is the isolation of a helical complex between single-stranded f1 DNA and its complementary RNA strand formed *in vitro* by the RNA polymerase. This complex gives a moderately well-defined diffraction pattern which closely resembles the A form of DNA (Milman *et al.*, 1967). It is of interest that, like double helical viral RNA, the diffraction pattern is insensitive to wide variations in humidity. Model building has shown that 2′-hydroxyl group cannot be accommodated into polynucleotide structures resembling the B form of DNA without considerable distortion, although there is little difficulty accommodating it in the A form. The rigidity of the hybrid helix may therefore directly reflect this steric hindrance.

4. *Transfer RNA*

Recent progress in understanding the tertiary structure of transfer RNA is described in Section V, C.

V. Structural Information on Transfer RNA

A. Primary and Secondary Structure

A great deal of information is now available concerning the nucleotide sequences of transfer RNA. This section will summarize the data currently available. The reader is referred to a recent review by Madison (1968) and to the original papers on this subject for more detailed information.

Table IV lists the complete nucleotide sequences of eleven transfer

TABLE IV

TRANSFER RNA SEQUENCES[a,b]

Sequence	tRNA (reference)
pG-G-G-C-G-U-G-U-G-G-C-G-C-G-U-A-G-U-C-G-G-U-A-G-C-G-C-G-C-U-C-C-C-U-U-I-G-C-I-ψ-G-G-G-A-G – A-G-U – C-U-C-C-G-G-T-ψ-C-G-A-U-U-C-C-G-G-A-C-U-C-G-U-C-C-A-C-C-A$_{OH}$	Alanine-tRNA (yeast) (Holley *et al.*, 1965)
pG-G-C-A-A-C-U-U-G-G-C-C-G-A-G – U-G-G-U-U – A-A-G-G-C-G-A-A-A-G-A-ψ-U-I-G-A-A-A-ψ-C-U-U-U-U-G-G-G-C-U-U-G-C-C-C-G-C-G-C-A-G-G-T-ψ-C-G-A-G-U-C-C-U-G-C-A-G-U-U-G-U-C-G-C-C-A$_{OH}$	Serine-tRNA (yeast) I, II (Zachau *et al.*, 1966)
pG-U-A-G-U-C-G-U-G-G-C-C-G-A-G – U-G-G-U-U – A-A-G-G-C-G-A-ψ-G-G-A-C-U-I-G-A-A-A-ψ-C-C-A-U-U-G-G-G-G-U-C-U-C-C-C-C-G-C-G-C-A-G-G-T-ψ-C-G-A-A-U-C-C-U-G-C-C-G-A-C-U-A-C-G-C-C-A$_{OH}$	Serine tRNA (rat liver) Staehelin *et al.* (1968)
pC-U-C-U-C-G-G-U-A-G-C-C-A-A-G-U-U-G-G-U-U-U-A-A-G-G-C-G-C-A-A-G-A-C-U-G-ψ-A-A-ψ-C-U-U-G-A – G-A-U – C-G-G-G-C-G-T-ψ-C-G-A-C-U-C-G-C-C-C-C-C-G-G-G-A-G-A-C-C-A$_{OH}$	Tyrosine-tRNA (yeast) (Madison *et al.*, 1966)
pG-C-G-G-A-U-U-U-A-G-C-U-C-A-G – U-U-G-G-G – A-G-A-G-C-G-C-C-A-G-A-C-U-G-A-A-Y-A-ψ-C-U-G-G-A – G-G-U C-C-U-G-U-G-T-ψ-C-G-A-U-C-C-A-C-A-G-A-A-U-U-C-G-C-A-C-C-A$_{OH}$	Phenylalanine-tRNA (yeast) (RajBhandary *et al.*, 1967)
pG-G-U-U-U-C-G-U-G-G-U-C-ψ-A-G – U-C-G-G-U-U-A-U-G-G-C-A-ψ-C-U-G-C-ψ-U-I-A-C-A-C-G-C-A-G-A – A-C-U – C-C-C-C-A-G-T-ψ-C-G-A-U-C-C-U-G-G-G-C-G-A-A-A-U-C-A-C-C-A$_{OH}$	Valine-tRNA (yeast) (Bayev *et al.*, 1967)
pG-G-U-U-U-C-G-U-G-G-U-C-ψ-A-G – U-U-G-G-U-C-A-U-G-G-C-A-ψ-C-U-G-C-ψ-U-I-A-C-A-C-G-C-A-G-A –C-C-C-C-A-G-T-ψ-C-G-A-U-C-C-U-G-G-G-C-G-A-A-A-U-C-A-C-C-A$_{OH}$	Valine tRNA (torula yeast) (Takemura *et al.*, 1968)
pG-G-U-G-G-G-G-Ü-U-C-C-C-G-A-G-C – G-G-C-C-A-A-A-G-G-G-A-G-C-A-G-A-C-U-C-U-A-A-A-ψ-C-U-G-C-C-G-U-C-A-U-C-G-A-C-U-U C-G-A-A-G-G-T-ψ-C-G-A-A-U-C-C-U-U-C-C-C-C-C-A-C-C-A-C-C-A$_{OH}$	(Tyrosine Su$_{III}$-tRNA (*E. coli*) (Goodman *et al.*, 1968)

```
                        O
                        Me                          *    +                      (C)(A)
pG-G-U-G-G-G-Ü-U-C-C-C-G-A-G-C - G-G-C-C-A-A-A-G-G-G-A-G-C-A-G-A-C-U-G-U-A-A-A-ψ-C-U-G-C-C-G-U-C-A-U-C-G-A-C-U-U-   C-G-A-A-G-G-T-ψ-C-G-A-A-U-C-C-U-U-C-C-C-C-C-A-C-C-A-C-C-A     Tyrosine-tRNA (E. coli) II/I
             X                                X   X               X       X                          X          X                                               X X X OH  (Goodman et al., 1968)
      5        10       15        20        25        30        35        40        45        50           60        65        70        75        80        85
                                                                                           (A)
                              Di                   O                                        7
                              H                    Me                                       Me
pC-G-C-G-G-G-G-Ü-G-G-A-G-C-A-G-C-C-U-G-G-U - A-G-C-U-C-G-U-C-G-G-G-C-U-C-A-U-A-A-C-C-C-G-A-A-   G-G-U      C-G-U-C-G-G-T-ψ-C-A-A-A-U-C-C-G-G-C-C-C-C-C-G-C-A-A-C-C-A     Formylmethionine-tRNA (E. coli)
             X                            X                X   X                                X      X                                             X X X OH  (Dube et al., 1968)
1     5        10       15        20        25        30        35        40        45                   50        55        60        65        70        75  77

                   (Di H U) Di2' DiDi                                                          7
                             H OMe H H                  +   +                                  Me
pG-G-C-U-A-C-G-Ü-A-G-C-U-C-A-G-U-U-G-G-U-U- -A-G-A-G-C-A-C-A-U-C-A-C-U-C-A-U-A-A-ψ-G-A-U-G-G-   G-G-X-   -C-A-C-A-G-G-T-ψ-C-G-A-A-U-C-C-C-G-U-C-G-U-A-G-C-C-A-C-C-A     Methionine-tRNA
             X                            X   X            X   X                                        X                                             X X X OH   (E. Coli)
1     5        10       15        20        25        30        35        40        45                   50        55        60        65        70        75  77   (Cory et al., 1968)
```

[a] After Madison (1968).

[b] Abbreviations used: A, adenosine or adenylic acid; C. cytidine or cytidylic acid; G, guanosine or guanylic acid; U, uridine or uridylic acid; Ψ, pseudouridine; p or-, phosphate residue, on the left of the nucleoside symbol they mean a 5′-phosphate, on the right a 3′-phosphate; the subscript OH is used to emphasize the presence of a 3′-hydroxyl group; Me, a methyl group whose position is indicated; OMe, 2′-O-methyl; Å, possibly a thioadenosine; A⁺, possibly an N-6 acetylated adenine derivative; IPA, N-$_6$-(Δ^2-isopentenyl) adenosine; AcC or C⁺, N-4 acetyl cytidine; 2MeG, N-2-methylguanosine; DiMeG, N-2-dimethylguanosine; I, inosine; T, ribothymidine; DiHU, 5,6-dihydrouridine; Ü, possibly 4-thiouridine; Ů, a mixture of uridine and dihydrouridine; Y, an unknown, highly fluorescent nucleoside; Pu, a purine nucleoside; Py, a pyrimidine nucleoside; X, an unknown nucleoside; G*, a modified guanine derivative.

RNA molecules which have been determined up until this time. These include alanine, phenylalanine, tyrosine, and valine from baker's yeast; two serine-tRNA's from brewer's yeast; valine-tRNA from torula yeast; methionine-, formylmethionine-, tyrosine (both I and II) -tRNA's, and the tyrosine suppressor (su^{+}_{III})-tRNA from *E. coli;* serine-tRNA from rat liver. The sequences have been arranged such that the terminal CCA ends and the anticodon sites are aligned. The lengths of these molecules vary from 76 residues for phenylalanine and valine-tRNA to 85 residues in the *E. coli* tyrosine- and su^{+}_{III}-tRNA's, and in the yeast and rat liver serine tRNA's. The anticodon sites are singly underlined. Sequences which are common, but which are not in the same position in all of the tRNA's, are doubly underlined. In all cases, the 3′-hydroxyl end is $CpCpA_{OH}$. In addition, there are seven positions containing similar nucleotides in almost every molecule, and this is shown by an X immediately below each sequence. Odd nucleotides, such as dihydrouracil dimethylguanine, 1-methyladenine, 5-methylcytosine, and others are seen to occur in similar areas of the molecule.

Figure 31 demonstrates how these tRNA molecules can be folded into a common cloverleaf arrangement. The upper limbs are double helical structures which contain predominantly Watson-Crick base pairs. The left-hand limb consists of three base pairs in the serine, tyrosine, valine, and su^{+}_{III} structures, and of four base pairs in the alanine-, methionine- formylmethionine-, and phenylalanine-tRNA's. The number of unpaired nucleotides in the left-hand loop varies from eight residues in the case of phenylalanine-tRNA to twelve in yeast tyrosine-tRNA. In all cases, there are seven unpaired nucleotide residues in the lower and right-hand loops and five base pairs in each lower and right-hand limb.

The nucleotide sequence -G-T-ψ-C-G- is found in the right-hand loop of all transfer RNA's except in yeast serine-tRNA-1 and in *E. coli* formylmethione-tRNA, where the sequence is modified to -G-T-ψ-C-A-, and this implies that the region has some common function for all these tRNA's. Another common finding are the bases immediately surrounding the anticodon; in all cases, uracil or pseudouracil is found to the left of the anticodon site, while adenine or a derivative of adenine appears immediately to the right. In between the lower and righthand limb there exists an additional nucleotide stretch whose length varies in the different tRNA's. It is seen that the extra length of the serine-tRNA, *E. coli* tyrosine-tRNA, and su^{+}_{III}-tRNA molecules is almost entirely accounted for by this extra protuberant "finger."

The cloverleaf model partly explains the sensitivity of certain nucleotide regions to nucleases and is particularly attractive because it fulfills the expectations of the adapter hypothesis suggested by Crick (1958).

It is clear, however, that detailed information pertaining to the secondary and tertiary structure of tRNA can only be obtained by detailed X-ray crystallographic structure analysis. The recent crystallization of transfer RNA provides the first step in attaining this goal (see Section V, C).

B. Codon-Anticodon Interaction: Crick's Wobble Hypothesis

The reader is referred to the excellent review by Speyer (1968) which summarizes current knowledge on the nature of the genetic code. It is now established that the code is a commaless triplet code in which three nucleotides code for a given amino acid. Since there are sixty-four possible codons (that is, nucleotide triplets), and only twenty amino acids, it has long been suspected that there was degeneracy (or redundancy) in the genetic code, i.e., more than one codon coded for a given amino acid. It has now been determined that this degeneracy reflects, almost exclusively, an ambiguity in reading the third base in each codon, and that the pattern of degeneracy obeys definite pairing rules. (An apparent exception to this rule may exist, since the codons AUG and GUG (and possibly UUG), when placed at or next to the 5′-terminal phosphate end of synthetic messenger RNA's, code for formylmethionine-tRNA. When these codons appear in the middle of the message, however, they code for methionine-tRNA.) Thus, for example, purified preparations of an alanine-tRNA contain the sequence IGC as its anticodon and recognizes the triplets GCU, GCC, and GCA, but not GCG. At a similar location in serine-tRNA, there is an IGA anticodon sequence, and this tRNA recognizes the triplets UCU, UCC, UCA, but not UCG. In tyrosine-tRNA, the sequence CψA recognizes the tyrosine codons, UAU and UAC, but not UAA or UAG. The general pattern of degeneracy that has been found is summarized in Table V.

TABLE V
General Pattern of Degeneracy in Genetic Code

tRNA anticodon (5′-hydroxyl end)	mRNA codon (3′-hydroxyl end)
U or pseudo-U	A or G
C	G
A	U
G	U or C
I	U, C, or A

These findings have led Crick (1966) to propose that the first two bases in each codon are read using the standard Watson-Crick antiparallel base-pairing configuration, while the third place may allow a little wobble and accept pairing combinations whose geometry is close

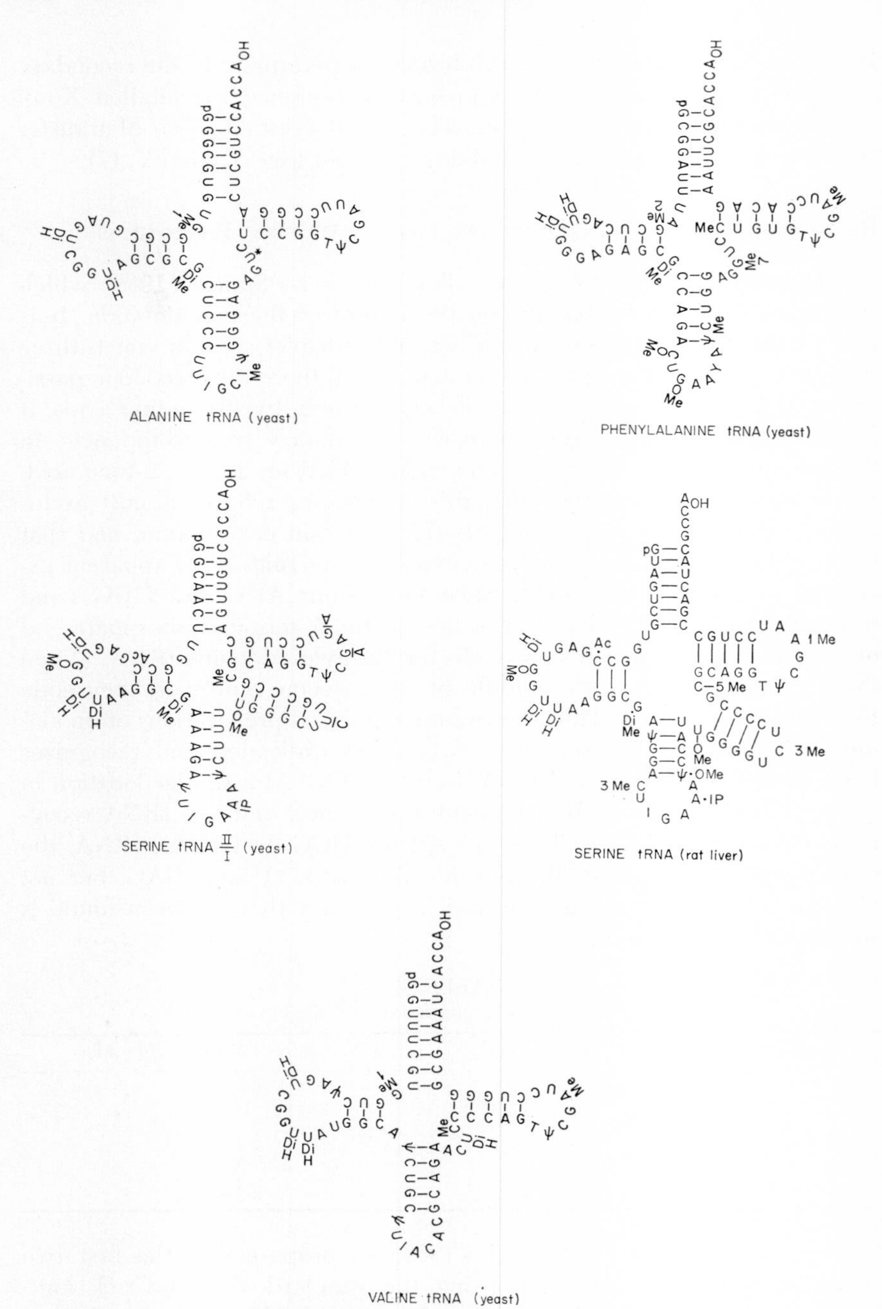

FIG. 31. Transfer RNA sequences arranged in the cloverleaf model. The anticodon sites are shown in the lowermost portion of each figure.

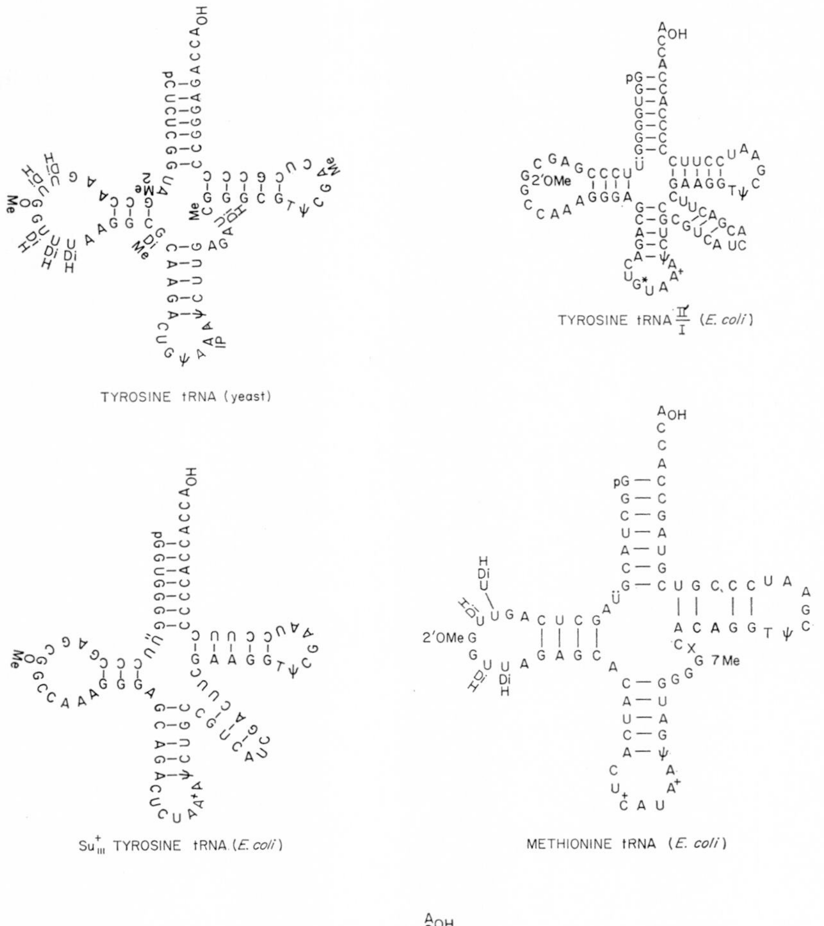

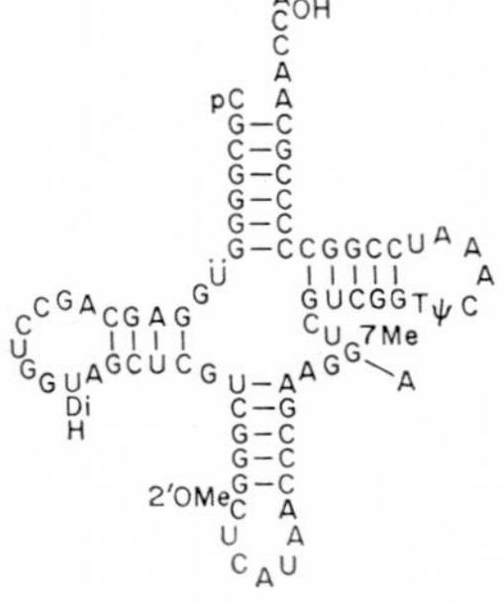

Fig. 31. *Continued.*

to, but not exactly the same as, the Watson-Crick geometry. These Crick wobble combinations are shown schematically in Fig. 32.

Although there is no physical evidence which directly bears on the guanine-uracil and hypoxanthine-uracil base-pairing configurations, the

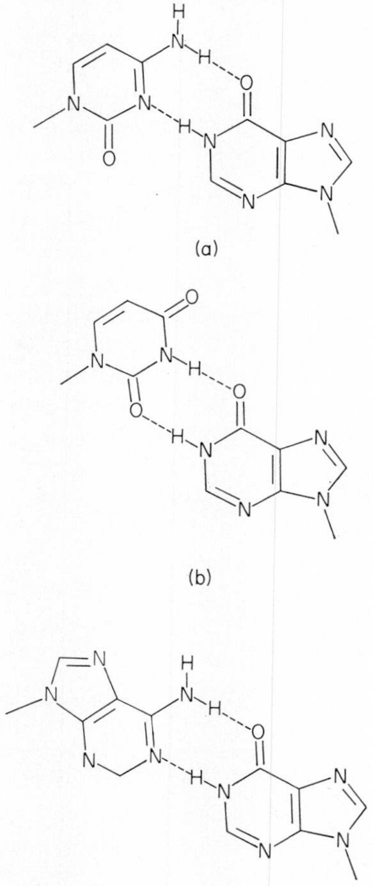

(a)

(b)

(c)

FIG. 32. Schematic diagram showing the Crick "wobble" combinations. (a) hypoxanthine–cytosine (b) hypoxanthine–uracil (the guanine–uracil pairing is similar to this) (c) hypoxanthine–adenine.

hypoxanthine-cytosine and hypoxanthine-adenine pairs have been found in synthetic polynucleotide complexes, as discussed earlier in Section IV, A. An adenine-hypoxanthine combination has also been seen in a cocrystallization study involving adenine and hypoxanthine derivatives cocrystallized as a hydrogen-bonded dimer. The configuration found in this study, however, differs from the Crick configuration in that hydrogen bonding involves the imidazole nitrogen, N-7, of adenine, rather than N-1. One wonders whether such a pairing configuration might possibly be used in codon-anticodon recognition. The configuration is an attractive possibility, since the distance between glycosidic carbon atoms and the direction of the glycosidic bonds more closely match the normal Watson-Crick base-pairing parameters and therefore the pairing involves less wobble than the Crick pairing. However, in order for this pairing to be used, the adenine residue on the messenger RNA must rotate into the *syn* conformation, as contrasted to the normal *anti* conformation of the purine nucleotide (the terms *syn* and *anti* refer to the relative orientation of the sugar residue to the base and are discussed in detail by Donohue and Trueblood, 1960). Although such a conformation may be less energetically favorable for steric reasons (see Haschemeyer and Rich, 1967), direct evidence that it is possible has been given by the deoxyguanosine-bromodeoxycytidine crystalline complex, described earlier, in which deoxyguanosine has been found to exist in the *syn* conformation. This, therefore, is a reasonable alternative to the Crick pairing, although no conclusions can be drawn as yet concerning this.

C. Tertiary Structure of Transfer RNA

Recently, several groups have reported success at crystallizing transfer RNA (Clark *et al.*, 1968; Hampel *et al.*, 1968; Kim and Rich, 1968b; Fresco *et al.*, 1968). Large single crystals have been obtained of phenylalanine-tRNA (*E. coli*) and formylmethionine-tRNA (*E. coli* and yeast) and preliminary X-ray data have become available. There appear to be two different crystalline modifications of *E. coli* formylmethionine-tRNA, an orthorhombic form (Clark *et al.*, 1968) and a hexagonal form (Kim and Rich, 1968b). The former crystals were obtained using dioxane as the precipitating agent, while the latter crystals have been grown from chloroform-water mixtures. Yeast formylmethionine-tRNA and *E. coli* phenylalanine-tRNA also crystallize from water-alcohol mixtures containing trace metal elements. Phenylalanine-tRNA crystallizes with hexagonal symmetry and it remains to be seen whether this type of crystal symmetry will occur in other transfer RNA samples. It is of interest that crude preparations of transfer RNA crystallize (Fresco *et al.*, 1968). If single crystals contain different transfer RNA species, then this could indicate that the tertiary structure of all these molecules is very similar. The complete X-ray analyses of these structures have begun, and the elucidation of the three-dimensional structure of these molecules is eagerly awaited.

VI. Structural Information on Ribosomal RNA

Since its original discovery in 1963, a low molecular weight ribosomal RNA with a sedimentation coefficient of 5S has been highly purified and sequenced using techniques developed by Sanger *et al.* (1965). Although the significance of this minor RNA component is not known, it is found in *E. coli* and in a mammalian carcinoma cell line, denoted KB (Forget and Weissman, 1967). Table VI shows the sequences of both 5 S RNA molecules. Although the detailed sequences of these molecules differ,

TABLE VI
NUCLEOTIDE SEQUENCES OF 5 S *E. coli* AND KB RIBOSOMAL RNA

E. coli 5 S RNA

pUGCCUGGCGGCCUUAGCGCGGUGGUCCCACCUGACCCCAUGCCGAACUCA
GAAGUGAAACGCCGUAGCGCCGAUGGUAGUGUGGGGUCUCCCCAUGCGA
GAGUAGGGAACUGCCAGGCAU$_{OH}$

KB cell 5 S RNA

pGUCUACGGCCAUACCACCCUGAACGCGCCCGAUCUCGUCUGAUCUCGGAA
GCUAAGCAGGGUCGGGCCUGGUUAGUACUUGGAUGGGAGACCGCCUGGG
AAUACCGGGUGCUGUAGGCUU(U)$_{OH}$

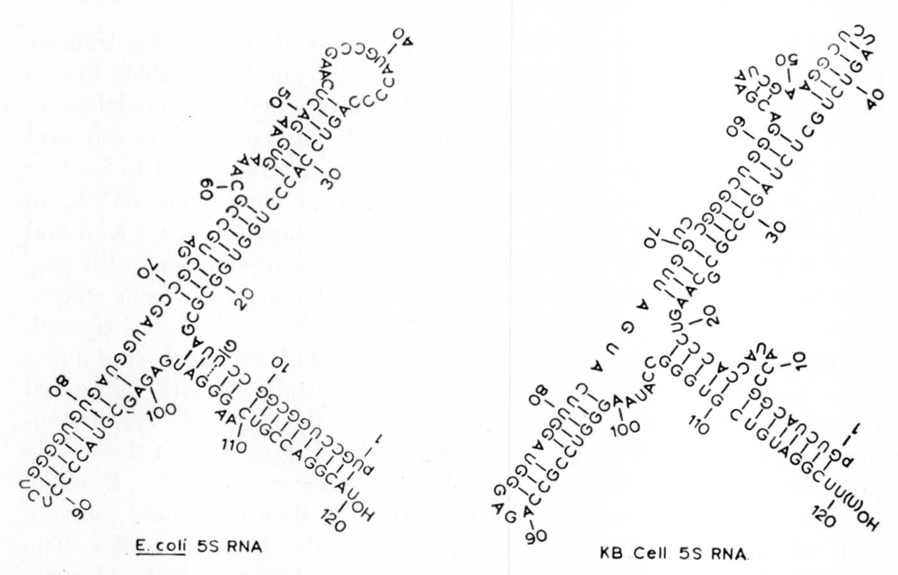

FIG. 33. Possible secondary structures for (a) *E. coli* 5 S ribosomal RNA (b) KB cell 5 S ribosomal RNA.

there are several similarities. In both cases there are approximately 120 residues arranged as a single linear molecule. These molecules have a large number of sequences in common (see Madison, 1968) and both can be folded into a common secondary structure (shown in Fig. 33). This suggests that these molecules may play some common role in *E. coli* and mammalian ribosomes, and further information about their function will be of great interest.

VII. Future Horizons

Before concluding this chapter, it seems worthwhile to say a few words about the probable future horizons which lay ahead in nucleic acid research in coming years. Kyogoku *et al.* (1968) have recently discovered that barbiturates, which are derivatives of 6-oxyuracil, interact very strongly and with great specificity with adenine and adenine derivatives in nonpolar solvents. Crystalline complexes have been isolated which contain these compounds crystallized as an intermolecular complex (Kim and Rich, 1968a), and it has been suggested that barbiturates exert their pharmacological action by interacting strongly and inactivating the adenine-containing coenzymes at the cellular membrane. It seems likely that a great deal of additional information can be obtained about the action of antibiotics and other drugs with similar studies. A good example might be actinomycin, which is known to interact with guanine residues on DNA. A crystalline complex containing actinomycin cocrystallized with a guanine-containing oligonucleotide would be of great interest in this connection, and attempts are currently being made to form such a complex.

Low angle scattering from ribosomes has given a surprising amount of X-ray diffraction information (Langridge and Holmes, 1962), and these pioneering efforts raise hopes that X-ray crystallography may one day succeed in elucidating the complete three-dimensional structure of the ribosome. The first major hurdle in this direction to be overcome is to form single crystals of ribosomes. Although this has not yet been possible, Byers (1967) has reported finding microcrystalline ribosomal aggregates in chick embryo cells. Similar reports of crystalline ribosomal aggregates *in vivo* have appeared (Crain *et al.*, 1964; Porte and Zahnd, 1961; Ghiara and Taddei, 1966), and this raises hopes that it may some day be possible to succeed in crystallizing ribosomes *in vitro*.

Another important area to be explored are the enzymes involved in nucleic acid metabolism. In particular, the *E. coli* DNA polymerase and RNA polymerase enzymes are of key importance, since they are intimately involved in the information transfer properties of the nucleic acids. Attempts to crystallize the *E. coli* DNA polymerase enzyme are in progress, and should these be successful, the X-ray structure analysis can begin on this important enzyme.

In the area of transfer RNA structure, an ever-growing list of nucleotide sequences for different transfer RNA molecules will almost certainly appear in future years, and this information will give insight into the function of different regions of the transfer RNA molecule in protein synthesis. The three-dimensional structure of transfer RNA, already suggested by the cloverleaf model, will find final verification by means of X-ray crystallography now that crystallization of transfer RNA has been achieved.

ACKNOWLEDGMENT

This work has been supported in part by the National Institutes of Health, United States Public Health Service, the National Science Foundation, the American Cancer Society, and an institutional grant to the University of Rochester from the American Cancer Society. Facilities for this work were supplied in part from the United States Atomic Energy Commission at the University of Rochester Atomic Energy Project and this paper has been assigned Report No. UR-49-1097.

REFERENCES

Akinrimisi, E. O., Sander, C., and Ts'o, P. O. P. (1963). *Biochemistry* **2**, 340.

Arnott, S., Wilkins, M. H. F., Hamilton, L. D., and Langridge, R. (1965). *J. Mol. Biol.* **11**, 391.

Arnott, S., Wilkins, M. H. F., Fuller, W., and Langridge, R. (1967a). *J. Mol. Biol.* **27**, 525.

Arnott, S., Wilkins, M. H. F., Fuller, W., and Langridge, R. (1967b). *J. Mol. Biol.* **27**, 535.

Arnott, S., Wilkins, M. H. F., Fuller, W., Venable, J. H., and Langridge, R. (1967c). *J. Mol. Biol.* **27**, 549.

Baklagine, Y. G., Volkenshtein, M. V., and Kondrashev, Y. D. (1966). *Zhu. Strukt. Khim.* **7**, 399.

Bayev, A. A., Venkstern, T. V., Mirsabekov, A. D., Krutilina, A. I., Li, L., and Axelrod, V. D. (1967). *Mol. Biol.* **1**, 754.

Blake, R. D., Massoulie, J., and Fresco, J. R. (1967). *J. Mol. Biol.* **30**, 291.

Brahms, J., Michelson, A. M., and Van Holde, K. E. (1966). *J. Mol. Biol.* **15**, 467.

Brahms, J., Maurizot, J. C., and Michelson, A. M. (1967). *J. Mol. Biol.* **25**, 465.

Brownlee, C. G., Sanger, F., and Barrell, B. G. (1967). *Nature* **215**, 735.

Byers, B. (1967). *J. Mol. Biol.* **26**, 155.

Clark, B. F. C., Doctor, B. P., Holmes, K. C., Klug, A., Marcker, K. A., Morris, S. J., Paradies, H. H. (1968). *Nature* **219**, 1222.

Cochran, W. (1951). *Acta Cryst.* **4**, 81.

Cochran, W., Crick, F. H. C., and Vand, V. (1952). *Acta Cryst.* **5**, 581.

Colthup, N. B., Daly, L. H., and Wiberley, S. E. (1964). "Introduction to Infrared and Raman Spectroscopy." Academic Press, New York.

Cory, S., Marcker, K. A., Dube, S. K., and Clark, B. F. C. (1968). *Nature* **220**, 1039.
Crain, S. M., Benitez, H., and Vatter, A. E. (1964). *Ann. N. Y. Acad. Sci.* **118**, 206.
Crick, F. H. C. (1958). *Symp. Soc. Exptl. Biol.* **12**, 138.
Crick, F. H. C. (1966). *J. Mol. Biol.* **19**, 548.
Davies, D. R. (1960). *Nature* **186**, 103.
Davies, D. R., and Baldwin, R. L. (1963). *J. Mol. Biol.* **6**, 251.
Davies, D. R., and Rich, A. (1959). *Acta Cryst.* **12**, 97.
Donohue, J., and Trueblood, K. N. (1960). *J. Mol. Biol.* **2**, 363.
Dube, S. K., Marcker, K. A., Clark, B. F. C., and Cory, S. (1968). *Nature* **218**, 232.
Fasman, G. D., Lindblow, C., and Grossman, L. (1964). *Biochemistry* **3**, 1015.
Felsenfeld, G., and Miles, H. T. (1967). *Ann. Rev. Biochem.* **36**, 407.
Felsenfeld, G., Davies, D. R., and Rich, A. (1957). *J. Am. Chem. Soc.* **79**, 2023.
Forget, B. G., and Weissman, S. M. (1967). *Science* **158**, 1695.
Freese, E. (1959). *J. Mol. Biol.* **1**, 87.
Fresco, J. R., Blake, R. D., and Langridge, R. (1968). *Nature* **220**, 1285.
Fuller, W., Wilkins, M. H. F., Wilson, H. R., and Hamilton, L. D. (1965). **12**, 60.
Fuller, W., Hutchinson, F., Spencer, M., and Wilkins, M. H. F. (1967). *J. Mol. Biol.* **27**, 507.
Ghiara, G., and Taddei, C. (1966). *Boll. Soc. Ital. Biol. Sper.* **42**, 784.
Giannoni, G., and Rich, A. (1964). *Biopolymers* **2**, 399.
Goodman, H. M., Abelson, J., Landy, A., Brenner, S., and Smith, J. D. (1968). *Nature* **217**, 1019.
Hackett, P. J., and Hanawalt, P. C. (1966). *Biochim. Biophys. Acta* **123**, 356.
Hamilton, L. D. (1968). *Nature* **218**, 633.
Hanawalt, P. C., and Ray, D. S. (1964). *Proc. Natl. Acad. Sci. U.S.* **52**, 125.
Hamlin, R. M., Jr., Lord, R. C., and Rich, A. (1965). *Science* **148**, 1734.
Hampel, A., Labanauskas, M., Connors, P. G., Kirkegard, L., RajBhandary, U. L., Sigler, P. B., and Bock, R. M. (1968). *Science* **162**, 1384.
Hartman, K. A., Jr., and Rich, A. (1965). *J. Am. Chem. Soc.* **87**, 2033.
Haschemeyer, A. E. V., and Rich, A. (1967). *J. Mol. Biol.* **27**, 369.
Haschemeyer, A. E. V., and Sobell, H. M. (1963). *Proc. Natl. Acad. Sci. U.S.* **50**, 872.
Haschemeyer, A. E. V., and Sobell, H. M. (1964). *Nature* **202**, 969.
Haschemeyer, A. E. V., and Sobell, H. M. (1965a). *Acta Cryst.* **18**, 525.
Haschemeyer, A. E. V., and Sobell, H. M. (1965b). *Acta Cryst.* **19**, 125.
Holley, R. W., Apgar, J., Everett, G. A., Madison, J. T., Marquisee, M., Merrill, S. H., Penswick, J. R., and Zamir, A. (1965). *Science* **147**, 1462.
Hoogsteen, K. (1959). *Acta Cryst.* **12**, 822.
Hoogsteen, K. (1963). *Acta Cryst.* **16**, 907.
Hoogsteen, K. (1968). "Molecular Associations in Biology" (B. Pullman, ed.). Academic Press, New York.
Howard, B. D., and Tessman, I. (1964). *J. Mol. Biol.* **9**, 364.
Howard, F. B., Frazier, J., Lipsett, M. N., and Miles, H. T. (1964). *Biochem. Biophys. Res. Commun.* **17**, 93.
Howard, F. B., Frazier, J., Singer, M. F., and Miles, H. T. (1966). *J. Mol. Biol.* **16**, 415.
Huang, W. M., and Ts'o, P. O. P. (1966). *J. Mol. Biol.* **16**, 523.
Jardetzky, O. (1964). *Biopolymers, Symp.* **1**, 501.
Kanner, L., and Hanawalt, P. C. (1968). *Biochim. Biophys. Acta* **157**, 532.

Katz, L., and Penman, S. (1966). *J. Mol. Biol.* **15**, 220.
Katz, L., Tomita, K., and Rich, A. (1965). *J. Mol. Biol.* **13**, 340.
Katz, L., Tomita, K., and Rich, A. (1967). *Acta Cryst.* **21**, 754.
Kim, S.-H., and Rich, A. (1967). *Science* **158**, 1046.
Kim, S.-H., and Rich, A. (1968a). *Proc. Natl. Acad. Sci. U.S.* **60**, 402.
Kim, S.-H., and Rich, A. (1968b). *Science* **162**, 1381.
Kim, S.-H., and Rich, A. (1969). *J. Mol. Biol.* **42**, 87.
Kornberg, A., Bertsch, L. L., Jackson, J. F., and Khorana, H. G. (1964). *Proc. Natl. Acad. Sci. U.S.* **51**, 315.
Küchler, E., and Derkosch, J. (1966). *Z. Naturforsch.* **21b**, 209.
Kyogoku, Y., Lord, R. C., and Rich, A. (1969). *Biochim. Biophys. Acta* **179**, 10.
Kyogoku, Y., Lord, R. C., and Rich, A. (1966). *Science* **154**, 518.
Kyogoku, Y., Lord, R. C., and Rich, A. (1967a). *J. Am. Chem. Soc.* **89**, 496.
Kyogoku, Y., Lord, R. C., and Rich, A. (1967b). *Proc. Natl. Acad. Sci. U.S.* **57**, 250.
Kyogoku, Y., Lord, R. C., and Rich, A. (1968). *Nature* **218**, 69.
Labana, L. L., and Sobell, H. M. (1967). *Proc. Natl. Acad. Sci. U.S.* **57**, 459.
Labana, L. L., and Sobell, H. M. (1967). Unpublished data.
Langridge, R., and Gomatos, P. J. (1963). *Science* **141**, 694.
Langridge, R., and Holmes, K. (1962). *J. Mol. Biol.* **5**, 611.
Langridge, R., and Rich, A. (1960). *Acta Cryst.* **13**, 1052.
Langridge, R., and Rich, A. (1963). *Nature* **198**, 725.
Langridge, R., Wilson, H. R., Hooper, C. W., Wilkins, M. H. F., and Hamilton, L. D. (1960). *J. Mol. Biol.* **2**, 19 and 38.
Langridge, R., Billeter, M. A., Borst, H., Burdon, A. R., and Weissmann, C. (1964). *Proc. Natl. Acad. Sci. U.S.* **52**, 114.
Lawley, P. D., and Brookes, P. (1962). *J. Mol. Biol.* **4**, 216.
Leng, M., and Felsenfeld, G. (1966). *J. Mol. Biol.* **15**, 455.
Lipsett, M. N. (1964). *J. Biol. Chem.* **239**, 1256.
Litman, R. M. (1957). Dissertation, University of California, Berkeley, California.
Madison, J. T. (1968). *Ann. Rev. Biochem.* **37**, 131.
Madison, J. T., Everett, G. A., and Kung, H. (1966). *Science* **153**, 531.
Marsh, R. E., Bierstedt, R., and Eichhorn, E. L. (1962). *Acta Cryst.* **15**, 310.
Marvin, D. A., Spencer, M., Wilkins, M. H. F., and Hamilton, L. D. (1961). *J. Mol. Biol.* **3**, 547.
Mathews, F. S., and Rich, A. (1964). *J. Mol. Biol.* **8**, 89.
Mazza, F., Sobell, H. M., and Kartha, G. (1969). *Acta Cryst.* (in press).
Meselson, M. and Stahl, F. W. (1966). *Phage and the Origins of Molecular Biology*, Cold Spring Harbor Laboratory of Quantitative Biology, pg. 246.
Michelson, A. M., Massoulie, J., and Guschlbauer, W. (1967). *Prog. Nucleic Acid Res. Mol. Biol.* **6**, 83.
Miles, H. T. (1964). (1964). *Proc. Natl. Acad. Sci. U.S.* **51**, 1104.
Miller, J. H., and Sobell, H. M. (1966). *Proc. Natl. Acad. Sci. U.S.* **55**, 1201.
Miller, J. H., and Sobell, H. M. (1967). *J. Mol. Biol.* **24**, 345.
Milman, G., Langridge, R., and Chamberlin, M. J. (1967). *Proc. Natl. Acad. Sci. U.S.* **57**, 1805.
Morgan, A. R., and Wells, R. D. (1968). *J. Mol. Biol.* 37, 63.
Nash, H. A., and Bradley, D. F. (1966). *J. Chem. Phys.* **45**, 1380.
O'Brien, E. J. (1963). *J. Mol. Biol.* **7**, 107.
O'Brien, E. J. (1967). *Acta Cryst.* **23**, 92.

Pimentel, G. C., and McClellan, A. L. (1960). "The Hydrogen Bond." Freeman, San Francisco, California.

Pitha, J., Jones, R. N., and Pithova, J. (1966). *Can. J. Chem.* **44**, 1044.

Pollak, M., and Rein, R. (1966). *J. Theoret. Biol.* **11**, 490.

Pople, J. A., Schneider, W. G., and Bernstein, H. J. (1959). "High Resolution Nuclear Magnetic Resonance." McGraw-Hill, New York.

Porte, A., and Zahnd, I. P. (1961). *Compt. Rend. Soc. Biol.* **155**, 1058.

Pullman, A., and Pullman, B. (1968). "Progress in Nucleic Acid Research and Molecular Biology" (J. N. Davidson and W. E. Cohn, eds.), Vol. 9, Academic Press, New York.

Pullman, B. (1968). "Molecular Associations in Biology" (B. Pullman, ed.) Academic Press, New York.

Pullman, B., and Caillet, J. (1967). *Theoret. Chim. Acta* **8**, 223.

Pullman, B., Claverie, P., and Caillet, J. (1966). *Proc. Natl. Acad. Sci. U.S.* **55**, 904.

Pullman, B., Claverie, P., and Caillet, J. (1967). *Proc. Natl. Acad. Sci. U.S.* **57**, 1663.

Radding, C., and Kornberg, A. (1962). *J. Biol. Chem.* **237**, 2877.

RajBhandary, U. L., Chang, S. H., Stuart, A., Faulkner, R. D., Hoskinson, R. M., and Khorana, H. G. (1967). *Proc. Natl. Acad. Sci. U.S.* **57**, 751.

Ralph, R. K., Connors, W. J., and Khorana, H. G. (1962). *J. Am. Chem. Soc.* **84**, 2265.

Rich, A. (1958). *Biochim. Biophys. Acta* **29**, 502.

Rich, A. (1959). *Symp. Mol. Biol., Univ. Chicago, 1959* p. 47. Univ. of Chicago Press, Chicago, Illinois.

Rich, A. (1960). *Proc. Natl. Acad. Sci. U.S.* **46**, 1044.

Rich, A., and Watson, J. D. (1954). *Nature* **173**, 995.

Rich, A., Davies, D. R., Crick, F. H. C., and Watson, J. D. (1961). *J. Mol. Biol.* **3**, 71.

Riley, M., Maling, B., and Chamberlin, M. J. (1966). *J. Mol. Biol.* **20**, 359.

Sakore, T. D., and Sobell, H. M. (1969a). To be published.

Sakore, T. D., and Sobell, H. M. (1969b). *J. Mol. Biol.* (in press).

Sakore, T. D., Tavale, S. S., and Sobell, H. M. (1969a). *J. Mol. Biol.* (in press).

Sakore, T. D., Mazza, F., Sobell, H. M., and Kartha, G. (1969b). *J. Mol. Biol.* (in press).

Sanger, F., Brownlee, C. G., and Barrell, B. G. (1965). *J. Mol. Biol.* **13**, 373.

Sasisekharan, V., and Sigler, P. (1965). *J. Mol. Biol.* **12**, 296.

Schachman, H. K., Adler, J., Radding, C. M., Lehman, I. R., and Kornberg, A. (1960). *J. Biol. Chem.* **235**, 3242.

Shoup, R. R., Miles, H. T., and Becker, E. D. (1966). *Biochem. Biophys. Res. Commun.* **23**, 194.

Smith, J. D., Abelson, J. N., Clark, B. F. C., Goodman, H. M., and Brenner, S. (1966). *Cold Spring Harbor Symp. Quant. Biol.* **31**, 479.

Sobell, H. M. (1966). *J. Mol. Biol.* **18**, 1.

Sobell, H. M. (1969). Unpublished data.

Sobell, H. M., Tomita, K., and Rich, A. (1963). *Proc. Natl. Acad. Sci. U.S.* **49**, 885.

Solie, T. N., and Schellman, J. A. (1968). *J. Mol. Biol.* **33**, 61.

Spencer, M. (1963). *Cold Spring Harbor Symp. Quant. Biol.* **28**, 77.

Spencer, M., and Poole, F. (1965). *J. Mol. Biol.* **11**, 314.

Speyer, J. F. (1968). *In* "Molecular Genetics" (J. H. Taylor, ed.), Part 2, p. 137. Academic Press, New York.

Spirin, A. S. (1963). *Dokl. Akad. Nauk SSSR* **44**, 1963.
Staehelin, M., Rogg, H., Baguley, R. C., Ginsberg, T., and Wehrli, W. (1968). *Nature* **219**, 1363.
Stevens, C. L., and Felsenfeld, G. (1964). *Biopolymers* **2**, 293.
Strelzoff, E. (1961). *Biochem. Biophys. Res. Commun.* **5**, 384.
Sulston, J., Lohrmann, R., Orgel, L. E., and Miles, H. T. (1968). *Proc. Natl. Acad. Sci. U.S.* **60**, 409.
Takemura, S., Mizutani, T., and Myazaki, M. (1968). *J. Biochem.* (Tokyo). **63**, 277.
Tavale, S. S., Sakore, T. D., and Sobell, H. M. (1969). *J. Mol. Biol.* (in press).
Thomas, G. J., Jr., and Kyogoku, Y. (1967). *J. Am. Chem. Soc.* **89**, 4170.
Tomita, K., and Rich, A. (1964). *Nature* **201**, 1160.
Tomita, K., and Rich, A. (1965). Unpublished data.
Tomita, K., Katz, L., and Rich, A. (1967). *J. Mol. Biol.* **30**, 545.
Tuppy, H., and Küchler, E. (1964a). *Biochim. Biophys. Acta* **80**, 669.
Tuppy, H., and Küchler, E. (1964b). *Monatsch. Chem.* **95**, 1677.
Ts'o, P. O. P., Melvin, I. S., and Olson, A. C. (1963). *J. Am. Chem. Soc.* **8**, 1289.
Voet, D., and Rich, A. (1969). *J. Am. Chem. Soc.* (in press).
Warner, R. C. (1957). *J. Biol. Chem.* **229**, 711.
Watson, J. D., and Crick, F. H. C. (1953). *Nature* **171**, 964.
Wilkins, M. H. F., and Randall, J. T. (1953). *Biochim. Biophys. Acta* **10**, 192.
Zachau, H. G., Dutting, D., and Feldman, H. (1966). *Z. Physiol. Chem.* **347**, 212.

III THE STRUCTURE AND DUPLICATION OF CHROMOSOMES

J. HERBERT TAYLOR

I. Introductory Remarks

Chromosome will be used as an all-inclusive term that refers to the structural and functional DNA-containing units of nuclei of higher cells as well as viruses, bacteria, and other cellular organelles of higher organisms, such as mitochondria and chloroplasts. The choice may not be logical, but the term is an evolving one and this image conforms to practice. To try to invent terms to signify real or supposed distinctions among the chromosomes of the various groups has limited value, and furthermore, attempts to introduce new terms for the genetic apparatus of viruses and bacteria (Ris and Chandler, 1963) have met with little success. Chromosomes (so named by Waldeyer in 1888 because of their intense staining capacity) became conceptually associated with linkage groups as the result of the work of Morgan (1911), Sturtevant (1913), and others in the first quarter of this century. When linkage studies were extended to bacteria and viruses in the late 1940's the concept of the chromosome as the physical basis for the observations was a natural one which has persisted. Admittedly the application of the term to the genetic apparatus of the smallest viruses would appear to be stretching the concept too far; perhaps molecule is applicable in such cases.

We know so little about the arrangement of DNA in the chromosomes and nuclei of most species, especially the higher cells, that subdivisions based on these variations are probably of little value. Nevertheless, the terms eucaryote and procaryote (Ris and Chandler, 1963) have found wide usage. Eucaryotes possess characteristic chromosomes, nuclei, and a typical mitotic apparatus. Nucleoli also appear to be characteristic of these nuclei. Procaryotes, on the other hand, include the organisms which do not possess a nucleus bounded by a nuclear membrane, do not divide by mitosis, i.e., do not have a spindle apparatus, and do not at any stage of division produce typical chromosome-like structures. By the latter the authors presumably mean condensed rod-shaped structures during some stage of division. The distinctions are, of course, clear at the extremes but like most other classifications, the differences become of dubious value at the borderlines. The presence or absence of histones associated with the DNA might be a further means of separating the group with chromosomes in the classic sense from those with DNA which does not condense into rod-shaped structures. However, the present knowledge of the factors involved in condensation make that also a dubious basis for distinctions. Perhaps the number and size of units of replication in chromosomes will prove to be a distinguishing feature between higher and lower forms. The regulatory unit in bacteria has been called a replicon (Jacob *et al.*, 1963) and the same name has been suggested for the regulatory units in chromosomes of higher cells (Taylor, 1963a). In the bacteria there appears to be one replicon per chromosome but in higher organisms the units appear to be of a smaller size and more numerous (Taylor, 1968; Huberman and Riggs, 1968). Perhaps the transition from a single replicon per chromosome to multiple replicons will prove to be an important evolutionary step which will serve to distinguish groups. For example, one would like to know what the maximum size is for a replicon to compete successfully in evolution. Do the chromosomes of bacteria such as *Escherichia coli* and *Bacillus subtilis* represent that stage in evolution?

We shall attempt to compare structure and replication in the various groups insofar as information is available. Differences and similarities will be noted and the knowledge gained from the smaller and presumably less complex chromosomes will be used in trying to visualize the more complex types. In addition, I would like to point out for the reader what will be obvious after reading the chapter. The following discussion is not a review of the papers on structure and duplication of chromosomes which have by now reached an enormous number, but an assessment of the state of knowledge based on the author's present views. Supplementary material and more extensive treatments of certain aspects of the problems may be found in reviews by Kaufmann *et al.* (1960), Moses (1964), Taylor (1962, 1963b and 1967), Thomas (1963), and Sueoka (1967), to mention a few.

II. Structure of the Genetic Apparatus

A. Viruses

The genetic apparatus of viruses appears to consist solely of DNA or, in some types, of RNA. In the smaller viruses the nucleic acid is single stranded in the mature virus particles, but becomes double-stranded during the reproductive cycle in all forms so far investigated. The single strand carries all of the genetic information or has the potential for generating it by replication in the proper cellular environment. The single strand may be the message as in the RNA phages Qβ and f2, i.e., the entering strand can be translated directly for protein synthesis by the cellular machinery. In other cases the entering strand may be the complement of the message; ϕX174 is an example. In the latter instance replication with the single strand as a template is the first event and only then is the genetic apparatus available for transcription of RNA. This RNA which is complementary to the new strand is the genetic message for proteins coded by the phage. One of the proteins is a specific DNA polymerase which produces the single-stranded DNA which is packaged in the maturation of new phage particles.

These small DNA viruses have nucleic acids with lengths (Table I) that vary from 1.58 μ for polyoma to about 1.86 μ for ϕX174 (see review

TABLE I
AMOUNTS OF DNA PER GENOME IN CELLS AND VIRUSES

Organism	μμg per genome	Length per genome (mm)	Chromosome number (n)	Reference
Viruses				
Polyoma		0.00158	1	Weil and Vinograd (1963)
φX174		0.00186	1	Chandler *et al.* (1964)
T7 phage		0.0122	1	Freifelder and Kleinschmidt (1965)
λ phage		0.0172	1	MacHattie and Thomas (1964)
T4 phage		0.0520 ± 6	1	Kleinschmidt (1967)
Fowl pox		0.0930 ± 6	1	Hyde *et al.* (1966)
Bacteria				
E. coli	0.45×10^{-2}	1.530	1	Lark (1966)
E. coli		1.400	1	Cairns (1963)
Algae				
Chlamydomonas	3×10^{-2}	10.200	8	Sager and Ishida (1963)
Fungi				
Yeast	1.8×10^{-2}	6.12	15	Laskowski *et al.* (1960)
Neurospora	4.6×10^{-2}	15.64	7	Horowitz and MacLeod (1960)
Higher plants				
Raphanus sativus	2.5	850.0	9	Baetcke *et al.* (1967)
Vicia faba	22.0	7,480.0	6	Baetcke *et al.* (1967)
Lilium longiflorum	53.0	18,020.0	12	Ogur *et al.* (1951)
Xanthorhiza	0.45	153.0	18	Rothfells *et al.* (1966)
Aquilegia	0.45	153.0	7	Rothfells *et al.* (1966)

Higher animals				
Invertebrates				
Insects				
Drosophila				
melanogaster	0.18	61.2	4	Rudkin (1964b)
Chironomus				
pallidivittatus	0.20	68.0	4	Edstrom (1964)
Vertebrates				
Fishes (various species)	0.7–2.8	238.0–952.0	24–30	Ohno and Atkin (1966)
Lungfishes				
Protopterus	50	17,000.0		Allfrey, Mirsky and Stern (1955)
Lepidosiren	124	42,160.0	19	Ohno and Atkin (1966)
Amphibia				
Frog (*Bufo*)	4.9	1666.0	13	Ohno and Atkin (1966)
Salamander (*Plethodon*)	24.6	8,364.0		Ohno and Atkin (1966)
Necturus maculosus	94.5	32,130.0	12	Ohno and Atkin (1966)
Reptiles				
Order Squamata	2.1–2.3	714.0–782.0	18–23	Atkin *et al.* (1965)
Order Crocodylia	2.9–3.1	986.0–1054.0	21	Atkin *et al.* (1965)
Birds (various species)	3.1–4.2	1054–1428.0		Ohno and Atkin (1966)
Mammals				
Chinese hamster	3.5	1190.0	11	Huberman and Riggs (1966)
Man	3.0	1020.0	23	Vendrely and Vendrely (1949)

by Kleinschmidt, 1967). Since DNA in the double-stranded form, B configuration, has 196 molecular weight units per angstrom, the molecular weight of these DNA's is a little over 3×10^6 daltons for the double-stranded forms. They have less than 5000 nucleotide pairs and could presumably code for a maximum of about 10 of the smaller polypeptides with 140–160 amino acid residues. The smaller RNA viruses may code for even fewer proteins. The minimum seems to be a specific RNA polymerase for their replication and perhaps two kinds of proteins involved in completing the coat.

The smaller DNA viruses seem to exist as rings both in mature particles and in the replicative forms, but the DNA of the larger viruses exists in the form of filaments in the mature particles. These filaments vary in length from about 6 μ for *B. subtilis* ϕ29 phage to 47–52 μ for the *E. coli* phages T2 and T4. Fowl pox virus has DNA molecules about 93 μ long (Table I).

Some of the DNA molecules from phages can form rings upon annealing *in vitro* and presumably do so in the infected host cell. *Escherichia coli* phage λ with a length of 16 μ is the most thoroughly studied. Hershey and Burgi (1965) have shown that the rings formed *in vitro* will open by partial denaturation of the DNA by heat, for example. The rings are formed and held together by complementary single-stranded regions at each end of the linear molecules. Other *E. coli* phages, T2, T4, T3, and T7, have terminal repetitions on the linear molecules, but these regions are double-stranded, and therefore, the phage chromosomes do not circularize *in vitro* as does λ (Ritchie *et al.*, 1967). If one strand is eroded back on each end by the action of exonuclease III, the DNA can then be caused to form circles or to link into linear or circular concatenations. On the other hand, randomly broken molecules which are partially digested in a similar way do not form circles. The length of the overlap region is 1–3% of the genome. In phages T2 and T4 the ends are not the same in all molecules, i.e., evidence from strand separation and reannealing (Ritchie *et al.*, 1967) as well as from studies of genetic recombination (Streisinger *et al.*, 1964) indicates that the chromosomes are circularly permuted. However, the chromosomes of phages λ, T3, and T7 are not circular permutations.

The circularized DNA molecules from phage λ can be sealed *in vitro* into covalently bonded circles by enzymes isolated from λ-infected *E. coli* cells (Gellert, 1967; Gefter *et al.*, 1967). The same type of reaction is assumed to occur in the cell, but mature phage particles contain only linear molecules. The circular DNA's would be protected from attack by exonucleases and it is supposed that this may be the significance of the ring forms. There is no convincing evidence for non-nucleotide linkers in any of the viral chromosomes. In fact, with the discovery of enzymes which close rings by forming phosphodiester linkages typical of those of regular polynucleotide chains, there is no reason to expect that any other type of linker will be found in DNA of viruses.

B. Bacteria

Genetic evidence derived from the sequence of transfer of markers during mating of *E. coli* first indicated that bacterial chromosomes may exist as rings. Among various strains of *E. coli* the first genetic locus to be transferred varied, but the sequence was similar, if not identical, in several strains which could be compared. Although it appeared that the ring could open during mating at a variety of places which were near the locus of the fertility factor (F^1), the chromosomes of vegetative cells behaved like rings (Jacob and Wollman, 1961). The chromosomes in various strains might be compared to the permuted chromosomes of phages T2 and T4 which exist in a population within any given strain. Later studies by Cairns (1963), using tritiated thymidine of high specific activity and autoradiography for visualizing single, fully extended chromosomes, showed that a physical ring-shaped chromosome existed in vegetative cells. The organization of these rings, which may be more than 1000 μ long within cells that are 1 or 2 μ in length, is still rather uncertain. Certainly they must be rather well packaged to occupy so little space and not become tangled. That they have one point of attachment at the nuclear membrane has been suggested from a variety of data (Jacob *et al.*, 1963) including electron micrographs in the case of *B. subtilis*. More recent and extensive autoradiographic studies of the segregation of labeled subunits of chromosomes in *E. coli* (Lark, 1966; Lark *et al.*, 1967) add strong evidence that a recently replicated chromosome becomes attached to a site in the cell (probably the cell surface) by means of the newest template strand. One template strand remains attached from the previous replication while the new one becomes attached by a mechanism which requires protein synthesis. Studies by Hanawalt and Ray (1964) and Ganesan (1965) have shown that when cells are lysed or broken the most recently replicated DNA is attached to the membrane. The hypothesis has been advanced by several of those who work on these problems that sorting occurs by the passage of the whole chromosome across or through such an attachment point during replication.

The existence of structural subunits in chromosomes of *E. coli* has been suggested by Massie and Zimm (1965) from the isolation of DNA with an average molecular weight of 250×10^6, i.e., about one eighth of a chromosome, by methods which involve almost no shear. Pietsch (1966) has found regions which bind phleomycin distributed along the bacterial chromosome. The number of sites is not certain, but there are several and he proposes that the sites consist of runs of AT polymer of at least 100 nucleotides. However, replication studies have revealed no large subunits and it seems best to assume at the present state of our knowledge that a chromosome of *E. coli* consists of a single molecule. Okazaki and associates (Okazaki *et al.*, 1967; Sakabe and Okazaki, 1966) have demonstrated small subunits which could have structural significance. After very short pulse labels, the labeled DNA can be shown to exist in pieces which separate from the bulk of the DNA by sedimentation through alkaline sucrose gradients. The newly synthesized DNA exists in pieces with sedimentation coefficients of 10–11 S, i.e., about 1000 to 2000 nucleotides per piece. He suggests that these pieces which are about the size estimated for cistrons are also units of replication. After formation the small units are rapidly linked into larger units presumably by DNA-linking enzymes which have been designated ligases.

Jacob *et al.* (1963) propose that a chromosome of *E. coli* consists of one replicon which would contrast with the situation in higher organisms where one must assume there are many replicons per chromosome (Taylor, 1963a and 1968; Huberman and Riggs, 1968). *Bacillus megaterium*, one of the largest bacteria, is reported to have 2.5×10^{-14} gm of DNA per haploid cell or about 5 *E. coli* equivalents. The organization and number of replicons is unknown. It would be interesting to know the maximum size that a replicon may attain, for as shown later the replicons of higher organisms appear to be considerably smaller than the whole *E. coli* chromosome.

C. Fungi and Algae

Well-defined chromosomes which go through the regular condensation cycle during meiotic divisions, at least, are found in *Neurospora* and other ascomycetes. *Neurospora* has about 4.6×10^{-14} gm of DNA per haploid set of chromosomes (Table I). This represents a length of 15.6 mm or about 2 mm per chromosome since the haploid number is 7. Therefore, each chromosome has about two times the length of DNA present in each *E. coli* chromosome. The organization of the DNA in the chromosome is unknown and there is no indication of how many replicons may be present in each. One difference does appear to be the existence of histone-like proteins in the cases where analyses of regular chromosomes have been made. *Escherichia coli* and other bacteria do not appear to have these proteins. Yeast chromosomes are smaller than those of *Neurospora,* for the haploid set (presumably 15–18, based on linkage groups) has about 1.8×10^{-14} gm of DNA. This would mean that each yeast chromosome is smaller than an *E. coli* chromosome, perhaps only one fourth or one fifth the size of the *E. coli* chromosomes.

Chlamydomonas has about 3×10^{-14} gm of DNA and probably 8 chromosomes in the haploid set. This alga then would have chromosomes of a size comparable to those of *E. coli.* However, they appear to undergo regular condensation during the mitotic cycle and perhaps are organized more like the chromosomes of higher plants. The behavior of chromosomes as small as these during mitosis is very difficult to investigate for they are near the limits of resolution of the light microscope.

Chromosomes of dinoflagelates appear to be somewhat like those of bacteria in that they lack histones. Giesbrecht (1966) produced electron micrographs which showed loops radiating in a fan-shaped arrangement from a region on the periphery of the circular cross-section. Oblique sections showed band-like structures somewhat like condensed metaphase chromosomes of higher organisms, but on a smaller scale.

Some green algae have moderate sized chromosomes and *Oedogonium,* a filamentous green algae, has rather large chromosomes with many if not all of the morphological features of chromosomes in higher plants and animals.

D. Higher Animals and Plants

Although we still have much to learn about the details of organization of DNA and other components in chromosomes of higher cells, one can say that such chromosomes appear to be remarkably similar in groups which diverged in evolution very early. In mechanics of condensation, coiling, and other visible changes over the cell cycle, chromosomes of plants such as *Vicia* and *Lilium* are very similar to those of salamanders and mammals. This statement should not be taken to indicate that variations do not occur, for there are some very interesting anomalies, such as localized or diffuse spindle attachments. However, these variations occur in both plants and animals and must be assumed to have arisen independently in evolution.

One of the most striking and puzzling differences, which again is not restricted to any one group, is the variation in amount of DNA per genome, especially in rather closely related forms. For example, among the higher plants, the family Ranunculaceae (buttercups) exhibits a range of chromosome complements, the DNA contents of which may vary 40-, or even 80-fold without significant changes in chromosome numbers (Rothfels *et al.*, 1966). Species of *Anemone* are reported to have 52.5×10^{-12} gm of DNA per diploid nucleus (2n = 16), while another genus of the same family, *Aquilegia* (2n = 14), has about 1.3×10^{-12} gm per nucleus. This difference indicates a 40-fold variation of DNA per genome. However, another genus, *Xanthorhiza,* has the same amount of DNA and a diploid complement of 36 chromosomes, two times that of *Aquilegia* and other near relatives. This means that *Xanthorhiza* is probably a tetrapoid with a base number of 9 chromosomes which would contain only 0.65×10^{-12} gm of DNA. The difference between 0.65×10^{-12} gm and 52.5×10^{-12} is about 80-fold. The value for *Xanthorhiza* reported in Table I was estimated on the assumption that the value obtained from measurements of root cells would be at least 30% too high. Because of the time spent in S and G2 (see Section IV) one may subtract at least 30% from the measured value to get about 0.45×10^{-12} gm of DNA. Since 10^{-12} gm of DNA would produce a DNA double helix about 340 mm long, each of the chromosomes of *Xanthorhiza* would have a little less than 10 mm of DNA.

Less dramatic variations may be noted in the families Leguminosae, Liliaceae, and Compositae among the higher plants. The Compositae, which are assumed to be one of the most recently evolved groups of higher plants, contains genera with the lowest number of chromosomes. *Haplopappus* with a haploid number of 2 (Jackson, 1957) and *Brachycome* also with a haploid number of 2 (Smith-White, 1968) represent one extreme. Although the author is not aware of any measurement of the amount of DNA per genome for either of these, the DNA content may not be as low as the case cited in the Ranunculaceae.

At the other extreme the Liliaceae probably have the highest amount of DNA per cell of any of the flowering plants. *Lilium* (n = 12) is reported to have about 53×10^{-12} gm of DNA per haploid complement (Ogur *et al.*, 1951). There are tetraploid species which presumably would have more than 200×10^{-12} gm per cell when DNA replication begins and 400×10^{-12} at its completion. By contrast in the same family some species of two other genera, *Crocus* and *Ornithogolum,* have only 3 chromosomes per genome, which are considerably smaller than the chromosomes of *Lilium.*

Among the vertebrates some amphibians and the lung fishes are notable for a high DNA content per cell. However, it is interesting to note that in more recently evolved groups such as birds and mammals the deviations are rather small. The birds have rather uniform amounts varying from 3.1 to 4.2 $\times$ 10^6 μg in the diploid complement, which is 44–59% of the mammalian complement. The reptiles have 60–89% of the DNA amounts typical of the mammalian complement (Atkin *et al.,* 1965). Most mammals have about 1 meter of DNA (3.0–3.5 $\times$ 10^{-6} mg) per haploid complement, i.e., about 800–900 *E. coli* equivalents of DNA. This amount is distributed among the 23 chromosomes of man, for example, and the 6 chromosomes of the wallaby, *Potorous* (Sharman and Barber, 1952). Determinations of the amount of DNA in the latter are not available but from the size of the chromosomes, the value must not be much lower than that in cells of placental mammals.

A lungfish (*Lepidosiren paradoxa*) is reported to have 38 chromosomes (diploid) with about 248 $\times$ 10^6 μg of DNA, i.e., about 35 times that of the mammals (Ohno and Atkin, 1966). This would mean it has nearly 800–900 *E. coli* equivalents per chromosome, or to put it in other terms, a mammalian haploid amount per chromosome. *Necturus* has about 28 times that of a mammal while the bullfrog (*Bufo*) has only 1.4 times the mammalian amount (Gall, quoted by Ohno and Atkin, 1966).

The structural arrangement, function, and evolution of chromosomes in creatures with such large amounts of DNA are interesting, unsolved problems. It must be arranged so that semiconservative distribution can occur at the level of the polynucleotide chains of DNA (Taylor *et al.,* 1957 and Prescott and Bender, 1963). The functional or nonfunctional state of what appears to be an unnecessarily large amount of DNA is not known. It is difficult to account for it unless the genetic information is highly redundant. Even in those species with a moderate amount of DNA in higher plants and animals, the genetic information is probably highly redundant, at least at certain loci. Since the DNA has been retained through long periods of evolution, it can be argued that it serves some functionally significant role. Otherwise much of it would have been lost.

It can also be argued that those species with the largest amounts of DNA are frozen with respect to evolution, perhaps because of the highly redundant genetic information. Certainly the lungfish is a relic from the Devonian period where fossil representatives of the subclass Crossopterygii have been found. As pointed out by Ohno and Atkins (1966), the lungfish is nearly as old as the class of bony fishes, *Pisces,* but appears to be more closely related to the class *Amphibia* than to these fishes. It is a survivor from this ancient period and the principal change since then may have been an increased redundancy of genetic information which fixed it in an evolutionary sense. The same argument could be made concerning other groups with the highest amounts of DNA. Groups which are more rapidly evolving probably have either retained lower amounts of DNA, or have invented mechanisms for reduction of redundancy.

Two principal views are held regarding the arrangement of the DNA in chromosomes which can explain these striking variations within rather closely related groups. One idea is that the chromosomes of higher forms are polyneme with respect to DNA strands, i.e., each chromosome has several to many DNA molecules and the DNA could double by a change in the degree of polynemy. The other idea is that the chromosome is somehow arranged so that additional copies of each unit of replication or function can be inserted simultaneously as tandem duplications, not once, but perhaps many times in the long process of evolution.

The first idea of polynemy would account for doublings and perhaps for other multiples of some basic amounts of DNA assumed to be present in each strand of a multistranded chromosome. The simplicity of the idea and some morphological evidence which indicates that large chromosomes may be multistranded are cited in favor of this idea (Rothfels *et al.*, 1966; Martin, 1968). On the other hand, the patterns of replication and distribution of chromosomal subunits are difficult to reconcile with such a model (Taylor, 1963b and 1959; Callan, 1967). In addition the evidence from studies of lampbrush chromosomes in amphibian oocytes provide strong evidence that these chromosomes are composed of one DNA duplex per chromatid (Gall, 1963a; Miller, 1965). A discovery by Keyl (1966) that tandem duplications of bands in species of *Chironomus* involve geometric doublings of DNA with ratios of 1:2, 4, 8, or 16 lends support to an increase by some process of tandem intercalation. Two species, *Chironomus thummi thummi* and *Chironomus thummi piger*, were studied; *thummi* has 27% more DNA than *piger* in its gametes, as well as its salivary gland in which the chromosomes are highly polytene. The difference is due, in a large measure at least, to tandem duplications of certain bands which may be equated with units of replication (replicons) which Keyl proposes to be circular, i.e., loops extending out from the axis of a chromosome. Irregular union at the base of one of these loops following replication sometimes results in a double-sized loop on one chromatid and a deficiency on its sister chromatid. Thereafter one chromosome would carry a double-sized loop at this locus.

We may extend this hypothesis and suppose that rarely the whole chromosome or complement of chromosomes undergoes such a process simultaneously. Such cells would have the same chance of survival as a tetraploid, but the DNA would double without a change in the number of chromosomes.

Redundancy would presumably fix a species evolutionarily only if it had not evolved a mechanism for mutant expression when multiples of the involved locus were present. For example, if 16, 32, 64, or more copies of a particular functional unit were present in a single genome, how would the random mutation of the various copies which are assumed to be self-replicating ever lead to a new functional unit of a uniform type? Callan and Lloyd (1960) have suggested one possible solution by supposing that such a cluster is composed of one "master" template and various numbers of "slave" templates from which RNA is transcribed. The only significant mutations or recombinants would be those which occurred in the master template which would, periodically during the life cycle, have a chance to correct the slaves to fit the master. Callan's (1967) choice for the time of correction was the long diplotene stage when lampbrush chromosomes are found in the oocyte of the amphibians, for example. Observations have indicated that a loop of a lampbrush chromosome and its matrix of attached particles move slowly by extending from one chromomere and retracting back into the one on the opposite side of a loop. Two types of evidence support such a polarized movement: (1) RNA, proteins, and other matrix materials of the loops accumulate in a polarized fashion on the loops; at one chromomere where a loop joins the axis the quantity of matrix is small, while at the other side it is large with a gradient between; and (2) observation of a giant granular loop of chromosome XII of *Trituris cristatus cristatus* (Gall and Callan, 1962) showed that labeling began on the side with little matrix and moved slowly (about 10 days) around the loop after a single injection of uridine-^{3}H. It was assumed that the loop moved with its matrix. Callan's (1967) hypothesis is that the movement is part of a process that allows each slave copy to be matched against the master and if different to be repaired by base substitution so that it corresponds to the master.

It is an attractive hypothesis which rationalizes a number of observations and would solve one of the outstanding puzzles in the genetics of higher organisms. However, there are no precise chemical mechanisms known which could explain such manipulations of macromolecules. The repair step could, of course, be comparable to that which is known to occur in cells of both bacteria and higher organisms, with the exception that the repaired strand would have to be limited to the slaves. A mechanism for this is probably not inconceivable, but there are no known examples of such asymmetrical repair.

III. Morphology and Fine Structures of Chromosomes

A. Metaphase Chromosomes

The structure of a metaphase chromosome is still a mystery in spite of the fact that one may find rather uniform and characteristic descriptions in various textbooks. A late prophase or metaphase chromosome consists of two cylindrical chromatids which appear to be cemented together along the surfaces that are in contact. At some stage in late metaphase or anaphase the two chromatids separate by a process which is completely obscure. No one knows the nature of the chemical bonds which hold them together or what is involved in the separation. Separation will often occur without the activity of the spindle, since in some cells, at least, the separation proceeds in the presence of colchicine. Most anaphase chromosomes appear as nearly cylindrical, but rather flexible, rods which appear to consist of a solid mass of fibrils. At least sections viewed with the electron microscope reveal no other structures. There is typically no membrane bounding the structure and no visible organization above the level of the 250 Å fibrils. Nevertheless most textbooks describe the anaphase chromosome as a helically coiled rod or two helically coiled strands with chromomeres along the axis of the strands. In some preparations a single rather irregular helically coiled strand is quite clear. This structural feature is readily demonstrated in anaphase I of meiosis where the two chromatids remain attached at the centromere and move as a unit to the poles. In large chromosomes such as those of *Lilium* two sets of coils are evident, a coiled coil. In other instances the same structure appears to consist of two strands twisted around each other in a plectonemic coil. These subunits have been called half-chromatids or subchromatids and are most convincingly demonstrated after air drying or fixation in hot water, treatment with hypotonic, slightly basic solutions before fixation, or by partial digestion with trypsin (Trosko and Wolf, 1965). However, a strong argument for their reality has been their demonstration in living cells (Bajer, 1965) and very clear demonstrations of double chromatids in anaphases and telophases from the protozoan, *Holomastigotoides* (Cleveland, 1949), without any unusual treatments. Cleveland found cells with single chromatids at anaphase and other cells with double chromatids at the same stage in the life cycle. Since no measurements of DNA were ever made for these different cells, it is not possible to know whether the difference was producd by an out-of-phase duplication of chromosomes and centromeres in some cells or a structural variation of chromosomes which had the same DNA content as those with single chromatids.

The fine structure of the fibrils is becoming a little clearer in spite of the fact that their organization into a chromosome is still obscure. Several models will be considered later for the arrangement of the fibrils into chromosomes, but first the nature of the fibrils will be considered. Ris (1957) has long contended that the fibrils contain 2 or 4 elementary strands. Originally he considered the 250 Å fibril to be composed of two 100–200 Å fibrils. These in turn were believed to consist of two 40 Å subunits, each of which presumably contained a double helix of DNA. Therefore, the fibrils were believed to contain four double helices. A modification of the method for spreading DNA on a protein film at a water-protein interface introduced by Kleinschmidt *et al.* (1961) has been useful for electron microscope studies of metaphase and interphase chromosomal fibrils (Gall, 1963b). Most of the investigators now agree that a variety of chromosomes, including both large and small ones, consist of fibrils with a diameter of 250–300 Å (Ris and Chandler, 1963; Wolfe, 1965; Wolfe and John, 1965; DuPraw, 1965; Wolfe and Hewitt, 1966 and Gall, 1966). The disagreements arise from the interpretation of the organization within these fibrils. As mentioned above, Ris and Chandler (1963) emphasized the multistranded concept, but more recently Ris (1967) has obtained evidence that the fibrils are basically two stranded (presumably two double helices of DNA). The 250 Å fibrils were found to contain many side branches which could be shown to be kinks in which the fibrils loop back and twist around each other. Therefore, the long 250 Å fibrils may be loops in which two 100 Å fibrils are twisted around each other in the way that closed rings of double-stranded DNA supercoil. However, the chromosomal fibrils contain protein in addition to DNA. When most of the protein was digested with pronase the 100 Å fibrils could be shown to contain a single 25 Å strand which stained with uranyl acetate and presumably was one DNA double helix.

DuPraw (1965) and Gall (1966) failed to find evidence for doubleness in the 250 Å fibrils. They assumed them to be formed by the coiling of a smaller fibril. DuPraw compared the 250 Å fibril to tobacco mosaic virus (TMV) with a coiled nucleic acid fibril in a protein coat. According to his observations, digestion with trypsin yielded a deoxyribonuclease-sensitive core which he interpreted as a single double helix of DNA. On the basis of the packing ratio which is much higher than for TMV he reasoned that the 250 Å fibril must be a coiled coil analogous to a chromatid which at the light microscope level sometimes exhibits a coiled coil. Electron scattering measurements (DuPraw and Bahr, 1968) have shown that the dry weight of the fibril is 6×10^{-20} gm per angstrom. Based on the determination of the ratio of DNA to protein (30:70) he calculated that each micron of a 230 Å fibril must contain 1.8×10^{-16} gm of DNA. This would be 56 μ of a Watson-Crick helix and the packing ratio would be 56:1. He concludes that no simple secondary coil can achieve such a ratio and that the structure is a coiled coil.

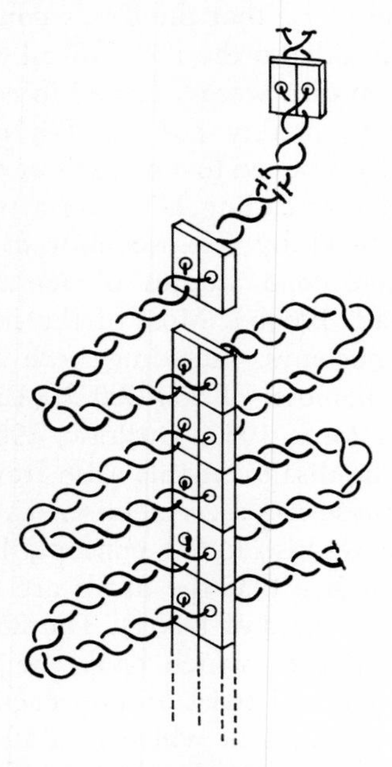

FIG. 1. A chromosome model constructed on the assumption that a chromatid or unreplicated chromosome contains a single DNA double helix held into a compact form by the aid of hypothetical replication guides attached to, but not necessarily interrupting, the DNA chains. The lampbrush chromosomes seen in meiotic prophase in some oocytes are the nearest approach to this skeleton model. At division stages when chromosomes are condensed, the loops would presumably be contracted into rings somewhat less than 1μ in diameter even in large chromosomes and the axis would be twisted and coiled. However, with a limited number of twists of the two ranks of rings a chromatid could sometimes present the appearance of two relationally coiled half-chromatids. The diagram shows only a short segment of a chromatid and the loops would be proportionally much larger than shown. Likewise, the replication guides would not be expected to be distinguishable even in the electron microscope. (From Taylor, 1966.)

The next problem to be considered is the possible organization of the fibrils into a chromosome. In this area the picture is still quite unclear, but a few models will be considered. The oldest concept is that which assumes the metaphase chromatid to be composed of a bundle of strands running longitudinally. This model has been suggested by Huskins (1937) and Nebel (1939) and elaborated by Kaufmann *et al.* (1960). Ris (1957) and Steffensen (1959) have presented models, the later with 32 or 64 strands indicated, and presented supporting evidence. The other extreme concept is that the chromosome is a tandem linkage of single DNA duplex molecules (Schwartz, 1955). This model has been elaborated by Freese (1958) and Taylor (1959) and later modifications eliminate the non-DNA linkers (Taylor, 1966; DuPraw, 1968). The strand would have to be folded, looped, or coiled in a very compact way to accommodate the dimensions of a metaphase chromosome. Some models assumed basically a series of loops (Schwartz, 1955; Taylor, 1966) while others assumed an irregular coil of the 250 Å fibril, or in later versions a folded fiber model which has strands folded or looped back at the ends. The model (Fig. 1) preferred by the author is some type of looped structure similar to a condensed lampbrush chromosome which will be described below. Since the lampbrush chromosome is found in the diplotene stage of meiosis and is capable of being converted into a regular metaphase chromosome later, it appears to serve as a model for the general structure of a chromosome.

B. Lampbrush Chromosomes

Lampbrush chromosomes were first described from studies of the oocytes of the shark, *Pristiurus*, by Ruckert (1892). The name was given because the loops off a central axis reminded him of a *Lampencylinderputzer*. If we were selecting a similar analogy today we would say bottle brush. That the bristles were actually loops became clear when similar chromosomes were isolated by micromanipulation from frog oocytes (Duryee, 1941 and 1950) and from *Trituris* (Callan, 1952; Gall, 1952).

The lampbrush chromosomes are bivalent, with homologs attached at one or more places (chiasmata), and reach their maximum development at diplonema (Fig. 2). The homologs are not visibly double as in many diplotene figures of other organisms. However, in short portions of some chromosomes the chromatids separate enough to be clearly distinguished (Callan and Lloyd, 1960). Lampbrush chromosomes are very large compared to regular diplotene chromosomes. Originally Duryee (1941) thought the whole loop was lost from the chromosome, since he found structures similar to the lateral loops free in the nuclear sap. After Callan (1955) and Gall (1956) established the nature of the loop connections with the chromomeres, this concept was changed and it was assumed that no free loops existed except by breakage. More recently, however, it has become clear that rings (loops) are derived from nucleoli at certain

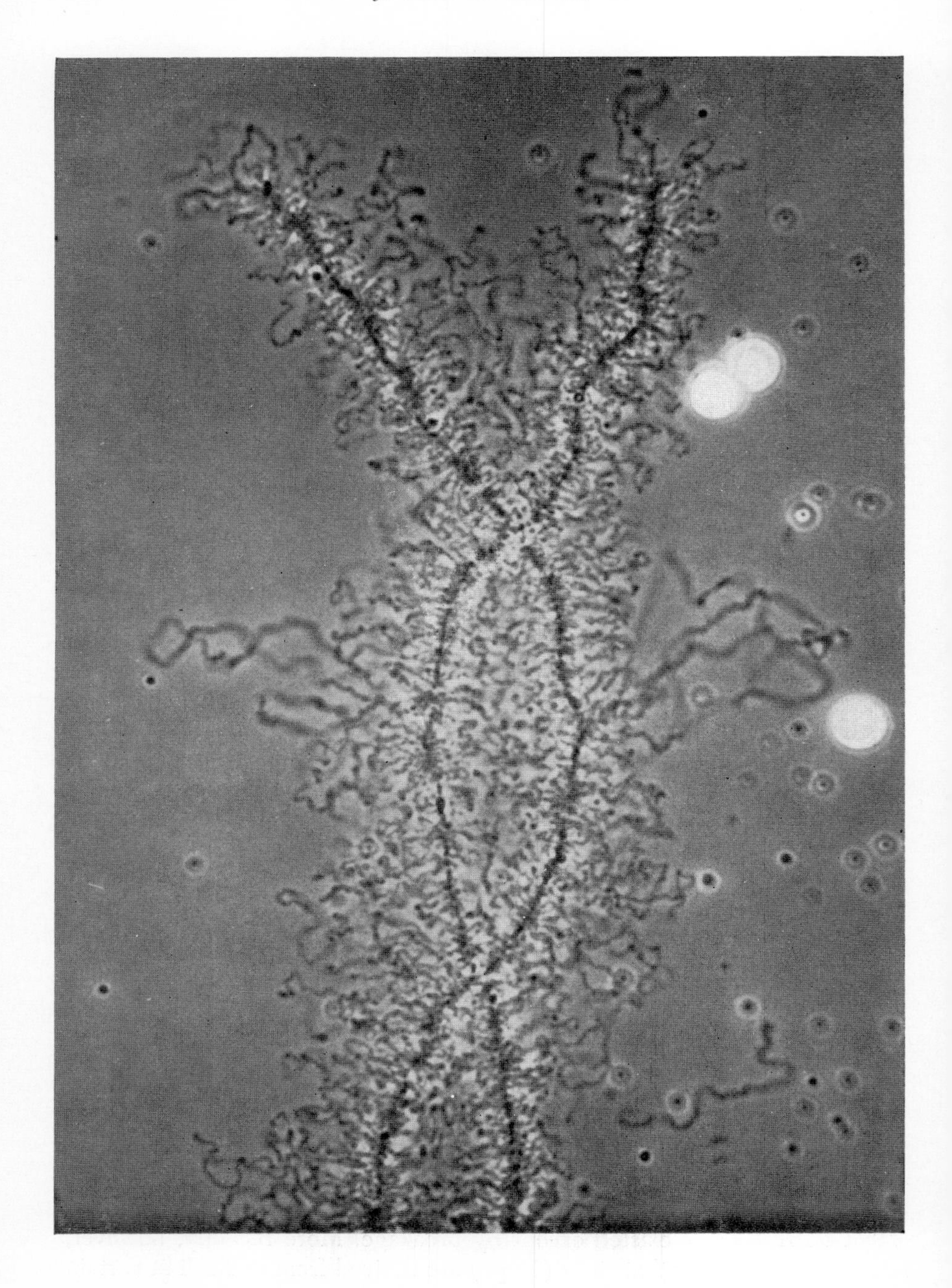

stages of maturation. The axial material of such rings is DNA, as shown by their breakage by deoxyribonuclease (Miller, 1966) and not by ribonuclease and proteases. The DNA in nucleoli appears to be copied from the chromosomal DNA which codes for ribosomal RNA. It is derived from DNA located in or near the nucleolar organizer regions of the chromosomes and is preferentially replicated in early stages of meiosis (Brown *et al.*, 1968). Brown *et al.* have estimated that 100–200 replicas of the master copy in the chromosomes must be formed to account for the DNA in the oocyte nuclei of *Xenopus*, the South African toad. An examination of a number of other cell types in which there is a wide range of rates of ribosomal RNA synthesis has failed to reveal any other instance of preferential synthesis of ribosomal cistrons. Ritossa *et al.* (1966) had previously reported a similar situation in a number of tissues from the chicken. The only exceptions known other than the oocytes of various animals are the salivary glands of the Sciaridae (Swift, 1962; Crouse and Keyl, 1968). Callan and Lloyd (1960) measured the lengths of chromosome V in more than 60 oocytes from *Trituris cristatus carnifex* which varied in diameter from 0.6 to 1.68 mm. They found a variation in length from 369–826 μ, but there was no correlation between sizes of oocytes, some of which were nearly mature, and the length of chromosome V. Since there are four other chromosomes longer than V, some bivalents may be more than a millimeter long. The lampbrush chromosomes are known to contract into regular-sized metaphase chromosomes before ovulation. Callan and Lloyd concluded that the contraction must be rather rapid and occurs at the very end of the period of yolk accumulation, since they did not observe shorter chromosomes in larger oocytes. A large amount of material is sloughed off the loops since the volume of the metaphase chromosome is much less than that of its diplotene counterpart.

The loops extend from chromomeres and when the chromosome is

FIG. 2. Photograph of an unfixed lampbrush bivalent chromosome from an oocyte of the newt, *Trituris viridescens*. Evidence indicates that a single DNA double helix extends through each loop which is a part of the chromatid axis. However, the main axis of each of the two homologous chromosomes forming the bivalent consists of two chromatids and, therefore, two DNA helices with many tightly coiled regions which are the chromomeres. A large amount of protein, RNA, and perhaps other components coat the DNA strands and make them visible to an observer with a light microscope. Magnification: × 650. (Original photograph contributed by J. G. Gall.)

stretched and broken, the break usually does not occur between chromomeres, but transversely across chromomeres in such a way as to separate pairs of loop insertions. The break is spanned by a pair of straightened-out loops (Callan, 1955 and Gall, 1956). Such observations lead to the concept that the loops are a part of the chromosomal linear axis. This view was strengthened by the demonstration that of various proteases and nucleases used for digestion of lampbrush chromosomes only deoxyribonuclease caused a rapid breakage of the loops and main axis (Callan and Macgregor, 1958). The other enzymes cause shrinkage but no detectable breakage. Later experiments by Gall (1963a) showed that the kinetics of breakage by pancreatic deoxyribonuclease indicated two polynucleotide chains in the loops and four in the main axis which is composed of two chromatids. This finding strengthens the view that a chromatid consists basically of a single double helix of DNA, because the loops are believed to be a continuation of the longitudinal axis of a chromatid. However, Ris (1967) has reported some observations which suggested to him that a lampbrush chromosome might be multistranded. He saw several strands bridging the gap between the insertions of some large loops. This led him to suggest that only one of several strands may be active in forming the loop at any chromomere. However, this suggestion does not seem to be in accord with the observation that there is an inverse correlation between the size of a chromomere and the size of a loop (Callan, 1955). Callan's observation, on the other hand, would indicate that a significant portion of the DNA in a chromomere is usually extended in the loop instead of one of several DNA fibrils.

We may summarize by repeating the analogy mentioned at the end of the last section. The lampbrush chromosome seems to be an unusual extension of a regular chromosome, many of the structural features of which it still retains. By retraction of the loops, perhaps by supercoiling and shortening of the axis along with the loss of loop matrix, these chromosomes would presumably become metaphase chromosomes. Each chromatid would be represented by a single DNA duplex, contracted into chromomeres or supercoiled into loops or coils visible with the light microscope.

C. Polytene Chromosomes

The polytene chromosomes, typically found in various dipteran tissues, differ from lampbrush chromosomes in that they clearly contain much more DNA than regular chromosomes of dividing cells. In addition they do not have a genetic continuity beyond the individual in which they develop. However, their morphological and genetic features can be predicted to reappear in the offspring in subsequent generations. Like the lampbrush chromosomes, in their fully developed form, they are restricted to a limited group of animals, the flies, mosquitoes, midges, and some ciliates, but available evidence indicates that they provide a functional and structural counterpart of typical mitotic chromosomes which has been valuable in many ways. They were first described by Balbiani in 1881, but their relation to the regular mitotic complement was not appreciated until Painter (1933) and Heitz and Bauer (1933) demonstrated the correlations. There followed an exciting period of investigation of their morphology, particularly their band patterns which were carefully mapped (for example, Bridges, 1935). The changes could frequently be correlated with genetically detectable phenomena: deletions, duplications, inversions, and reciprocal exchanges as first demonstrated by Painter (1933 and 1934). The correlations were studied in great detail in one of the most fruitful endeavors in which cytologists and geneticists ever collaborated. After many of the cytogenetic problems had been solved, there was a period of reduced interest and activity in investigations of these giant chromosomes in spite of the fact that many of the most basic questions of their structure and function remained unanswered.

Developments beginning in the early 1950's were soon to cast the structures in a new role in a study of the functional counterpart of a regular chromosome. Beermann (1950, 1952 and 1956) began studies on the constancy of the band patterns which has been so useful in cytogenetic studies. The interest in this matter grew out of arguments and questions concerning observed variations in band patterns during development and particularly by a comparison of different tissues such as salivary gland, Malpighian tubules, midgut, etc. Brewer and Pavan (1955) also initiated studies of developmental changes in the giant chromosomes of *Rhynchosciara*, which has proved to be a very favorable material for such studies. Beermann's contribution was to show that bands were constant except for specific differences between some tissues and between some stages of development in the same tissues. Even when the patterns appeared to vary, homologies could usually be shown by careful comparisons at several stages in development. It became clear that morphological changes, puffs, for example, occur by the alteration of specific bands and that these changes typically occur only in certain tissues and at specific stages in larval development. One period of striking alteration was the premolting stage in chironomids. Clever and Karlson (1960) showed that the puffing patterns associated with molting could be prematurely induced by injection of ecdyson. Later studies showed that specific puffs would appear in less than 1 hour after injection of the hormone. If the concentration was sufficient, molting would proceed. Otherwise the small puffs would subside after a time.

The most important point for our discussion was the observations of Beermann and Bahr (1954) and later studies (Beermann, 1968) which showed that the bands expanded into submicroscopic fibrils which are continuous along the length of the chromosomes. The larger strands in the polytene chromosomes which must consist of bundles of chromonemata separate into finer and finer fibrils in the large puffs which in *Chironomus* are called Balbiani rings. These studies and other supporting evidence has led to the view that each chromonema may be represented by a single DNA double helix which runs through the length of the chromosomes (Swift, 1962). Bundles of these which represent doublings of the original two chromonemata produce 1024, 2048, and even higher multiples of the original DNA. Whether the amounts are exact multiples of the mitotic complements or of some reduced amount has been debated. However, Rudkin's (1964) studies indicate that parts of the chromosome complement cease replication rather early in development. The analogy between the structure of chromonemata in puffs of polytene chromosomes to the loops of lampbrush chromosomes has been suggested by Beermann and Bahr (1954) and everything learned about them since indicates that the analogy is a valid one.

IV. Synthesis of DNA over the Cell Cycle

The genetic equivalence of cells produced by mitosis provides the basis for the concept that chromosomes are reproduced once each division cycle and usually segregated so that both daughter cells get identical genomes. The rare exceptions to this usual mode of distribution which coincides with visible nondisjunction of a particular chromosome allowed Bridges (1916) to obtain convincing evidence that the genetic factors were located on the chromosomes. The genetic equivalence of nuclei has also been demonstrated by transplantation experiments of Briggs and King (1957), although these and later experiments (King and DiBerardino, 1967) showed that development is frequently limited when differentiated nuclei are substituted for the zygote nucleus in developing eggs. Whether these limitations are due to gross chromosomal abnormalties or to more subtle changes is not yet clear. Certainly changes in chromosomes occur during differentiation, but whether these changes are fully reversible remains to be established.

The rate and timing of chromosome duplication over the cell cycle varies with the growth conditions and with the stages of embryonic development in higher forms. For example, the 15 x 10^{-6} μg of DNA of each cell of a cleaving frog egg must be replicated in less than 1 hour at 18°C (Pollister and Moore, 1937), but in the young frog the *S* phase (DNA synthetic period of the cell division cycle) is probably 6–12 hours at the same temperature.

For the purposes of discussion the cell cycle will be divided into the four stages used by Howard and Pelc (1953), namely, *M* or mitotic stages; *G1*, the gap before DNA synthesis begins; S, the DNA synthetic stage; and *G2*, the gap between the end of DNA replication and mitosis. The most variable stage appears to be *G1*, although Lajtha (1963) prefers to call the stage following *M* and preceding *G1*, the *GO* stage, especially when cells such as those in adult liver may not divide for months. Some cells do not have the *G1* or *G2* stages, or if they do, these are shortened until they are hardly recognizable. Bacteria such as *E. coli* growing in a rich medium have replication of DNA proceeding in nearly all cells. However, the fact that a small percentage of cells are immune to killing by thymine starvation indicates that there is a short period when these cells are not in S. Cells starved of an essential amino acid eventually stop synthesis of DNA. After a culture increases its DNA content by about 50% nearly all cells have reached the stage with completed chromosomes (comparable to *G1* or *G2*) and are immune to thymineless death (Lark, 1963). However, if the amino acid is supplied and thymine withheld the cells proceed into a stage from which they can not recover.

A series of studies of replication with particular attention to the number and distribution of conserved units of DNA in *E. coli* (Lark, 1966) has shown that these cells contain two identical chromosomes, both of which are replicating at the same time when growing in a medium which supports growth with a generation time of 40 minutes. At slower rates of growth in less enriched medium (generation time of 70 minutes) each cell contains two chromosomes but only one is replicating at any time. At still slower rates, about 120 minutes, each cell contains one chromosome which is replicated throughout the cell cycle. However, when cells are slowed to a generation time of 180 minutes or longer, the single chromosome appears to replicate in the first half of the cell cycle and is therefore approaching the situation in cells of higher forms which have developed beyond the early embryonic stages.

In the meiotic cycle in higher forms the DNA is also replicated in the interphase preceding the first meiotic prophase (Swift, 1950; Taylor and McMaster, 1954; Taylor, 1957 and 1967; Rossen and Westergaard, 1966). In a few species the synthesis extends into early prophase, but appears to be essentially complete before meiotic pairing and crossing-over occur (Abel, 1965; Henderson, 1966; and Peacock, 1968). However, a small, but detectable synthesis has been reported to occur in meiotic prophase (Wimber and Prensky, 1963; Hotta *et al.*, 1966). No synthesis occurs preceding the second division of meiosis. However, Chiang and Sueoka (1967) have reported a rather unusual cycle of synthesis in *Chlamydomonas* in connection with meiosis. Vegetative cells and gametes are reported to have an amount of DNA which they call 2D. The zygote has 4D and a round of replication occurs after synapsis and crossing-over, but during meiosis, to produce a germinating zygospore with 8D. This zygospore then yields 8 zoospores, without further synthesis, which have the D amount of DNA. A round of replication then restores the 2D amount to the zoospores before they become vegetative cells. Since these conclusions are based on rounds of replication rather than measurements of amounts of DNA per cell, it is conceivable that a part of the interpretation is in error. However, it may be similar in some respects to the behavior described for the ascomycete, *Neottiella rutilans*, by Rossen and Westergaard (1966). In this fungus the two haploid nuclei which are to fuse later and form the primary ascus nucleus, i.e., the gametic nuclei, have their DNA replicated before sexual fusion (caryogamy) and no further synthesis of DNA occurs until meiosis is complete and 4 nuclei are formed. An increase of DNA occurs before the next mitotic division and the formation of 8 ascospores. Then another DNA doubling occurs before the ascospores mature. If the interpretation of Chiang and Sueoka (1967) is correct, the difference would be that in Chlamydomonas the usual postmeiotic synthesis occurs while the zygospores still have one nucleus, i.e., before the two meiotic divisions. Diplochromosomes, which would presumably be produced by such an early synthesis, rarely occur spontaneously in mitotic cycles when two DNA replications occur in one interphase, but such behavior as a regular phenomenon of meiosis is unknown except for its possible occurrence here.

V. Autoradiographic Studies and DNA Transfer Experiments on Mechanisms of DNA Replication

A. Review of Some Autoradiographic Experiments on DNA Replication

Following the proposal of a specific structure for DNA by Watson and Crick (1953a) and a suggested mechanism for its replication (1953b) by what Delbrück and Stent (1957) later called a semiconservative mechanism of replication, several attempts were made to obtain evidence for or against the hypothesis. One of the first indications that the replication might be occurring in the way proposed by Watson and Crick was obtained by observing the segregation of chromosomes labeled with tritiated thymidine (Taylor *et al.*, 1957). Thymidine is a nucleoside which Friedkin *et al.* (1956) had shown to be readily used by cells for the synthesis of DNA. W. L. Hughes (Taylor, 1960b) was able to label the nucleoside at high specific activity with tritium (^3H). Tritium, which had been used very infrequently up to that time for autoradiography, has a great advantage because the low energy β-particles emitted are stopped within a fraction of a micron of their source in a photographic emulsion. This means that the position of labeled atoms can be resolved to the dimensions of a part of a chromosome.

Vicia faba was chosen for the first experiments because it has large chromosomes and the cell cycle was rather well known from the experiments of Howard and Pelc (1953). At 18–20°C the cell cycle is completed in 20–24 hours. DNA replication extends over a 6–8 hour period and the *G2* stage is also 6–8 hours. The division requires 1–2 hours and that leaves 6–8 hours for *G1*. Roots exposed to thymidine-^3H for 8 hours were washed and transferred to a solution of colchicine for 10 hours. Another group remained in colchicine for 34 hours. Roots were then fixed, stained by the Feulgen reaction, squashed on microscope slides, and coated with stripping film. After appropriate exposure the autoradiographs were developed and examined under the light microscope. Chromosomes and silver grains appeared in nearly the same optical plane; therefore, the distribution of labeled and unlabeled chromatids could be followed (Fig. 3).

Cells from the 12-hour sample had 12 chromosomes typical of the diploid nuclei from *Vicia* roots. Both chromatids were labeled (Fig. 3a). However, the 34-hour sample had some cells with 24 chromosomes, as well as some with 12 and a few with 48 chromosomes. Those with 24, which were labeled with tritium had been synthesized with thymidine-^3H available, but must have continued one full cycle, i.e., out of the first blocked metaphase into *G1* and through another *S* phase without thymidine-^3H, and were again blocked in metaphase at the time of fixation. The chromosomes in these cells with 24 metaphase chromosomes showed a remarkable segregation of the labeled DNA. At any one level along the chromosomes it was all in one chromatid or all in the other (Fig. 3b). The occasional switch points which occurred were assumed to be the result of sister chromatid exchanges.

It was evident that DNA was conserved in two subunits (Fig. 3c) of

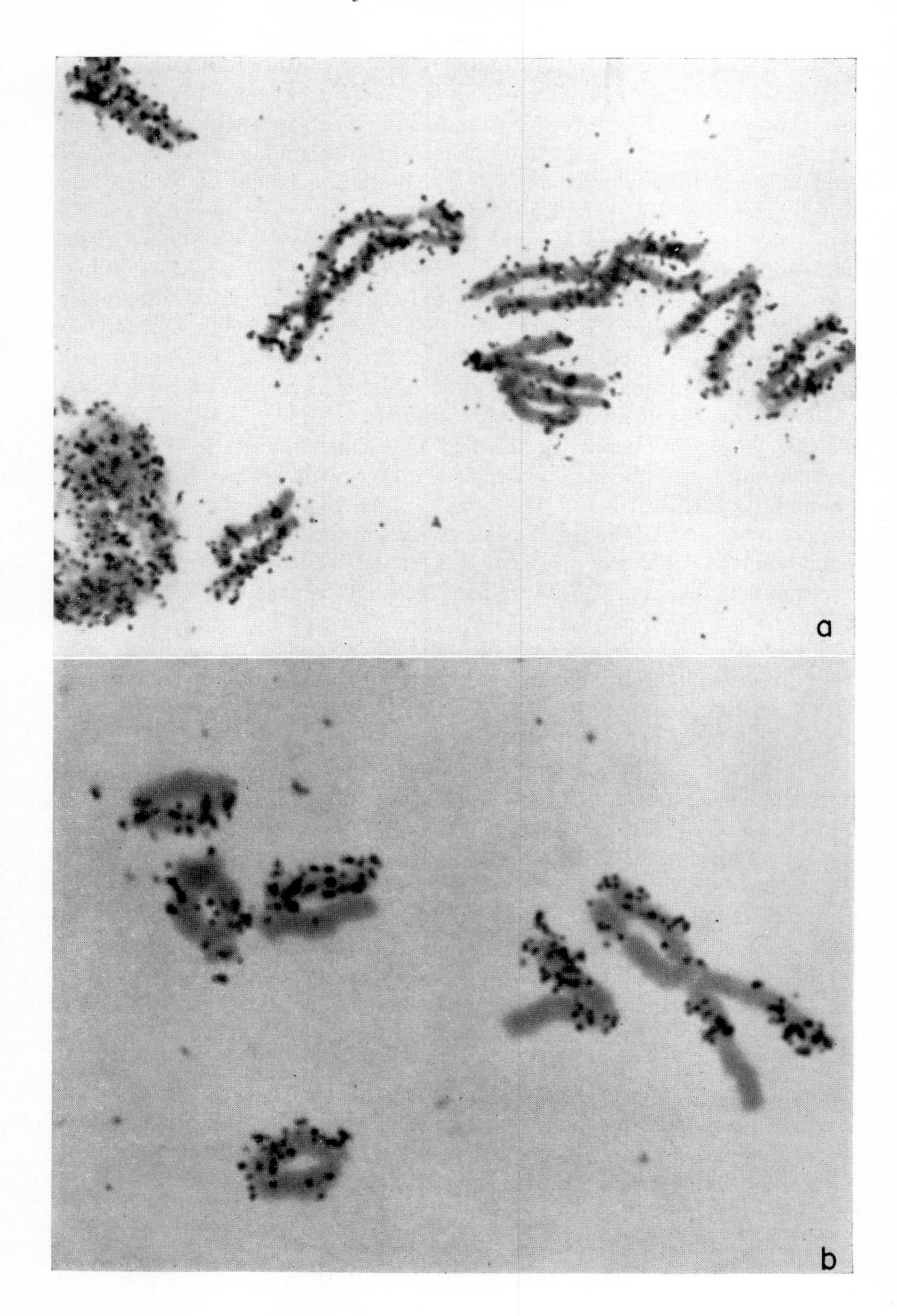
a
b

each chromosome (excluding sister chromatid exchange) and segregated in a semiconservative manner as predicted by Watson and Crick for the DNA duplex. It certainly appeared that each subunit of a chromatid might contain one or many DNA chains which were somehow able to unwind and separate from the other chain or chains in the other subunit.

However, so little was known of the structure of a chromatid that the behavior of the individual DNA chains could not be definitely inferred from the behavior of whole chromosomes. Alternative explanations were possible, but it seemed that a system for such an elaborate and precise segregation of DNA chains would hardly have been maintained over the ages of evolution unless it had some meaning and significance for the mechanisms of heredity. The only one that made sense was the segregation of complementary strands of DNA. Nevertheless, science is not built on the assumption that nature is necessarily logical, simple, and expedient in the selection of modes of evolution. It was possible that nature is capricious and the chromosome could possibly be any one of the following: (1) A bundle of DNA chains somehow linked as a unit, but presumably existing as duplexes, with another bundle of chains, which somehow separated from their complementary chains at each duplication and assorted as a conserved unit. (2) The conserved unit was a DNA duplex or a bundle of these which somehow reproduced another duplex or bundle, which for reasons unknown remained attached to the original conserved group for one or more division cycles. (3) The unreplicated

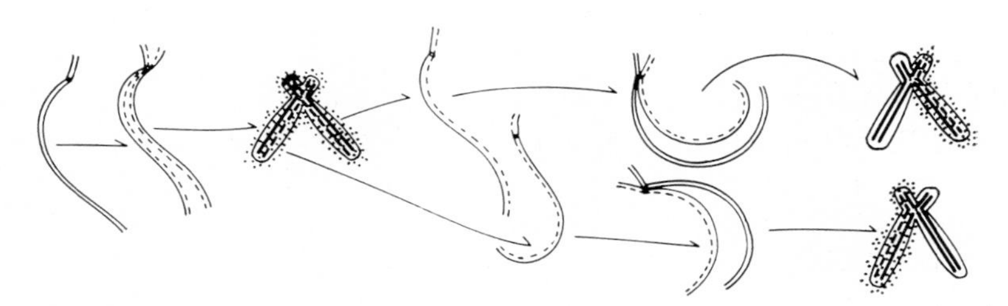

FIG. 3. (a) Autoradiogram of *Vicia* chromosomes after one replication cycle in which thymidine-^{3}H was incorporated into the DNA. (From Taylor, 1962.) (b) Autoradiogram of a part of the chromosome complement of a *Vicia* root cell which had incorporated thymidine-^{3}H during one replication cycle and then replicated once without the radioactive label. The labeled subunits of DNA have segregated into separate chromatids, except for sister chromatid exchanges. (From Taylor, 1962.) (c) Diagram showing a distribution of subunits which explains the segregation of labeled DNA seen in (a) and (b). Broken lines represent labeled subunits and unbroken lines represent unlabeled subunits. (From Taylor *et al.*, 1957.)

chromosome was a single giant DNA molecule or a tandemly linked group of DNA molecules which replicated semiconservatively as predicted by Watson and Crick. We never doubted the correctness of either the first or third possibilities, but we knew others would and the doubts were soon expressed in print.

Plant and Mazia (1956) had already published results on segregation of thymidine-^{14}C in whole anaphase nuclei of *Crepis* roots which appeared to be at variance with the equal distribution of label at the first division after incorporation. LaCour and Pelc (1958) found contradictory autoradiographic results with *Vicia.* Peacock (1963) agreed that segregation was regular at the division following labeling (X_1), but found some isolabeling of chromatids at the second division (X_2) when segregation should be all or nothing on the simplest hypothesis. Finally Prescott and Bender (1963) published completely confirmatory results for the semiconservative hypothesis using Chinese hamster cells in tissue culture. In the meantime other pertinent experiments were being reported. The same type of semiconservative segregation was reported for DNA in *Crepis* root cells (Taylor, 1958a), *Bellevalia* root cells (Taylor, 1958b), and human cells (HeLa) in culture (Taylor, 1960a). J. G. Gall (personal communication, 1958) had confirmed the results in cells of onion roots. A later report by Zweidler (1964) that there is a cell generation delay in the segregation of label in chromosomes of the onion roots is probably in error.

Two experiments were soon reported which made it very likely that a chromatid arm was a single DNA duplex or some very precise and tandemly linked group of molecules. One was the report by Meselson and Stahl (1958) that DNA replication was semiconservative at the molecular level in *E. coli* and the other was the finding that the sister chromatid exchanges in large chromosomes were restricted in such a way as to indicate that the conserved subunits of a chromatid very likely had different polarities similar to the two chains of the DNA duplex (Taylor, 1958b).

Meselson and Stahl (1958) grew *E. coli* for 14 successive cell generations in a medium in which the nitrogen source was the heavy isotope ^{15}N. Cells were washed and switched abruptly to a medium with the usual isotope of nitrogen, ^{14}N. The increase in the population of cells continued at an exponential rate. Samples of cells were withdrawn at intervals and their DNA isolated and centrifuged in a solution of cesium chloride with a mean density which was similar to that of the suspended DNA molecules. When centrifuged at 44, 770 rpm for about 20 hours the cesium ions form a linear concentration gradient along the direction of the centrifugal force and the DNA floats and forms a band at a position corresponding to its mean density. In this way the DNA-^{15}N and DNA-^{14}N could be separated. Since the DNA solutions were spun in a Model E Spinco centrifuge with a window in the rotor, the spinning compartment with the solution of DNA could be photographed with ultraviolet light. The bands of DNA absorb ultraviolet light strongly at wavelengths around 260 mμ and show up as dark bands across the spinning compartment.

After transfer to the ^{14}N medium a lighter band of DNA very quickly appeared and at the end of one cell generation nearly all of the DNA had been converted to a lighter species. By two cell generations about one half of the DNA had been converted to a still lighter species while the other half remained at the position of the intermediate band. On continued growth the intermediate band remained, but the light band continued to increase in amount. This behavior is that predicted for semiconservative replication of the DNA duplex. The intermediate band would be formed by the first replication of DNA-^{15}N when a conserved ^{15}N chain would serve as a template for building each ^{14}N chain. The lightest band would be expected to show up when these hybrid molecules, composed of one light chain and one heavy chain, replicated. The light chain would serve as a template for another light chain to produce the DNA banding at the lightest position, while its complementary heavy chain would reform a hybrid molecule. A further test of the hybrid nature of the molecules in the intermediate region was obtained by heating this DNA to 100°C for 30 minutes and then rebanding it in the centrifuge. Now it separated into two bands of different densities as expected, since heating DNA was thought to cause collapse of the duplex and separation of its chains.

Confirmatory results were soon reported for other organisms: an alga, *Chlamydomonas* (Sueoka, 1960), mammalian cells in culture (Djordjevic and Szybalski, 1960; Simon, 1961; Chun and Littlefield, 1961), and higher plant cells (Filner, 1965.) However, doubts were again raised that the particles (molecules) in solution were simple DNA duplexes and that heating could separate the chains (Cavalieri and Rosenberg, 1961). They determined molecular weights of bacterial and mammalian DNA's before and after denaturation and performed experiments on kinetics of reduction of molecular weight which indicated to them that denatured DNA was still double stranded. Since heating reduced the molecular weight by about one half without making the DNA single-stranded by their criterion, they proposed that the conserved unit was the Watson-Crick double helix which never separated during replication. Baldwin and

Shooter (1963) provided evidence that hybrid DNA produced during replication was not a four-strand structure in which the two new strands formed a Watson-Crick helix. They took advantage of the fact that DNA, in which bromouracil is substituted for thymine, denatures at a lower pH than regular thymine-containing DNA. They produced a hybrid and a fully substituted DNA by growing a thymine-requiring mutant of *E. coli* in bromouracil. The two density species were isolated in a preparative centrifuge by banding in an appropriate cesium chloride gradient. Samples of the hybrid DNA were banded in CsCl in the analytical centrifuge in carefully buffered solutions. At a pH which would denature bromouracil DNA, the fully substituted DNA would undergo a density shift typical of denatured DNA in CsCl. At a higher pH the thymine-containing DNA would undergo a similar shift in density. The hybrid, if composed of two paired strands of bromouracil DNA and two paired strands of thymine-containing DNA, would undergo the shift in two steps. However, it failed to show the two-step transition which made the model of Cavalieri and Rosenberg (1961) seem very unlikely. However, the demonstration of the mass per unit length of phage DNA and *E. coli* DNA by electron microscopy (Kleinschmidt, 1967), by autoradiography (Cairns, 1961), and by better methods of molecular weight comparisons in the centrifuge (Burgi and Hershey, 1963; Studier, 1965) finally showed that DNA of phages and bacteria were two-stranded molecules. This information along with the demonstration of semiconservative replication in phage λ (Meselson and Weigle, 1961), for example, gave a clear indication for separation of strands during replication. No such information is available for DNA of higher cells, but there is no reason to believe replication is basically different from that in bacteria or viruses. However, it must be admitted that there are many details of DNA replication in all types of cells, and especially higher cells, which have not yielded to any of the methods so far applied. Perhaps it will be worthwhile to review the evidence for semiconservative replication in higher cells; the Chinese hamster cells in culture serve as a good model. As mentioned above, Prescott and Bender (1963) demonstrated semiconservative segregation at the chromosomal level by the use of thymidine-^{3}H in experiments similar to those described above for *Vicia* root cells.

B. A Density Transfer Experiment to Illustrate Semi-conservative Replication

The procedures for demonstrating semiconservative replication by the use of the preparative centrifuge can be illustrated by an extension of some experiments designed to measure rates of chain growth in Chinese hamster cells (Taylor, 1968). DNA replication in mammalian cells can be followed by the use of a density label, bromodeoxyuridine, which is an analog of thymidine (Djordjevic and Szybalski, 1960). The heavy nucleoside is readily phosphorylated by most mammalian cells and although the DNA containing such an analog is not quite normal, the mechanism of replication can be revealed. Because of the complexity of the medium required for growing mammalian cells, density labeling with ^{15}N, ^{13}C, and ^{2}H to a level required to repeat the Meselson-Stahl experiment described above has never been accomplished. However, transfer experiments can be performed with bromouracil as a density label if one does not require two complete rounds of replication with the analog.

Rapidly growing Chinese hamster cells, when transferred to a medium with bromodeoxyuridine and an inhibitor of thymidylate synthesis, will begin nearly complete substitution of the analog immediately and produce a density hybrid DNA with a density of about 1.74 or more compared to 1.70 for the unsubstituted DNA. A clean separation of the hybrid from unsubstituted DNA is usually not achieved because the DNA does not necessarily break at the transition points when it is fragmented in preparation for banding in CsCl (Taylor, 1968). However, if the cells are synchronized at the beginning of the *S* phase and then allowed to substitute bromodeoxyuridine for several hours, the separation of substituted and unsubstituted DNA is quantitative (Taylor and Miner, 1968).

In the experiments to be described the DNA was isolated by a procedure which yielded a product with a high molecular weight and relatively free of protein. It was sheared to a uniform molecular weight ranging between 10 and 20 x 10^{-6} daltons before banding in CsCl. This size particle forms a rather narrow band and flows out of the tube to form a nearly symmetrical distribution when removed by puncturing the bottom of the tube (Fig. 4). DNA was added to a buffer solution (0.05 *M* Na_2HPO_4 and 0.015 *M* sodium citrate at pH 7.4) and then CsCl was added to produce a density appropriate for banding DNA's to be centrifuged. Gradients were formed by centrifuging 3–4 ml of the solution (density 1.72–1.75) at 37,500 rpm for 48 hours in a swinging bucket or fixed angle rotor. During this period the gradient forms with a density varying from about 1.65 at the top to about 1.85 at the bottom of the tube. Enough DNA may be added (15–20 μg/3–4 ml) so that the principal bands can be detected by the optical density of diluted fractions. The bromodeoxyuridine may be labeled with either ^{14}C or ^{3}H if small amounts of the hybrid or fully substituted DNA are to be detected. Figure 4 shows the hybrid DNA which was formed in 3 hours after transferring a culture of cells synchronized at the beginning of the S phase to a medium containing bromodeoxyuridine. The new DNA banded at a position which indicated a density in the range 1.74–1.75, while the unreplicated DNA had a density of about 1.70. After 22.5 hours in culture with the

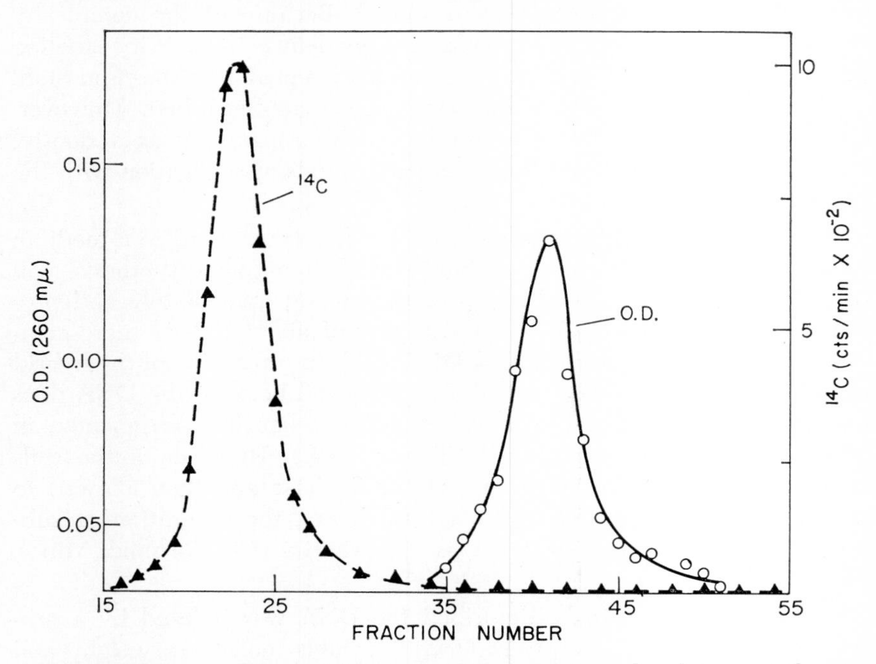

FIG. 4. Density profile of DNA in a CsCl gradient produced by centrifugation in a Spinco 40.2 rotor at 39,000 rpm for 48 hours; 60 fractions were dripped from the bottom of the tube containing 3 ml and assayed for radioactivity to detect the bromodeoxyuridine-^{14}C and optical density at 260 mμ to detect the remainder of the unsubstituted DNA. The cells from which the DNA was extracted were synchronized at the beginning of the S-phase and grown for 3 hours in a medium with 10^{-5} *M* bromodeoxyuridine. (From Taylor and Miner, 1968.)

analog, nearly all of the DNA was converted to either a density hybrid or a DNA which banded at a position (Fig. 5; fractions 21–24) indicating complete substitution in all strands of the DNA. This type of transfer experiment illustrates the mode of replication predicted on the basis of the Watson-Crick mechanism of replication. First the hybrid DNA double helices are formed and later when these replicate one half of the second generation double helices have both strands substituted with the density label and the other half remain hybrid (Fig. 5). Figure 5 was produced by growing a 6-hour culture of cells for 22.5 hours in a medium with bromodeoxyuridine.

Another way to follow the replication of hybrid DNA which does not require continued growth in bromodeoxyuridine is illustrated in Fig. 6.

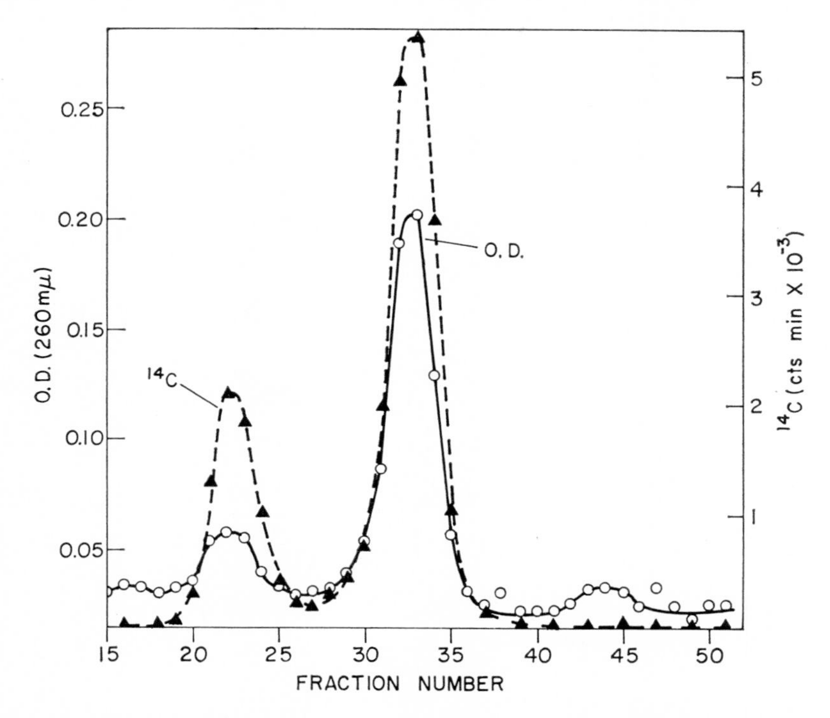

FIG. 5. Density profile of DNA in a CsCl gradient produced by centrifugation in a Spinco SW 39 rotor at 37,500 rpm for 48 hours. The DNA was labeled by growing the cells in 10^{-5} *M* bromodeoxyuridine-^{14}C for 22.5 hours along with an inhibitor of thymidylate synthesis. Three density species of DNA are detectable: Fractions 42–46 mark the position of unsubstituted DNA; fractions 30–35, the position of the density hybrid DNA; and fractions 20–25, the position of the fully substituted DNA. (From Taylor, 1968.)

A 48-hour nearly stationary phase culture of Chinese hamster cells was transferred to new medium. After 4 hours when most of the cells were reattached to the surface and in the *G1* phase 2×10^{-5} *M* bromodeoxyuridine and 10^{-6} *M* fluorodeoxyuridine were added to the medium for 18 hours. The fluorinated nucleoside inhibits the synthesis of thymidylate and forces the complete substitution of bromouracil for thymine in the new strands of DNA. After 18 hours when the DNA was all in the density hybrid form, the cells were transferred to a medium with 10^{-5} M thymidine for 8 hours. This interval should allow all cells which had begun the S phase in bromodeoxyuridine to complete synthesis so that no growing ends of chains containing the density label are available for attachment of new DNA. The cells were changed to conditioned medium

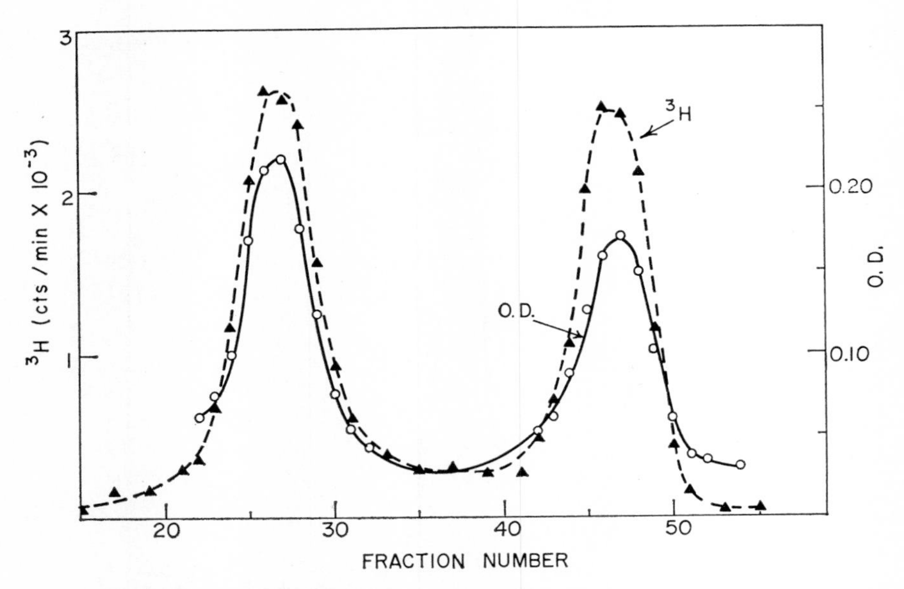

FIG. 6. Profile of DNA produced by centrifugation as described for Fig. 4. Two density species can be detected; they were produced by allowing bromouracil-substituted density hybrid DNA to replicate with the incorporation of thymidine-^{3}H to detect the new DNA. Note that nearly equal amounts of new ^{3}H-labeled density hybrid and unsubstituted ^{3}H-labeled DNA are produced as predicted by the semiconservative replication scheme in which each chain is conserved and incorporated into different molecules during replication. (Taylor, unpublished data.)

(medium in which cells had been grown previously) without thymidine for 40 minutes to deplete the thymidylate pool and then 10^{-5} M thymidine-^{3}H and 10^{-6} M fluorodeoxyuridine was added for 10 minutes. During this 10-minute period only cells containing density hybrid DNA should be in the S phase. The result is that predicted by the semiconservative replication of DNA. Equal amounts of thymidine-^{3}H appeared in hybrid molecules and in molecules without bromodeoxyuridine (Fig. 6).

To show that the new DNA chains of higher density are not linked tandemly with the template, the length of chains can be reduced by shear and the DNA rebanded. When this was done and as long as the DNA was not denatured, the average density did not increase above that of the longer pieces of hybrid (Taylor and Miner, 1968). However, to show that the ^{3}H-labeled DNA was actually only hydrogen bonded to the bromouracil-containing DNA, a sample of the labeled DNA was heated to 100° for 5 minutes in the buffer used for dissolving CsCl. Formaldehyde (final concentration 0.35%) was added and the mixture

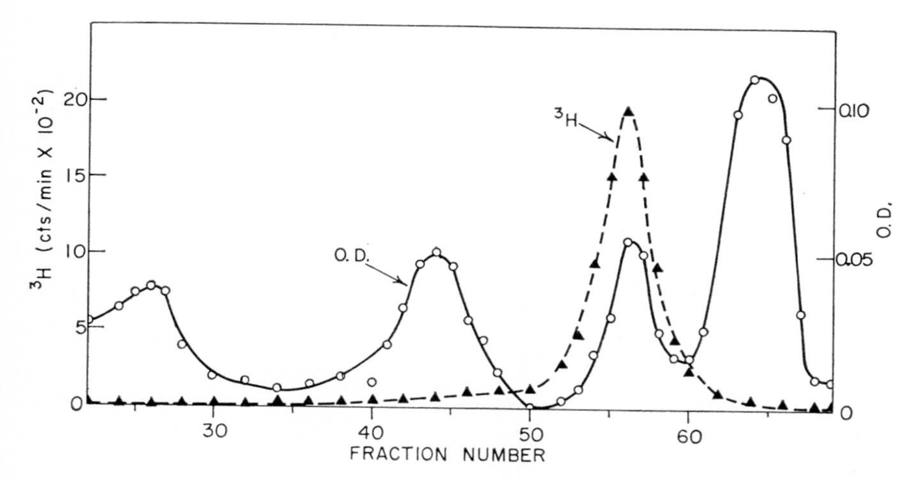

FIG. 7. Profile of DNA in a CsCl gradient prepared as described for Figs. 4 and 6. The profile contains unsubstituted, native DNA (fractions 61–67), denatured, unsubstituted DNA (fractions 54–58), native, bromouracil-substituted, density, hybrid DNA (fractions 42–46), and denatured bromouracil-substituted DNA (fractions 23–27). (Taylor, unpublished data.)

chilled in ice water. The formaldehyde reacts with the amino groups on the bases and prevents the strands from reannealing. Bromouracil hybrid DNA and native unlabeled DNA were added as density markers. CsCl was added to produce a density of 1.745 and the DNA was banded by centrifugation for 48 hours. The first optical density peak on the right is native, unsubstituted DNA; the second is denatured, unsubstituted DNA; the third is native, hybrid, density DNA; and the fourth is denatured bromouracil DNA. The significant result was that no detectable radioactivity remained associated with either the hybrid or denatured bromouracil-containing DNA. All of it is in the band of denatured, unsubstituted DNA (Fig. 7).

C. Polarity Differences in the Conserved Subunits of Chromosomes

From many observations, it is known that chromosomes break and exchange parts without steric restrictions. Spontaneous changes, which produce duplications, deletions, inversions, and reciprocal translocations; radiation and chemically induced breakage and reunion; and the sister chromatid exchanges observed in autoradiographic studies all support this concept. There are apparently no restrictions on the union of chromatids—homologous and nonhomologous regions unite—and a chromatid shows no polarity when inversions are produced or in any other reunion. This has been taken to mean that the linear components which hold chromosomes together are similar or identical along many if not all places in chromosomes of related organisms. We now know from the work of Callan and Macgregor (1958) and Gall (1963a) that DNA is the component that is primarily, if not solely, responsible for the linear integrity of chromosomes. Of a variety of enzymes, various proteases, ribonuclease, and deoxyribonuclease, only deoxyribonuclease severed the main axis and the loops of lampbrush chromosomes in oocytes of salamanders.

It appeared that an investigation of the pattern of sister chromatid exchanges might be useful for revealing the nature of the conserved subunits within chromatids. The DNA duplex proposed by Watson and Crick (1953a) has polynucleotide chains of reversed polarity. Supporting evidence for this model was obtained by Josse *et al.* (1961) from studies of *in vitro* replication. Nearest neighbor frequencies of the four nucleotides were found to be compatible with a chain synthesized on a template of opposite polarity rather than on one of the same polarity.

If the subunits were single polynucleotide chains, the patterns of exchanges predicted would be quite different from those predicted if each subunit consisted of a DNA duplex or some other linear component with the same polarity.

If the four subunits in a replicating chromosome were free to assort independently and were without polarity in breakage and reunion, semiconservative segregation at the chromosome level would be very irregular. Since it is quite regular, in some species at least—*Bellevalia* (Taylor, 1958b) and Chinese hamster cells (Prescott and Bender, 1963)—we may rule out that possibility. Even in *Vicia* where some odd segregation may be occurring at the second (X_2) division following incorporation of thymidine-^{3}H, the overall segregation is regular enough to exclude independent segregation of subunits. If segregation were independent even occasionally, the first division (X_1) would be irregular in that the sister chromatids would not always be labeled at the same loci. LaCour and Pelc (1958) reported that labeling was irregular, but Woods and Shairer (1959) and Peacock (1963) did not agree. Even the irregular segregation at the X_2 division reported by Peacock (1963) was variable among his various experiments. The *Vicia* material has been reinvestigated with better preparations and more extensive analyses (see below).

We may then consider the type of exchanges which would occur if the two subunits nearly always assort as one new and one old for the whole length of each chromosome, but can break and unite to produce sister chromatid exchanges. It is necessary to examine X_2 divisions in tetraploid cells where all of the products of replication and the original subunits are present and analyzable. Now if an original and template subunit assort together and are of opposite polarity the pattern of exchanges is predictable. Each exchange before the X_1 division will produce two visible exchanges at X_2 at the same locus on identical chromosomes (descendants from the same original chromosomes)—these have been called twin exchanges (Taylor, 1958b). In addition, all exchanges occurring during the S phase between the X_1 and X_2 divisions will produce a single visible exchange at X_2. Therefore, if the frequencies of exchange are equal before and after the X_1 division, the number of twin exchanges should equal the number of single exchanges. However, if the frequency of twins (two exchanges from a single event) is compared to the frequency of singles (one exchange from a single event) the ratio expected would be 1:2 because there are half as many chromosomes before X_1 as after that division.

On the other hand, if the subunits are alike and free to rejoin at random, but each two in a chromatid assort as a unit, the predicted ratio of twins to singles is 1:10. The reasoning behind that prediction is that exchanges before X_1 will yield four possible recombinations of labeled and unlabeled subunits (Fig. 8). One will show up as a twin at X_2, two will show up as singles at X_2, and one will not produce a switch point of label. After X_2 there will be two times as many chromosomes and with an equal chance of exchange (each will produce a visible switch point) before and after X_1, there should be eight singles for each four events before X_1. Summing up, there should be a ratio of one twin to each ten singles visible at X_2. The original data presented by Taylor (1958b and 1959) supported the hypothesis that the two subunits of a chromatid have opposite polarity and therefore are probably composed of single polynucleotide chains. Each chromatid is a highly folded and coiled DNA double helix and each chromosome after replication and before anaphase contains two such DNA double helices, i.e., has four subunits or poly-

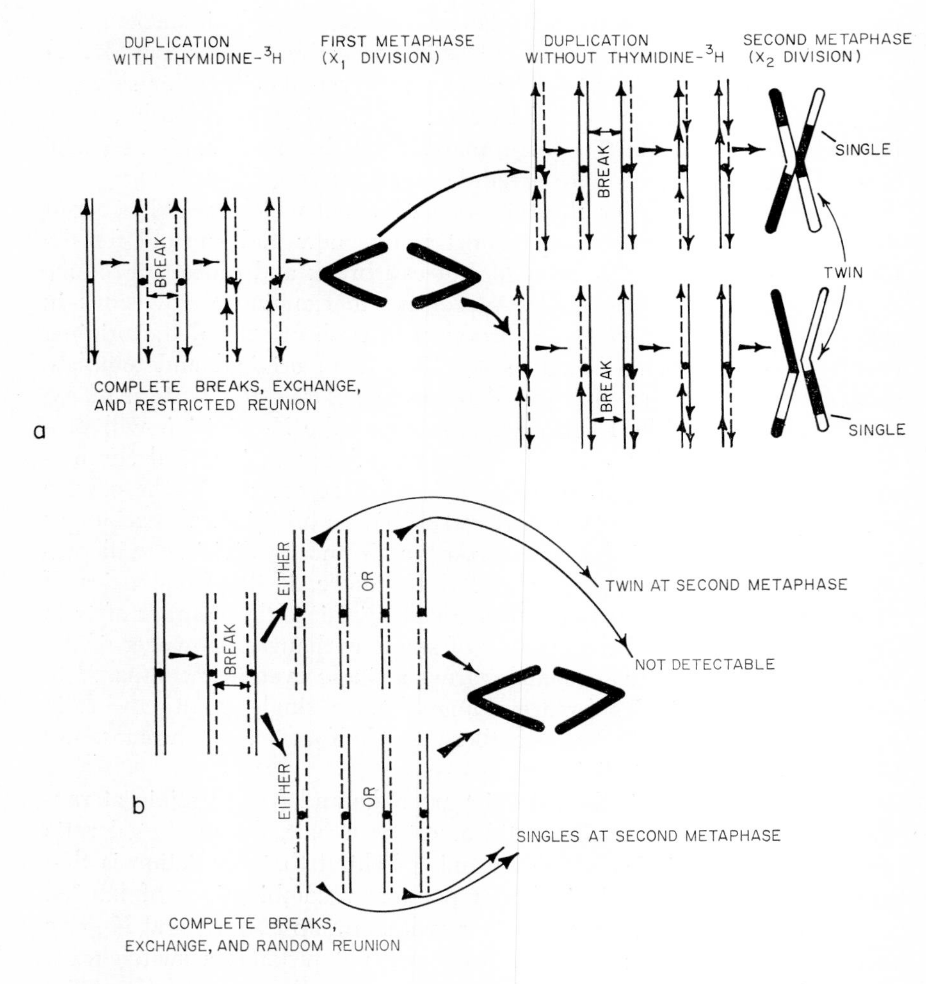

FIG. 8. Diagram showing the predicted results of sister chromatid exchange when incorporation of thymidine-^{3}H occurs during one cycle of replication and the pattern of labeled DNA is examined two divisions later, the X_2 division. The dashed lines represent labeled DNA subunits and the solid lines represent unlabeled subunits. The chromatids in solid black represent regions over which silver grains would appear in an autoradiogram and those in outline the unlabeled segments. (a) The predicted results when the exchange is restricted to subunits of the same polarity. (b) The predicted results if there are no restrictions on the way the subunits within a chromatid reunite (see text for a detailed analysis of the frequencies of single and twin exchanges predicted.)

nucleotide chains, which are the units that break and are rejoined during sister chromatid exchange.

Data presented later for Chinese hamster cells by Walen (1965) also support this model, although her data were analyzed for the purpose of showing the orienting influence of the centromeres at the half-chromatid level. Another report on human chromosomes by Herreros and Gianelli (1967) also supports the model. However, the size of the chromosomes makes the distinction between singles and twins difficult. More recent and extensive data on the marsupial chromosomes (*Potorus*) presented by Peacock and Bruen (personal communication, 1968) also confirm the earlier reports. In addition, Peacock and Bruen were able to disprove an alternate hypothesis suggested as a problem in Stahl's book (1964). Stahl pointed out that if one assumed breakage of all four subunits at nearly identical sites and reunion at random (no polarity), with the restriction that subunits in the one chromatid could not exchange with subunits in the other chromatid, the ratio of one twin to two singles would also be predicted. Although these are a very unlikely set of restrictions, it is interesting that the hypothesis can be eliminated as an explanation for the frequency of twins.

Recently Heddle (1968) has critically examined the quantitative aspects of the data reported in the literature on the ratios of twin and single exchanges. He concluded that the margin of error is large enough to cast considerable doubt on the ability to distinguish between the two predicted ratios, one twin to two singles and one twin to ten singles. While this conclusion is probably correct for medium-sized chromosomes, it is probably not correct for the larger chromosomes when the frequency of sister chromatid exchanges is below one per division cycle. Geard and Peacock have made a rather extensive study of the sister chromatid exchanges in *Vicia*. The analysis showed that the frequency of twin to single exchanges is 1:2 as originally predicted by Taylor (1958). They also considered the potential errors in classification discussed by Heddle (1969). Their conclusion was that such errors can be adjusted for and that confidence can be placed in the finding on one twin to two singles. Therefore, the data clearly support the postulate that the two replication subunits of a chromosome are dissimilar.

D. Transfer Experiments and Segregation of Chromosomal DNA in Meiotic Divisions

The premeiotic DNA replications have long been suspected of being different from those preceding regular mitosis. While the duration of replication may be different in the last premeiotic interphase (Callan and Taylor, 1968), other differences, if any exist, have not yet been discovered. Both at the chromosomal level (Taylor, 1965) and the molecular level (Chiang and Sueoka, 1966) the synthesis appears to follow the regular semiconservative mode of replication. These demonstrations were delayed several years after the original autoradiographic and density labeling studies were made because of technical difficulties. In the case of autoradiographic studies, the labeling at the premeiotic interphase and the preceding interphases is difficult to obtain in many organisms. Because the prophase I is so long and its length depends on so many variables, it proved difficult to find the few labeled cells and to determine the time when they became labeled. Finally, spermatocytes of the grasshopper, *Romalea,* were sufficiently labeled and found in division II as well as division I (Taylor, 1965). By injecting many animals and examining a few at daily intervals it could be established that the first labeled cells to arrive at anaphase I and metaphase II were labeled in both chromatids. However, 1 and 2 days later a few cells appeared at these stages with only one half of the chromatids labeled. Since it was difficult to make a complete analysis of all chromosomes in anaphase I groups, the next best analysis appeared to be metaphase II dyads. Nevertheless, a few anaphase I figures gave convincing indications that one half of the length of all chromatids were usually labeled in some complements which contained the most radioactivity. Metaphase II dyads are identical to their anaphase I counterparts in distribution of labeled DNA, because no synthesis of DNA and probably no exchange between chromatids occur in the interkinesis between the first and second division of meiosis. At metaphase II, some dyads were found in which one chromatid was labeled along the whole length while the other was unlabeled (Fig. 9a). Because of the exchanges produced by crossing-over, other dyads were asymmetrically labeled in ways that could not be accounted for by sister chromatid exchanges (Fig. 9b and c). These observations actually provided the first proof that physical exchange of material occurred during meiosis. Thus the proof was finally obtained that reciprocal exchanges of chromatid segments between homologous chromosomes occurred in meiosis. All other observations could be explained on the basis of some copy choice mechanism such as Belling (1931; 1933) originally proposed. Moreover, the number of exchanges was equal to that predicted if each chiasma represents an exchange between nonsister chromatids. The original conclusion that the exchanges were distributed at random among chromosomes while chiasmata were not was interpreted to indicate that the two events might not be correlated in a 1:1 fashion (Taylor, 1965).

However, in a reanalysis of the data by Miss Alice L. Andersen of Dr. David Perkin's laboratory, Stanford University (personal communication), it could be shown that the exchanges predicted from the reported chiasmata distributions were not significantly different from the exchange frequencies observed. Therefore, the original conclusion was incorrect and there remains no reason to doubt a correlation between the two events.

Transfer experiments at the premeiotic DNA synthesis proved difficult because of the technical difficulties of labeling such cells and of obtaining enough synchronized cells for density gradient centrifugation. The experiments carried out with *Chlamydomonas* (Chiang and Sueoka, 1966 and 1967) show semiconservative replication of DNA in the mitotic cycle and during a replication of chromosomal DNA which occurs during the germination of zygotes to form zoospores, but after crossing-over is thought to occur. Methods were available for synchronizing vegetative cells and for obtaining nearly synchronous maturation of zygotes in which meiosis occurs. However, the gametes turned out to have two times as much DNA as the *G1* vegetative cells, i.e., the premeiotic DNA replication occurred before gametes fused. In this respect *Chlamydomonas* has a cycle similar to *Neottiella*, an ascomycete, in which premeiotic DNA replication occurs before nuclear fusion (Rossen and Westergaard, 1966). Following meiosis in this ascomycete one postmeiotic DNA replication and then mitosis occurred so that asci have 8 nuclei. *Chlamydomonas* appears to have a similar behavior except that the equivalent of the postmeiotic DNA replication occurs before the postmeiotic four nucleate stage (see discussion of paper by Chiang and Sueoka, 1967). The chromosomes appear to have their DNA replicated in late prophase I of meiosis rather than after the divisions are complete. This replication was shown to be semiconservative and is followed by the production of 8 zoospores from each zygote. Therefore, the last premeiotic synthesis of DNA does not occur and cannot be studied in zygotes of *Chlamydomonas.* It occurs in the maturation of gametes from vegetative cells and was not specifically studied in the transfer experiments of Chiang and Sueoka (1966 and 1967). However, the synthesis of chromosomal and two cytoplasmic DNA's were shown to be semiconservative, and there is no reason to think that the last replication in the vegetative cells before gamete formation is different from other premitotic replications in vegetative cells. Nevertheless, semiconservative replication in the premeiotic DNA replication remains to be demonstrated by density transfer experiments. Certainly premeiotic interphase and meiotic prophase are still relatively unknown stages in terms of molecular events.

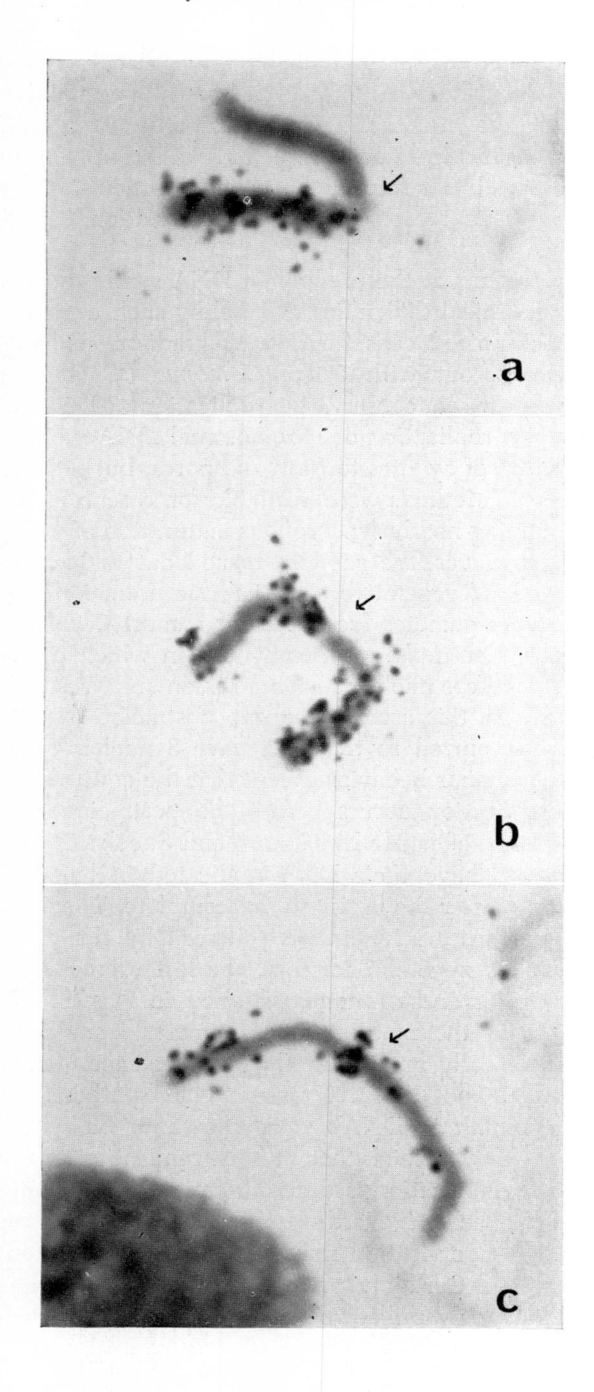
a
b
c

Hotta *et al.* (1966) have demonstrated what appears to be significant synthesis of DNA during the pachytene stage in *Lilium* microsporocytes. Even though it is less than 1% of that synthesized during premeiotic DNA replication, it could prove of importance in recombination events. However, if the few exchanges which result in chiasmata represent all of the changes in DNA during meiosis it is difficult to see how the events will ever be detected by isotopic labels, either radioactive or density.

Several sets of data now indicate that some events affecting recombination, which can be disturbed by high temperature, occur in the early meiotic prophase stages. Abel (1965) studied the effect of heat treatments on intergenic recombinations in the liverwort, *Sphaerocarpus*, and found the frequency to be influenced by treatments given when sporocytes were in leptotene stages. Henderson (1966) and Peacock (1968) showed that heat treatment could reduce chiasma frequency in grasshopper spermatocytes when applied 2–3 days after premeiotic DNA synthesis, i.e., in zygotene or early pachytene stages. These experiments took advantage of thymidine-^{3}H to label and time the arrival of cells at some later stage of maturation, and therefore could definitely place the events disturbed by heat after the regular premeiotic S phase. However, the same types of experiments, but without the benefit of autoradiographic techniques, were conducted earlier. These indicated that chiasmata frequency could be affected by treatments as late as the early pachytene stages (Straub, 1937; 1938). It is still not clear whether the events affected involve DNA metabolism, i.e., completion of a small

---

FIG. 9. Autoradiographs of individual chromosomes (dyads) from metaphase II in a secondary spermatocyte of *Romalea* labeled with thymidine-^{3}H at the interphase preceding the last spermatogonial division prior to meiosis. Arrows indicate the position of the terminal centromeres which are the only attachment points for the two chromatids at this stage. The parental chromosomes of these dyads entered the replication immediately preceding meiotic prophase with one DNA subunit in each chromatid labeled. After this premeiotic replication each chromosome (composed of two chromatids) should have one labeled and one unlabeled chromatid except for chromatid interchanges. (a) A chromosome (dyad) from metaphase II in which no segmental interchange is apparent. (b) A dyad with one reciprocal switch point proximal to the centromere and a nonreciprocal one toward the distal end of the left chromatid. The nonreciprocal switch points are presumably the result of a reciprocal exchange with the other dyad or homologous pair of chromatids which would be in another secondary spermatocyte. The dyad (pair of chromatids) in the photograph has three visible switch points. The transition from a labeled to an unlabeled region at the arrow is the centromere and is not considered a switch point. (c) A dyad with two nonreciprocal switch points in one chromatid. The other chromatid (right of arrow) was not considered to have visible switch points even though a few silver grains appear to be associated with it. (From Taylor, 1965.)

amount of synthesis or possibly breakage and repair, or some other phenomena associated with pairing of chromosomes.

VI. Patterns of Chromosomal Replication

The first indication that variations occurred in the replication patterns of the DNA within a chromosome were obtained by autoradiographic studies with thymidine-^{3}H in *Crepis* root tip cells (Taylor, 1958a). The observations indicated that in contrast to those of *Vicia* and *Bellevalia*, the chromosomes of *Crepis* had the ends of chromosome arms replicated earlier in the S phase than the regions around the centromeres. Later Lima-de-Faria (1959) showed that the X chromosome of the grasshopper continued replication of its DNA after most of the remainder of the complement had completed synthesis. An examination of patterns of replication in chromosomes of the Chinese hamster (Taylor, 1960c) revealed the first out-of-phase replication of mammalian chromosomes which was later found to be correlated with genetic suppression. The long arm of the X and the whole Y chromosome are replicated in the last half of the S phase while the short arm of the X chromosome is replicated in the first half in somatic cells of the male, as shown in Fig. 10 (Hsu *et al.*, 1964). In females the whole of one X and the long arm of the other X are replicated late. Most of the length of two of the smaller autosomes is late in replication. Later one of the X chromosomes of the human female was shown to be late in finishing replication (Morishima *et al.*, 1962; German, 1962). A suggestion had already been made by Lyon (1961) that one of the X chromosomes of the mouse had its genetic loci suppressed in somatic cells on the basis of variegation-type position effects produced when coat-color genes were translocated to an X chromosome (Russell, 1961, 1963; Cattanach, 1961). The early condensation of one X chromosome in the somatic cells of several mammals had also been shown by Ohno *et al.* (1959, 1961). Therefore it appeared that late replication, early condensation in prophase, the heteropycnotic appearance of chromosomes in interphase (the Barr body in mammalian cells, for example), and genetic suppression were correlated phenomena. More direct evidence for the genetic suppression was obtained by Davidson *et al.* (1963) for at least one locus on the human X chromosome. They found a female with two electrophoretic variants of the enzyme, glucose-6-phosphate dehydrogenase, designated A and B, which were under control of a sex-linked genetic locus. They obtained explants of skin from this individual. Cultures of these skin cells had both types of enzyme molecules, but when the cells were cloned, each clone yielded only one variant of the enzyme. This indicated that the locus on one X was suppressed in any individual cell. The suppressed locus could be either the one yielding the A or B form of the enzyme depending upon which X chromosome was suppressed in a particular cell. The suppression or inactivation presumably occurred rather early in embryological development, but at a stage in the embryo when many cells were present. Once established the suppression appears to be either nonreversible or nearly so. Kinsey (1967) studied the replication pattern in early embryos of the rabbit and found that late replicating X chromosomes first appeared in late blastula at about the time that sex chromatin (Barr bodies) could be observed.

Recently autoradiographic studies (Flickinger *et al.* 1967) of early embryonic division in the frog, *Rana pipiens,* have revealed that patterns of differential replication could first be demonstrated in frog embryos after gastrula stages. At the neurula and tailbud stages early and late replicating DNA could be demonstrated in a number of the chromosomes, but at earlier stages labeling occurred simultaneously in all parts of the complement.

That late replication patterns are not limited to sex chromosomes was shown by the early autoradiographic studies (Taylor, 1958, 1960c) as well as by numerous similar studies since, but that a large class of DNA in most cells may fit into this category is indicated by the study by Mueller and Kajiwara (1966), who used radioactive and density labels to demonstrate that nearly one half of the DNA in HeLa cells replicates early or late each cell cycle and that the same regions of the DNA are early or late in each cycle. Cells were synchronized in culture and labeled for a short interval with thymidine-^{3}H so that only early replicating DNA was radioactive. Cells were allowed to go through two or three additional division cycles and then blocked before DNA synthesis by treatments that depleted thymidylate. These cells were then released by supplying bromodeoxyuridine in place of thymidine as a density label. Early replicating DNA was then identified by its acquisition of the density label. The radioactive DNA was again labeled in the first half of the S phase, at least within the limits of the error of the method which was rather large because of the limitations imposed by imperfectly synchronized cells.

The only striking instance of a variation in the pattern of synthesis in mammalian chromosomes of different tissues has been reported by Utakoji and Hsu (1965). They found that the patterns in the X and Y chromosomes of the spermatogonial cells of the Chinese hamster are just

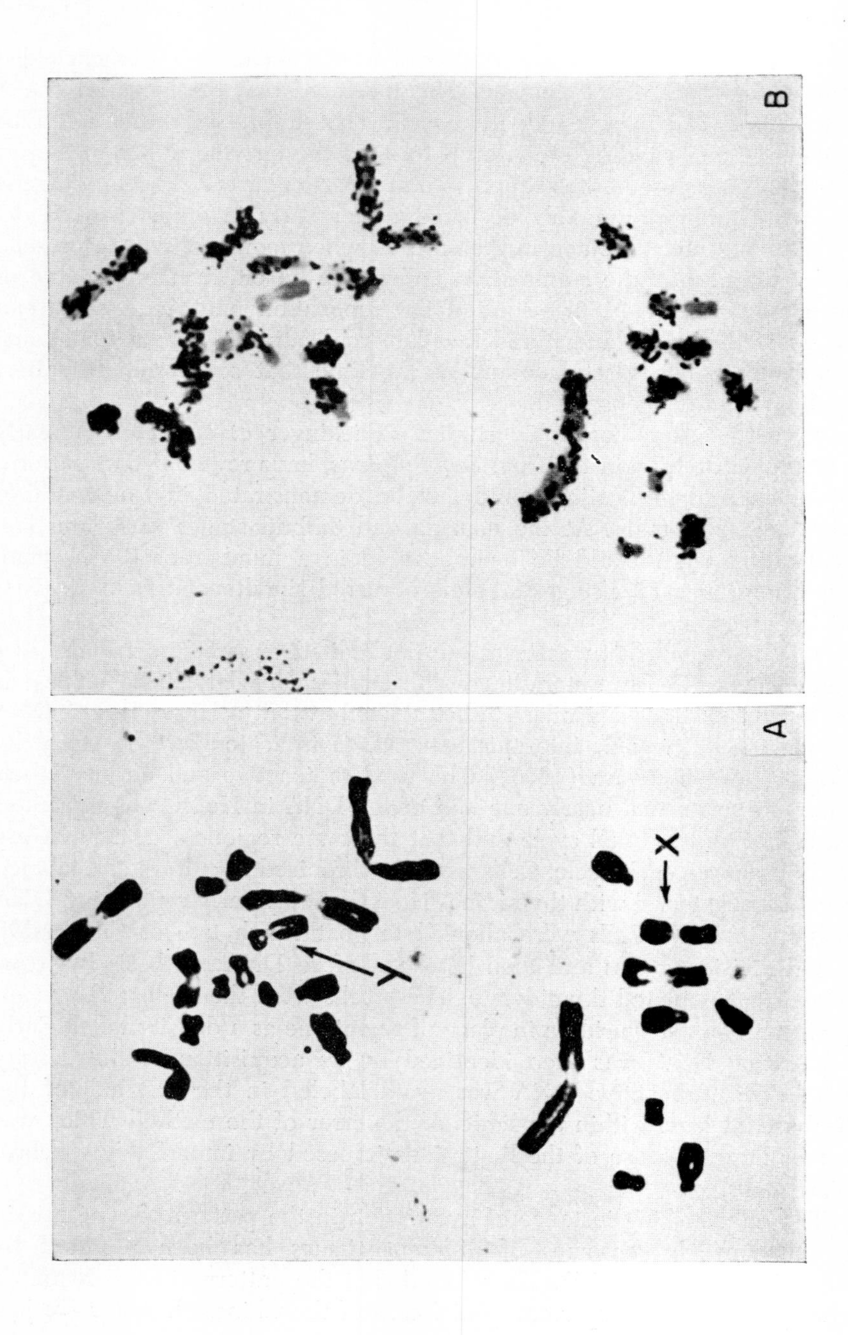
B
A
Y
X

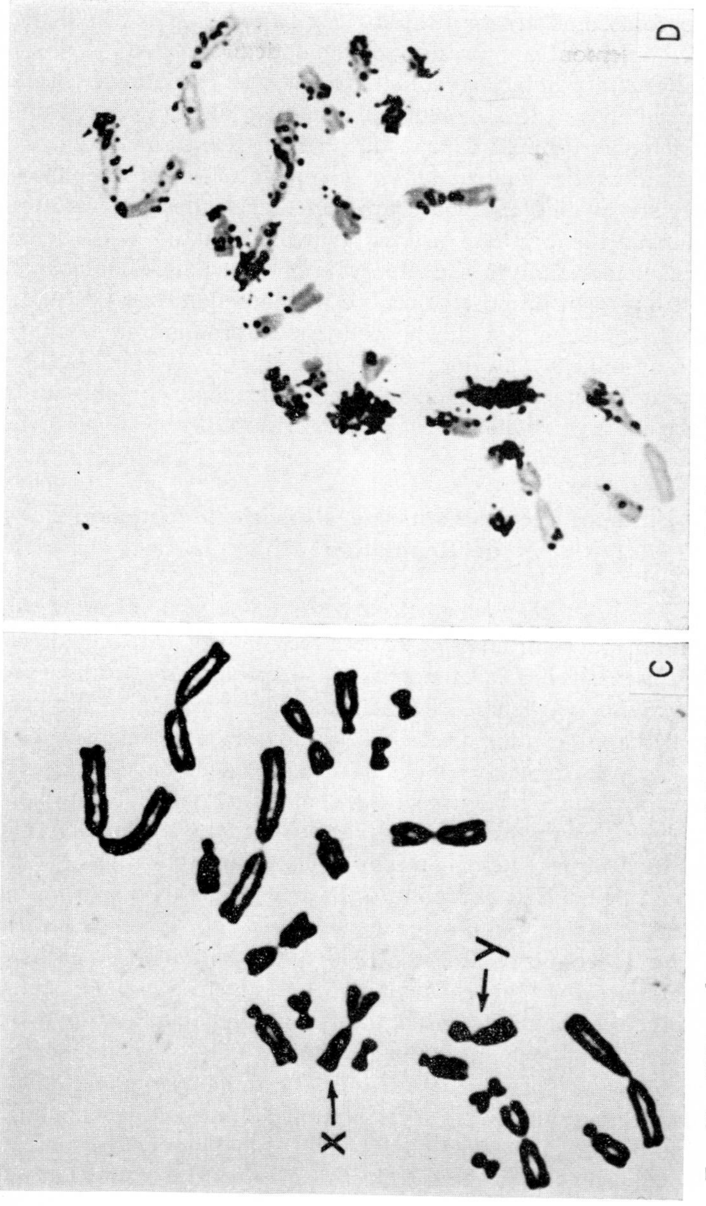

Fig. 10. DNA replication patterns of the Chinese hamster revealed by labeling with thymidine-^{3}H. Left figures show the chromosome complement before autoradiographs were prepared and right figures show the distribution of silver grains over radioactive areas. A and B show labeling during the first 5 minutes of the *S* phase. C and D show labeling during the last 45 minutes of the *S* phase. Notice the reverse labeling patterns of early and late stages, particularly in the X and Y chromosomes. (From Hsu *et al.*, 1964.)

the reverse of that in the somatic cells. Both the long arm of the X and the whole Y chromosome are replicated early in these cells where these parts of the complement are presumably genetically active. There are many other instances of heteropycnotic chromosomes or parts of chromosomes which are believed to be genetically inactive in the tissue in question or at least have a limited expression of their genetic material.

The mechanism of the control of DNA replication is still unknown, but some clues are available. Since it appears that patterns or variations of replication over the S phase are somehow correlated with genetic activity at a particular time in the life cycle or in a particular differentiated tissue, it has been proposed that the phenomenon may be the key to a major control mechanism during cellular differentiation in most if not all higher organisms (Taylor, 1964; Hsu, 1964). It then becomes important to search for the mechanism of control and to understand the structure of the units of replication and transcription in chromosomes.

VII. Units of Replication and Possible Mechanism of Regulation

Although it has been clear since the early studies of replication with tritiated thymidine that chromosomes must have more than one site for initiation, only recently has it been possible to predict the number and size of the units with any degree of precision. One approach to the problem aside from the autoradiographic observations is to measure rates of chain growth in replicating DNA. Several measurements have been made by autoradiography and by the use of the centrifuge and a density label. Autoradiographs of single extended DNA helices heavily labeled with thymidine-^{3}H by Cairns (1966) and by Huberman and Riggs (1968) gave an estimate of 0.5–2 μ per minute for mammalian cells in culture compared to 20–50 μ per minute for *Escherichia coli.* By the use of a density label for short pulses Taylor (1966, 1968) obtained an estimate of 1–2 μ per minute for Chinese hamster cells in culture. In a similar approach, Painter *et al.* (1966) used the density label to estimate that there were 5,000–10,000 growing points in human (HeLa) cells. Perhaps the most direct measurement of the size of the units of replication was obtained by Huberman and Riggs (1968) from autoradiographic studies. Partially synchronized cells were obtained by blocking DNA synthesis with fluorodeoxyuridine for 12 hours. These cells were then released with thymidine-^{3}H of high specific activity for 30 minutes. Autoradiographs of the extended helices of DNA showed replicating segments spaced about 20–100 μ apart; the average size was considered to be about 30 μ for the Chinese hamster cell. HeLa cells treated in a similar way produced labeled pieces about one half as far apart. One other interesting observation indicated the pattern of chain growth. When cells were removed from thymidine and grown for an additional 45 minutes, the precursors were diluted slowly enough so that most labeled segments showed a tapering off of grains at each end of the original segment labeled in the first 30 minutes. They interpreted these pictures to indicate that replication units were produced by initiation at a point with chains growing in both directions.

Recent attempts to verify the model by isolation of the replicating pieces in alkaline sucrose gradients (Taylor, 1968) indicate that the labeled single strands increase in length at a rate of 1 μ per minute. If the units are growing from both ends, this indicates a chain growth of 0.5 μ per minute. Synchronized cells were obtained by blocking at metaphase with colcemid. Cells released from the colcemid block were further synchronized by treatment before the beginning of the *S* phase with fluorodeoxyuridine and then released into *S* by supplying thymidine-^{3}H or thymidine-^{14}C. At intervals of 10 minutes cells were lysed with as little shear as possible. The lysates were layered on an alkaline sucrose gradient and spun for an appropriate interval. The labeled single DNA chains separated from their templates and their length could be estimated from their sedimentation rate. In 10 minutes the chains reached a length of about 10 μ; in 20 minutes a length of 18–20 μ; and by 30 minutes, they had an average length of 30–32 μ. After a lag period these pieces, which are probably units of replication, became linked into pieces about 100–125 μ long. The size could be limited by the shear involved in lysing cells and spreading lysates on the gradients, but the present evidence indicates that the large particles are significant subunits of chromosomes (Taylor, 1968). Their molecular weights were estimated to be about 200×10^{6} daltons, which is about 100 μ of bihelical DNA. The subunits denature in alkaline (pH 11.8) sucrose gradients but the two polynucleotide chains did not separate in these gradients at low ionic strength. However, the chains separated in 4 *M* sodium trichloroacetate which is known to denature DNA (Hamaguchi and Geiduschek, 1962); the high ionic strength, or at least some property of this solvent, allowed the separation of the chains into subunits with a sedimentation constant of about 57 which is equivalent to a molecular weight of over 100×10^{6} daltons. Phage T4 was used as a reference marker in the gradients. The chromosome subunits are about two times the length of phage T4 DNA.

Evidence is accumulating that the larger units of DNA in chromosomes of both bacteria and higher cells are replicated by growth of short segments of DNA which are then linked by ligases. The first indication came from a report by Sakabe and Okazaki (1966) that the DNA produced by a pulse label with thymidine-^{3}H appeared as short segments which separated from the template when the DNA was centrifuged in an alkaline sucrose gradient after denaturation at pH 12.0. These short, labeled segments became linked to much larger pieces if the pulse-labeled cells were incubated in the presence of unlabeled thymidine for a few minutes. In a later paper, Okazaki *et al.* (1967) reported that the particles produced by a 10-second pulse at 20°C had a sedimentation coefficient of 11 S while the remainder of the DNA had a higher sedimentation rate (47 S). Within 10 minutes nearly all of the labeled DNA had become associated with a 38 S component. The same type of behavior was noted for *E. coli* infected by T4. They also reported that pulse-labeled, native DNA centrifuged on neutral sucrose gradients yielded the short labeled segments which appeared to be single stranded when tested by elution from hydroxyapatite columns. These findings have been confirmed by Oishi (1968a) and Oishi (1968b). Earlier studies (Taylor, 1968; Taylor and Miner, 1968) on incorporation of bromodeoxyuridine in chromosomes of higher cells had indicated small subunits of replication in addition to the presumed larger subunits of 100 μ or more. More recently the small subunits similar to those reported for bacteria have been demonstrated by centrifugation and chromatography on hydroxyapatite (Schandl and Taylor, 1969). The results indicate a growth mechanism which produces either a cluster of growing points or initiation of short segments in tandem sequence. The short chains produced are easily removed from the template in the first moments after initiation, but as soon as longer chains are formed, the regular double-stranded, hybrid DNA (new and old strands) can be demonstrated.

Figures 11 and 12, which show profiles of DNA's in an alkaline sucrose gradient, illustrates the progressive increase in size of DNA segments after a pulse label of Chinese hamster cells. Cells in log phase growth were rinsed for 15–30 seconds with medium containing 10^{-5} *M* fluorodeoxyuridine, transfered to a medium with thymidine-^{3}H (10^{-5} *M*; sp. act. 15 c/m*M*) for 1 minute or less; after rinsing two times very quickly in a medium with 10^{-4} *M* unlabeled thymidine, they were either lysed or returned to the original medium to which was added 10^{-5} *M* unlabeled thymidine. The pulse label of 1 minute or less duration yielded only small particles (4–8 S, molecular weight of less than 5×10^5 daltons). A seven minute chase after a one minute pulse produced pieces covering a wider spectrum of sizes but with a peak of 26 S (mean molecular weight of 10–12 $\times$ 10^6 daltons). Within one hour the labeled DNA was associated with chains showing two peaks, one at 54 S and one at 72 S. If these 54 S pieces are randomly coiled single polynucleotide chains, the average molecular weight is nearly 100 $\times$ 10^6 daltons.

The nature of the small subunits and their distribution on the template is still rather obscure. However, the concept that polynucleotide chains must be growing from either end at replicating forks (Cairns, 1963; Cairns and Davern, 1967; Huberman and Riggs, 1968) has no compelling evidence to support it. Even the report by Bishop *et al.* (1967) that indicated chains of Qβ-RNA could be replicated in either direction may have an alternate interpretation (August *et al.*, 1968). The appearance of the short pieces makes it quite possible that the growth of chains is always by addition to the 3′-OH ends of chains on a template with the opposite polarity. This is, of course, the specific property of the original polymerase discovered by Kornberg (see Bessman, 1963 for a review of its properties).

The possible role of a primer for *in vivo* synthesis is still obscure, although it has never been demonstrated that any DNA polymerase can initiate chain growth on a template by joining two free nucleoside triphosphates. Oligonucleotides have been shown to increase the rate of polymerization on complementary templates *in vitro* (Bollum, 1967). One of the most interesting is the demonstration of a requirement for a heat-stable factor from *E. coli* for the replication *in vitro* of intact rings of the phage ϕX (Goulian, 1968). The factor proved to be oligodeoxynucleotides, which presumably act as primers by base pairing with the circular template which provides no free ends.

The mechanism of regulation of replication is a related problem, the details of which still elude us. In *E. coli* the initiation requires at least two steps which involve the utilization of amino acids (Bird *et al.*, 1968). The first step is sensitive to chloramphenicol, which presumably means that polypeptide synthesis is required. However, the second step occurs in the presence of high concentrations of chloramphenicol, and therefore remains more obscure. Initiation of DNA replication also appears to require protein synthesis in higher cells (Mueller and Kajiwara, 1966) although the more complex media required by higher cells do not allow the exclusion of protein synthesis to the extent possible in the studies with bacteria.

The replicon hypothesis of regulation proposed by Jacob *et al.* (1963) assumes that there is a replication locus, analogous to an operator locus in operons. The hypothesis also presumes a regulator substance. In *E. coli* there would presumably be one replicon per chromosome, but it is clear that in higher cells each chromosome would have many replicons, perhaps 10,000 per genome if the subunits 100 μ in length turn out to be replicons.

Would each of these have a replicator locus? If so, would each have a different regulator substance? The alternative could be that a whole

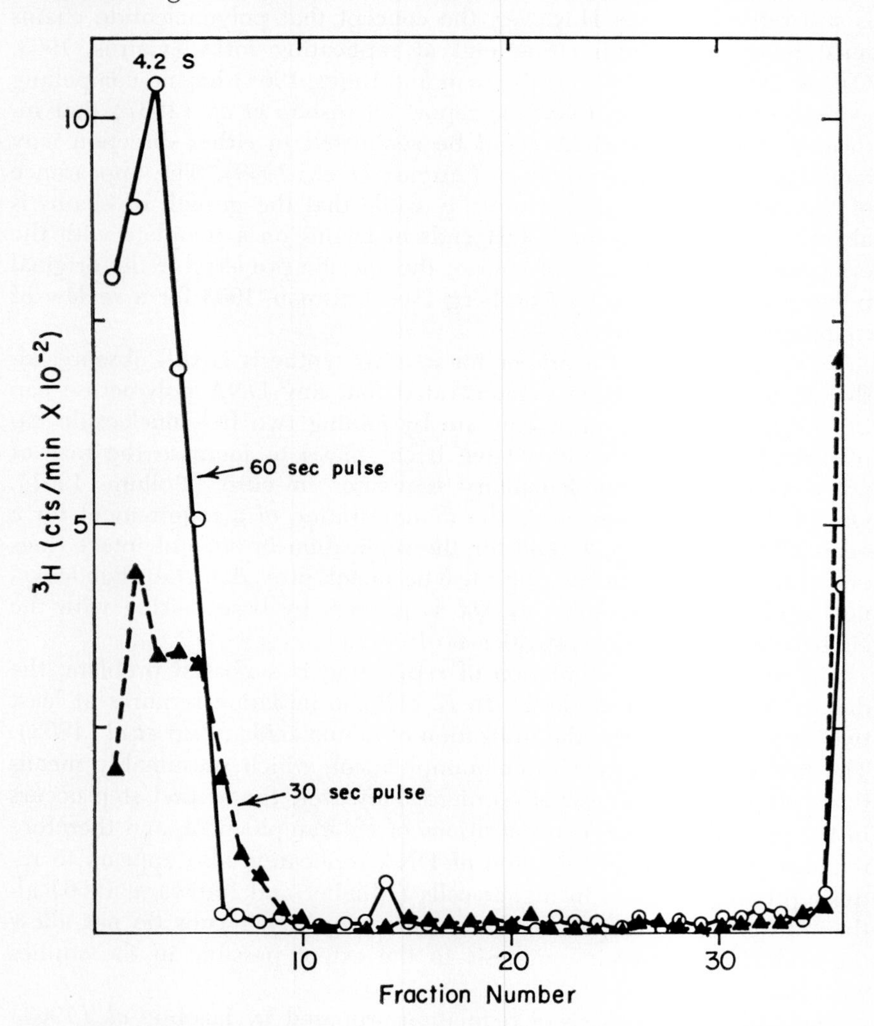

FIG. 11. Sedimentation profiles of two DNA preparations in alkaline sucrose gradients. Cells were pulse labeled for 30 and 60 seconds, respectively, and lysed directly on the glass surface to which they were attached. Centrifuged at 26,500 rpm for 9 hours (From Schandl and Taylor, 1969).

class of DNA replicons, late or early replicating, would have a common regulator substance. Perhaps there are a number of subclasses, but these two entities (early and late replicating sectors on chromosomes) have been clearly demonstrated in various higher cells since the initial report

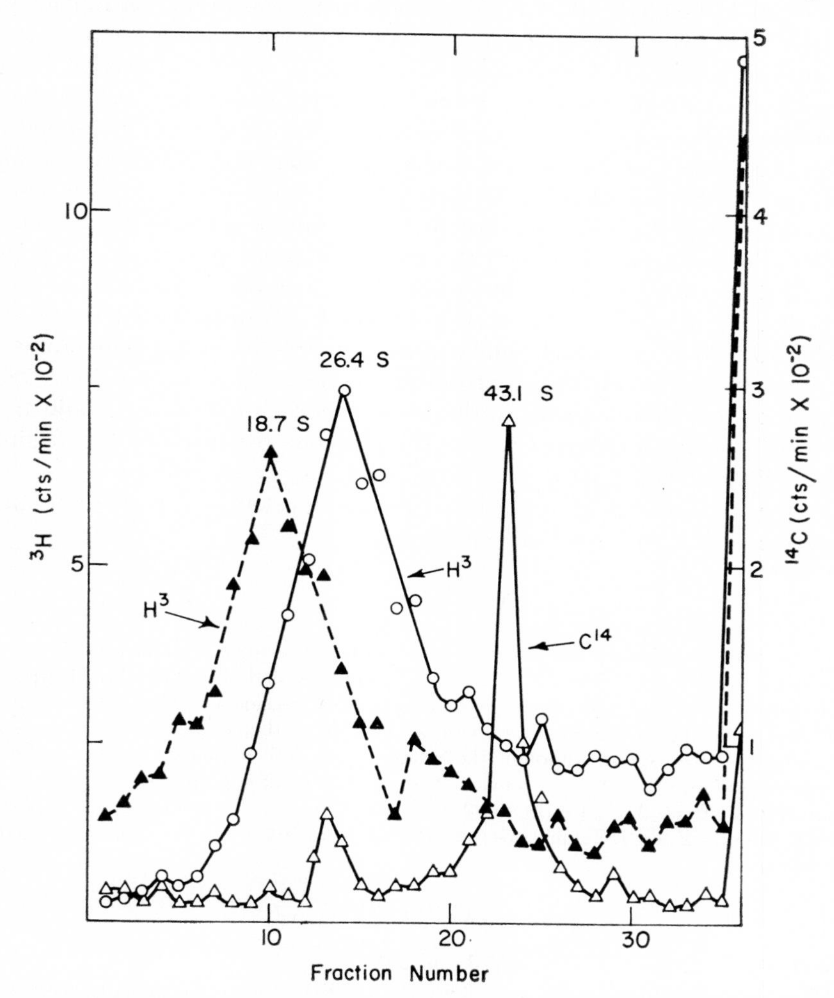

FIG. 12. Sedimentation profiles of two DNA preparations in alkaline sucrose gradients. Cells were pulse labeled for 30 seconds with ^{3}H-thymidine, rinsed with medium containing unlabeled thymidine and grown for 2 and 7 minutes, respectively. ^{14}C-labeled phage T4 was lysed along with the cells as a marker. Centrifuged at 26,500 rpm for 12 hours. –▲– –▲–, 2 minute chase; —○—○—○—, 7 minute chase; —△—△—△—, T4 phage DNA (From Schandl and Taylor, 1969).

for the Chinese hamster cell (Taylor, 1960c). Although it has been demonstrated that the pattern of replication may change during development, the pattern is remarkably stable in most adult tissues. How are such changes in pattern likely to be coded? It seems to the author that the DNA could be changed by a process similar to transformation in bacteria —not from externally added DNA, but from intracellular sources. It is conceivable that DNA replication is initiated by a primer (Taylor, 1966) and that whole blocks of replicating units might be transformed simultaneously by a slight variant of the regular primer which is presumed to function in replication. Primer in the sense used here is relatively short pieces of DNA (single strands) generated at some site in the cell which has a sequence complementary to the initiation sites in the chromosomes. These would pair with the chromosome at the specific sites and replication would continue by adding of monomers to the ends of such primers as guided by the template strands. Transformation could be effected in the absence of regular primer molecules by the generation of primers differing only slightly from the sequence at initiation sites. Once incorporated, the new sites in the transformed chromosomes would now normally accept a different primer from the one preferred by the original site. Without going into great detail, the suggestion will be made that changes in patterns of replication during development may sometimes involve such transformation phenomena.

Acknowledgments

The author wishes to acknowledge the assistance of members of his laboratory, particularly Mr. Emil Schandl who provided unpublished data, and Mr. Joseph F. Metcalf who assisted with the experiments on DNA replication described in Section V, B. Dr. Wolfgang Beerman was kind enough to read and make suggestions concerning the manuscript, although the author takes full responsibility for any deficiencies or errors which may remain. The original work reported was supported in part by Contract AT (401-1) 2690 with the Division of Biology and Medicine, U. S. Atomic Energy Commission, and Grant GB-6568 from the National Science Foundation.

REFERENCES

Abel, W. O. (1965). *Z. Verebungsl.* **96**, 228.

Allfrey, V. G., Mirsky, A. E., and Stern, H. (1955). *Advan. Enzymology* **16**, 411.

Atkin, N. B., Mattinson, G., Becak, W., and Ohno, S. (1965). *Chromosoma* **17**, 1.

August, J. T., Banerjie, A. K., Eoyang, L., de Fernandez, M. T. F., and Hori, K. (1968). *Cold Spring Harbor Symp. Quant. Biol.* **23** (in press).

Baetcke, K. P., Sparrow, A. H., Nauman, C. H., and Schwemmer, S. S. (1967). *Proc. Natl. Acad. Sci. U.S.* **58**, 533.

Baldwin, R. L. and Shooter, E. M. (1963). *J. Mol. Biol.* **7**, 511.
Bajer, A. (1965). *Chromosoma* **17**, 291.
Beermann, W. (1950). *Naturwissenschaften* **37**, 543.
Beermann, W. (1952). *Chromosoma* **5**, 139.
Beermann, W. (1956). *Cold Spring Harbor Symp. Quant. Biol.* **21**, 217.
Beermann, W. (1968). Mendel Centennial Symp. (Ft. Collins, Colo., 1965) Univ. of Wisconsin Press, (in press).
Beermann, W. and Bahr, G. F. (1954). *Exptl. Cell Res.* **6**, 195.
Belling, J. (1931). *Univ. Calif. Publ. Bot.* **16**, 311.
Belling, J. (1933). *Genetics* **18**, 388.
Bessman, M. J. (1963). *In* "Molecular Genetics" Part I (J. H. Taylor, ed.), p. 1. Academic Press, New York.
Bird, R., Renger, H., and Lark, K. G. (1968). *Cold Spring Harbor Symp. Quant. Biol.* **23** (in press).
Bishop, D. H. L., Pace, N. R., and Spiegelman, S. (1967). *Proc. Natl. Acad. Sci. U.S.* **58**, 1790.
Bollum, F. J. (1967). *In* "Genetic Elements" (D. Shugar, ed.) Academic Press, New York.
Brewer, M. E. and Pavan, C. (1955). *Chromosoma* **7**, 371.
Bridges, C. B. (1916). *Genetics* **1**, 1 and 107.
Bridges, C. B. (1935). *J. Heredity* **26**, 60.
Briggs, R., and King, T. J. (1957). *J. Morph.* **100**, 269.
Brown, D. D., Weber, C. S., and Sinclair, J. H. (1968). Carnegie Inst. Yearbook **66**, 8.
Burgi, E., and Hershey, A. D. (1963). *Biophy. J.* **3**, 309.
Cairns, J. (1961). *J. Mol. Biol.* **3**, 756.
Cairns, J. (1963). *J. Mol. Biol.* **6**, 208.
Cairns, John (1966). *J. Mol. Biol.* **15**, 372.
Cairns, J. and Davern, G. I. (1967). *J. Cell Physiol.* **70** (suppl. 1), 65.
Callan, H. G. (1952). *Symp. Soc. Exp. Biol.* **6**, 243.
Callan, H. G. (1955). Symp. on Fine Structure of Cells, I. U. B. S., series B, **21**, 89.
Callan, H. G. (1967). *J Cell. Sci.* **2**, 1.
Callan, H. G. and Lloyd, L. (1960) *Phil. Trans. Roy. Soc. B* **130**, 324.
Callan, H. G. and Macgregor, H. C. (1958). *Nature* **181**, 1479.
Callan, H. G. and Taylor, J. H. (1968). *J. Cell Sci.* **3**, 615.
Cattanach, B. M. (1961). *Z. Verebungslehre* **92**, 165.
Cavalieri, L. F. and Rosenberg, B. H. (1961) *Biophys. J.* **1**, 323.
Chandler, B., Hayashi, M., Hayashi, M. N., and Spiegelman, S. (1964). *Science* **143**, 47.
Chiang, K. S. and Sueoka, N. (1966). *Genetics* **54**, 327.
Chiang, K. S. and Sueoka, N. (1967). *J. Cell Physiol.* **70**, (Suppl. 1), 89.
Chun, E. H. L. and Littlefield, J. W. (1961). *J. Mol. Biol.* **3**, 668.
Cleveland, L. R. (1949). *Trans. Amer. Phil. Soc. N.S.* **39**, 1.
Clever, U. and Karlson, P. (1960). *Exptl. Cell Res.* **20**, 623.
Crouse, H. V. and Keyl, H. (1968). *Chromosoma* **25**, 357.
Davidson, Ronald G., Nitowsky, H. M., and Childs, Barton. (1963). *Proc. Natl. Acad. Sci. U.S.* **50**, 481.

Delbrück, M. and Stent, G. S. (1957). *In* "Symposium on the Chemical Basis of Heredity" (W. D. McElroy and B. Glass, eds.), p. 699. Johns Hopkins Univ. Press, Baltimore.
Djordjevic, B. and Szybalski, W. (1960). *J. Exptl. Med.* **112**, 509.
DuPraw, E. J. (1965). *Nature* **206**, 338.
DuPraw, E. J. (1968). "Cell and Molecular Biology," Academic Press, New York.
DuPraw, E. J. and Bahr, G. F. (1968). Abstr. Publ. 3rd Internatl. Congress Histochem. and Cytochem. New York.
Duryee, W. R. (1941). *In* "University of Pennsylvania Bicentennial Conference on Cytology, Genetics and Evolution," p. 129. University of Penn. Press, Philadelphia.
Duryee, W. R. (1950). *Ann. N.Y. Acad. Sci.* **50**, 920.
Edstrom, J. E. (1964). "Role of Chromosomes in Development" (M. Locke, ed.), Academic Press, New York.
Filner, P. (1965). *Exptl. Cell Res.* **39**, 33.
Flickinger, R. A., Freedman, M. L., and Stambrook, P. J. (1967). *Dev. Biol.* **16**, 457.
Freifelder, D. and Kleinschmidt, A. K. (1965). *J. Mol. Biol.* **14**, 271.
Freese, E. (1958). *Cold Spring Harbor Symp. Quant. Biol.* **23**, 13.
Friedkin, M., Tilson, D. and Roberts, D. (1956). *J. Biol. Chem.* **220**, 627.
Gall, J. G. (1952). *Exptl. Cell Res.* (Suppl.) **2**, 95.
Gall, J. G. (1956). *Brookhaven Symp. Biol.* **8**, 17.
Gall, J. G. (1958). Personal communication; Department of Zoology, University of Minnesota, Minneapolis.
Gall, J. G. (1963a). *Nature* (London) **198**, 36.
Gall, J. G. (1963b). *Science* **139**, 120.
Gall, J. G. (1966). *Chromosoma* **20**, 221.
Gall, J. G. and Callan, H. G. (1962). *Proc. Natl. Acad. Sci. U.S.* **48**, 562.
Ganesan, A. T. (1965). *Biochem. Biophys. Res. Comm.* **18**, 824.
Geard, C. R., and Peacock, W. J. (1969). *Genetics* (in press).
Gefter, M. L., Becker, A., and Hurwitz, J. (1967). *Proc. Natl. Acad. Sci.* **58**, 240.
Gellert, M. (1967). *Proc. Natl. Acad. Sci. U.S.* **57**, 148.
German, J. L. (1962). *Trans. New York Acad. Sci.* Series II **24**, 395.
Giesbrecht, P. (1966). *Proc. 6th Intern. Conf. Electron Microscopy*, Kyoto, p. 341.
Goulian, M. (1968). *Cold Spring Harbor Symp. Quant. Biol.* **23** (in press).
Hanawalt, P. C. and Ray, D. S. (1964). *Proc. Natl. Acad. Sci. U.S.* **52**, 125.
Hamaguchi, K. and Geiduschek, E. P. (1962). *J. Am. Chem. Soc.* **84**, 1329.
Heddle, J. A. (1968). *J. Theoretical Biol.* (in press).
Heitz, E. and Bauer, H. (1933). *Z. Zellforsch. Mikroskop. Anat.* **17**, 67.
Henderson, S. A. (1966). *Nature* **211**, 1043.
Herreros, B., and Giannelli, F. (1967). *Nature* **216**, 286.
Hershey, A. D. and Burgi, E. (1965). *Proc. Natl. Acad. Sci. U.S.* **53**, 325.
Horowitz, N. H. and MacLeod, H. (1960). *Microbiol. Genet. Bull.* **17**, 6.
Hotta, Y., Ito, M. and Stern, H. (1966). *Proc. Natl. Acad. Sci. U.S.* **56**, 1184.
Howard, A. and Pelc, S. R. (1953). *Heredity* **6** (Suppl.), 261.
Hsu, T. C. (1964). *In* "Biology of Cells and Tissues in Culture" (E. N. Willmer, ed.), p. 397. Academic Press, New York.
Hsu, T. C., Schmid, W. and Stubblefield, E. (1964). *In* "The Role of Chromosomes in Development" (M. Locke, ed.), p. 83. Academic Press, New York.

Huberman, J. A. and Riggs, A. D. (1966). *Proc. Natl. Acad. Sci. U.S.* **55**, 599.
Huberman, J. A., and Riggs, A. D. (1968). *J. Mol. Biol.* **32**, 327.
Huskins, C. L. (1937). *Cytologia, Fujii Jub. Vol.* **1015**.
Hyde, J. M., Randall, C. C., and Gafford, L. G. (1966). *Proc. 6th Intern. Conf. Electron Microscopy,* Kyoto, 193.
Jackson, R. C. (1957). *Science* **126**, 1115.
Jacob, F., Brenner, S., and Cuzin, F. (1963). *Cold Spring Harbor Symp. Quant. Biol.* **28**, 329.
Jacob, F. and Wollman, E. L. (1961). "Sexuality and the Genetics of Bacteria". Academic Press, New York.
Josse, J., Kaiser, A. D., and Kornberg, A. (1961). *J. Biol. Chem.* **236**, 864.
Kaufmann, B. P., Gay, H. and McDonald, M. (1960). *Internatl. Rev. Cytol.* **9**, 77.
Keyl, H. G. (1966). "3. Wissenschaftliche Konferenz der Gesellschaft Deutscher Naturforscher und Arzte, Semmering bei Wein 1965" (P. Sitte, ed.) p. 53. Springer-Verlag, Heidelberg.
King, T. J., and Di Berardino, M. A. (1967). *Dev. Biol.* **15**, 102.
Kinsey, J. D. (1967). *Genetics* **55**, 337.
Kleinschmidt, A. K. (1967). *In* "Molecular Genetics" Part II, p. 47. (J. H. Taylor, ed.). Academic Press, New York.
Kleinschmidt, A. K., Wang, D., Plescher, C., Hellman, W., Haass, J., Zahn, R. K., and Hagedorn, A. (1961). *Z. Naturforsch.* **16b**, 730.
LaCour, L. F. and Pelc, S. R. (1958). *Nature* **182**, 506.
Lajtha, L. G. (1963). *J. Cell. Comp. Physiol.* **62** (Suppl. 1), 143.
Lark, K. G. (1963). *In* "Molecular Genetics" Part I (J. H. Taylor, ed.) p. 175. Academic Press, New York.
Lark, K. G. (1966). *Bacteriol. Rev.* **30**, 3.
Lark, K., Eberle, H., Consigle, R., Minocha, H., Chai, N. and Lark, C. (1967). "Chromosome Segregation and the Regulation of DNA Replication". Organizat. Biosyn. Bicent. Symp.
Laskowski, W. Lochmann, E., Wacker, A. and Stein, W. (1960). *Zeitschr. f. Naturf.* **15b**, 730.
Lima-de-Faria, A. (1959). *J. Biophys. Biochem. Cytology* **6**, 457.
Lyon, M. E. (1961). *Nature* **190**, 372.
MacHattie, L. A. and Thomas, C. A., Jr. (1964). *Science* **144**, 1142.
Martin, P. G. (1968). *In* "Replication and Recombination of Genetic Material" (W. J. Peacock and R. D. Brock, eds.). p. 93. Australian Academy of Science, Canberra.
Massie, H. R. and Zimm, B. H. (1965). *Proc. Natl. Acad. Sci. U.S.* **54**, 1636.
Meselson, M. and Stahl, F. W. (1958). *Proc. Natl. Acad. of Sci. U.S.* **44**, 671.
Meselson, M. and Weigle, J. J. (1961). *Proc. Natl. Acad. Sci. U.S.* **47**, 857.
Miller, O. L. (1965). *Natl. Cancer Inst. Monogr.* **18**, 79.
Miller, O. L., Jr. (1966). *Natl. Cancer Inst. Monogr.* **23**, 53.
Morgan, T. H. (1911). *Science,* N.S. **34**, 384.
Morishima, A., Grumbach, M. M. and Taylor, J. H. (1962). *Proc. Natl. Acad. Sci. U.S.* **48**, 756.
Moses, M. J. (1964). *In* "Cytology and Cell Physiology" (G. Bourne, ed.). Academic Press, New York.
Mueller, G. C. and Kajiwara, K. (1966). *Biochem. Biophy. Acta* **114**, 108.
Nebel, B. R. (1939). *Bot. Rev.* **5**, 563.

Ogur, M., Erickson, R. O., Rossen, G. U., Sax, K. and Holden, C. (1951). *Exptl. Cell Res.* **2**, 73.
Ohno, S. and Atkin, N. B. (1966). *Chromosoma* **18**, 455.
Ohno, S., Kaplan, W. D., and Kinosita, R. (1959). *Exptl. Cell Res.* **18**, 415.
Ohno, S. and Weiler, C. (1961). *Chromosoma* **12**, 362.
Oishi, M. (1968a). *Proc. Natl. Acad. Sci. U.S.* **60**, 329.
Oishi, M. (1968b). *Proc. Natl. Acad. Sci. U.S.* **60**, 691.
Okazaki, R., Okazaki, T. Sakabe, K., and Sugimoto, K. (1967). *Jap. J. Med. Sci. and Biol.* **20**, 255.
Painter, R. B., Jermany, D. A., and Rasmussen, R. E. (1966). *J. Mol. Biol.* **17**, 47.
Painter, T. S. (1933). *Science* **78**, 585.
Painter, T. S. (1934). *Genetics* **19**, 175.
Peacock, W. J. (1963). *Proc. Natl. Acad. Sci. U.S.* **49**, 793.
Peacock, W. J. (1968). *Genetics* (in press).
Pietsch, P. (1966). *J. Cell Biol.* **31**, 86A.
Plant, W. and Mazia, D. (1956). *J. Biophys. and Biochem. Cytol.* **2**, 573.
Pollister, A. W. and Moore, J. A. (1937). *Anat. Rec.* **68**, 489.
Prescott, D. M. and Bender, M. A. (1963). *Exptl. Cell. Res.* **29**, 430.
Ris, H. (1957). *In* "Symp. Chem. Basis of Heredity" (W. D. McElroy and B. Glass, eds.). p. 23. The Johns Hopkins Press, Baltimore.
Ris, Hans (1967). *In* "Regulation of Nucleic Acid and Protein Biosynthesis" (V. V. Koningsberger and L. Bosch, eds.) p. 11. Elsevier Publishing Co., Amsterdam.
Ris, H. and Chandler, B. L. (1963). *Cold Spring Harbor Symp. Quant. Biol.* **28**, 1.
Ritchie, D., Thomas, C. Jr., MacHattie, L. and Wensink, P. (1967). *J. Mol. Biol.* **23**, 365.
Ritossa, F. M., Atwood, K. C., Lindsey, D. L., and Spiegelman, S. (1966). *Natl. Cancer Inst. Monogr.* **23**, 449.
Rossen, J. M. and Westergaard, M. (1966). *C. R. Trans. Lab. Carlsberg* **35**, 233.
Rothfels, K., Sexsmith, E., Heimburger, M. and Krause, M. O. (1966). *Chromosoma* **20**, 54.
Ruckert, J. (1892). *Anat. Anz.* **7**, 107.
Rudkin, G. T. (1964). *In* "Genetics Today," Proc. XI Intern. Congress of Genetics, The Hague, p. 359. Pergamon Press, New York.
Russell, L. B. (1961). *Science* **133**, 1795.
Russell, L. B. (1963). *Science* **140**, 976.
Sager, R. and Ishida, M. R. (1963). *Proc. Natl. Acad. Sci. U.S.* **50**, 725.
Sakabe, K. and Okazaki, R. (1966). *Biochim. Biophys. Acta* **129**, 651.
Schwartz, D. (1955). *J. Comp. Cell. Biol.* **45** (Suppl. 2) 171.
Schandl, E. K., and Taylor, J. H. (1969). *Biochem. Biophys. Res. Comm.* **34**, 291.
Sharman, G. B. and Barber, H. N. (1952). *Heredity* **6**, 345.
Simon, E. H. (1961). *J. Mol. Biol.* **3**, 101.
Smith-White, S. (1968). *Chromosoma* **23**, 359.
Stahl, F. W. (1964). "The Mechanics of Inheritance". p. 79 Prentice-Hall, Englewood Cliffs.
Steffensen, D. (1959). "A Comparative view of the Chromosome" Brookhaven Symp. Biol. No. 12. Brookhaven National Lab., Upton, New York.
Straub, J. (1937). *Ber. d. dtsch. bot. Ges.* **55**, 160.
Straub, J. (1938). *Zeit. f. Bot.* **32**, 225.
Streisinger, G., Edgar, R. S., and Denhardt, G. H. (1964). *Proc. Natl. Acad. Sci. U.S.* **51**, 775.

Studier, F. W. (1965). *J. Mol. Biol.* **11**, 373.
Sturtevant, A. H. (1913). *J. Exptl. Zool.* **14**, 43.
Sueoka, N. (1960). *Proc. Natl. Acad. Sci. U.S.* **46**, 83.
Sueoka, N. (1967). *In* "Molecular Genetics" (J. H. Taylor, ed.) Part II, Chapt. 1. Academic Press, New York.
Swift, H. (1950). *Proc. Natl. Acad. Sci. U.S.* **36**, 643.
Swift, H. (1962). *In* "The Molecular Control of Cellular Activity" (J. M. Allen, ed.) p. 73. McGraw-Hill, New York.
Taylor, J. H. (1957). *Amer. Nat.* **91**, 209.
Taylor, J. H. (1958a). *Exptl. Cell Res.* **15**, 350.
Taylor, J. H. (1958b). *Genetics* **43**, 515.
Taylor, J. H. (1959). *Pric. X Internatl. Cong. Genetics, Vol.* **1**, 63, Montreal 1958. University of Toronto Press.
Taylor, J. H. (1960a). *Amer. Sci* **48**, 365.
Taylor, J. H. (1960b). *Adv. Biol. Med. Physics* **7**, 107.
Taylor, J. H. (1960c). *J. Biophysic. and Biochem. Cytol.* **7**, 455.
Taylor, J. H. (1962). *Intern. Rev. Cytol.* **13**, 39.
Taylor, J. H. (1963a). *J. Cell. Comp. Physiol.* **62** (Suppl. 1), 73.
Taylor, J. H. (1963b). "Molecular Genetics" (J. H. Taylor, ed.) Part I, p. 65. Academic Press, New York.
Taylor, J. H. (1964). *Symp. Intern. Soc. for Cell Biol.* **3**, 175. Academic Press.
Taylor, J. H. (1965). *J. Cell Biol.* **25**, 57.
Taylor, J. H. (1966). *In* "3 Wissenschaftliche Konferenz der Gesellschaft Deutscher Naturforscher und Arzte, Semmering bei Wien, 1965 "(P. Sitte, ed.) p. 9. Springer-Verlag, Heidelberg.
Taylor, J. H. (1967). *Encyclopedia of Plant Physiology* **18**, 344. (W. Ruhland, ed.) Springer-Verlag, Berlin.
Taylor, J. H. (1968). *J. Mol. Biol.* **31**, 579.
Taylor, J. H. (1969). *Proc. XIII Internatl. Cong. Genetics,* Tokyo, 1968. (in press).
Taylor, J. H. and McMaster, R. D. (1954). *Chromosoma* **6**, 489.
Taylor, J. H. and Miner, P. (1968). *Cancer Res.* **28**, 1810.
Taylor, J. H., Wood, P. S., and Hughes, W. L. (1957). *Proc. Natl. Acad. Sci. U.S.* **43**, 122.
Thomas, C. A. (1963). *In* "Molecular Genetics" (J. H. Taylor, ed.) Part I, Chapt. III. Academic Press, New York.
Trosko, J. E. and Wolf, S. (1965). *J. Cell Biol.* **26**, 125.
Utakoji, T. and Hsu, T. C. (1965). *Cytogenetics* **4**, 295.
Vendrely, R. and Vendrely, C. (1949). *Experientia* **5**, 327.
Waldeyer, W. (1888). *Arch. Mikr. Anat.* **32**, 1.
Walen, Kirsten H. (1965). *Genetics* **51**, 915.
Watson, J. D. and Crick, F. H. C. (1953a). *Nature* **171**, 737.
Watson, J. D. and Crick, F. H. C. (1953b). *Nature* **171**, 964.
Weil, R. and Vinograd, J. (1963). *Proc. Natl. Acad. Sci. U.S.* **50**, 730.
Wimber, D. E. and Prensky, W. (1963). *Genetics* **48**, 1731.
Wolfe, S. L. (1965). *J. Ultrastruct. Res.* **12**, 104.
Wolfe, S. L. and Hewitt, G. M. (1966). *J. Cell Biol.* **31**, 31.
Wolfe, S. L. and John, B. (1965). *Chromosoma* **17**, 85.
Woods, P. S., and Schairer, M. V. (1959). *Nature* **183**, 303.
Zweidler, A. (1964). *Arch. Julius Klaus-Stiftung Vererbungsforsch. Sozialanthropol. Rasenhyg.* **39, 54.**

IV LINKAGE AND RECOMBINATION AT THE MOLECULAR LEVEL

WALTER F. BODMER AND ANDREW J. DARLINGTON

I. Introduction

The formal analysis of recombination in higher organisms has suggested two fundamentally different models for the reassortment of linked genetic markers. The problem has been to arrive at a compromise between the classic hypotheses of copy choice and breakage and reunion. These are distinguished according to whether a parental chromosome strand contributes to the recombinant a part of its substance or merely, via replication, the information it carries. There are two essential features of any model of recombination, namely, synapsis, which is the alignment of homologous chromosome segments, and exchange which is the recombining of the information or substance carried by the synapsed homologs. Since genetic maps presumably represent the linear structure of deoxyribonucleic acid (DNA) molecules, recombination must be an interaction between such molecules. Recombination data from higher organisms throw little light on the molecular mechanisms of the process, especially in the absence of a clear picture of the structure of their chromosomes in relation to DNA structure. Systems are needed in which the physicochemical properties of DNA molecules can be correlated with their genetic properties. So far, bacteria and their phages are the only organisms in which this is feasible. Our aim is, therefore, to review the data from these organisms which throw some light on the molecular mechanism of recombination and to try, briefly, to relate this information to recombination as studied in higher organisms. Such a review cannot be comprehensive but is, we hope, representative. Somewhat greater emphasis is placed on data from bacterial transformation, since those from other systems have been adequately reviewed recently by a number of authors (see Meselson, 1966; Taylor, 1967).

II. Correlation of the Genetic and Physical Chromosome Maps

Conventional linkage maps depend, in general, on the analysis of the whole progeny of a suitable cross. As genetic distances decrease the method becomes laborious and, beyond a certain point, impossible. However, if microorganisms are used in such a way that their genotypes may be assayed by selective techniques, the resolving power of genetic mapping may be enormously increased, as illustrated by the classic work of Pritchard (1955, 1960) with *Aspergillus* and Benzer (1959, 1961) on the structure of the rII region of phage T4. Both these workers showed that the functional genetic unit could be broken down into a large number of mutable sites separable by recombination. The mutants with which Benzer worked map in a small region of the phage genome, the rII region, and differ from the wild-type in plaque morphology and in being unable to form plaques on strain K of *Escherichia coli* though they will do so on strain B. Thus, in a cross between two rII mutants, the total number of progeny of a mixed infection may be estimated by plating on strain B and the number of wild-type recombinants by plating at a higher concentration on strain K. Benzer isolated a very large number of rII mutants, many of which were deletions identified as such by the fact that they would not form wild-type recombinants with several other mutants. He was able to assign mutants to groups on the basis of whether they gave wild-type recombinants when crossed with a series of deletion mutants. They were finally assigned to positions on a linear map according to the frequency with which they produced wild-type recombinants in combination with other rII mutants. This fine structure analysis established in great detail the genetic map of a large number of mutable sites within a single functional unit and showed that it was strictly linear. From the standpoint of the mechanism of recombination the important questions are, first, what does this linear map represent, and second, what distances separate sites that may recombine?

The chromosome of the T-even phages, T2, T4, and T6, consists of an uninterrupted DNA double helix containing approximately 1.8×10^3 nucleotide pairs (Davison *et al.*, 1961; Tomizawa and Anraku, 1965; C. A. Thomas, 1966). The rII region, therefore, presumably occupies a small section of this molecule, whose size may be estimated from the fraction of the genetic map which it occupies. Assuming map distances (basically recombination fractions over short distances) to be strictly proportional to physical distances, this gives an estimate of 1000 base pairs, within which Benzer located 308 mutable sites separable by recombination. Thus, it seems likely that the lowest recombination frequency observed by Benzer is that occurring between adjacent nucleotide pairs. More precise information on the minimum distance between recombinant sites has been obtained by Henning and Yanofsky (1962). They showed that different mutants of the tryptophan synthetase A protein in *E. coli*, resulting in different amino acid substitutions at the same site in the protein molecule, could recombine, and that the recombinant form had a new amino acid at this same site. Thus, recombination had occurred within a single codon consisting of only three nucleotide pairs.

There are no comparable results obtainable for more complex organisms, though fine structure maps have been made in fungi (Pritchard, 1955; Case and Giles, 1958), and to a limited extent in *Drosophila* (Chovnick, 1966) and mice, for the H2 region (Stimfling and Richardson, 1965). There is no reason to suppose that the problems of constructing fine structure maps in higher organisms are not due to the experimental difficulties imposed by the very low recombination fractions involved. The work on fine structure analysis lays the basis for assuming a direct correlation between the physical map and the DNA molecule. This has been elegantly demonstrated in a more direct way in experiments using the *E. coli* phage λ. Kaiser and Hogness (1960) established a system of DNA transformation for this phage which makes it possible to correlate very directly the genetic map of λ obtained by conventional means with a physical map. Thus, Kaiser (1962), Radding and Kaiser

(1963), Hogness and Simmons (1964), and Hogness *et al.* (1966) have been able to correlate the genetic transforming activities obtained from fragments of DNA molecules formed by stepwise breaking into halves, quarters, and smaller pieces with the position of markers on the genetic map. Their work provides a direct verification of a one-to-one relationship between a genetic map and the structure of the associated DNA molecule.

The problem with which we are concerned is how genetic markers on homologous molecules come to be located on the same molecule. As mentioned above, two basically different models have been considered. The first supposes that material recombination occurs by breakage and joining of homologous molecules. This is the theory that was proposed originally by Janssens (1909) to account for the observed correlation between meiotic chiasmata and genetic recombination. The second model, that of copy choice, supposes that the genetic information, but not the material, is recombined; the recombinant molecule is synthesized at the time of recombination and contains information derived from both parental molecules. This model was proposed by Belling (1931) and revived by Hershey and Rotman (1949) to account for certain aberrant phenomena in phage recombination which will be discussed later (see also J. Lederberg, 1955). Neither of these models explains how homologous DNA molecules pair prior to recombination, whatever the exchange mechanism may be. It is clear that pairing must be very precise since the functional integrity of the recombined structure is of vital importance. The only known process with the requisite precision is the base pairing of the complementary strands of a DNA double helix.

III. Molecular Studies of Recombination in Bacteriophage λ

Bacterial viruses have proved to be very suitable for molecular studies on recombination. Since the phage particle is stable and has a relatively very large DNA content, its density is highly correlated with that of the DNA it contains. As in bacterial transformation, therefore, the physical and the biological properties of DNA molecules may be assayed. The major technique that has been used to analyze DNA preparations is that of equilibrium density-gradient centrifugation in cesium chloride (CsCl) solutions, originally applied to the analysis of DNA by Meselson *et al.* (1957). This depends on the establishment of a stable gradient in density when very concentrated cesium chloride solutions are centrifuged at high speed. DNA added to such a gradient will come to equilibrium as a narrow band at a position corresponding to its own density. If the DNA has been isolated from organisms grown in media containing heavy isotopes, for example, of nitrogen, hydrogen, or carbon (^{15}N, ^{2}H, or ^{13}C), its density will differ from that of unlabeled DNA. Fully labeled, half-labeled, and unlabeled molecules expected on the assumption of semi-conservative DNA replication have been satisfactorily resolved in cesium chloride density gradients (Meselson and Stahl, 1958). Exactly analogous results may be obtained with intact phage particles (Meselson and Weigle, 1961). When phages derived from bacteria growing in heavy medium are used to infect a culture growing in light unlabeled medium, the resultant lysate shows three peaks in a cesium chloride density gradient. The size and position of these peaks corresponds with that expected of phages containing conserved, semiconserved, and new DNA double helices.

Since the genotype of phages which have been subjected to this treatment may readily be determined by appropriate selective plating if suitable genetic markers are used, this technique is capable of answering the question: Do recombinant phages contain DNA derived from the parent particles or is there DNA synthesis after the infection which gives rise to the recombinants? These are the two contrasting models discussed above. Meselson and Weigle (1961) and Kellenberger *et al.* (1961) have clearly shown the occurrence of breakage and reunion, in the case of phage λ.

The genetic map of λ phage is short, about 15 map units. In a cross between, for example, λ++ and λ*cmi* all four types, parental (λ++, λ*cmi*) and recombinant (λ*c*+, λ+*mi*) can readily be assayed (see Fig. 1).

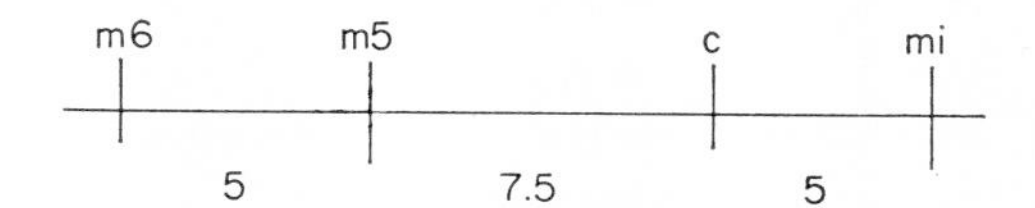

FIG. 1. Genetic map of bacteriophage λ.

If one of the parents carries an isotopic density label, the cross is performed in unlabeled medium and the progeny analyzed on a density gradient, the copy choice hypothesis predicts that only parental types should contain fully labeled material. However, the breakage-reunion hypothesis predicts, first, that labeled material should be found in recombinant phages, and second, that the amount found in different recombinant types should correspond to the parental contribution expected on the basis of the map positions of the particular markers used. This is precisely what Meselson and Weigle found. Figure 2 shows an analysis of the progeny of the cross [λ++ (heavy) $\times$ λcmi (light)]. The heavy parental type phage (++) appears in three peaks corresponding to heavy conserved, hybrid semiconserved, and light new progeny. The light parental type (*cmi*), on the other hand, is still found predominantly

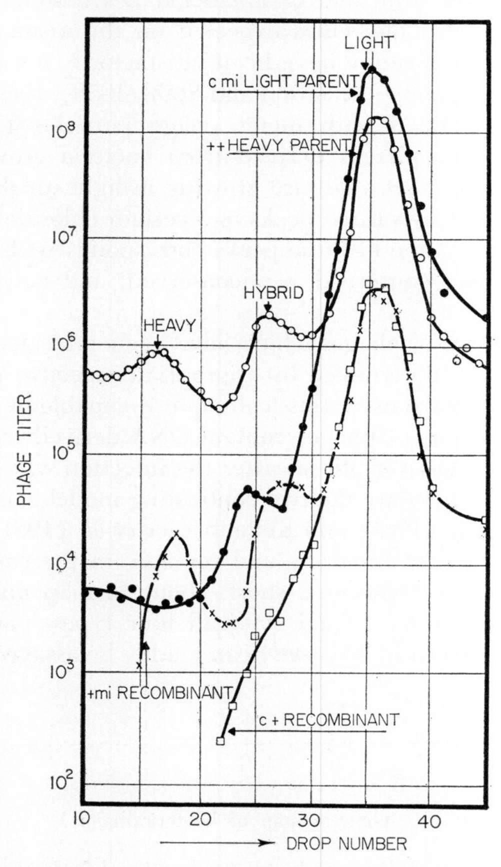

FIG. 2. Cross of λ++-$^{13}C^{15}N$ (heavy) $\times$ λcmi-$^{12}C^{14}N$ (light). Titer of the progeny phages of different genotypes in the drops collected after cesium chloride density-gradient centrifugation (Redrawn with permission from Meselson and Weigle, 1961.)

in the light region. One of the recombinant types, *c*+, does not appear in significant numbers outside the light fractions, while the other, +*mi* appears in two peaks close to the positions of heavy and semiconserved hybrid phage particles, as well as in the light position. If recombinants resulted from breakage and rejoining of parental strands, *+mi* should have received approximately 85% of a heavy strand and *c+* only 15%, which accounts for the very small numbers of these latter recombinants in the denser regions of the gradient. This interpretation was confirmed by the reciprocal cross in which the *λcmi* parent was heavy. The result obtained was exactly complementary, *c+* recombinants appeared near the heavy and hybrid regions while *+mi* recombinants did not appear outside the light region to any appreciable extent.

Comparable results were obtained in parallel experiments by Kellenberger *et al.* (1961), using mutants which have different densities as genetic markers and using radioactive phosphorus to follow the contribution of parental DNA to the recombinant types.

These important and elegant experiments, which are models on which many subsequent experiments on recombination at the molecular level have been based, show that genetic recombinants contain parental DNA and that the extent of the parental contribution is consistent with a breakage-reunion theory without appreciable involvement of DNA synthesis. Since the bulk of the recombinants does not contain DNA derived from the input phages (the light *cmi* recombinant peak is 100 times larger than either the heavy or the hybrid recombinant peak), other mechanisms of recombination cannot be excluded by these experiments. The light parental types are correspondingly numerous. The fact that labeled parental DNA appears, in appropriate crosses, in one recombinant type and not the other eliminates the possibility that the parental label is incorporated by breakdown and random resynthesis. The finding of a recombinant peak containing more than 50% parent heavy DNA also suggests that recombination may occur without appreciable replication. Under normal conditions at low multiplicities of infection, all molecules do replicate. Only when the infection occurs at a high multiplicity does an appreciable fraction of phage, whose DNA has not been replicated, appear. Further experiments (Meselson, 1964) involving crosses where both parents carried a density label confirmed that no appreciable amount of DNA synthesis is needed for the formation of recombinants. The recombinants were found at the density position of fully labeled DNA. Furthermore, to show that this association between DNA molecules of different origins is stable, phage from the heavy region were grown for one cycle in unlabeled bacteria at a low multiplicity of infection to ensure that no further recombination occurred. The distribution of hybrid and light particles among the progeny was the same for the recombinant as for the parental type.

IV. Bacterial Transformation

The process of DNA mediated bacterial transformation, originally described by Griffith in 1928 and later clarified in a classic paper by Avery *et al.* (1944), offers an excellent opportunity for studying the physical and chemical properties of DNA molecules in relationship to their biological activity. While the true genetic nature of the phenomenon was not at first apparent, the discovery of recombination between genetic markers (Ephrussi-Taylor, 1951; Hotchkiss and Marmur, 1954) established it as a system of genetic recombination analogous to bacterial conjugation and transduction, and phage recombination. The formal genetics of transformation has been considerably amplified in recent years by the demonstration of linkage between different markers in a variety of systems, especially in *Bacillus subtilis* (Nester and Lederberg, 1961; Nester *et al.*, 1963; Anagnostopoulos and Crawford, 1961; Ephrati-Elizur *et al.*, 1961; Goodgal and Herriott, 1957), the development of genetic fine structure mapping (Hotchkiss and Evans, 1958; Lacks, 1965; Sicard and Ephrussi-Taylor, 1965), and the correlation of maps constructed by transformation with those obtained in other ways (O'Sullivan and Sueoka, 1967; Dubnau *et al.*, 1967).

We shall review mainly those aspects of the experimental work on transformation which bear directly on an understanding of the molecular mechanism of recombination. The interpretation of the relevant experiments does, however, depend on an understanding of the basic features of the transformation process. Comprehensive reviews will be found in Ravin (1961), Hayes (1964), Schaeffer (1964), and Spizizen *et al.* (1966).

Transformation was first described in *Pneumococcus* and subsequently in a few other species of bacteria, notably, *Hemophilus influenzae* (Alexander and Leidy, 1951) and *Bacillus subtilis* (Spizizen, 1958). The ability of a culture to be transformed depends on a special physiological state called competence, which is empirically determined and still incompletely understood. The proportion of competent cells in a culture to be transformed is quite variable. It can approach 100% in *Pneumococcus*, but is probably in most cases less than 20% in *B. subtilis*. The routine procedures for preparing DNA (e.g., Marmur, 1961) break the chromosome, most probably at random, into some 30 to 100 fragments. The proportion of transformants for a particular marker increases linearly with the DNA concentration and the time of contact between DNA and bacteria, provided both are small. This suggests that a one-hit mechanism mediates association between donor DNA molecules carrying the relevant marker and the recipient bacteria (R. Thomas, 1955; Fox and Hotchkiss 1957). Most pairs of genetic markers chosen at random will be on different donor DNA molecules, since these are usually very small compared to the size of the total genome. When recipient cells having, for example, a double nutritional requirement corresponding to a pair of genes $x^- \ y^-$, are mixed with DNA from $x^+ \ y^+$ cells, double transformants ($x^+ \ y^+$) occur with a frequency which is approximately proportional to the product of the frequencies of the two singly transformed classes, $x^- \ y^+$ and $x^+ \ y^-$. The proportion of such double transformants thus varies quadratically with the DNA concentration, when this is low. Such pairs of markers are said to be unlinked. Occasional pairs of requirements are, however, found for which the frequency of double transformants is comparable to that of singles and also varies linearly with the DNA concentration. These pairs of genetic markers must be sufficiently close to each other on the bacterial chromosome to occur frequently on the same molecule, and so are said to be linked. The distinction between a linear and a quadratic response of double transformants with respect to DNA concentration is the main criterion for linkage in transformation (see Bodmer, 1967, for review).

The use of donor DNA prepared from cells grown in the presence of ^{32}P or other radioisotopes allows one to trace the physical fate of the donor DNA in the recipient cell. Thus, Lerman and Tolmach (1957) and others have found that the amount of DNA, as measured by ^{32}P counts, irreversibly bound to the recipient cell [that is not removed by a combination of extensive washing and deoxyribonuclease (DNase) treatment] is roughly proportional to the number of transformants. Under optimal conditions, the number of cell equivalents of DNA irreversibly bound is almost the same as the number of transformants, suggesting almost complete efficiency of the utilization of bound DNA for the production of transformants (see Goodgal and Herriott, 1957; Bodmer and Ganesan, 1964). The irreversible binding by recipient cells of ^{32}P counts from donor DNA demonstrates the physical incorporation of DNA into cells but says nothing about its fate after entry into the cell, and, in particular, about the nature of the association between donor and recipient DNA. Only a small fraction of the donor DNA is generally incorporated into the recipient cells, which is one of the most severe technical limitations to the use of transformation as a molecular probe for recombination mechanisms.

The first significant study of the fate of donor DNA following its incorporation into competent recipient cells was that of Fox and Hotchkiss (1960). Their approach, which is the basis of much of the subsequent work in this field, was to determine the donor activities in the DNA re-extracted from transformed cells at various times after the addition of donor DNA. This procedure is analogous to the study of phage maturation using premature lysis of the infected bacteria. The test scheme for such an uptake experiment is indicated in Fig. 3. A pair of linked genetic markers (a^+, a^- and b^+, b^-) is used in such a way that the donor, recipient, and recombinant transforming activities present in the re-extracted DNA can be selectively identified by using a^-, b^- cells as recipients for transformation with this material. Typically, a large excess of DNase is added at various times after the addition of donor DNA to recipient cells, in order to inactivate all of the donor DNA not incorporated. The recipient cells are then washed carefully to remove all traces of the degraded unincorporated DNA and the DNA prepared from these washed transformed cells is used to transform the a^-, b^- test cells.

Using drug resistance markers (rather than nutritional requirements) in *Pneumococcus,* Fox (1960) showed that donor activity was not initially recovered in the reisolated DNA and did not reach its maximum value until at least 5 to 6 minutes after the addition of donor DNA. Linkage between donor and recipient genetic markers, as determined by the appearance of a^+, b^+ transformants, was rapidly established (maximal level at 12 to 15 minutes) following this apparent eclipse. Voll and Goodgal (1961) in *Hemophilus* and Venema *et al.* (1965) in *B. subtilis* also showed the rapid establishment of linkage between donor and recipient markers. Inhibition of DNA synthesis during transformation did not apparently prevent this rapid integration process, suggesting that physical association of donor and recipient DNA's occurred without appreciable DNA synthesis. Attainable levels of inhibition are, however, rarely adequate to rule out preferential synthesis in the neighborhood of the integrated region. (Note that in our usage, "incorporation" refers to the absorption or entry of donor DNA with or into the recipient cells, while "integration" refers to the processes leading to the establishment of linkage between donor and recipient genetic markers and so the fixing of the transformed state.)

An initial attempt to determine more directly whether physical integration of donor DNA into the recipient genome accompanied genetic integration was made by Fox (1963) using the phenomenon of ^{32}P suicide. ^{32}P disintegrations cause breaks in the DNA backbone which inactivate transforming activity. Physical integration was suggested by the fact that a significant fraction of the donor activity still decayed even in extracts made after more than 60 minutes growth at 37°C following a 15-minute period of contact between cells and donor DNA.

More definitive evidence for the physical integration of donor DNA in the *B. subtilis, Pneumococcus,* and *Hemophilus* transformation systems comes from the use of heavy and radioactive isotopes as pioneered in recombination studies with phage [Szybalski (1961), Bodmer and Ganesan (1964), Fox and Allen (1964), Pene and Romig (1964), and Notani and Goodgal (1966)]. When donor DNA is prepared from cells grown in the presence of heavy isotopes (e.g., ^{13}N and ^{2}H) and carries a differential radioactive label (e.g., ^{3}H) from that of the recipient (e.g., ^{32}P), then donor and recipient DNA's can be physically separated in a CsCl density gradient by their difference in buoyant density (pycnography) and their constituent atoms can be identified by differential counting of the two isotopes using a scintillation counter.

	DONOR DNA	RECIPIENT CELLS
GENETIC MARKERS	a^+ b^-	a^- b^+
RADIOACTIVE LABEL	^{3}H	^{32}P (or ^{14}C)
DENSITY LABEL	^{15}N ^{2}H (HEAVY)	^{14}N ^{1}H (LIGHT)

FIG. 3. General scheme for uptake experiments to determine the fate of donor DNA in transformation. a+, a− and b+, b− refer to pairs of linked genetic markers (usually nutritional requirements), of which a+ and b+ can be selectively identified. Using a− b− cells as recipients for transformation by reextracted DNA: a− b+ transformants are mainly derived from DNA from the a− b+ recipient cells; a+ b− transformants are derived from the original donor, and possibly recombinant DNA; a+ b+ transformants must come from recombinant molecules.

Consider, for example, the experiments of Bodmer and Ganesan with *B. subtilis,* which used the labeling scheme shown in Fig. 3. Following 30 minutes' contact between donor DNA and recipient competent cells, an excess of DNase was added to stop transformation, the cells were washed extensively to remove donor DNA not taken up, and a DNA extract was prepared from the resulting transformed cells. The fractionation of the untreated, sheared (to reduce the molecular weight by about one half), and heat-denatured extracts in cesium chloride density gradients is illustrated in Fig. 4. Donor DNA is distinguished from the recipient by its heavier density ($\rho = 1.753$ as compared with 1.703) and by ^{3}H as opposed to ^{32}P counts. The expected position of fully heavy DNA is indicated by H. In parallel control experiments using light instead of heavy donor DNA, no density differences between ^{3}H and ^{32}P counts were observed. This shows that the density differences shown in Fig. 4

are due solely to the incorporated heavy isotopes and not to any conformational peculiarities of donor-recipient DNA complexes. The data illustrate the following points:

1. The donor DNA in the untreated sample bands almost in the

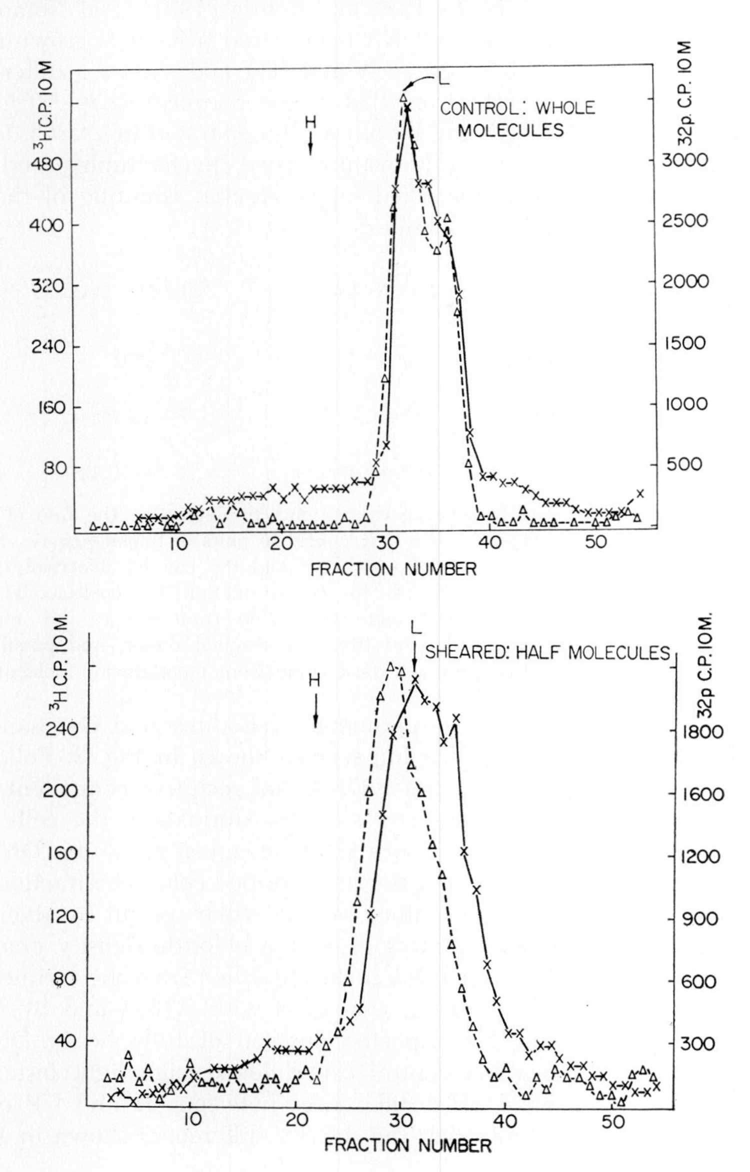

same position as the recipient DNA. This can only occur if the proportion of donor DNA in the donor-recipient complex is small (see Fig. 5).

2. Shearing to reduce the molecular weight by about one half results in a relative increase in the density of those molecules which carry donor DNA. Since the specific activity of the recipient (^{32}P) must be much less than that of the donor (^{3}H), in order for counts from the former not to swamp the latter, the recipient DNA associated with donor-carrying molecules remains undetected. No DNA with the same density as the original donor is produced. The observed density shift implies that the material from the donor DNA associated with the recipient molecule is not uniformly dispersed throughout those molecules but rather that it remains concentrated in a few discrete regions. The proportion of donor DNA in any molecule was small, so that the number of donor regions per molecule must also have been small. If the average number is less than 1, shearing in half should approximately double the density difference between donor-carrying and recipient molecules (see Fig. 5); otherwise, shearing would have little or no effect.

3. Heat denaturation, rendering the molecules single-stranded, also

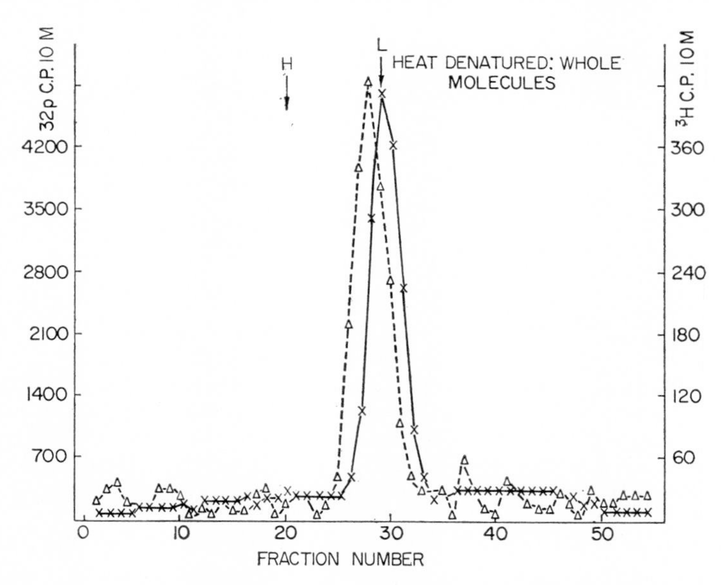

FIG. 4. A. Cesium chloride density-gradient analysis of DNA extracted from transformed *Bacillus subtilis* cells. The labeling scheme is shown in Fig. 3. Heavy and light positions in the gradient are marked H and L, respectively. - - - - - △ ^{3}H; ——x—— ^{32}P. (From Bodmer and Ganesan, 1964.) B. As in A, but the DNA has been exposed to mechanical shear. C. As in A, but heat-denatured.

causes a relative shift in the density of donor-carrying molecules and produces no material with the original donor density. This shows that the bonds linking donor and recipient are resistant to heat denaturation and so are probably covalent. As indicated in Fig. 5, the simplest interpreta-

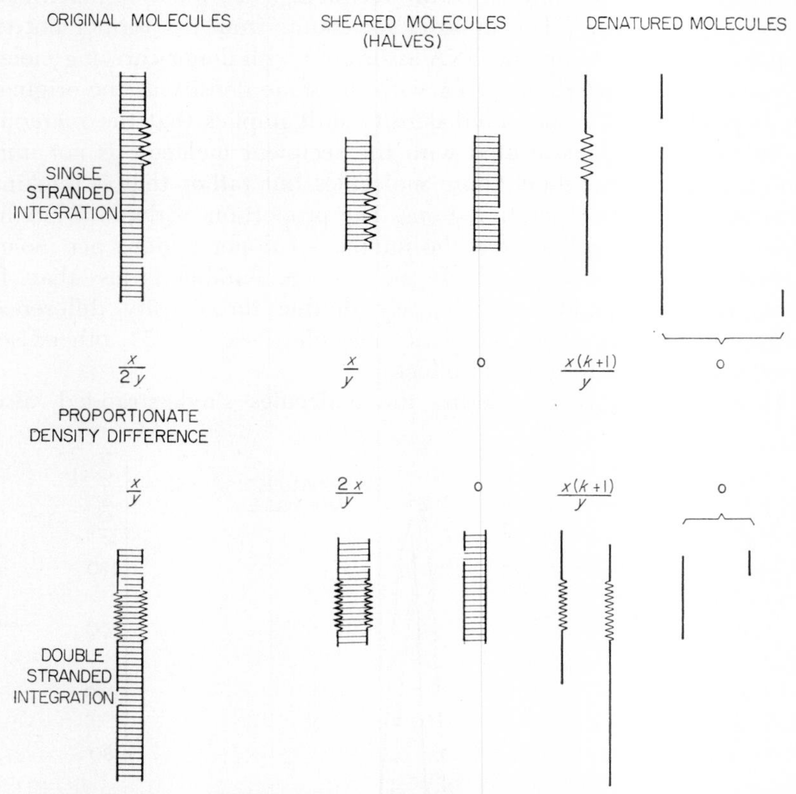

FIG. 5. Illustration of the effects of shearing and denaturation on the density of complexes of heavy donor and light recipient DNA formed during transformation. ———— recipient DNA (average length y); ∿∿∿∿ donor DNA (average inserted length x); k is the mean number of preexisting single-strand breaks. The possibility of breaks occurring within the donor segment is ignored.

tion of the density shift is that donor DNA is integrated into only one of the two recipient strands. This interpretation is, unfortunately, complicated by the occurrence of single-strand breaks in "normal" DNA preparations. These mimic the effects of shearing and so lead to an expected increase in the relative density of donor-carrying fragments after denaturation even if integration is into both strands. The magnitude of the observed density shift in relation to the estimated number of single-strand breaks was, however, large enough to suggest integration of only one strand of the donor DNA.

Similar experiments by Fox and Allen (1964), using *Pneumococcus*, gave essentially the same results, showing, in addition, that fairly extensive sonication of the donor-recipient complex produces at most material of hybrid density and none with donor density, indicating more directly that integration is single stranded.

There are at least four other lines of evidence which support the hypothesis of single-stranded integration.

1. Notani and Goodgal (1966) have shown that the density of the donor-recipient complex obtained using a heavy donor DNA and a light recipient culture is not changed after one generation's growth in light medium following transformation. If integration were double stranded, the density differential of the complex should be reduced by one half after one replication in a normal light medium.

2. Following heat treatment of a transformed pneumococcal culture to 5% survival, Guerrini (quoted by Fox, 1966) found that 36 out of 44 transformant colonies contained cells of the parental type, whereas colonies isolated after one or more cell doublings were predominantly pure. The heat treatment was enough to more or less exclude the survival of more than one nucleus per cell. These experiments thus suggest a hybrid structure for the transformants, at least with respect to the genetic information.

3. Efficient transformation by single-stranded DNA has been demonstrated in *Hemophilus* by Postel and Goodgal (1966, 1967) and Goodgal and Postel (1967). They also showed that the mechanism of integration seemed to be the same as for double-stranded DNA. However, this is not conclusive evidence that normal integration occurs in this way. Roger *et al.* (1966) have presented suggestive evidence that both sense and nonsense single strands are capable of giving rise to transformants, based on the physical separation of the two types of strands. Chilton (1967) has demonstrated transformation by both complementary strands in *B. subtilis* single-stranded transformation.

4. Lacks (1962) and Lacks *et al.* (1967) have provided clear-cut evidence in *Pneumococcus* that ^{32}P counts from added donor DNA are found predominantly in single-stranded DNA early after addition of donor DNA to the recipient culture. At later times these counts gradually move over into double-stranded material. On the basis of these results Lacks suggested that donor DNA is rendered single-stranded as it enters its cell, so that, in agreement with Fox's observation (1960) of an eclipse

immediately following DNA uptake, no donor-transforming activity is recovered from inside the cell until the incorporated single strand has been integrated into the recipient genome. Evidence for some denatured DNA at early times after the addition of donor DNA was also found by Fox and Allen (1964) and Bodmer and Ganesan (1964). However, in *B. subtilis,* Bodmer and Ganesan (1964), Bodmer and Laird (1968), and Notani and Goodgal (1966) in *Hemophilus* found native DNA clearly associated with washed transformed cells at early times after DNA addition, apparently ruling out Lack's hypothesis in their systems. The possibility that the single-stranded material found at early times is a product of the integration process or the extensive DNase treatment used to terminate transformation, rather than a precursor, has not been ruled out.

A direct estimate of the size of the integrated region (x) can be obtained from a knowledge of the donor-recipient density difference of untreated molecules ($\rho - \rho_1$) and the average size (y) of the isolated molecules. Thus from Fig. 5, assuming single-stranded integration:

$$x = 2y \frac{\rho - \rho_1}{\rho_2 - \rho_1}$$

where ρ is the observed density of reisolated material; ρ_1 is the density of light DNA; ρ_2 is the density of heavy DNA; y, $\rho - \rho_1$, and $\rho_2 - \rho_1$ can be determined experimentally and x can be calculated. Bodmer and Ganesan (1964) gave an estimate of 5.6×10^6 molecular weight or 8.4×10^3 nucleotide pairs for the size of the integrated region, while Bodmer (1966), using somewhat more accurate data based on sizes and densities of single strands, gave an estimate of 4.9×10^5 single-stranded molecular weight corresponding to only 1500 nucleotide pairs. Fox and Allen's data (1964) for *Pneumococcus* suggest a size of 4×10^6–10^7 molecular weight (6–15 x 10^3 nucleotide pairs) for the integrated region, while Fox (1966) quotes a figure of 2000 to 3000 nucleotide pairs. Notani and Goodgal (1966) obtained an estimate for *Hemophilus* which is about 1.2×10^7 molecular weight (1.8×10^4 nucleotide pairs). The wide range of these various estimates may reflect size differences between the donor DNA's which were used. Single-strand breaks have been shown to inhibit transformation and cotransfer of linked markers (Bodmer, 1966; Strauss *et al.*, 1966), suggesting that only the regions between the breaks on any given single strand can be integrated. On the assumption that the aromatic linkage group (Nester *et al.*, 1963) covered some 10 to 15 cistrons with an average length of 10^3 nucleotide pairs, Bodmer and Ganesan (1964) suggested that a DNA giving 50% cotransfers for these markers should lead, on the average, to an integrated region of 5–8 x 10^3 nucleotide pairs. Ephrussi-Taylor and Gray (1966) suggest, on the basis of genetic fine structure studies and assuming that their lowest recombination fraction of 1.0 x 10^{-4} corresponds to recombination between adjacent nucleotide pairs, that the mean length of a donor region corresponds to a single-stranded molecular weight of 3 x 10^5 or to only 450 nucleotide pairs. Lacks (1965) has also given an estimate of the size of the donor fragment, based on fine structure recombination data, of 2000 nucleotide pairs. It seems likely that the direct estimates which range from 1.5 x 10^3 to 1.5 x 10^4 nucleotide pairs are more reliable than estimates based on a recombination data since there is no means of correlating physical and genetic distance, and also measurements based on target theory (Litt *et al.*, 1958; Lerman and Tolmach, 1959).

However, Ephrussi-Taylor and Gray point out that the physicochemical data are insufficient to distinguish between integration of an uninterrupted single strand and multiple crossovers in a region of integration. They suggest that 20 crossovers per donor DNA molecule would account for the high frequency of recombination which they observe between markers which may map in adjacent nucleotides in *Pneumococcus*. However, in *B. subtilis* evidence from crosses involving several linked markers indicates that multiple crossovers do occur within a single site of integration, but they are much less frequent than Ephrussi-Taylor and Gray's hypothesis would predict (Darlington and Bodmer, 1968).

The uptake experiments discussed above, while establishing physical integration of donor DNA into the transformant genome, leave open the question of the extent to which DNA synthesis may be involved in the integration process. Inhibition of DNA synthesis in *Pneumococcus* by 5-fluorouracil (Fox and Hotchkiss, 1960) and thymine deprivation of a thymine-requiring strain of *B. subtilis* (Farmer and Rothman, 1965) indicate that integration can take place in the absence of appreciable overall DNA synthesis in the competent culture. This does not, however, rule out the possibility of relatively extensive DNA synthesis associated specifically with the region of integration. The involvement of DNA synthesis during uptake and integration was investigated by Bodmer (1965), using a competent nonleaky, thymine-requiring mutant which was transferred to a medium containing tritiated 5-bromouracil (5–BU) at the time of addition of donor DNA. This allows the selective identification, in a cesium chloride density gradient, of DNA which was synthesized during the transformation process. The density of donor and recombinant DNA isolated from such transformed cells is then a direct measure of its association with DNA newly synthesized during transformation. The results of such an experiment are shown in Fig. 6 (see also Bodmer and Laird, 1968). The ^{3}H counts in the hybrid region (mainly fractions 20–27) identify the DNA of the recipient culture which was replicated following the addition of donor DNA before ter-

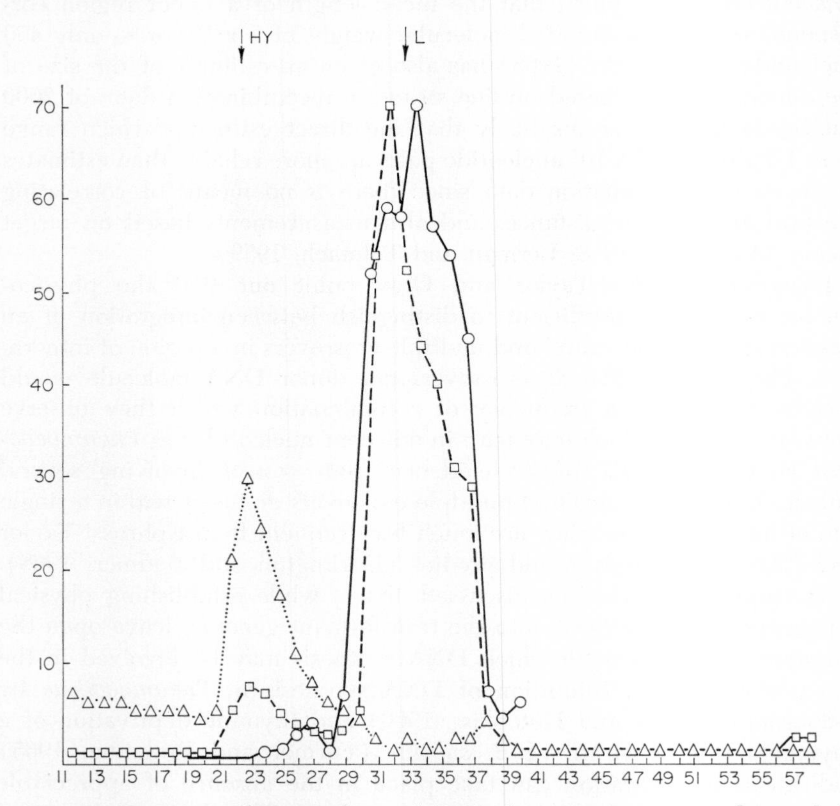

FIG. 6. Density gradient analysis of DNA from a competent culture of *B. subtilis* to which ^{3}H-labeled bromouracil (△ . . . . △) has been added at the time of transformation. DNA of the recipient culture (□ – – – □) is labeled with C^{14} and the donor marker (○——○) is try+. Hybrid and light positions in the gradient are marked HY and L, respectively.

mination of transformation (approximately 5%). Less than 5% of the try_2^+ transformants representing donor and recombinant activity, are in this hybrid position. This provides direct evidence that extensive DNA synthesis is not needed for the integration of donor DNA and is, perhaps, one of the most direct demonstrations that copy choice as originally proposed is not a part of the recombination process, at least in transformation. An amount of DNA synthesis corresponding to about 10% of the length of the average isolated molecule cannot be excluded.

The possibility that integration during transformation involved repair DNA synthesis along the gaps within a recombinant region which are not covered by the incoming donor DNA fragment, was specifically suggested by Bodmer and Ganesan (1964) and Fox and Allen (1964). A UV radiation-sensitive and mitomycin C-sensitive mutant of *B. subtilis* which also gives greatly reduced transformation levels, and seems to be analogous to the recombination deficient mutants of *E. coli* (see below), has been described by Okubo and Romig (1966). This finding strongly supports the suggestion that repair DNA synthesis is a feature of integration during transformation. The extent of synthesis expected may be quite small, and so well within the range tolerated by the data on replication during transformation. However, the possibility that repair synthesis may be detected by a direct approach, using heavy density labeling, needs to be further investigated.

Lacks and Hotchkiss (1960) and Iyer and Ravin (1962) reported considerable differences in the relative frequencies with which different genetic markers are integrated in *Pneumococcus*, and this phenomenon has since been studied in detail (see Ephrussi-Taylor and Gray, 1966; Lacks, 1965; Chen and Ravin, 1966b). In all cases mutations with widely differing integration efficiencies are found within a single cistron, and both Lacks and Ephrussi-Taylor and Gray suggest that low efficiency may be associated with a tendency for the donor or recipient DNA sequence associated with some genetic markers to be preferentially excised and the resulting gaps filled by repair DNA synthesis. This is exactly the mechanism originally suggested by Holliday (1962, 1964) for gene conversion and subsequently shown to be the basis for the removal of thymine dimers from UV-irradiated DNA (Setlow and Carrier, 1964). A similar mechanism has been proposed by Hogness (1966) to account for anomalous results of DNA-mediated infection experiments using hybrid DNA with different genetic markers on the two strands in the λ phage system. If the excision is at all extensive, neighboring markers on the same strand as that carrying the excised marker may be affected, resulting in an apparent partial suppression of their transforming activity: Ephrussi-Taylor and Gray observed exactly the same phenomenon in pneumococcal transformation and it could account for anomalous marker effects reported by Ravin and Iyer (1962). Lacks (1965) found four distinct efficiency classes, which he related to the base pair transitions that can occur at a single nucleotide pair site, but this hypothesis is apparently not borne out by Ephrussi-Taylor's data. So far no such large differences in marker integration efficiency have been reported in *B. subtilis.*

The major molecular features of integration during transformation seem to be reasonably well established, at least in outline. Single-stranded portions of intact donor DNA replace approximately homologous sections of the recipient DNA. Repair DNA synthesis fills in the gaps between the ends of the donor fragments and the recipient strand following which a ligase (Gellert, 1967; Weiss and Richardson, 1967) enzyme forms the

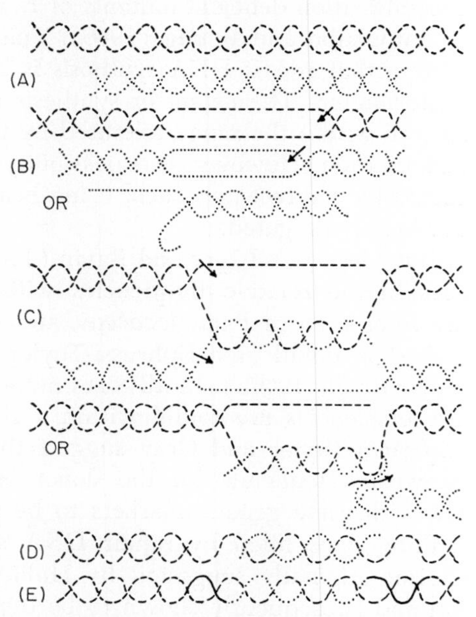

FIG. 7. Schematic model of integration of donor DNA during bacterial transformation. — — — — recipient; . . . . donor; ———— repair.

phosphodiester bonds needed to link donor and recipient strands. It is evident that the precision of the insertion event must depend on base pairing, and the relative inefficiency of transformation crosses between different bacterial species is evidence for this (Schaeffer, 1956; Chen and Ravin, 1966a). Ravin and Chen (1967) have further shown that linkage between markers is reduced in heterospecific as compared with homospecific transformation, probably reflecting the smaller average size of the integrated region due to impaired homology.

A schematic model for integration based on these ideas is shown in Fig. 7. An obvious possible modification involves the use of only one isolated strand for integration, as suggested by Lacks (1962). The molecular basis for most of these events now seems fairly well established *in vitro* and will be discussed at greater length below.

Two major questions remain essentially unsolved. First, what is the mechanism by which the relatively enormous donor DNA molecule attaches to the bacterial cell and maneuvers its way relatively close to the correct position of the recipient genome? Second, what physico-chemical models can be constructed to account for the partial unwinding of DNA, the exchanges of base-pairing partners, and the controlled breakage of the parental DNA strands?

An answer to the first question may rest partly on an understanding of the physiology of the competent state. A biosynthetic latency of competent *B. subtilis* cells, reflected in their increased resistance to penicillin, was demonstrated by Nester and Stocker (1963) and Nester (1964). A parallel latency in DNA synthesis, resulting in delayed replication of a donor genetic marker, was shown by Bodmer (1965) and Bodmer and Laird (1968). When a competent culture of *B. subtilis* is grown in 5-BU following transformation, donor DNA when first replicated appears at an anomalous density intermediate between hybrid and light. This was originally interpreted to indicate that integration of donor DNA occurred at a preexisting stationary replicating point of DNA synthesis. The discovery that the replicating point is membrane associated (Ganesan and Lederberg, 1965) suggested that it might be in a position in association with the bacterial surface which makes it readily accessible for contact with donor DNA in solution around the cell. However, an initial attempt to verify this hypothesis directly, by seeking for donor markers integrated into a replicating point labeled by density, gave negative results (Laird, 1966; Bodmer and Laird, 1968). Plausible alternative explanations for the anomalous intermediate density material are that DNA synthesis is precociously initiated at an integrated region, or that DNA synthesis is temporarily suspended when it arrives at the integration region. An early DNase-sensitive stage for the donor DNA in the recipient cell was suggested by R. Thomas (1955) and has been clearly demonstrated more recently by Laird *et al.* (1968; see Bodmer and Laird, 1968). Perhaps regions of the recipient genome which are being transcribed are readily available for integration of donor DNA.

The possibility that some recombination in higher organisms occurs at a chromosomal replication point was suggested by Bodmer (1965) on the basis of the results with transformation. This has also been discussed by Grell and Chandley (1965) for *Drosophila* and Holliday (1965) for *Ustilago maydis*. It should be emphasized that single-stranded integration is not in conflict with evidence from phage, and presumably higher organisms, which suggests recombination involves breakage and joining of double-stranded molecules. The amount of donor DNA inserted in a transformation event is small, and Bodmer and Ganesan (1964) and Meselson (1964) suggest that events occurring during transformation may be analogous to those taking place at a point of exchange in other systems where there may be extensive regions of hybrid material.

V. Bacterial Conjugation, Transduction, and the Attachment and Release of Prophages

A. Bacterial Conjugation

When cultures of appropriate strains of *E. coli* are mixed, a part of the genome of cells of one strain, the male donor Hfr (high frequency of recombination), is transferred linearly into the cells of the other strain (F^-), the female recipient (see Jacob and Wollman, 1961). Following this process, recombination may occur between the resident and the introduced genetic material, resulting in the formation of stable recombinants. Siddiqi (1963) obtained evidence that recombination during conjugation occurred by breakage and reunion. He showed that ^3H label from the male donor DNA was incorporated into recombinant exconjugants. Bresler and Lanzov (1967) have done similar experiments using ^{32}P suicide of donor DNA in recombinants. Oppenheim and Riley (1966) have investigated the fate of ^3H-labeled Hfr DNA injected into F^- cells grown in density-labeled medium. They demonstrated an association between the resident and the incoming DNA, using the technique of density-gradient centrifugation on the DNA of the mated culture. Tritium label was found in the density position of hybrid and heavy DNA. Since ^3H label appeared in the heavy region, some of the incoming DNA must have become associated with unreplicated recipient DNA. They found that the association between donor and recipient material was disrupted by denaturation, suggesting that few, if any, covalent bonds were formed. Recently it has been shown that covalent bonds are formed if the culture is incubated for a long enough time for some replication to occur after conjugation (Oppenheim and Riley, 1967).

B. Transduction

Transfer of genetic markers from one bacterial strain to another may be mediated by phage infection, as shown with *Salmonella* and its phage

P22 by J. Lederberg *et al.* in 1951. This was called transduction (see Hayes, 1964, for review). The details of this process are not relevant to a discussion of recombination; it merely provides another means by which bacterial genetic material may be transferred between strains. Recombination may (or may not) subsequently occur between the recipient and the newly introduced genetic markers.

C. The Attachment and Release of Prophages

Many bacterial strains harbor bacteriophage genomes which are inactive but which can, on appropriate treatment, give rise to viable infective phage particles. Such strains are said to be lysogenic for the phage in question (Lwoff and Guttman, 1949). It has been shown in many cases that the phage genome, or prophage, is located at a specific point on the bacterial chromosome (E. M. Lederberg and Lederberg, 1953). Various models have been proposed to account for the process by which this integration takes place, such as the formation of a branched structure (Jacob and Wollman, 1961). Campbell (1962) suggested that integration occurs by a single recombinational event between a circular bacterial chromosome and a circularized phage genome. A single event thus gives rise to a phage genome linearly inserted into the circular bacterial genome. Franklin *et al.* (1965) showed that deletions in the region of an integrated prophage could extend into it for varying distances, thus confirming that insertion is linear.

Hershey *et al.* (1963) and MacHattie and Thomas (1964) showed that DNA molecules prepared from purified λ phage can be induced to form circles *in vitro*, presumably by base pairing of complementary single-stranded ends. This hypothesis was elegantly confirmed by Strack and Kaiser (1965), who showed that *in vitro* circle formation could be inhibited if the single-stranded cohesive sites at the ends were made double-stranded by treatment with *E. coli* DNA polymerase. Bode and Kaiser (1965) and Young and Sinsheimer (1964) (see also Vinograd and Lebowitz, 1966) showed that soon after phage λ injects its DNA into the bacteria, the DNA takes up the form of a covalently linked circle. These circles can be broken open by nucleases which cause single-strand breaks. An important requirement for phage replication and maturation must therefore be an enzyme activity which joins the final bond in the noncovalently linked circles to make them covalent. Such an enzyme has recently been demonstrated *in vitro* in *E. coli* extracts by Gellert (1967) and is probably an important part of the recombination process. Thus the mechanism by which prophage integration and release take place is a useful system for studying the molecular events associated with recombination.

Ptashne (1965) has shown that phage λ may be released from the prophage site without DNA replication. If density-labeled *E. coli* cells lysogenic for λ are infected with another phage, a few λ particles are released in the subsequent burst. Some of the released particles have the density of conserved phage, consistent with a breakage mechanism of detachment. Prell (1965) has obtained similar results using "host-controlled modification" to mark the phage DNA, and zygotic induction to introduce the phage genome into the host. Zygotic induction occurs when a phage genome, integrated into the chromosome of a male cell, is transferred during mating into a female cell sensitive to that phage. The phage multiplies and lyses the cell (see Hayes, 1964). The progeny after zygotic induction contained particles with the modification characteristics of the cell from which the prophage originated. This experiment, unlike that of Ptashne, could only show that at least one strand of the prophage DNA was incorporated into progeny phage, since semiconservatively replicated phage genomes may have the modification characteristics of the parent particle conserved on a parental DNA strand.

VI. Processes Leading to Recombination in Bacteriophages

Phage T4 undergoes much more recombination per chromosome than phage λ; the length of its linkage map is about 800 map units. The amount of recombination taking place makes this a suitable system for analyzing the intermediate structures formed in the process. When the infecting phage carries a radioactive label, the radioactivity is widely distributed among the progeny as might be expected if breakage and reunion were responsible for recombination. Kozinski (1961) analyzed DNA extracted from T4 infected cells at various times after infection, in order to study the nature of the recombinant DNA molecules. If *E. coli* cells, growing in a density-labeled medium, are infected with light phage labeled with ^{32}P, no radioactive material is found at the heavy position of the density gradient after 3 minutes' infection. After 17 minutes, however, almost all the ^{32}P is found in the heavy region. If this DNA from the heavy region is subjected to ultrasonic vibrations, which break it into small fragments, and again analyzed on a density gradient, the radioactive peak is moved to the position expected of hybrid (½ heavy and ½ light) molecules. This is consistent with the exchange of DNA fragments between phage DNA molecules before or after replication.

The average size of the parental contribution may be estimated by comparing the position of the peak of infective centers at the heavy position in the gradient and the position of the radioactivity peak (Kozinski and Kozinski, 1963). This gives a figure of 5–10%, which is slightly increased as the multiplicity of infection is increased. In order to determine the nature of the association between input and newly synthesized phage DNA's, the parental DNA peak was subjected to denaturation by heat and alkali. If the separated strands are made up of covalently linked contributions between parental and progeny DNA they should not separate when denatured and further analyzed in a density gradient. Kozinski *et al.* (1963) found no separation and therefore concluded that the joint between recombined molecules must be a covalent one. 5-FUdR (fluorodeoxyuridine), which inhibits the synthesis of DNA, did not prevent the fragmentation of parental DNA (Kozinski *et al.*, 1963). However, chloramphenicol (CAP), the main effect of which is to inhibit protein synthesis, does inhibit fragmentation if added sufficiently early after phage infection. DNA replication in the presence of chloramphenicol is reduced but not completely inhibited. If CAP is added after 5 minutes of infection, DNA synthesis continues at a reduced rate but no fragmentation is observed. This effect cannot be due to the reduced size of the DNA pool since FUdR, which reduces the DNA pool size more markedly, does not inhibit fragmentation. If CAP is added 9 minutes after infection or later, no inhibition of fragmentation is observed. These results suggest that phage-specified, or phage-induced, enzyme or enzyme system which catalyzes the breakage and rejoining of DNA molecules is synthesized under normal conditions beginning 6–7 minutes after infection. The formation of this enzyme, or enzymes, would therefore be inhibited if chloramphenicol were added sufficiently early after infection.

Tomizawa and Anraku (1964a,b) have done experiments on a very similar system but with DNA synthesis more rigorously inhibited by KCN. They used parental phages, one of which was labeled with ^{32}P and the other with the density label 5-BU incorporated in place of thymine. When KCN was added 5 minutes after infection, radioactivity was found in the denser regions of the gradient, indicating an association between the incoming parental DNA molecules, as found also in the absence of KCN by Kozinski and co-workers. This association is disrupted by hydrodynamic shear, just as Kozinski found that his complexes were disrupted by ultrasonic vibration. However, if the complexes formed in the presence of KCN are subjected to denaturation by heating, the labels from the different parents are observed to separate when analyzed in a density gradient. This indicates that no covalent association is formed between DNA molecules under these conditions. It seems likely that the KCN treatment is inhibiting the final repair of the joint between the two recombined DNA molecules, which depends on pairing between single strands of different origins. It is interesting to consider whether breakage of DNA molecules occurs before or after the association which eventually gives rise to recombinants. If random fragmentation is followed by association of fragments of different origins, then infection with a single phage particle in the presence of KCN, when no replication occurs, should often not give rise to any plaques when the infected bacteria are plated out after removal of KCN. It is likely that the fragments produced from a single phage particle could not efficiently reassort to form a complete chromosome. Tomizawa's results indicate such cells do give rise to plaques with a high efficiency.

Anraku and Tomizawa (1965a) showed that similar complexes are formed in the presence of 5-FUdR, as was found by Kozinski and Kozinski (1963). However, FUdR does not completely inhibit DNA synthesis. They were able to show that whereas at 15–20 minutes after infection in the presence of FUdR noncovalently linked complexes were formed, at a later time (30–45 minutes) there was already some covalent bond formation between the components of the complex. This was also observed in the few progeny phages that are produced in the presence of FUdR. Since FUdR limits availability of thymidine, inhibition by FUdR suggests that repair involves incorporation of new nucleotides rather than merely the joining of adjacent nucleotides.

Another means of examining complex formation in the absence of DNA synthesis is by using conditional lethal mutants which are defective in DNA synthesis. Anraku and Tomizawa (1965b) have used amber mutants of phage T4. Amber mutants are a class of suppressible mutants: They will not grow in *E. coli* strain B but will in strain CR63. The mutant site leads to polypeptide chain termination in the wild-type strain but is mistranslated as an amino acid in CR63 (see Brenner *et al.*, 1965). These suppressible mutants occur throughout the phage genome (see Epstein *et al.*, 1963). Anraku and Tomizawa examined the fate of DNA derived from pairs of differentially labeled phage carrying amber mutations using *E. coli* K which did not allow their growth. They compared two phage strains, one of which had a multiple block in DNA synthesis and steps leading to it, and the other having a defect in a single gene leading to the production of abnormally high levels of DNA polymerase and other early enzymes. They found that the noncovalently linked complexes were formed by the multiple mutant strain unable to synthesize DNA while covalently linked complexes were formed by the strain which was able to synthesize DNA.

Tomizawa *et al.* (1966) later found an amber mutant of phage T4, located in gene *32*, which prevents complex formation between differentially labeled parental DNA molecules. This mutant is also unable to synthesize DNA. Thus, although DNA synthesis itself is apparently not necessary for the formation of these joint complexes, recombination and DNA synthesis must have some steps in common. The nature of the defect in this mutant is not known. On infection of *E. coli* K it was found to induce normally the synthesis of DNA polymerase, deoxycytidinetriphosphatase, and deoxyribonuclease. They suggest that the function defective in this mutant may be needed for opening of double-standard DNA structures.

The experiments of Tomizawa and co-workers suggest models of recombination involving some association of homologous DNA molecules by base pairing in regions of hybrid overlap, followed by repair of single-stranded gaps and formation of covalent bonds, which are essentially the same as those already discussed for other systems. It should be emphasized that neither these experiments nor those of Kozinski and co-workers correlate the physical structures presumed to be intermediates in recombination with recombinant genetic activity. Only the experiments of Meselson and others on phage λ and the work on transformation, discussed previously, decisively link the physicochemical with the genetic information.

VII. Phage Heterozygosis and High Negative Interference

A. Phage Heterozygotes

Most bacteriophage particles carry only a single allele of a given genetic marker (i.e., are haploid). Occasionally, however, particles are found which carry different alleles of the same gene. A T4 particle carrying both the rII and rII^+ alleles may be detected by the formation of a mottled, or heterogeneous plaque following infection at a low multiplicity to ensure that no plaques are derived from more than one particle. By analogy with higher organisms these particles were called heterozygotes. Levinthal in 1954 observed that heterozygous particles of *E. coli* phage T2 were generally recombinant for markers situated on either side of the locus which was heterozygous. Since heterozygosity was found to extend over only a small fraction of the phage genome (only closely linked pairs of markers were found to be heterozygous in the same phage particle), it seemed probable that these regions represented a special structure associated with recombination. He suggested that the heterozygosity was between DNA strands in a region of hybrid overlap, at a point where rejoining of recombinant

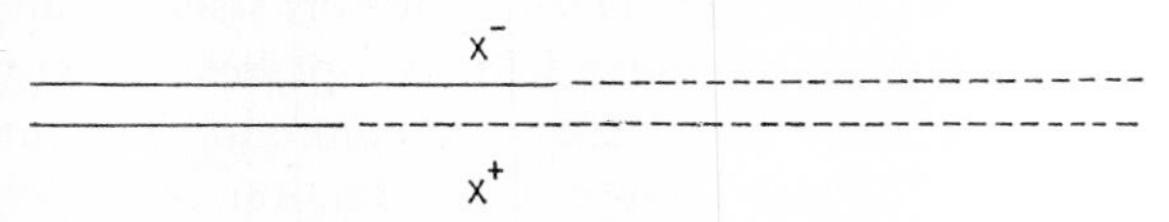

FIG. 8. Levinthal's model of phage heterozygotes.

strands had occurred. The subsequent literature on phage heterozygotes has been confused and it is now apparent that there are two kinds of heterozygotes in phage T2 and T4. Nomura and Benzer (1961) observed that the frequency of heterozygotes in T4 was lower in crosses involving deletions (0.4%) than in those involving point mutations (1.4%). This suggests that a certain type of heterozygote cannot be formed by a deletion mutant, a property that is expected of an internal overlap heterozygote. A possible explanation for heterozygotes involving deletions arose from the finding of Streisinger *et al.* (1964) that the linkage map of T4 phage is circular. Heterozygosity in these particles might, therefore, be due to terminally redundant regions. These two

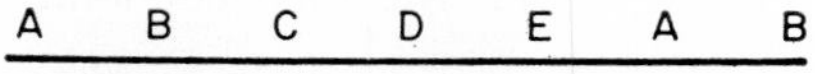

A B C D E A B

FIG. 9. Genetic structure of terminal redundancy heterozygote.

hypothetical types of heterozygotes should behave differently on replication. The redundant type should increase in frequency, whereas the overlap type should segregate out. Therefore, when DNA synthesis is inhibited the loss of overlap heterozygotes should be prevented and their number should increase. Inhibition of DNA synthesis should, on the other hand, have no effects on the terminally redundant heterozygotes. This prediction was tested on deletion and point mutant heterozygotes by Séchaud *et al.* (1965). They found that in T4-infected bacteria, under conditions of reduced DNA synthesis in the presence of 5-fluorodeoxyuridine (FUdR), the number of deletion heterozygotes did not increase with time following mixed infection, whereas the number of heterozygotes involving point mutations did increase. T4 particles vary in DNA content and hence in density. An obvious prediction of the terminal redundancy hypothesis is that large particles are more likely to be heterozygous for a given marker. This was confirmed by Doermann and

Boehner (1964) for some T4 heterozygotes using density-gradient centrifugation.

The circularity of the T2 phage chromosome and the general occurrence of terminal redundancy has recently been verified directly by Thomas and co-workers (C. A. Thomas, 1966; MacHattie *et al.*, 1967). They showed that T2 phage DNA can be made to form circles *in vitro* by DNA annealing coupled with partial degradation of one strand at each end of the molecule.

In bacteriophage λ heterozygotes are not formed by terminal redundancy. Kellenberger *et al.* (1962) showed that the density of λ heterozygotes was indistinguishable from that of normal phage particles. This phage is, therefore, more suitable than T2 or T4 for the study of supposed internal heterozygotes, although these occur at very low frequencies. However, Kellenberger *et al.* (1962) found that their frequency could be increased 100-fold by ultraviolet treatment of the infected bacteria. They confirmed Levinthal's (1954) and Trautners' observations (1958) that heterozygotes were frequently recombinant for outside markers. Sixty-two percent of heterozygotes were recombinant for a pair of outside markers separated by only 15 map units, from which they estimated the average length of the heterozygous region as 1.5 map units. Thus, this type of heterozygote (1) is frequently formed in association with one or more recombination events; (2) contains no more DNA than other phage particles; (3) is destroyed during DNA replication; and (4) involves only small regions of DNA. These properties strongly support a model of internal overlap derived by recombination of outside markers as originally proposed by Levinthal. However, this may not be the only structure that can give rise to heterozygotes. How do heterozygotes nonrecombinant for outside markers arise? Recently Berger (1965) has analyzed the genetic constitution of T4 heterozygotes formed when DNA synthesis in the infected bacteria is inhibited by FUdR, which, as mentioned before, increases the number of point mutant heterozygotes formed in a phage cross. He found that a much larger proportion of these induced heterozygotes (about 50%) were not recombinant for outside markers. The most plausible explanation of this finding is that under these conditions some heterozygote formation occurs by the incorporation of small single-stranded regions into the DNA, as originally suggested by Kellenberger *et al.* (1962).

It could be argued that these structures are artifacts, induced by FUdR and not occurring under normal conditions. However, if they did occur naturally at low frequency they would never be distinguishable from heterozygotes formed from DNA molecules which previously or subsequently underwent a second recombinational event in this region. It is quite feasible that such nonrecombinant structures would be selectively preserved in the presence of FUdR. They only involve breaks in one strand of the DNA and therefore might survive better than a structure involving breaks in both strands under conditions of reduced availability of thymidine, and presumably reduced DNA repair. These nonrecombinant phage heterozygotes are analogous to the gene conversion tetrads found in some fungi, which, also, are not associated with recombination of outside markers.

B. High Negative Interference

It has been observed in a number of organisms that recombination between two closely linked markers increases the probability of one or more further exchanges occurring very close to it (Pritchard, 1955; Chase and Doermann, 1958). Chase and Doermann, for example, using three factor crosses involving rII mutants of phage T4, found that over the very small distances involved within the rII region, the probability of a second exchange decreased as the map distance involved increased. Exactly analogous observations were made in *Aspergillus* by Pritchard (1955, 1960). High negative interference has been found wherever genetic techniques allow it to be detected.

Several hypotheses have been proposed to account for this phenomenon. Pritchard (1960) suggested that DNA molecules, or chromosomes, were only effectively paired during recombination over very small regions. Therefore, within an "effective pairing segment" several exchanges could occur and this would satisfactorily account for the observed clustering. Thus multiple exchanges may be a characteristic product of what is virtually a single recombination event. Holliday (1968) argues that true negative interference, an increase in the exchange frequency for closely linked markers, does not occur in fungi; all the observed results can be accounted for by gene conversion due to repair in the region of an exchange.

Edgar (1961) found that the number of recombinants produced in a cross between two closely linked T4 rII mutants was greatly increased if the progeny were allowed to grow for one cycle in *E. coli* B, which allows growth of rII mutants and wild types, and then the infected B cells spread on indicator plates of *E. coli* K. A comparable increase was not observed in a reconstruction experiment in which B cells mixedly infected with two rII mutants were spread on K. This suggests that a structure is formed in the original mixed infection which is not itself recombinant but is capable of giving rise to recombinants at a high frequency. These structures might be nonrecombinant heterozygotes of the type

$$\frac{r_1 \quad +}{+ \quad r_2},$$

which is the result of an overlap covering the entire region under consideration; they could equally be terminal redundancy heterozygotes. Edgar suggests models of how nonrecombinant heterozygotes may give rise to recombinants. However, an alternative explanation for these results cannot be ruled out: It is possible that, for spatial reasons, markers entering a cell on the same heterozygous DNA molecule have a higher probability of giving rise to recombinants than markers entering on two different strands, that is, mating pools derived from different particles do not completely mix.

The general theory that multiple recombinants between closely linked markers arise from events within a single overlap predicts that such recombinants should also be recombinant for outside markers, whether or not the number of exchanges selected within the overlap region was single or double. This was tested with rII mutants of phage T4 by Steinberg and Edgar (1962). They selected r^+ double recombinants from crosses of the type shown in Fig. 10 and examined them for the

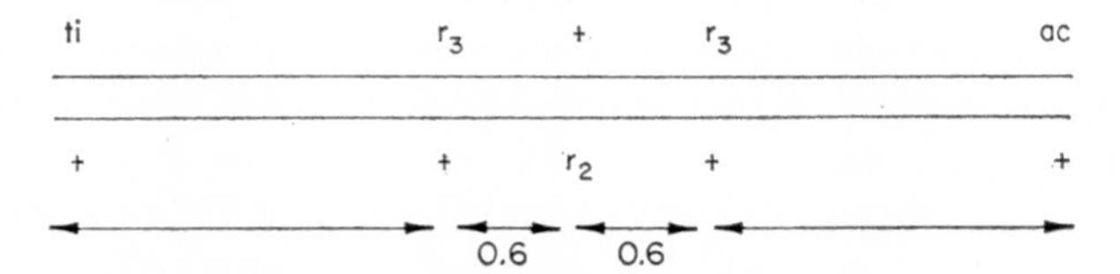

FIG. 10. Steinberg and Edgar's experiment.

markers ti and ac. These markers were found and largely in the parental configuration, which is that predicted by classic genetic theory. This is in apparent conflict with the model based on multiple exchanges only occurring within overlaps formed during the normal recombination process and is in conflict with Edgar's earlier data indicating that recombinants for closely linked markers arise from heterozygotes.

Steinberg and Edgar's data could be explained, however, if the rare double recombinants, unlike single recombinants, arise from the insertion of a small single-stranded piece of DNA, as proposed by Kellenberger *et al.* (1962) and by Berger (1965) for phage heterozygote formation in the presence of FUdR. This event would clearly preserve the linkage of outside markers and would account for high negative interference when two exchanges are selected. This model is analogous to the models which have been proposed to explain integration of donor DNA during bacterial transformation (see Fig. 7). However, it can only account for the increased number of triple exchanges observed over small regions, if multiple switches within regions of effective pairing are invoked, as proposed by Pritchard. The switches need only involve single-stranded regions where outside markers are not recombined.

VIII. Recombination-Deficient Mutants

The possibility that gene conversion in fungi (see Chapter V) might, at least in part, be explained by a recombination mechanism which involves removal and replacement of portions of DNA in a hybrid region (that is, repair) was first suggested by Holliday in 1962. Molecular models for recombination incorporating some repair DNA synthesis have been elaborated by Meselson (1964, 1966), Whitehouse (1963), and Holliday (1964). Boyce and Howard-Flanders (1964) and Setlow and Carrier (1964) suggested that the repair synthesis possibly occurring during recombination was analogous to that demonstrated to take place following recovery from ultraviolet (UV) irradiation. This was dramatically confirmed by Clark and Margulies (1965) who showed that mutants selected as deficient in recombination were also abnormally sensitive to ultraviolet irradiation. A considerable number of recombination-deficient mutants have now been described, mostly in *E. coli* (Clark and Margulies, 1965; Howard-Flanders and Theriot, 1966; van der Putte *et al.*, 1966), but also in *B. subtilis* (Okubo and Romig, 1966) and other species. Several means of selection have been used to isolate these mutants. Clark and Margulies (1965) used a direct method. They looked for F^- colonies which failed to give recombinants when inoculated onto a lawn of Hfr cells on a medium on which only recombinants would grow. van der Putte *et al.* (1966) selected recombination-deficient strains by their sensitivity to UV light. Sensitivity to other agents which damage DNA such as mitomycin C (Okubo and Romig, 1966) and X rays (Howard-Flanders and Theriot, 1966) has also been used though only a fraction of mutants sensitive to such agents are defective in genetic recombination.

Bacteria and other organisms are killed by exposure to ultraviolet light. The action spectrum of this effect, with a peak at a wavelength about 2600Å which is close to the absorption spectrum of thymine, and the fact that UV is mutagenic, strongly suggests this is a specific effect on DNA. Wacker *et al.* (1960) and Beukers and Berends (1960) showed that ultraviolet light induced the formation of dimers between adjacent pyrimidine residues on the same strand of a DNA molecule. Setlow and Setlow (1962) later showed that the loss of transforming activity of UV-treated DNA was correlated with the formation of thymine dimers. Though there may be other lesions induced by ultraviolet light, it has been shown (see Witkin, 1966) that pyrimidine dimers are the most numerous. Mechanisms exist which reverse ultraviolet damage. The killing effect of ultraviolet on *E. coli* is much reduced by post irradiation treatment with visible light, a phenomenon known as photoreactivation. Rupert (1962) showed that UV-irradiated transforming DNA could be reactivated by an extract of yeast or *E. coli* in the presence of light. The yeast enzyme splits thymine dimers *in situ* in the presence of light, thus restoring the structure and biological activity of the DNA (Cook, 1967).

This is not the only mechanism by which microorganisms are able to repair lesions in their genetic material. Setlow and Carrier (1964) and Boyce and Howard-Flanders (1964) showed that if a culture of *E. coli* previously grown in the presence of tritiated thymidine is irradiated with UV light and subsequently incubated in the dark, some radioactive material is made acid-soluble. This acid-soluble material was shown to consist of oligonucleotides containing thymine dimers. The time at which virtually all thymine dimers had been removed was observed to be approximately that at which observable DNA synthesis resumed after inhibition by the ultraviolet light treatment. Thus there must be an enzymic mechanism which recognizes thymine dimers and removes them from the DNA in the dark. Setlow and Carrier also showed that an UV-sensitive strain of *E. coli* was unable to excise thymine dimers. Since DNA molecules cannot contain extensive gaps, we must suppose that the gaps formed by excision are repaired. There is now good evidence for such repair processes. Pettijohn and Hanawalt (1964a,b) have elegantly analyzed the incorporation of new material into the DNA of a wild-type bacterial culture following ultraviolet irradiation. They found that DNA isolated from unirradiated cells after a short period of growth in 5-bromouracil (5-BU) was largely at the light and hybrid positions in a cesium chloride density gradient, though a significant amount of material was found at intermediate positions. This was separated by sonication into fragments which banded at the hybrid and light positions. If, on the other hand, a culture was exposed to 5-BU following UV irradiation, bromouracil was also incorporated into DNA but breakage of the DNA by sonication did not produce any material of predominantly hybrid density. This is consistent with the incorporation of 5-BU at a large number of small widely distributed sites. Presumably similar processes are involved in the repair of gaps after genetic recombination but this has not yet been directly demonstrated.

The precise nature of the process defective in recombination-deficient mutants is at present obscure. Howard-Flanders and Theriot (1966) showed that their mutants were able to excise thymine dimers from DNA and, furthermore, that incubation after exposure to ultraviolet light caused extensive breakdown of DNA in most of these strains. This also occurs after UV irradiation, though to a lesser extent, in the wild type. However, Howard-Flanders and Theriot (1966) estimated that, whereas 200–500 nucleotides are excised per dimer in the wild type, rec^- strains degraded 20–40 times as much as this. Clark *et al.* (1966) have tested one of these mutants for all known exo- and endonuclease activities, DNA phosphatase, and DNA polymerase, but found no significant differences from the wild type. It is likely that some process which limits DNA breakdown in the wild-type strain, or is concerned with the initiation of resynthesis, is defective in these strains. These hypotheses would satisfactorily account for recombination deficiency, since the breakage necessary for recombination would lead to lethal damage in these strains and recombinant molecules would not be formed.

IX. The Enzymology of DNA Breakage and Repair

All the available evidence from microorganisms indicates that recombination at the level of the DNA molecule takes place by some combination of breakage of DNA strands, homologous base pairing, repair of single-strand gaps, and final formation of covalent bonds between recombination fragments (see Fig. 11). Each of these processes (or analogs of them) has now been observed *in vitro*.

Enzymes that break DNA strands, the deoxyribonucleases, have been known for a long time, though their function in the living cell has remained obscure. DNases are broadly classified into two categories: the exonucleases, which attack the molecule at a free end and degrade it one nucleotide at a time, and the endonucleases, which break bonds internal to the molecule. A wide variety of specificities is found among both exo- and endonucleases, even within a single cell (see Lehman, 1963). Endonucleases may tend to break one or both DNA strands, may have a preference for either native or denatured DNA, and in addition may attack preferentially certain sequences of DNA. The breakage of DNA strands prior to their recombination must be a highly organized process. If the model shown in Fig. 11 applies, the position of each break relative to that in the complementary strand, and to those in the homologous double helix, is very important. If the breaks in complementary strands are too close, base pairing will fail to hold the molecule together: An irreparable double-strand break will result. If they are too far apart it may be impossible to open up the molecule to allow exchange to take

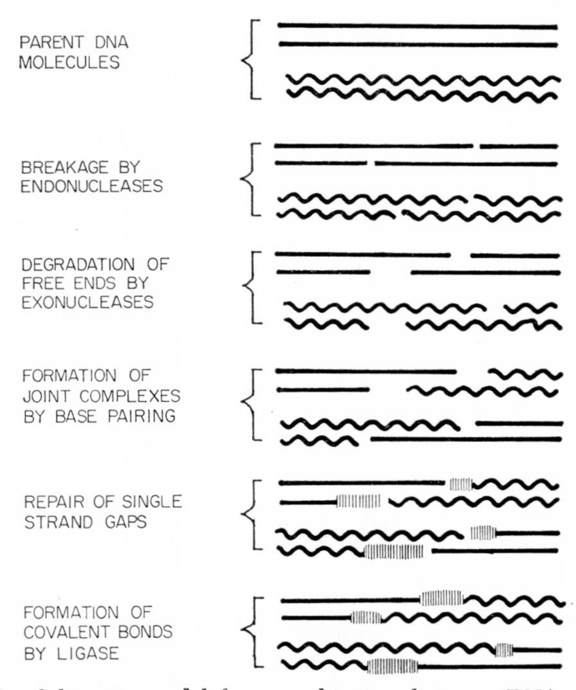

FIG. 11. Schematic model for recombination between DNA molecules.

place. It is conceivable that an enzyme complex is able to make all four breaks simultaneously, or in some way to measure accurately the distance between successive breaks. This kind of mechanism might account for the opening up of phage λ circular molecules, though a set of sequence-specific nucleases could be responsible for making the necessary breaks.

Following the formation of single strand breaks by endonucleases, exonucleases will mediate the excision of single-stranded sections of the DNA in the neighborhood of a break. The key example of such an enzyme is the *E. coli* exonuclease III described by Richardson *et al.* (1964b). This is the enzyme that was used by MacHattie *et al.* (1967) to form circles of T4 phage DNA and which forms the basis for the demonstration of *in vitro* repair (Richardson *et al.*, 1964a). It is clear that partial degradation following breakage is only necessary if the breaks in the homologous DNA molecules are not exactly aligned. It is not known if they are or not, but the correlation between ability to repair ultraviolet damage and the ability to recombine strongly suggest that they are not.

Marmur and Lane (1960) first demonstrated the annealing of two single strands of DNA by appropriate heat treatment. This process has since been much studied (for review, see Marmur *et al.*, 1963) and clearly involves recognition of homologous base pairs with ultimate re-formation of the DNA double helix from two single strands, The *in vitro* annealing process, which is probably most closely analogous to joining of DNA molecules during recombination, is the formation of DNA circles of λ and T phage from linear DNA molecules by base pairing of overlapping, terminally redundant sequences as studied by Hershey, Kaiser, Thomas, and their colleagues. Recently, C. A. Thomas (1966) has calculated that the minimum length of polynucleotide needed for DNA molecules to recognize each other during the annealing process may be as little as 12 nucleotide pairs. He further speculates that if synapsis occurs among freely recombining DNA molecules, repetitious sequences longer than the minimum recognition length would "confuse" the synapsis of DNA molecules. Though little, or nothing, is really known concerning the mechanisms that align two DNA molecules for synapsis, it seems most unlikely that large portions of the genome are simultaneously available for the pairing process.

The *in vitro* repair of partially single-stranded DNA templates by DNA polymerase was demonstrated by Richardson *et al.* (1964a). They showed that a partially single-stranded DNA, prepared by limited digestion of each strand with *E. coli* exonuclease III, could be restored to its native fully double-stranded form by the action of *E. coli* DNA polymerase. In fact, there is the distinct possibility that the *E. coli* DNA polymerase as originally described by Kornberg and co-workers is the repair rather than the replication enzyme.

Recently enzymes have been discovered which will mend single-strand breaks in DNA (Gellert, 1967; Weiss and Richardson, 1967; Olivera and Lehman, 1967; Gefter *et al.*, 1967). Gellert (1967) showed, as mentioned above, that in normal *E. coli* extracts there is an enzyme which will convert noncovalent λ circles formed *in vitro* into covalent circles. Weiss and Richardson (1967), on the other hand, have purified from extracts of T4-infected *E. coli* an enzyme which will mend breaks inserted by the action of DNase I, which inserts single-stranded breaks at random. They have in addition shown that a T4 amber mutant which cannot synthesize DNA is deficient with respect to this enzyme.

The processes described above include all the ingredients needed for the *in vitro* formation of recombinant DNA molecules. The mechanism is almost the same as that originally proposed for formation of T4 phage heterozygotes by Levinthal in 1954.

X. Recombination in Higher Organisms

Although, as discussed in Section I, the formal analysis of genetic recombination in higher organisms has provided the background for the construction of models of the recombination process in microorganisms, very little direct evidence for molecular mechanisms of recombination is available from higher organisms. No convenient system has yet been discovered which readily allows the corrlelation between physicochemical and genetic recombinant structures to be studied. In fact, a major block to any study is the lack of any convincing evidence concerning the molecular structure of the chromosome. Two techniques, so far, have been useful: the use of inhibitors of metabolic activities and autoradiography. Thus, following his classic studies on the semiconservative replication of plant chromosomes, using autoradiography, Taylor (1965) has obtained suggestive evidence for breakage and reunion of semiconserved subunits during meiosis in the grasshopper. A major point of discussions concerning molecular mechanisms of recombination in higher organisms has been the possible involvement of DNA synthesis. Studies of the timing of DNA synthesis in relation to meiosis throw some light on this question. Most indicate that DNA synthesis is complete before chiasmata are visible, namely, during interphase or early prophase (Howard and Pelc, 1951; Taylor, 1957). Pritchard (1960) and others have, however, argued that cytologically visible chiasmata may be reflections of a recombination event which took place much earlier and which, if it only occurs in short effective pairing regions, might be almost impossible to visualize. Bodmer (1965), as discussed above, Holliday (1965), and Grell (1965) have argued that recombination might actually take place at a growing point of DNA synthesis, this being the analog of the effective pairing region proposed by Pritchard. Thus, Grell and Chandley (1965) showed that heat shocks during times corresponding to the period of DNA synthesis increased recombination frequencies in *Drosophila.* Holliday (1965) similarly, showed that mitotic crossing-over in *Ustilago* was only stimulated by ultraviolet if irradiation was carried out during the period of

DNA synthesis. In addition he showed that the further a marker was from the centromere the earlier during the period of DNA synthesis did UV irradiation stimulate crossing-over in regions proximal to that marker. This was explained by a sequential association of DNA synthesis with recombination, as also suggested independently by Bodmer (1965). Convincing evidence, however, that recombination need have no connection with normal DNA synthesis was provided by Rossen and Westergaard (1966). These authors showed that in the ascomycete, *Neottiella rutilans*, replication was completed before sexual fusion, that is, before the diploid primary ascus was established. At the time of replication the homologous chromosomes were in fact in two separate haploid nuclei. Sueoka *et al.* (1967) have also shown, using isotopic density labeling, that replication during meiosis in *Chlamydomonas reinhardi* is strictly semiconservative and probably occurs after crossing-over has taken place. Nevertheless, at least some recombination may still take place during DNA synthesis. In particular, gene conversion, which is often not associated with recombination of outside markers, may to a large extent occur at the growing point. Thus, Holliday (1968) presents convincing evidence that gene conversion for nitrate-requiring mutants of *Ustilago* is predominantly stimulated by UV treatment during the period of DNA synthesis, as was also found for mitotic crossing-over. He suggests, in addition, a model for conversion involving repair which is analogous to the models discussed by Hogness *et al.* (1966), also providing evidence that FUdR increases the frequency of intergenic crossing-over in *Aspergillus* but does not affect intragenic recombination. While it seems likely that recombination in higher organisms has an ultimate basis analogous to that discussed for microorganisms, the problem is clouded by the lack of knowledge concerning chromosome structure and organization during meiosis.

XI. Conclusions

At the molecular level there are models which can account for the main features of genetic recombination as it occurs in bacteria and their viruses. However, a major problem yet to be resolved is how recognition occurs between homologous regions before recombination takes place. Presumably, once homologous regions are aligned, breaks may be introduced in a controlled way, and gaps in the recombinant molecules repaired by enzyme systems similar to those described above. Satisfactory molecular models for the way in which double-stranded DNA molecules could partially unwind and exchange base-pairing partners have not yet been constructed. Annealing provides an explanation for the recognition process but says nothing about the structural organization needed to hold the pairing DNA molecules in the proper configuration. The problem in higher organisms is, of course, accentuated by the complexity of their chromosome structure. It certainly is hard to conceive of any mechanism which could allow pairing simultaneously at all points along a chromosome. Some sort of effective pairing regions, as postulated by Pritchard (1960), must exist and possibly also a mechanism for moving paired regions along the chromosome. Studies with microorganisms may well have provided the basis for an understanding of what takes place within such a paired region, but have so far left almost untouched the understanding of the way in which the paired regions are formed.

ACKNOWLEDGMENTS

We should like to thank Drs. Lawrence Okun and Dale Kaiser for their helpful comments on the manuscript. This work was supported, in part, by a Public Health Service Research Career Program award (to Walter Bodmer, GM 35002) and a research grant from the National Science Foundation.

REFERENCES

Alexander, H. E., and Leidy, G. (1951). *J. Exptl. Med.* **93**, 345.
Anagnostopoulos, C., and Crawford, I. P. (1961). *Proc. Natl. Acad. Sci. U.S.* **47**, 378.
Anraku, N., and Tomizawa, J. (1965a). *J. Mol. Biol.* **11**, 501.
Anraku, N., and Tomizawa, J. (1965b). *J. Mol. Biol.* **12**, 805.
Avery, O. T., MacLeod, C. M., and McCarty, M. (1944). *J. Exptl. Med.* **79**, 137.
Belling, J. (1931). *Univ. Calif. (Berkeley) Publ. Botany* **16**, 153.
Benzer, S. (1959). *Proc. Natl. Acad. Sci. U.S.* **45**, 1607.
Benzer, S. (1961). *Proc. Natl. Acad. Sci. U.S.* **47**, 403.
Berger, H. (1965). *Genetics* **52**, 239.
Beukers, R., and Berends, W. (1960). *Biochim. Biophys. Acta* **41**, 550.
Bode, V. C., and Kaiser, A. D. (1965). *J. Mol. Biol.* **14**, 399.
Bodmer, W. F. (1965). *J. Mol. Biol.* **14**, 534.
Bodmer, W. F. (1966). *J. Gen. Physiol.* **49**/6, 233.
Bodmer, W. F. (1967). *Proc. 5th Berkeley Symp. Statistics Probability,* Univ. of California Press, p. 377.
Bodmer, W. F., and Ganesan, A. T. (1964). *Genetics* **50**, 717.
Bodmer, W. F., and Laird, C. D. (1968). *In* "Replication and Recombination of the Genetic Material," pp. 184–205, Canberra.
Boyce, R. P., and Howard-Flanders, P. (1964). *Proc. Natl. Acad. Sci. U.S.* **51**, 293-300.
Brenner, S., Stretton, A. O. W., and Kaplan, S. (1965). *Nature* **206**, 994.
Bresler, S. E., and Lanzov, V. A. (1967). *Genetics* **56**, 117.
Campbell, A. (1962). *Advan. Genet.* **11**, 101.
Case, M. E., and Giles, N. H. (1958). *Cold Spring Harbor Symp. Quant. Biol.* **23**, 119.
Chase, M., and Doermann, A. H. (1958). *Genetics* **43**, 332.
Chen, K., and Ravin, A. W. (1966a). *J. Mol. Biol.* **22**, 109.

Chen, K., and Ravin, A. W. (1966b). *J. Mol. Biol.* **22**, 123.
Chilton, M. D. (1967). *Science* **157**, 817.
Chovnick, A. (1966). *Proc. Roy. Soc.* **B164**, 198.
Clark, A. J., and Margulies, A. D. (1965). *Proc. Natl. Acad. Sci. U.S.* **53**, 451.
Clark, A. J., Chamberlin, M., Boyce, R. P., and Howard-Flanders, P. (1966). *J. Mol. Biol.* **19**, 442.
Cook, J. S. (1967). *Photochem. Photobiol.* **6**, 97.
Darlington, A. J., and Bodmer, W. F. (1968). *Genetics* **60**, 681.
Davison, P., Freifelder, F. D., Heder, R., and Levinthal, C. (1961). *Proc. Natl. Acad. Sci. U.S.* **47**, 1123.
Doermann, A. H. (1965). *Proc. 11th Intern. Congr. Genet., The Hague, 1963* Vol. **2**, p. 69. Pergamon Press, Oxford.
Doermann, A. H., and Boehner, L. (1964). *J. Mol. Biol.* **10**, 212
Dubnau, D., Goldthwaite, C., Smith, I., and Marmur, J. (1967). *J. Mol. Biol.* **27**, 163.
Edgar, R. S. (1961). *Virology* **13**, 1.
Ephrati-Elizur, E., Srinivasan, P. R., and Zamenhof, S. (1961). *Proc. Natl. Acad. Sci. U.S.* **47**, 56.
Ephrussi-Taylor, H. (1951). *Cold Spring Harbor Symp. Quant. Biol.* **16**, 445.
Ephrussi-Taylor, H., and Gray, T. C. (1966). *J. Gen. Physiol.* **49/6**, 211.
Epstein, R. H., Bolle, A., Steinberg, C. M., Kellenberger, E., De La Tour, E. B., and Chevalley, R. (1963). *Cold Spring Harbor Symp. Quant. Biol.* **28**, 373.
Farmer, J. L., and Rothman, F. (1965). *J. Bacteriol.* **89**, 262.
Fox, M. S. (1960). *Nature* **187**, 1004.
Fox, M. S. (1963). *J. Mol. Biol.* **6**, 85.
Fox, M. S. (1966). *J. Gen. Physiol.* **49/6**, 183.
Fox, M. S., and Allen, M. K. (1964). *Proc. Natl. Acad. Sci. U.S.* **52**, 412.
Fox, M. S., and Hotchkiss, R. D. (1957). *Nature* **179**, 1322.
Fox, M. S., and Hotchkiss, R. D. (1960). *Nature* **187**, 1002.
Franklin, N. C., Dove, W. F., and Yanofsky, C. (1965). *Biochem. Biophys. Res. Commun.* **18**, 910.
Ganesan, A. T., and Lederberg, J. (1965). *Biochem. Biophys. Res. Commun.* **18**, 824.
Gefter, M. L., Becker, A., and Hurwitz, J. (1967). *Proc. Natl. Acad. Sci. U.S.* **58**, 240.
Gellert, M. (1967). *Proc. Natl. Acad. Sci. U.S.* **57**, 148.
Goodgal, S. H., and Herriot, R. M. (1957). *In* "The Chemical Basis of Heredity" (W. D. McElroy and B. Glass, eds.), p. 336. Johns Hopkins Press, Baltimore, Maryland.
Goodgal, S. H., and Postel, E. H. (1967). *J. Mol. Biol.* **28**, 261.
Grell, R. F. (1965). *Natl. Cancer Inst. Monograph* **18**, 215.
Grell, R. F., and Chandley, A. C. (1965). *Proc. Natl. Acad. Sci. U.S.* **53**, 1340.
Griffith, F. (1928). *J. Hyg.* **27**, 113.
Hayes, W. (1964). "The Genetics of Bacteria and Their Viruses." Wiley, New York.
Henning, U., and Yanofsky, C. (1962). *Proc. Natl. Acad. Sci. U.S.* **48**, 183.
Hershey, A. D., and Rotman, R. (1949). *Genetics* **34**, 44.
Hershey, A. D., Burgi, E., and Ingraham, L. (1963). *Proc. Natl. Acad. Sci. U.S.* **49**, 748.
Hogness, D. S. (1966). *J. Gen. Physiol,* **49**/6, 29.

Hogness, D. S., and Simmons, J. R. (1964). *J. Mol. Biol.* **9**, 411.
Hogness, D. S., Doerfler, W., Egan, J., and Black, L. (1966). *Cold Spring Harbor Symp. Quant. Biol.* **31**, 129.
Holliday, R. (1962). *Genet. Res.* **3**, 472.
Holliday, R. (1964). *Genet. Res.* **5**, 282.
Holliday, R. (1965). *Genet. Res.* **6**, 104.
Holliday, R. (1968). *In* "Replication and Recombination of the Genetic Material," pp. 157–183, Canberra.
Hotchkiss, R. D., and Evans, A. H. (1958). *Cold Spring Harbor Symp. Quant. Biol.* **23**, 85.
Hotchkiss, R. D., and Marmur, J. (1954). *Proc. Natl. Acad. Sci. U.S.* **40**, 55.
Howard, A., and Pelc, S. R. (1951). *Exptl. Cell Res.* **2**, 178.
Howard-Flanders, P., and Theriot, L. (1966). *Genetics* **53**, 1137.
Iyer, V. N., and Ravin, A. W. (1962). *Genetics* **47**, 1355.
Jacob, F., and Wollman, E. L. (1961). "Sexuality and the Genetics of Bacteria." Academic Press, New York.
Janssens, F. A. (1909). *Cellule* **25**, 389.
Kaiser, A. D. (1962). *J. Mol. Biol.* **4**, 275.
Kaiser, A. D., and Hogness, D. S. (1960). *J. Mol. Biol.* **2**, 392.
Kellenberger, G., Zichichi, M. L., and Epstein, H. T. (1962). *Virology* **17**, 44.
Kellenberger, G., Zichichi, M. L., and Weigle, J. J. (1961). *Proc. Natl. Acad. Sci. U.S.* **47**, 869.
Kozinski, A. W. (1961). *Virology* **13**, 124.
Kozinski, A. W., and Kozinski, P. B. (1963). *Virology* **20**, 213.
Kozinski, A. W., Kozinski, P. B., and Shannon, P. (1963). *Proc. Natl. Acad. Sci. U.S.* **50**, 746.
Lacks, S. (1962). *J. Mol. Biol.* **5**, 119.
Lacks, S. (1965). *Genetics* **53**, 207.
Lacks, S., and Hotchkiss, R. D. (1960). *Biochem. Biophys. Acta* **39**, 508-517.
Lacks, S., Greenberg, B., and Carlson, K. (1967). *J. Mol. Biol.* **29**, 327.
Laird, C. D. (1966). Ph.D. Dissertation, Stanford University.
Laird, C. D., Wang, L., and Bodmer, W. F. (1968). *Mutation Res.* **6**, 205-209.
Lederberg, E. M., and Lederberg, J. (1953). *Genetics* **38**, 51.
Lederberg, J. (1955). *J. Cellular Comp. Physiol.* **45/2**, 95.
Lederberg, J., Lederberg, E. M., Zinder, N. D., and Lively, E. R. (1951). *Cold Spring Harbor Symp. Quant. Biol.* **16**, 413.
Lehman, I. R. (1963). *Progr. Nucleic Acid Res.* **2**, 83.
Lerman, L. S., and Tolmach, L. J. (1957). *Biochim. Biophys. Acta* **26**, 68.
Lerman, L. S., and Tolmach, L. J. (1959). *Biochim. Biophys. Acta* **33**, 371.
Levinthal, C. (1954). *Genetics* **39**, 169.
Litt, M., Marmur, J., Ephrussi-Taylor, H., and Doty, P. (1958). *Proc. Natl. Acad. Sci. U.S.* **44**, 144.
Lwoff, A., and Guttman, A. (1949). *Compt. Rend.* **229**, 679.
MacHattie, L. A., and Thomas, C. A., Jr. (1964). *Science* **144**, 1142.
MacHattie, L. A., Ritchie, D. A., Thomas, C. A., Jr., and Richardson, C. C. (1967). *J. Mol. Biol.* **23**, 355.
Marmur, J. (1961). *J. Mol. Biol.* **3**, 208.
Marmur, J., and Lane, D. (1960). *Proc. Natl. Acad. Sci. U. S.* **46**, 453.
Marmur, J., Rownd, R., and Schildkraut, C. L. (1963). *Progr. Nucleic Acid Res.* **2**, 231.
Meselson, M. (1964). *J. Mol. Biol.* **9**, 734.

Meselson, M. (1966). *In* "Heritage from Mendel" (R. A. Brink, ed.), pp. 81-104. Univ. of Wisconsin Press, Madison, Wisconsin.
Meselson, M., and Stahl, F. W. (1958). *Proc. Natl. Acad. Sci. U.S.* **44**, 671.
Meselson, M., and Weigle, J. J. (1961). *Proc. Natl. Acad. Sci. U.S.* **47**, 857.
Meselson, M., Stahl, F. W., and Vinograd, J. (1957). *Proc. Natl. Acad. Sci. U.S.* **43**, 581.
Nester, E. W. (1964). *J. Bacteriol.* **87**, 867.
Nester, E. W., and Lederberg, J. (1961). *Proc. Natl. Acad. Sci. U.S.* **47**, 52.
Nester, E. W., and Stocker, B. A. D. (1963). *J. Bacteriol.* **86**, 785.
Nester, E. W., Schafer, M., and Lederberg, J. (1963). *Genetics* **48**, 529.
Nomura, M., and Benzer, S. (1961). *J. Mol. Biol.* **3**, 684.
Notani, N., and Goodgal, S. H. (1966). *J. Gen. Physiol.* **49/6**, 197.
Okubo, S., and Romig, W. R. (1966). *J. Mol. Biol.* **15**, 440.
Olivera, B. M., and Lehman, I. R. (1967). *Proc. Natl. Acad. Sci. U.S.* **57**, 1426.
Oppenheim, A. B., and Riley, M. (1966). *J. Mol. Biol.* **20**, 331.
Oppenheim, A. B., and Riley, M. (1967). *J. Mol. Biol.* **28**, 503.
O'Sullivan, A., and Sueoka, N. (1967). *J. Mol. Biol.* **27**, 349.
Pene, J. J., and Romig, W. R. (1964). *J. Mol. Biol.* **9**, 236.
Pettijohn, D. E., and Hanawalt, P. C. (1964a). *J. Mol. Biol.* **8**, 170.
Pettijohn, D. E., and Hanawalt, P. C. (1964b). *J. Mol. Biol.* **9**, 395.
Postel, E. H., and Goodgal, S. H. (1966). *J. Mol. Biol.* **116**, 317.
Postel, E. H., and Goodgal, S. H. (1967). *J. Mol. Biol.* **28**, 247.
Prell, H. H. (1965). *J. Mol. Biol.* **13**, 329.
Pritchard, R. H. (1955). *Heredity* **9**, 343.
Pritchard, R. H. (1960). *Genet. Res.* **1**, 1.
Ptashne, M. (1965). *J. Mol. Biol.* **11**, 90.
Radding, C. M., and Kaiser, A. D. (1963). *J. Mol. Biol.* **7**, 225.
Ravin, A. W. (1961). *Advan. Genet.* **10**, 61.
Ravin, A. W., and Chen, K. (1967). *Genetics* **57**, 851.
Ravin, A. W., and Iyer, V. N. (1962). *Genetics* **47**, 1369.
Richardson, C. C., Inman, R. B., and Kornberg, A. (1964a). *J. Mol. Biol.* **9**, 46.
Richardson, C. C., Lehman, I. R., and Kornberg, A. (1964b). *J. Biol. Chem.* **239**, 251.
Ris, H., and Chandler, B. L. (1963). *Cold Spring Harbor Symp. Quant. Biol.* **28**, 1.
Roger, M., Beckmann, O., and Hotchkiss, R. D. (1966). *J. Mol. Biol.* **18**, 174.
Rossen, J. M., and Westergaard, M. (1966). *Compt. Rend. Trav. Lab. Carlsberg* **35**, 233.
Rupert, C. S. (1962). *J. Gen Physiol.* **45**, 703.
Schaeffer, P. (1956). *Ann. Inst. Pasteur* **91**, 192.
Schaeffer, P. (1964). *In* "The Bacteria" (I. C. Gunsalus and R. Y. Stanier, eds.), Vol. 5, pp. 87-153. Academic Press, New York.
Séchaud, J., Streisinger, G., Emrich, J. Newton, J., Lanford, H., Reinhold, H., and Stahl, M. M. (1965). *Proc. Natl. Acad. Sci. U.S.* **54**, 1333.
Setlow, R. B., and Carrier, W. L. (1964). *Proc. Natl. Acad. Sci. U.S.* **51**, 226.
Setlow, R. B., and Setlow, J. K. (1962). *Proc. Natl. Acad. Sci. U.S.* **48**, 1250.
Sicard, A. M., and Ephrussi-Taylor, H. (1965). *Genetics* **52**, 1207.
Siddiqi, O. H. (1963). *Proc. Natl. Acad. Sci. U.S.* **49**, 589.
Spizizen, J. (1958). *Proc. Natl. Acad. Sci. U.S.* **44**, 1072.

Spizizen, J., Reilly, B. E., and Evans, A. H. (1966). *Ann. Rev. Microbiol.* **20**, 371.
Steinberg, C. M., and Edgar, R. S. (1962). *Genetics* **47**, 187.
Stimpfling, J. H., and Richardson, A. (1965). *Genetics* **51**, 831.
Strack, H. B., and Kaiser, A. D. (1965). *J. Mol. Biol.* **12**, 36.
Strauss, B. S., Wahl-Synek, R., Reiter, H., and Searashi, T. (1966). *G. Mendel Mem. Symp. Mutational Process, Prague, 19* Abstr., pp.39–49.
Streisinger, G., Edgar, R. S., and Denhardt, G. H. (1964). *Proc. Natl. Acad. Sci U.S.* **51**, 775.
Sueoka, N., Chiang, K., and Kates, J. (1967). *J. Mol. Biol.* **25**, 47.
Szybalski, W. (1961). *J. Chim. Phys.* **58**, 1098.
Taylor, J. H. (1957). *Am. Naturalist* **91**, 209.
Taylor, J. H. (1965). *J. Cell Biol.* **25**, 57.
Taylor, J. H., ed. (1967). "Molecular Genetics," Part 2, Chapter 3. Academic Press, New York.
Thomas, C. A., Jr. (1966). *J. Gen. Physiol.* **49/3**, 143.
Thomas, R. (1955). *Biochim. Biophys. Acta* **18**, 467.
Tomizawa, J., and Anraku, N. (1964a). *J. Mol. Biol.* **8**, 508.
Tomizawa, J., and Anraku, N. (1964b). *J. Mol. Biol.* **8**, 516.
Tomizawa, J., and Anraku, N. (1965). *J. Mol. Biol.* **11**, 509.
Tomizawa, J., Anraku, N., and Iwama, U. (1966). *J. Mol. Biol.* **9**, 247.
Trautner, T. A. (1958). *Z. Vererbungslehre* **89**, 266.
van der Putte, P., Zwenk, H., and Rorsch, A. (1966). *Mutation Res.* **3**, 381.
Venema, G., Pritchard, R. H., and Venema-Schröder, T. (1965). *J. Bacteriol.* **89**, 1250.
Vinograd, J., and Lebowitz, J. (1966). *J. Gen. Physiol.* **49**, 103.
Voll, M. J., and Goodgal, S. H. (1961). *Proc. Natl. Acad. Sci. U.S.* **67**, 505.
Wacker, A., Dellweg, H., and Weinblum, D. (1960). *Naturwissenschaften* **47**, 477-480.
Weiss, B., and Richardson, C. C. (1967). *Proc. Natl. Acad. Sci. U.S.* **57**, 1021.
Whitehouse, H. L. K. (1963). *Nature* **199**, 1034.
Witkin, E. M. (1966). *Science* **152**, 1345.
Young, E. T., and Sinsheimer, R. L. (1964). *J. Mol. Biol.* **10**, 562.

V LINKAGE AND RECOMBINATION AT THE CHROMOSOME LEVEL

STERLING EMERSON

I. Introduction

Fragments of rather precise information about mechanisms of recombination between linked genetic markers are currently being produced by ingenious applications of new and old techniques in organisms having diverse genetic systems. There is justification for the belief that a collation of information so far obtained in studies of different systems by different methods may lead to a better understanding of how recombination occurs at all levels of organization of the genetic material.

Our knowledge of recombinational processes in eucaryotic (nucleated) organisms has been derived largely by the use of purely genetic techniques. The precision of mitotic and meiotic processes (Chapter III) endows genetic analyses with great accuracy. The complex organization of the genetic material in the chromosomes of eucaryotes, with the accompanying inaccessibility of the DNA, has limited the use of those biophysical methods which have proved so successful in studies of genetic recombination in species with more simply organized material (Chapter IV). Consequently, the material to be reviewed here is that resulting from genetic methods. Special attention will be given to the kinds, extent and accuracy of information so obtained and to the validity of inferences drawn therefrom regarding occurrences at the molecular level as well as to grosser features of chromosome behavior.

II. Intergenic Crossing-Over

A. Linkage Groups and Chromosomes

The characteristics of the genetic material and its structured organization as inferred from observed linkage and recombination between inherited traits closely reflect the cytological structure and meiotic behavior of chromosomes. The conclusion that a common process is being examined by two methods, genetic and cytological, is now too well established to require a detailed discussion—a mere listing of the kinds of evidence upon which this conclusion is based should be sufficient in this review.

Proof that chromosomes are the physical carriers of genetic determiners was obtained by Bridges (1916). Occasional "exceptional" daughters which had inherited sex-linked traits from the maternal parent only were found to have three sex chromosomes—a Y chromosome from the paternal parent and two X chromosomes of maternal origin resulting from nondisjunction at meiosis. Such nondisjunction occurred invariably in a strain of *Drosophila* studied by L. V. Morgan (1922) in which physical attachment of the two X chromosomes results in strict mother-to-daughter inheritance of all normally sex-linked traits.

The linear order of genes in linkage maps was established by Sturtevant (1915) by an algebraic treatment of recombination data obtained from hybrids heterozygous at three or more linked loci. Only a unique linear order of loci would satisfy the condition that the frequency of coincident crossovers in two intervals does not exceed the products of the frequencies of total crossovers in the two intervals concerned. That the order of genes in the physical chromosomes is identical with that determined for genetic maps was verified by correlated cytological and genetic mapping of the positions of chromosome breaks (in transloca-tions, inversions, deletions, etc.). Complete agreement in gene order (but only approximate agreement in map lengths between genes) was first obtained for somatic chromosomes—from ganglion nuclei (Dob-zhansky, 1930) and salivary gland nuclei (Painter, 1933; Bridges, 1935), both in *Drosophila*—and later for meiotic prophase chromosomes in other organisms, especially in flowering plants and fungi.

Evidence that a physical exchange between regions of homologous chromosomes accompanies genetic crossing-over was obtained almost simultaneously by Stern (1931) in *Drosophila* and by Creighton and McClintock (1931) in maize. Recombination between cytologically visi-ble differences at two locations in pairs of homologous chromosomes was observed to accompany genetic crossing-over.

B. Chiasmata and Genetic Exchanges

1. *The Chiasmatype Interpretation*

Following synapsis between homologous chromosomes, as seen in cyto-logically favorable material, there occurs a relaxation in the tightness of pairing between chromosomes and it can be seen that each chromosome consists of two sister chromatids. At intervals along the length of a chro-mosome pair there are chiasmata in which one pair of tightly associated chromatids exchanges partners with the paired chromatids of the homol-ogous chromosome. Three such chiasmata in a single pair of chromosomes are illustrated in Fig. 1a, as they appear in cytological preparations. The chiasmatype interpretation of this figure, proposed by Janssens (1909) and now adopted by most cytologists (see Darlington, 1932; also Chap-ter III of this volume), is that early intimate associations on both sides of a chiasma are between sister chromatids (that is, between daughter chromatids of each homologous chromosome); consequently, an exchange of segments between two nonsister chromatids must be responsible for the production of each chiasma. This interpretation of chromatid pair-ing and exchange as applied to the three chiasmata in Fig. 1a is dia-gramed in Fig. 1b, and perhaps more clearly in Fig. 1c in which the chromosomes have been untwisted.

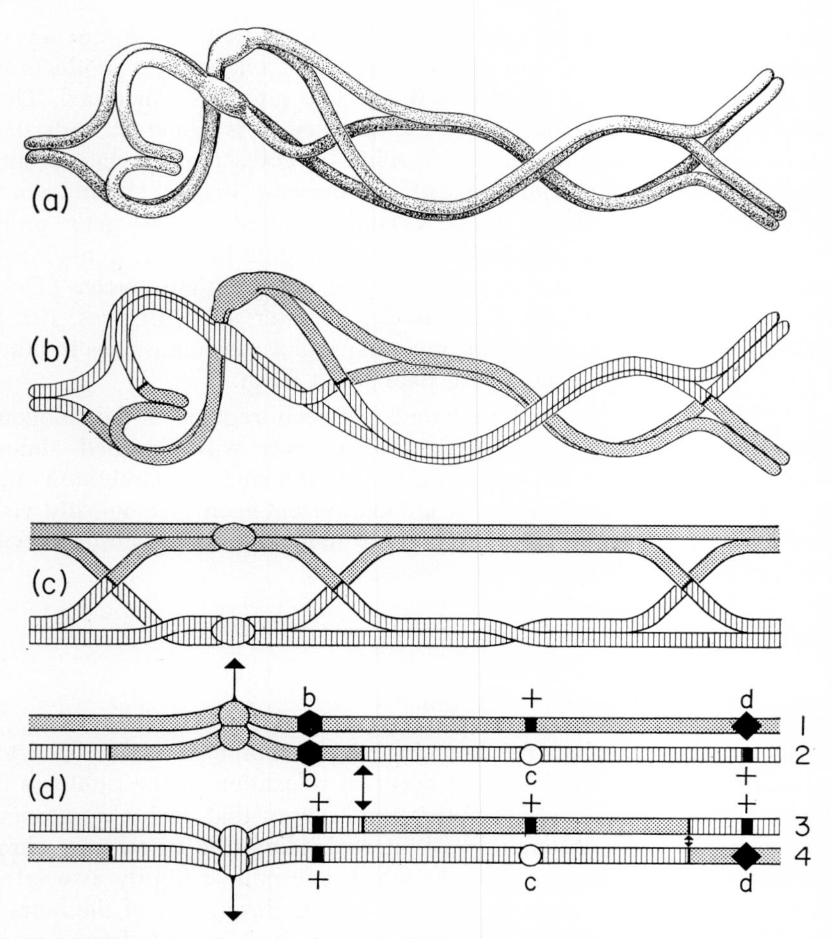

FIG. 1. Chiasmata. (a) A pair of diplotene chromosomes with three chiasmata, as seen in cytological preparations. (After Janssens, 1909.) (b) Interpretation of relation of sister chromatids to chiasmata: Sister chromatids (daughter products of a single homolog) are indicated by the same type of shading; homologous chromatids by different shading patterns. (c) Similar to (b), but with the twists in the chromosomes straightened out to show the individuality of the chromatids. (d) Expected genetic consequences of chiasmata shown in (a) as interpreted in (c), following terminalization of chiasmata: b, c, and d, recessive alleles at three loci in the right chromosome arm; +, the wild-type allele of each, distinguished by their positions; arrows are attached to centromeres and indicate segregation to opposite poles in anaphase of the first meiotic division; chromatids 1 and 2 separate in one second division figure, 3 and 4 in the other, each being recovered in a separate meiotic product.

Direct cytological proof that pairing on each side of a chiasma is between sister chromatids is difficult to obtain inasmuch as homologous chromatids as well as sister chromatids ordinarily have identical morphologies. In examples in which the two homologous chromosomes have recognizable structural differences it can be (and has been) argued that structural heterozygosity itself may prevent the close pairing of nonsister chromatids. (Somewhat later in meiotic prophase there is close association between nonsister chromatids as a result of "chiasma terminalization": Darlington, 1932.) It is noteworthy that the chiasmatype interpretation is in harmony with the known characteristics of genetic crossing-over, many of which were predicted by the chiasmatype model.

2. *Genetic Exchange*

Genetic consequences of the chiasmatype model are illustrated diagrammatically in Fig. 1d in which the location of three genes is indicated by their recessive alleles, *b, c,* and *d,* whose dominant alleles are designated by +'s. Verified predictions from the chiasmatype model about characteristics of intergenic crossing-over include the following: (1) Crossing-over occurs between chromatids, not between whole chromosomes. (2) Two and only two chromatids take part in crossing-over at any one point in the length of the chromosome (i.e., at any one exchange). (3) Reciprocal crossovers between heterozygous genes flanking the exchange are produced in all instances.

There are some limitations to the detection of chromatid exchanges by genetic methods. (1) An exchange will be undetected unless a heterozygous gene is present on each side of it, as in the exchange between *b*/+ and *c*/+ and that between *c*/+ and *d*/+ in the diagram—the leftmost exchange illustrated would be undetected because of no heterozygous gene to the left of it. (Exchanges between the centromere and a heterozygous gene can be detected by tetrad analysis but the chromatids involved cannot necessarily be identified.) (2) The exact point at which an exchange occurs is not evident from genetic tests. All that is indicated is that an exchange (and not necessarily only one) has occurred somewhere between the flanking marker genes. (3) If only random products of meiosis are recovered (as in eggs and sperm of animals and in megaspores and microspores of most higher plants) the potential genetic information is still more incomplete. The product receiving chromatid 1 in the diagram (Fig. 1d) fails to show that any exchange has occurred; that receiving chromatid 2 reveals a single exchange between *b*/+ and *c*/+, and that in chromatid 4 a single exchange between *c*/+ and *d*/+.

Only the product which receives chromatid 3 shows that two exchanges have occurred, and it does not tell whether the same two chromatids were involved in both exchanges or (as illustrated) that only one chromatid was involved in both, but with different sister chromatids of the homologous chromosome in the two regions. Maximum genetic information is obtained only when all four products of single meiotic events are recovered and their common origin identified, as in tetrad analysis.

C. Analysis of Random Meiotic Products

In relatively few organisms it is possible to recover other than random meiotic products. Samples of discharged sperm of animals, or of pollen of higher plants, consist of random mixtures of products derived from many different spermatocytes or pollen mother cells, whereas on the female side three of the four products of each individual meiosis are shunted into nonreproductive cells (polar bodies or nondeveloping megaspores) and only one product directly becomes, or indirectly engenders, the egg nucleus.

Much of our knowledge of mechanisms by which crossing-over occurs was derived from studies of random meiotic products. Things so learned include: (1) genes are carried by the cytologically visible chromosomes; (2) the genes are arranged in linear orders in the chromosomes; (3) the frequency of recombination between linked genes is a (complex) function of their physical distance from one another in the chromosome: (4) recombination occurring in one interval is associated with less frequent recombination in adjoining intervals than occurs in the absence of recombination in the first region; (5) interallelic recombination is correlated (in a complex manner, to be discussed later) with recombination between genes lying on either side of (flanking) the locus undergoing interallelic recombination. Points on which random isolates do not give direct evidence include: (1) that crossing-over involves an exchange between two of four chromatids; (2) information relating to chromatid interference—the choice of chromatids involved in neighboring exchanges; (3) determination of the division at which segregation has occurred; (4) the nature of correlations between interallelic recombination and gene conversion.

D. Half-Tetrad Analysis

In organisms in which it is possible to recover only random products of meiosis it may be possible to recover two of the four chromatids involved in exchanges within a single product. Such was the case in Bridges' studies (1916) of X-chromosome nondisjunction in *Drosophila,* in which two chromatids occasionally failed to separate from one another during meiosis. Such disjunctional failure was found to occur regularly in a strain of *Drosophila* later shown to have the two homologous X chromosomes attached to a single centromere (L. V. Morgan, 1922). The sporadic occurrence of genetic exchange unaccompanied by chromosome reduction in somatic cells [first observed in *Drosophila* (Stern, 1936) and now known to occur fairly commonly in fungi in which there is normally a diploid stage or in which selection for sporadic diploidy is possible (Roper, 1966)] also regularly results in products carrying two of the four chromatids which had taken part in recombinational events. The recovery of two meiotic products within a single gamete, or of two products of crossing-over in a single somatic cell, results in more information about recombinational processes that can be gained from random meiotic products which are completely haploid.

1. *Attached-X Chromosomes*

Current concepts of the manner in which intergenic crossing-over takes place stem most directly from studies of attached-X chromosomes of *Drosophila* (Anderson, 1925).* Results from these studies showed conclusively that crossing-over occurs at the 4-strand stage. They also strongly suggested that only two strands (chromatids) are involved in any one exchange and that the chromatids (exclusive of sister chromatids) take part in each exchange pretty much at random. Because of the historical importance of the attached-X studies to the development of current genetic concepts the nature of the evidence derived from them will be examined in some detail.

Double exchanges of essentially the same kind previously used to illustrate crossing-over in unattached meiotic chromosomes (Fig. 1d) are represented diagrammatically in Fig. 2a and b. The two homologous X chromosomes are attached to a single centromere which, as in unattached chromosomes, divides only once during the two meiotic divisions —during the division at which homologous centromeres normally segregate all elements attached to the undivided X-chromosome centromere pass to one pole, none to the other; during the equational division the two daughter centromeres, each attached to two chromatids which are nonsisters from the centromere to the most proximal exchange, pass to opposite poles. (In most instances, attached-X females have received a

* See historical note, Addendum I, A.

Y chromosome from their father; and disjunction occurs between the Y centromere and that of the attached-X.)

The first exchange (that most proximal to the centromere) may take place between two homologous chromatids which are attached to different daughter centromeres, as in Fig. 2a, or which are attached to the same one, as in Fig. 2b, with recognizably different consequences. The

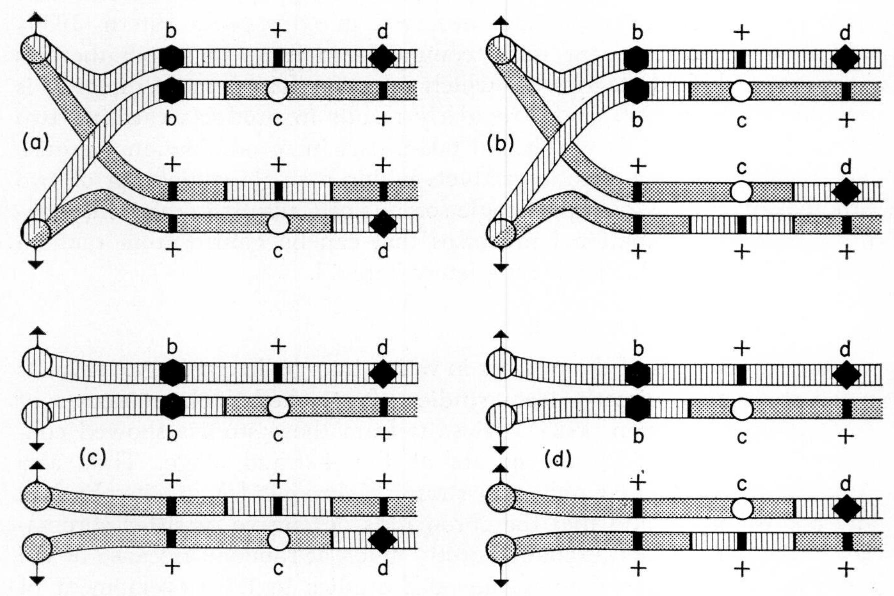

FIG. 2. The recovery of half-tetrads of products of crossing-over. (a) and (b) Attached-X-chromosomes (or attached left or right arms of autosomes) of *Drosophila melanogaster*: Two homologous chromosome arms are attached to a common centromere and cannot separate at the meiotic division in which homologous centromeres ordinarily separate—the separation indicated is equivalent to the separation between sister centromeres (and sister chromatids proximal to the first exchange) in the second meiotic division involving normal (unattached) pairs of homologous chromosomes. The diagrams are essentially identical to that in Fig. 1d except that two attached chromatids segregate together and enter a single meiotic product: chromatids 1 and 3 to one product, 2 and 4 to another, in (a); 1 associated with 4, and 2 with 3, in (b). (c) and (d) Mitotic recombination in diploid nuclei: No reduction division occurs, but sister centromeres of each homolog disjoin as in normal somatic mitoses, one daughter chromatid of each homolog being distributed to each daughter nucleus—two of the chromatids of a chromosome pair undergoing exchange are recovered in each daughter nucleus. Chromatid 1 is included in the same nucleus as chromatid 3, and 2 with 4, following one orientation of the two pairs of sister centromeres on the mitotic spindle (c), or 1 with 4 and 2 with 3 following another orientation (d).

latter results in a reciprocal crossover, so called because the two complementary products of the exchange pass together to one daughter cell. The former kind of first exchange (Fig. 2a), results in a nonreciprocal crossover—one crossover and one noncrossover chromatid are recovered in either daughter nucleus. Exchanges resulting in reciprocal crossovers can thus be detected in only half of the products, those resulting in nonreciprocal crossovers in both.

The actual detection and identification of reciprocal and nonreciprocal crossovers in attached-X chromosomes requires one further generation of testing, and some differences exist in the ways in which identifications are made. In the examples illustrated in Fig. 2, the constitution of the attached-X chromosomes before crossing-over may be represented b + d/+ c +, the reciprocal crossover between b/+ and +/c would then be b c +/+ + d, the nonreciprocal either b c +/+ c +, or b + d/ + + d, depending upon which product was recovered (compare with Fig. 2a and b, disregarding the exchange between +/c and d/+). Both the parental (noncrossover) and reciprocal crossover are phenotypically wild type and heterozygous at all three loci. Nonreciprocal crossovers are homozygous at the two distal loci, with phenotypes characteristic of either mutant c or mutant d. Phenotypic expression alone is insufficient for the identification of the region of exchange in nonreciprocal crossovers —phenotype c can also result from an exchange between the centromere and b/+ to give + c +/+ c +; phenotype d can also result from an exchange between +/c and d/+ to give b + d/+ c d. A further generation of testing will show whether or not recessive allele b is still present in flies of phenotype c, or recessive allele c in flies of phenotype d, thus definitely identifying the regions in which the exchanges had occurred.

The detection of double exchanges also varies with the type of exchange occurring nearest the centromere. Only a single kind of double exchange, a 3-strand double, is diagrammed in Fig. 2a and b—diagrams of all possible kinds may be seen in Figs. 22 and 28 in Addendum II. The kinds of products recovered from each of the eight possible different double exchanges are summarized in Table I. It should be noted that, with a reciprocal first exchange, one product of a 2-strand double can be identified definitely as arising from that kind of exchange, whereas the other product is a noncrossover; one product of each 3-strand double can also be identified as definitely arising from a 3-strand double exchange, whereas the other is a single crossover in the more distal region; but neither product of a 4-strand double can be recognized as a product of double crossing-over, each having a single reciprocal crossover, one in the proximal and one in the distal region. With a nonreciprocal first

TABLE I
Exchanges Detected in Analysis of Half-Tetrads[a]

Double exchange type	Attached-X (first exchange)	Mitotic (recovered strands)	Both exchanges detected							One exchange detected				None
			$\frac{(1\,2)}{(1\,2)}$	$\frac{(1\,2)}{(1)}$	$\frac{(1)}{(1\,2)}$	$\frac{(1\,2)}{—}$	$\frac{—}{(1\,2)}$	$\frac{(1)}{(2)}$	$\frac{(2)}{(1)}$	$\frac{(1)}{(1)}$	$\frac{(2)}{(2)}$	$\frac{(2)}{—}$	$\frac{—}{(2)}$	$\frac{—}{—}$
2-strand	Reciprocal	1,3 or 2,4	$\frac{b\ c\ d}{+\,+\,+}$	—	—	—	—	—	—	—	—	—	—	$\frac{b+d}{+\ c\ +}$
	Nonreciprocal	1,4 or 2,3	—	—	—	$\frac{b\ c\ d}{+\ c\ +}$	$\frac{b+d}{+\,+\,+}$	—	—	—	—	—	—	—
3-strand	Reciprocal	1,3 or 2,4	—	—	$\frac{b\ c\ +}{+\,+\,+}$	—	—	—	—	—	—	—	$\frac{b+d}{+\ c\ d}$	—
	Nonreciprocal	1,4 or 2,3	—	—	—	—	$\frac{b+d}{+\,+\,+}$	$\frac{b\ c\ +}{+\ c\ d}$	—	—	—	—	—	—
3-strand	Reciprocal	1,3 or 2,4	—	$\frac{b\ c\ d}{+\,+\ d}$	—	—	—	—	—	—	—	$\frac{b\,+\,+}{+\ c\ +}$	—	—
	Nonreciprocal	1,4 or 2,3	—	—	—	$\frac{b\ c\ d}{+\ c\ +}$	—	—	$\frac{b\,+\,+}{+\,+\ d}$	—	—	—	—	—
4-strand	Reciprocal	1,3 or 2,4	—	—	—	—	—	—	—	$\frac{b\ c\ +}{+\,+\ d}$	$\frac{b\,+\,+}{+\ c\ d}$	—	—	—
	Nonreciprocal	1,4 or 2,3	—	—	—	—	—	$\frac{b\ c\ +}{+\ c\ d}$	$\frac{b\,+\,+}{+\,+\ d}$	—	—	—	—	—

[a] Numerals in parentheses in the heading indicate the intervals in which crossovers are present in the two recovered chromatids: (1) is proximal, between b/+ and +/c; (2) is distal, between +/c and d/+; the parental constitution is $\frac{b+d}{+\ c\ +}$. Mitotic strands are numbered as in Fig. 2c and d. The alleles present in recovered chromatids are shown in the body of the Table.

exchange, on the other hand, both products of each type of double exchange are recognizable double crossovers, but the number of strands involved is not completely determinable—one product of a 3-strand double is indistinguishable from products of 2-strand doubles, the other from products of 4-strand doubles.

Inasmuch as the data from attached-X chromosomes of *Drosophila* are the principal basis for the widely held conclusion that chromatids are involved completely at random at each successive region of exchange, the data concerned are examined critically in Addendum II, B. That analysis shows that the inference is probably no more than approximately correct.

2. *Mitotic Recombination*

The products of somatic recombination are of the same kinds as those described for attached-X chromosomes, their essential difference being solely in their mode of origin, as illustrated in Fig. 2c and d. There is no reduction division following somatic crossing-over. Instead, sister chromatids separate mitotically over the region extending from the centromere to the most proximal exchange (in each chromosome arm if centromeres are located more-or-less centrally instead of terminally as illustrated). In Fig. 2c the orientation of sister centromeres on the mitotic spindle is such that chromatids 1 and 3 pass to one daughter cell, chromatids 2 and 4 to the other, resulting in genotypic constitutions identical with the nonreciprocal proximal crossover illustrated in Fig. 2a. In Fig. 2d the distribution of sister centromeres to daughter cells is such that chromatids 1 and 4 are included in one, 2 and 3 in the other, resulting in the same genotypic constitutions as those derived from the reciprocal crossover illustrated in Fig. 2 b.

Following mitotic recombination in somatic tissues of multicellular organisms it is obviously impossible to determine genotypic constitutions, and only those recombinants which are homozygous for recessive alleles can be detected. In unicellular organisms with normal diploid phases, such as the yeast *Saccharomyces cerevisiae*, meiosis can be induced in clones derived from a cell resulting from such recombination, thus permitting genotypic analyses. Information obtained from such analyses are exactly comparable to that obtained from attached-X chromosomes of *Drosophila*, except that it is usually impossible to know the exact time at which recombination occurred, or the total frequencies of recombination in the population of dividing cells.

The frequencies of somatic recombination can be greatly increased by ultraviolet irradiation or by the use of alkylating agents—at least in

Saccharomyces cerevisiae (James and Lee-Whiting, 1955; Roman and Jacob, 1958) and in *Ustilago maydis* (Holliday, 1961). A further advantage associated with the use of inducers of somatic recombination is that the time of occurrence of induced recombination is known, and techniques are sometimes possible which permit reasonably accurate frequency determinations (Luzzati, 1965). There is no complete assurance, however, that induced mitotic recombination occurs in a manner identical with that taking place spontaneously.

The available information suggests that in most species somatic recombination is not generalized, but occurs sporadically in single chromosome arms. In *Saccharomyces*, however, there is evidence that the cells in which somatic recombination takes place are in a stage closely resembling meiotic prophase—many chromosome arms undergo recombination in an individual cell (Wilkie and Lewis, 1963; Hurst and Fogel, 1964). Wilkie and Lewis conclude that chromosome disjunction following mitotic recombination is of the meiotic type—a reductional separation of homologs at the centromere region—but with diploidy maintained either through suppression of the equational separation of sister chromatids, as occurs in the second meiotic division, or by some other mechanism. Hurst and Fogel, on the other hand, conclude that chromosome disjunction immediately following somatic recombination is strictly mitotic in nature, as diagrammed in Fig. 2a and b above. The observations reported in the two papers just cited are also discordant—parts of the pertinent data are discussed in greater detail in Addendum II, C, from which I conclude that there is no support for the conclusion that chromosome disjunction is of the meiotic type.

E. Analysis of Tetrads and Octads

Organisms in which all four products of the meiotic divisions can be recovered and identified as having a common meiotic origin offer opportunities for obtaining rather precise information about meiotic events related to genetic recombination. Situations permitting genetic analysis of such tetrads are known to occur in some of the algae, in many of the higher fungi, in some mosses and liverworts, and even in some flowering plants in which pollen grains remain associated in tetrads, though technical difficulties have prevented their exploitation. Among these organisms, ascomycetes with strictly linearly ordered eight-spored asci, such as *Neurospora crassa* and *Sordaria fimicola,* yield direct information about a larger number of meiotic events than do the others. Hence, the situation obtaining in such organisms will be used as the primary example.

1. *Cytogenetic Background*

a. *Nuclear History during Meiosis.* Features in the life cycle of linearly ordered eight-spored ascomycetes are illustrated diagrammatically in Fig. 3 and outlined in the legend to that figure. Important considera-

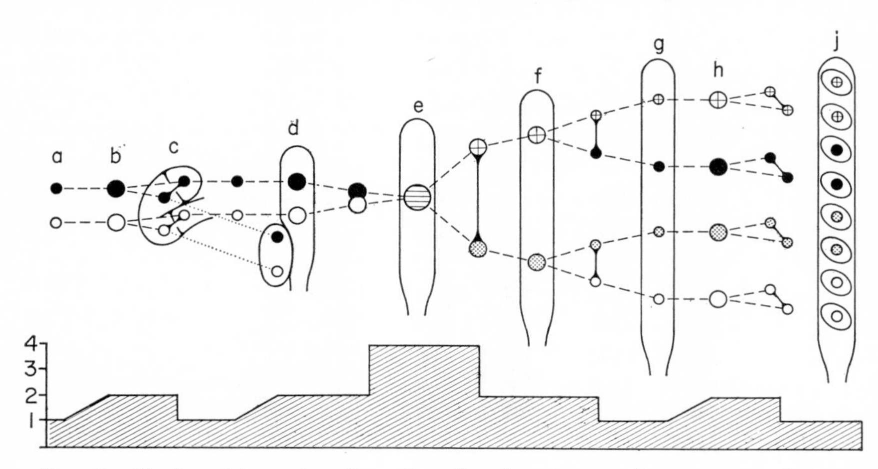

FIG. 3. Nuclear history in a linearly ordered eight-spored ascomycete. (a) Haploid nuclei of opposite mating type in the terminal hook of a fruiting hypha; (b) the same after DNA replication; (c) last premeiotic mitosis in the terminal hook (crozier); (d) two gamete nuclei, prior to fusion, but after completion of the last premeiotic DNA replication, in the ascus initial (the cell containing sister nuclei of the two gamete nuclei may form a second crozier, or even more, by the same process); (e) fusion, or zygote, nucleus—the only diploid nucleus in the entire life cycle; (f) the two nuclei resulting from the first meiotic division, well-separated in the ascus; (g) tetrad of nuclei resulting from the second meiotic division, in which the two spindles are tandem and parallel to the long axis of the ascus; (h) the same nuclei after the first postmeiotic DNA replication; (j) eight haploid nuclei resulting from the third division in the ascus (first postmeiotic mitosis), each cut out in a separate ascospore (in *Neurospora crassa* one further mitosis occurs before the spores mature). The dotted lines show the lines of descent of the nuclei in each ascospore. Based on cytological studies of *Neurospora* by McClintock (1945) and Singleton (1953). Histogram at the bottom is a diagrammatic representation of the amounts of DNA per nucleus at the stages pictured directly above, based on studies of *Neottiella rutilans* by Rossen and Westergaard (1966).

tions are: (1) Chromosomes of different parental origin do not coexist within a nucleus until the two haploid gamete nuclei fuse to form the zygote nucleus in the developing ascus (Fig. 3e). (2) The zygote nucleus is the only nucleus in the entire life cycle in which a full complement of chromosomes from both parents is present. (3) Complete haploidy is reestablished at the end of the second meiotic division in each of the four resultant nuclei. (4) Positions of spores in the linear array within the ascus accurately portray their lineages, permitting identification of the meiotic division at which segregation has occurred (spore patterns characteristic of first and second meiotic division segregations are illustrated in Fig. 4). Complete genetic identity between sister spores resulting from mitotic divisions of the tetrad of meiotic products is an assurance that segregation was completed by the end of the second meiotic division. These attributes, taken together, permit accurate inferences relative to meiotic events to be drawn from genetic observations.

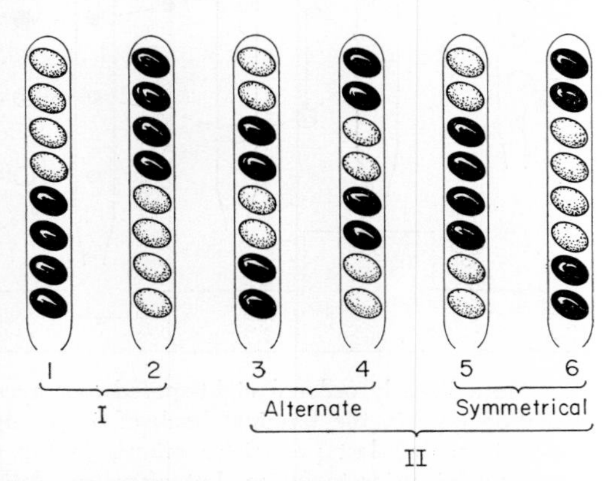

FIG. 4. Segregation patterns involving a single heterozygous locus. Spores carrying the mutant allele are represented as being white, those with the wild-type allele as black I: Patterns resulting from segregation at the first meiotic division, of which 1 and 2 represent different polarities. II: Patterns resulting from segregation at the second meiotic division, with 2, 3, 4, and 5, resulting from different combinations of polarity in the two second-division spindles.

b. *Timing of DNA Synthesis Relative to Meiosis.* The presence of a single diploid nucleus which immediately undergoes meiosis has long appeared to offer unique possibilities for determining the time of DNA synthesis relative to genetic recombination. Biochemical idiosyncrasies in the fungi (at least in *Neurospora*) have made direct labelling of chromosomes with tritiated thymidine impossible, but, taking advantage of the unusually large nuclei in *Neottiella rutilans,* Rossen and Westergaard (1966) have obtained reasonably accurate measurements by spectrophotometric methods. Whereas their method was sufficiently accurate to permit identification of doubling and quadrupling, and of equivalent reductions, in DNA content per nucleus, it was not sufficient to identify the exact onset and end of duplication. The histogram at the bottom of

Fig. 3 actually represents more than the authors report. Together with their cytological studies and the fine-structure studies of Westergaard and von Wettstein (1966), it is definitely shown that the last premeiotic DNA replication occurs before the gamete nuclei fuse, hence before any opportunity for genetic recombination. Their data permit the strong inference that the tetrads of nuclei resulting from meiosis have, to begin with, the DNA content characteristic of unreplicated haploid nuclei, and then undergo a replication preparatory to the first postmeiotic mitosis; actual measurements showed a complete range between unreplicated and replicated haploid amounts. The nuclei of young ascospores were shown definitely to have the amount of DNA per nucleus characteristic of unreplicated haploid nuclei.

It is unfortunate that, up to the present time at least, it has proved impossible to follow DNA replication in an ascomycete adaptable to genetic studies. Nuclei of nearly all ascomycetes are too small to permit accurate measurements by the method used by Rossen and Westergaard, and it has so far been impossible to culture *Neottiella rutilans*. In the unicellular alga, *Chlamydomonas reinhardii*, which produces eight unordered products following the first postmeiotic mitosis, DNA replication has been observed to be completed in the nuclei of the haploid gametes before fusion of the gamete cells (Chiang *et al.*, 1965; Sueoka *et al.*, 1967).

2. *Genetic Analyses*

a. *Genetic Characteristics.* A number of long-standing genetic postulates relative to the independent assortment of genes carried by different (nonhomologous) chromosomes can be directly tested in species with linearly ordered ascospores. The most extensive data known to me which make use of ordered eight-spored asci are discussed in relation to these postulates in Addendum II,C,1. These data from *Neurospora crassa* are in exceedingly close agreement with the interpretation that the assortment of genes on independent chromosomes is completely random both with respect to the poles of meiotic spindles to which alleles are distributed and to the distribution of alleles of one gene in relation to the distribution of alleles of another gene on a different chromosome.

b. *Chromatid Interference.* In genetically well-marked crosses, in which distances between heterozygous loci are sufficiently small that the possibility of double crossing-over between loci is negligible, tetrads of meiotic products, whether ordered or not, permit exact identifications of all types of double exchanges; there is no need for comparative analyses such as are required in handling half-tetrad data (Addendum II, B). This advantage, however, has not yet led to a clearer understanding of factors affecting chromatid interference. Lindegren and Lindegren (1939, 1942) were the first to report discrepancies from the expected 1:2:1 ratio of 2-strand, 3-strand, and 4-strand double exchanges in tetrad analyses. Because the number of observations reported was not large, because of a possibility (now largely discredited) that overlapping spindles in the second meiotic division might account for some observed discrepancies, and especially because of the generally accepted interpretation of random chromatid involvement (based on attached-X data), the Lindegrens' observations were not duly appreciated by most geneticists interested in recombination. Since then their results have been supported and extended by a number of investigators.

Reviews by Perkins (1955) and Emerson (1963) showed that significant deviations from expected frequencies of different types of double exchange were very commonly encountered, and that heterogeneity between different experiments was very great. The data summarized in the latter review showed the greatest discrepancy to be between 2-strand and 4-strand doubles—there were 615 and 476, respectively, with probability of less than 0.1% of so great a divergence from equality being due to sampling errors. The discrepancy between 3-strand doubles and the sum of 2-strand and 4-strand doubles was not great (977 and 1030, respectively—probability 24%), but the heterogeneity between experiments was very great (probability less than 0.1%). More recently published data continue to show discrepancies which are so far unaccounted for by any plausible interpretation. Especially striking are the data of Prakash (1964): Among 68 asci in his control experiments, 60% were 2-strand doubles, 22% 3-strand, and 18% 4-strand; in experiments using EDTA treatments, in a total of 275 asci, 24.4% were 2-strand, 22.2% 3-strand, and 53.4% 4-strand. These trends, a great deficiency in 3-strand doubles, a great excess of 2-strand doubles in the control, and of 4-strand doubles in treated material, were remarkably consistent whether the map distance between exchanges was small or large. The absence of a consistent relationship between relative frequencies of the different types of double exchanges and distance between exchange seems to be general in *Neurospora* crosses.

To the best of my knowledge, direct searches for causes of the vast heterogeneity observed in chromatid interference in *Neurospora crassa* (and also in *Saccharomyces cerevisiae*) have not been seriously attempted. It seems to me that such searches might result in information important to an understanding of recombination mechanisms.

III. Gene Conversion and Intragenic Recombination*

Recombination between different mutant alleles in a single locus is observationally of two kinds: reciprocal and nonreciprocal. Whether or not the two are fundamentally different is still not certain. So far there is no operational criterion by which reciprocal intragenic recombination can be distinguished from reciprocal crossing over between linked loci. Nonreciprocal recombination, on the other hand, involves aberrant segregation at one or more mutant sites—that is, it involves gene conversion.

A. Gene Conversion†

Gene conversion differs from the more usual kind of meiotic segregation in ways which are not evident in studies of random meiotic products. Tetrad analysis is essential, and an analysis of octads (or the use of some genetic method whereby sister products of the first postmeiotic mitosis can be identified) is necessary for the detection of some conversion patterns. Hence, the following discussion will deal almost entirely with the eight-spored ascomycetes.

1. *Patterns of Conversion at Single Mutant Sites*

The most commonly observed result of gene conversion is the recovery of three meiotic chromatids carrying the allele derived from one parent, and only one chromatid with the alternative allele from the other parent. The ratios then observed are 3+:1m and 1+:3m in tetrads, 6+:2m and 2+:6m in octads. Two additional aberrant ratios, 5+:3m and 3+:5m, are observable in octads in a number of instances. In these, two members of one spore pair are not identical, indicating that postmeiotic segregation has occurred during the ontogeny of one spore pair of the ascus. A fifth conversion pattern involves postmeoitic segregation in the production of two spore pairs, resulting in $4+:4m^{p}$ segregation which can be distinguished from normal $4+:4m^{m}$ (meiotic) segregation only when pairs of sister spores can be identified.

* The term gene, originally defined as a region of a chromosome within which crossing over does not occur, is now more commonly defined in terms of a region with a unitary function. The term cistron will here be used only in examples in which alleles are known to affect the same function. The term gene will be used more loosely to imply unity of one sort or another, including a unit of recombination based on observed coincidence of conversion at different mutant sites. Intragenic recombination will refer to the recombination between different alleles of such a loosely defined gene.

† See historical note, Addendum I.

Each of these five patterns has been observed in *Sordaria fimicola* (Kitani *et al.,* 1962), *Neurospora crassa* (Case and Giles, 1964), and *Ascobolus immersus* (Emerson, 1968). They are most clearly seen in the spore-color mutants of *S. fimicola* (Fig. 5), with its linearly ordered asci. (Only one of a number of variations in exact patterns of spores within asci has been illustrated for each conversion ratio. With an alternate orientation on the first division spindle, the four spores illustrated at the tip of the ascus would occur at the base. Alternative orientations of chromatids on second division spindles would reverse the positions of the tip

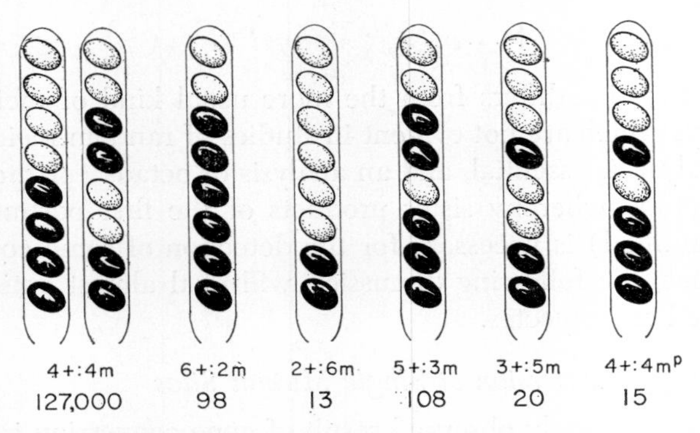

FIG. 5. Normal and aberrant segregation patterns produced in a cross between the gray-spored mutant and wild type of *Sordaria fimicola,* which has linearly ordered, eight-spored asci. Observed numbers of each pattern are listed below the segregation ratio occuring in each ascus. (After Kitani *et al.,* 1962.)

and/or basal spore pairs. Similarly, a different orientation of meiotic half-chromatids on spindles in the first post-meiotic mitosis would change the relative positions of members of unlike spore pairs.)

2. *Frequencies of Conversion Patterns*

There is considerable variation in total frequencies of conversion between species studied, between loci within a species, and even between mutant sites within a locus. A similar variation is observed in the relative frequencies of the different conversion patterns, though perhaps less so between alleles than between loci within a species. Selected examples illustrating such differences are summarized in Table II. It is noteworthy that each of the five conversion patterns is the most frequent in one example or another—those shown in bold-faced type in the table.

A few other remarks should be made concerning the examples summarized in the table. (1) All data are from crosses in which there was

TABLE II

VARIATIONS IN FREQUENCIES OF DIFFERENT CONVERSION PATTERNS AMONG SPORE-COLOR MUTANTS IN EIGHT-SPORED ASCOMYCETES

	Converted asci		Relative frequencies (%) of different conversion patterns[b]				
Example[a]	No.	%	6+:2m	5+:3m	4+:4m^p	3+:5m	2+:6m
1	99	0.20	31.3 (21.7–44.3)	**35.4** (24.8–49.3)	13.1 (7.1–22.5)	14.1 (7.9–23.6)	6.1 (2.2–13.1)
2	96	0.23	7.3 (2.9–15.0)	27.0 (18.0–39.5)	**36.4** (25.7–50.8)	23.9 (15.4–36.6)	5.2 (1.7–12.1)
3	97	0.23	6.2 (2.3–13.5)	28.8 (19.5–41.4)	**48.4** (35.8–64.3)	12.4 (6.5–21.6)	4.1 (1.1–10.5)
4	100	0.22	3.0 (0.6–8.8)	9.0 (4.1–17.1)	35.0 (24.7–48.8)	**40.0** (29.9–54.3)	13.0 (7.0–22.0)
5	49	0.20	0.0 (0.0–7.5)	**42.8** (27.1–65.5)	38.8 (24.1–60.2)	18.4 (8.6–34.7)	0.0 (0.0–7.5)
6	53	0.23	1.9 (0.05–10.5)	7.6 (1.3–19.5)	30.2 (17.4–49.1)	**52.9** (35.7–76.0)	7.6 (1.3–19.5)
7	47	0.22	4.3 (0.5–15.4)	23.4 (11.9–41.5)	23.4 (11.9–45.5)	**42.6** (26.8–65.6)	6.4 (1.3–18.7)
8	6	0.08	0.0 (0.0–61.7)	—	Undetected	—	**100** (38.3–100)
9	162	0.73	3.1 (0.1–6.4)	—	Undetected	—	**96.9** (93.6–99.9)
10	644	2.11	**51.2** (46.8–56.6)	—	Undetected	—	48.8 (43.4–54.2)
11	169	0.50	**75.4** (68.0–82.8)	—	Undetected	—	24.6 (17.2–32.0)
12	74	0.26	**58.1** (40.5–70.3)	—	Undetected	—	41.9 (29.7–59.5)
13	139	4.0	15.8 (10.1–23.7)	**64.0** (51.8–78.4)	Undetected	17.3 (11.5–25.9)	2.9 (0.8–7.2)
14	571	8.4	**45.0** (39.8–50.3)	10.2 (7.7–13.1)	Undetected	23.8 (20.3–25.0)	21.0 (17.5–25.0)
15	991	12.3	22.1 (19.3–24.9)	17.5 (15.8–20.3)	1.2 (0.6–2.1)[c]	**41.4** (37.4–54.2)	17.9 (15.2–20.4)
16	2661	17.7	**71.8** (68.7–74.4)	3.2 (2.6–3.9)	Undetected	4.8 (4.1–5.9)	20.2 (18.5–21.9)
17	439	18.6	**72.0** (64.7–79.3)	1.1 (0.5–2.9)	0.9 (0.2–2.3)[c]	0.7 (0.1–2.0)	25.1 (20.8–30.1)
18	693	18.7	**46.8** (42.2–51.1)	16.3 (13.6–19.5)	0.9 (0.3–1.8)[c]	16.7 (13.9–20.1)	19.3 (16.2–22.8)

[a] Examples 1 to 7, respectively, alleles g_1, h_2, h_{2a}, h_3, h_{3a}, h_4, and h_{4b}, at the gray-spored locus in *Sordaria fimicola* (Kitani and Olive, 1967b); 8 to 10, respectively, alleles 1604, 63, and 137, of series 46 of *Ascobolus immersus* (Rossignol, 1964); 11 and 12, respectively, alleles 77 and 775 of series Y of *Ascobolus immersus* (Kruszewska and Gajewski, 1967); 13, mutant w-6 × +, 14, mutant w-62 × +, 15, double mutant w-62 gr-1 × ++, 16, w-10(P) × +, 17, w-10(P) × gr-1, and 18, w-78(p) × gr-1, among which w-10 and w-78 are presumptive alleles and gr-1 an independently segregating marker (Emerson and Yu-Sun, 1967, 1968).

[b] Figures in parentheses are 95% confidence limits based on frequencies among total asci, method of W. L. Stevens (Fisher and Yates, 1957, Table VIII_I). Figures in bold face indicate the most frequent pattern.

[c] Postmeiotic 4+:4m^p segregation in unordered asci was detected by the use of a second, unlinked, ascospore character.

a single heterozygous allele at the converting locus. (2) Corrected frequencies have been used for the examples from Rossignol (1964) and from Kruszewska and Gajewski (1967) in which genotypic tests frequently do not confirm phenotypic determinations. The correction involves multiplying the numbers obtained from phenotypic scoring by the fraction of tested phenotypes which were substantiated by genetic tests. (3) Our own data have not been so corrected because, in the particular samples chosen, phenocopies have proved to be of negligible frequency, and because normal rates of spontaneous mutation to new spore-color mutants are very low in comparison to the high conversion rates occurring. Many spore-color mutants in the Pasadena strain of *Ascobolus* do have the more usual low conversion frequencies (Yu-Sun, 1966).

3. *Relation of Conversion to Recombination of Flanking Markers*

It has long been known (e.g., Mitchell, 1955) that, in a population selected for wild-type intragenic recombinants, crossing-over between loci which are relatively close to the converting locus, and lying one on each side of it, is greatly enhanced over that occurring in the general unselected populations. Frequencies of recombination between flanking markers in populations of wild-type intragenic recombinants often approximate 50%, no matter how closely linked the flanking markers. A similar increase in recombination between flanking markers is found to be associated with gene conversion in monoallelic crosses, though the published data on this point are not as extensive. The most thoroughly examined case so far reported involves the gray-spored locus of *Sordaria fimicola* (Kitani *et al.*, 1962; Kitani and Olive, 1967a,b).

This *Sordaria* example is especially interesting because all five kinds of conversion ratios (with the possible exception of 2+:6m) occur in reasonably high frequencies in crosses involving one or another allele, and because the frequency of flanking marker recombination varies with the conversion ratio occurring. In the pooled results from crosses of all alleles to wild type, flanking marker recombination is greatest in $4+:4m^p$ asci, 52.1% (standard error $\pm$ 3.8%), followed in order by 5+:3m with 46.9 $\pm$ 3.1%, 3+:5m with 41.7 $\pm$ 2.4%, 2+:6m with 39.5 $\pm$ 7.9%, and 6+:2m which has only 27.8 $\pm$ 3.9%. As shown in Table III, variations in responses of the different alleles is sufficiently small to warrant pooling their results, with the exception of the 3+:5m asci, in which the probability that the observed heterogeneity between alleles is due to sampling errors is only 7%. The statistical validity of the differences observed between conversion patterns is also indicated in this table. The relation of 2+:6m asci to flanking marker recombination does not differ signifi-

TABLE III
PROBABILITIES THAT OBSERVED DIFFERENCES IN FLANKING MARKER RECOMBINATION FREQUENCIES ARE DUE TO SAMPLING ERRORS

	6+:2m	5+:3m	4+:4m^p	3+:5m	2+:6m
			A. Heterogeneity between alleles[a]		
	0.72	0.40	0.42	0.07	0.30
			B. Differences between patterns[b]		
	6+:2m	5+:3m	4+:4m^p	3+:5m	2+:6m
2+:6m	0.17	0.37	0.17	0.80	—
3+:5m	0.01	0.27	0.05	—	0.80
4+:4m^p	0.001	0.29	—	0.05	0.17
5+:3m	0.001	—	0.29	0.27	0.37
6+:2m	—	0.001	0.001	0.01	0.17

[a] Probabilities based on χ^2 with 4 degrees of freedom, except 6+:2m and 2+:6m with 1 degree of freedom.
[b] One degree of freedom.

cantly from that of any other pattern, whereas that of 6+:2m asci does differ significantly from all others except 2+:6m.

4. *Number of Chromatids Taking Part in Conversion*

a. *Inference That Two Chromatids Interact.* In those instances in which conversion has resulted in 4+:4m^p segregation, two of the four meiotic chromatids have been affected—both became hybrid at the mutant site concerned by the time they separated in the second meiotic division (Fig. 6). In other instances, those resulting in 5:3 and 6:2 segregations, only one chromatid has an altered genotype: In 5:3 segregations it is hybrid at the converting site, in 6:2 segregations it is completely altered to the allelic state carried by the two sister chromatids homologous to it. Inasmuch as conversion of one allelic state to another occurs only when the acquired state is coexistent with that which is altered, some role in conversion must be ascribed to the genetic constitution of one or both sister chromatids of the homologous chromosome. The most direct inference (relative to 6:2 segregations) is that two homologous chromatids take part in some interaction which results in the alteration of the mutant site in one of them, but leaves no evidence of alteration at the corresponding site in the other. Genetic evidence, however, does not make this inference inescapable.

In instances in which flanking marker recombination accompanies conversion, chromatids which have become heterozygous in the process have also, as a rule, taken part in the intergenic exchange. This is most definitely shown in examples of 4+:4m^p segregation, in which those two chromatids becoming heterozygous at the converting site are both recombinant for flanking markers (Fig. 7) in nearly every instance (Kitani and Olive, 1967b). In instances in which 5:3 segregation is accompanied by flanking marker recombination, that chromatid which became heterozygous at the converting site is also usually recombinant for flanking markers, e.g., in 30 of 31 asci segregating 5+:3m (Kitani *et al.*, 1962), with the single exceptional case compatible with the frequency of such crossing-over in the absence of conversion at the included locus. If one

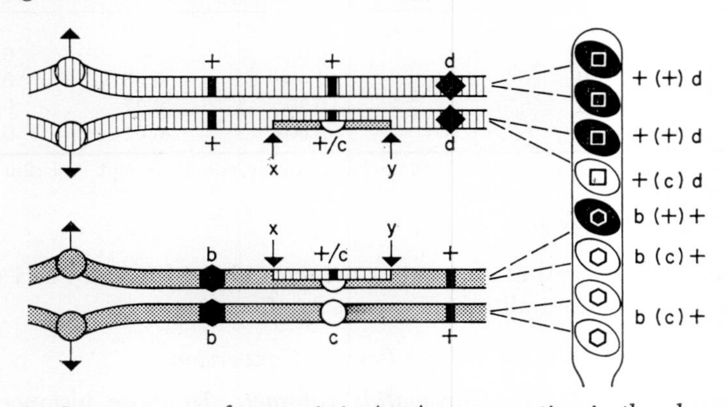

FIG. 6. Interpretation of postmeiotic 4+:4mp segregation in the absence of recombination between flanking markers. At the right, observed spore pattern in the ascus with genotypic constitutions of the spores. Broken lines indicate the origins, during the first postmeiotic mitosis, of members of each spore pair from particular chromatids which are represented as just entering anaphase of the second meiotic division. Arrows attached to centromeres indicate the poles to which chromatids will pass on the second division spindles—the upper pair have separated from the lower during the first meiotic division. Chromatids, or regions thereof, derived from one parent have vertical-line shading, those from the other parent are stippled. The locus undergoing conversion, c/+, is heterozygous at only one site; it governs ascopore color. The proximal flanking marker is b/+, the distal flanking marker d/+.

accepts the inference that two chromatids have interacted in conversion, the second recombinant chromatid, unaltered at the converting site in 6:2 segregations, is that which had interacted with the chromatid becoming heterozygous.

b. *Possible Involvement of More Than Two Chromatids.* There is strong evidence from tetrad and half-tetrad analyses of intergenic crossing-over (see Section II and Addendum II,B and D) in support of the inference that two, and only two, of the four meiotic chromatids are involved in genetic exchange in any one short interval of a genetic map. Similarly, from tetrad analyses of gene conversion one can infer that no more than two chromatids are involved at any one mutant site; though there are sporadic observations that suggest the possible, occasional involvement of more than two chromatids. In some mutant crosses in which 6+:2m conversion patterns predominate, rare occurrences of 8+:0m patterns are encountered (personal observation); and rare occurrences of 7+:1m have been observed in crosses in which the predominant patterns are 5+:3m and 6+:2m (Kitani *et al.*, 1962; Kitani and Olive, 1967b). Asci in which more than six members of the octads carry

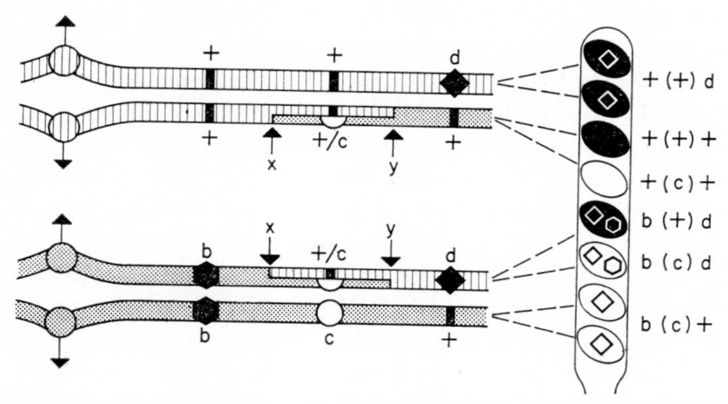

FIG. 7. Interpretation of postmeiotic 4+:4mp segregation when there is recombination between flanking markers. Conventions as in Fig. 6.

one allele are probably sufficiently infrequent to be accounted for by some occurrence which is independent of conversion, such as mutation —their occurrence is almost certainly sufficiently infrequent to justify omitting them from considerations of the nature of the conversion process at this time. There is, however, a report of one completely analyzed ascus (Case and Giles, 1964) in which three chromatids were definitely involved—in this ascus of a triallelic cross, one meiotic chromatid was heterozygous (at the end of the second meiotic division) at the middle and distal mutant sites and was also reciprocally recombinant with a second chromatid between the proximal and middle sites, and a third chromatid was heterozygous at the distal site (see Fig. 16, in Section III,B,3,a).

B. INTRAGENIC RECOMBINATION

1. *Observations on Wild-Type Recombinants*

Intragenic recombination has been most intensively studied in connection with fine-structure mapping of alleles within a gene, often within a cistron, where the relation between genetic maps and complementation maps is of especial interest (Catcheside and Overton, 1958; Woodward *et al.*, 1958; Leupold, 1961; de Serres, 1963; Catcheside *et al.*, 1964; Fincham, 1966). For such fine-structure mapping the usual practice is to determine the frequencies of wild-type recombinants among random meiotic products in crosses of one allele to another (i.e., when mutant alleles are in repulsion—in a *trans* configuration). The very low frequencies in which interallelic recombination ordinarily occurs makes it almost imperative to use selective screening techniques, as in selecting for prototrophic recombinants between auxotrophic alleles, if one is to obtain significant numerical data. The method has, in general, proved satisfactory for establishing the linear order of allelic sites within a gene, and its relation to the order of other genes on the chromosome map (Jessop and and Catcheside, 1965; Catcheside, 1966), though additivity between regions in map distances so obtained is often far from perfect (Holliday, 1964).

On the other hand, with one exception, use of this selective technique has not been fruitful in furnishing information about the nature of the intragenic recombination process. The exception has to do with the correlation between intragenic recombination and recombination between flanking markers, which was first observed by this method (Mitchell, 1955). Further, it was soon found that a bias exists in frequencies of association of wild-type interallelic recombinants with the two non-parental combinations (Freese, 1957a,b), and sometimes with the two parental combinations (Murray, 1963), of flanking markers. These polarities in association of wild-type intragenic recombinants with different combinations of flanking markers have been postulated to reflect unequal frequencies of conversion at the two mutant sites concerned (Murray, 1963; Jessop and Catcheside, 1965). Direct evidence that such a relationship does exist has been obtained by genetic analysis of tetrads containing wild-type recombinants.

2. *Tetrads and Octads with Wild-Type Recombinants*

a. *Polarity with Respect to Flanking Markers.* Very extensive analyses have recently been made by Fogel and Hurst (1967) of tetrads of *Saccharomyces cerevisiae* which contained one wild-type recombinant between alleles of the *histidine-1* locus, with *threonine-3* and *arginine-6* as flanking markers. The genetic map of this region is illustrated in Fig. 8, which shows the relative distances between alleles, based on the frequencies of wild-type recombinants, their linear order with respect to the flanking markers, and the relationships of alleles involved in different crosses. Their data from 834 recombinant asci, in which all four spores were completely identified, are summarized in Table IV.

Three different origins of wild-type interallelic recombinants were observed: one from reciprocal recombination between alleles; one from conversion of the more proximal mutant allele (1) to wild type (+); and one from conversion of the more distal mutant allele (2) to wild type. Each of these three origins was found to be associated with strongly biased distributions of the flanking markers, the bias being different in each.

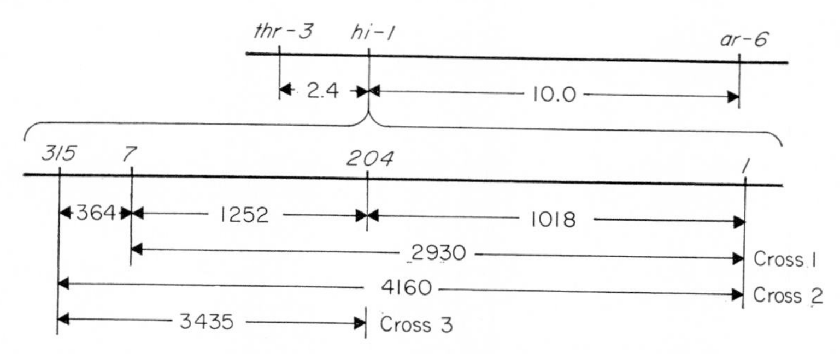

FIG. 8. Top heavy line: Portion of the chromosome map showing the position of the *histidine-1* locus to that of the proximal, *threonine-3,* and distal, *arginine-6,* flanking markers; distances between loci are shown in standard map units. Lower heavy line: Map showing positions of alleles *315, 7, 204,* and *1,* of the *histidine-1* locus; distances between allelic sites are shown as the number of asci per million containing wild-type recombinant spores. (After Fogel and Hurst, 1967.)

The reciprocal recombinants between alleles are perhaps the most interesting and enlightening, as they constitute the only large sample of asci of this sort so far reported. Over 95% of the asci in this group involve recombination between flanking markers, and in 92.3% both wild-type (+ +) and double mutant (1 2) interallelic recombinant chromatids were recombinant for flanking markers, with the specific associations expected from reciprocal crossing over between the two *hi-1* alleles (Fig. 9). A second noteworthy point is that the total frequency of reciprocal interallelic recombination varied significantly ($P < 0.001$) depending upon which two alleles were intercrossed: 14.8% (95% confidence interval 11.7–18.3%) in cross 1, involving alleles *7* and *1*; 8.0% (4.9–12.3%) in cross 2, involving alleles *315* and *1*; and 1.6% (0.3–7.0%) in cross 3, involving alleles *315* and *204*. Reference to the fine-structure map (Fig. 8) shows that reciprocal recombination is most frequent in crosses including the region between alleles *204* and *1*. On the other hand, reciprocal interallelic recombination does occur when that interval is excluded (cross 3), making it difficult to assume that reciprocal recombination occurs at only one point in the fine-structure map. In this connection it is noteworthy that, among all asci resulting from conversion, the fraction in which conversion occurred at the more

TABLE IV

TETRAD PATTERNS IN ASCI WITH WILD-TYPE INTRAGENIC RECOMBINANTS[a]

Tetrad genotypes at the *hi-1* locus with respect to flanking marker combinations[b]				Observed numbers of asci			
Parental		Recombinant		Cross 1	Cross 2	Cross 3	Total
+–d	b–+	b–d	+–+				
+ (++) d + (1 +) d	b (1 2) + b (+ 2) +	—	—	2	1	0	3
+ (++) d	b (1 2) +	b (+ 2) d	+ (1 +) +	1	0	0	1
+ (1 +) d	b (1 2) +	b (++) d	+ (+ 2) +	3	0	0	3
+ (1 +) d	b (+ 2) +	b (++) d	+ (1 2) +	65	17	2	84
Total reciprocal intragenic recombinants				71	18	2	91
+ (++) d + (1 +) d	b (+ 2) + b (+ 2) +	—	—	173	81	64	318
+ (++) d	b (+ 2) +	b (+ 2) d	+ (1 +) +	17	5	6	28
+ (++) d	b (+ 2) +	b (1 +) d	+ (+ 2) +	1	5	1	7
+ (1 +) d + (+ 2) d	b (++) + b (+ 2) +	—	—	1	0	0	1
+ (+ 2) d	b (++) +	b (1 +) d	+ (+ 2) +	2	0	0	2
+ (1 +) d	b (+ 2) +	b (++) d	+ (+ 2) +	113	66	29	208
+ (+ 2) d	b (+ 2) +	b (++) d	+ (1 +) +	2	0	0	2
+ (1 +) d	b (+ 2) +	b (+ 2) d	+ (++) +	41	20	8	69
Total conversion of (1) to (+)				350	177	108	635
+ (1 +) d + (1 +) d	b (++) + b (+ 2) +	—	—	17	9	8	34
+ (1 +) d	b (++) +	b (+ 2) d	+ (1 +) +	0	2	0	2
+ (1 +) d	b (+ 2) +	b (++) d	+ (1 +) +	41	15	10	66
+ (1 +) d	b (+ 2) +	b (1 +) d	+ (++) +	2	3	1	6
Total conversion of (2) to (+)				60	29	19	108
Grand total				481	224	129	834

[a] From study by Fogel and Hurst (1967) on the *histidine-1* locus of *Saccharomyces cereviseae*. Cross, + (1 +) d × b (+ 2) +.

[b] Alleles of *hi-1* enclosed in parentheses; b and d are flanking markers.

distal allele is the same in all crosses—14.6% (11.3–18.4%), 14.1% (9.8–19.7%), and 15.0% (9.7–22.6%), respectively, for crosses 1, 2, and 3.

Convertants at the two mutant sites were exactly opposite in their relative frequencies of association with two parental combinations of flanking markers: Among (1 +) to (+ +) convertants, the ratio of one parental combination to the other is 353:3; among (+ 2) to (+ +) convertants it is 0:34. As illustrated in Fig. 10a and b, this is the result to be expected, because conversion has occurred in chromatids of opposite parental constitution in the two instances.

When conversion is accompanied by recombination between flanking markers, the situation is less straightforward. Among convertants (1 +)

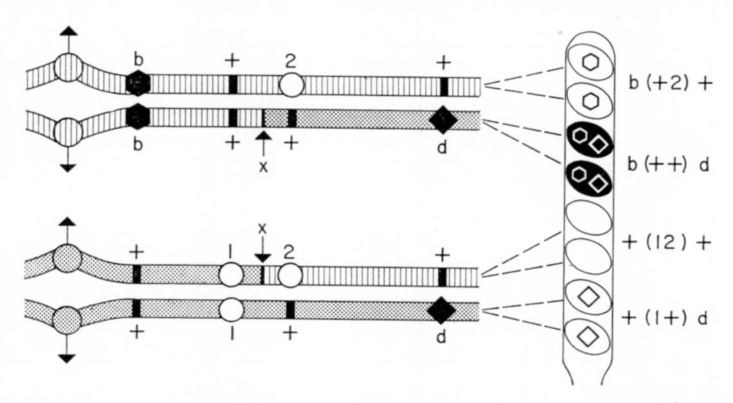

FIG. 9. Reciprocal interallelic recombination, resulting in two wild-type spores (+ +), two single-mutant spores of each parental type (1 +) and (+ 2), and two double-mutant spores (1 2), with respect to alleles at the center locus. Both wild-type and double-mutant types are interallelic recombinants and were present in the two chromatids which are recombinant with respect to the flanking markers b/+ and c/+. Conventions as in Fig. 6.

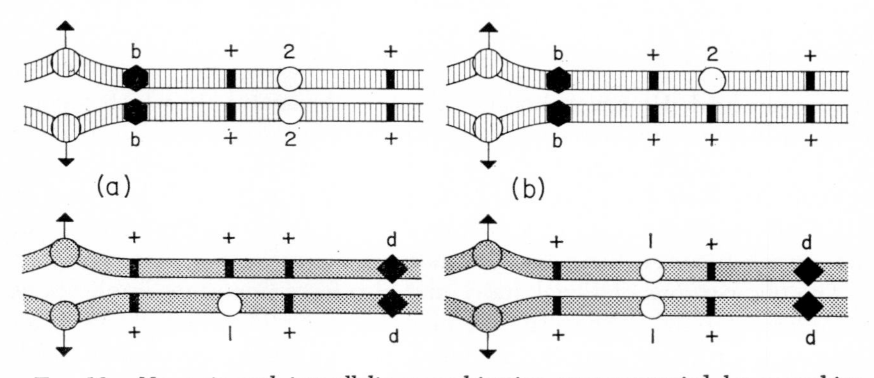

FIG. 10. Nonreciprocal interallelic recombination unaccompanied by recombination between flanking markers. (a) Conversion of allele 1 to + occurs in a chromatid with the parental flanking marker combination + —— d. (b) Conversion of allele 2 to + occurs in a chromatid with the parental flanking marker combination b—— +.

to (++) the observed frequencies of b–d and +–+ flanking marker combinations are 210 and 69, respectively; among (+ 2) to (++), 66 and 6. The predominant recombinant combination is the same whichever allele has been converted; but numerical differences are significant at the 1% level—frequencies of the less frequent class are 24.7% (19.7–30.1%) for conversion at the proximal site, 8.3% (3.2–17.4%) for conversion at the distal site. What this difference is due to is open to question. One might infer that the majority class required crossing-over proximal to allele (1) when that allele was converted, and distal to allele (2) when

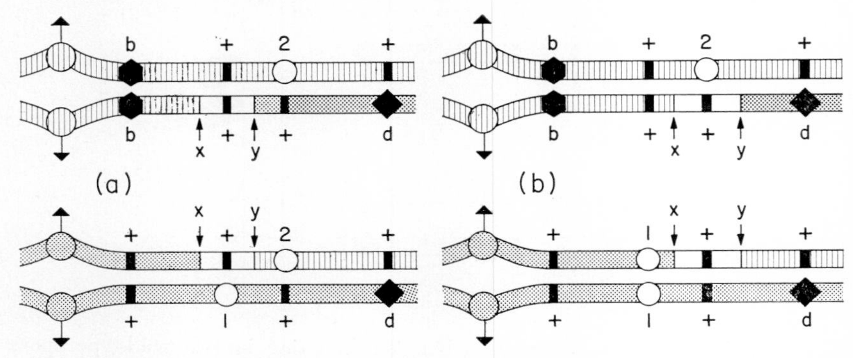

Fig. 11. Nonreciprocal interallelic recombination accompanied by recombination between flanking markers. Conversion of allele 1 to + (a) and of allele 2 to + (b) occur in the recombinant chromatid with flanking marker combination b —— d. (a) Because both chromatids recombinant for flanking markers have the + allele at mutant site 1, it is impossible to distinguish between crossing-over at x, proximal to site 1, and at y, between mutant sites. (b) Both chromatids recombinant for flanking markers carry the + allele at mutant site 2; crossing over could have occurred between mutant sites, at x, or distal to site 2, at y.

it was involved. However, because the two chromatids in which recombination occurs are homozygous for the converted allele, it is equally possible, as indicated in Fig. 11a and b, that recombination occurred between the two allelic sites in both cases—at y instead of at x in Fig. 11a, and at x instead of at y in Fig. 11b. Genetic observations do not distinguish between these two alternatives.

The observations of Fogel and Hurst also have important implications for interpretations of fine-structure map order relative to interlocus order when fine-structure mapping is based on frequencies of random wild-type interallelic recombinants. For example, one recombinant combination of flanking markers is contributed to by the major fraction of recombinants accompanying reciprocal interallelic recombination as well as those accompanying conversion of the proximal allele, whereas a similar contribution to the other flanking marker recombination comes only from conversion to the distal allele.

b. Polarity Within a Locus: The Polaron. Additional information about interrelations between alleles has been gained from the extensive studies of interallelic recombination between white-spored mutants of *Ascobolus immersus* by Rizet and his students (Rizet *et al.*, 1960a, b; Lissouba, 1960; Lissouba *et al.*, 1962; Rizet and Rossignol, 1963; Rossignol, 1964). Very strong directional biases in frequencies of conversion, and restriction of reciprocal recombination to limited regions, have been established to occur in at least two loci.

The strongest evidence for polarity in gene conversion was obtained in their series 46. Data relating to seven alleles (those concerning which there is the most information) are summarized in Fig. 12. Four relationships within this series have been demonstrated, viz.:

(1) Functional tests for allelism are available for six of the seven mutants—all but mutant 46. Negative complementation in heterokaryotic ascospores was obtained in all combinations of mutants 1216, 137, and 277, and between each of these with mutants 1604, 63, and W.

(2) As can be seen in Fig. 12, on the basis of frequencies of 2+:6m asci, only one map order is possible for mutants 1604, 63, 46, W, and 1216. Additivity, however, is far from perfect, as longer regions have recombination frequencies larger than the sums of component smaller regions: recombination between 1064 and 1216 is 12.3 (10.1–14.7) per thousand, whereas the sum of regions 1604 to 46 and 46 to 1216 is only 3.07 (2.43–3.99), and the sum of the four smallest intervals only 1.38 (1.00–2.03). This is a fairly general characteristic of interallelic recombination, which Holliday (1964) has termed map expansion; it has been interpreted (Emerson, 1966a; Emerson and Yu-Sun, 1967) to result from "concident" conversion at closely linked mutant sites (origin C in Section III,C,3,c).

The map position of mutant 137 relative to W and 1216 is somewhat ambiguous. Whether the order is W–1216–137, or W–137–1216, recombination between the two outside mutants would be less than the sum of that occurring in the two intervals. Recombination between 137 and each of 46, 63, and 1604, however, is significantly greater in each instance than recombination between 1216 and the same three mutants, giving some assurance that the order shown in Fig. 12 is correct. There is no strong support from recombination frequencies for positioning mutant 277 at the extreme right of the map—this has been done because it differs from the others relative to reciprocal interallelic recombination.

(3) In all intragenic crosses except those involving mutant 277, tested

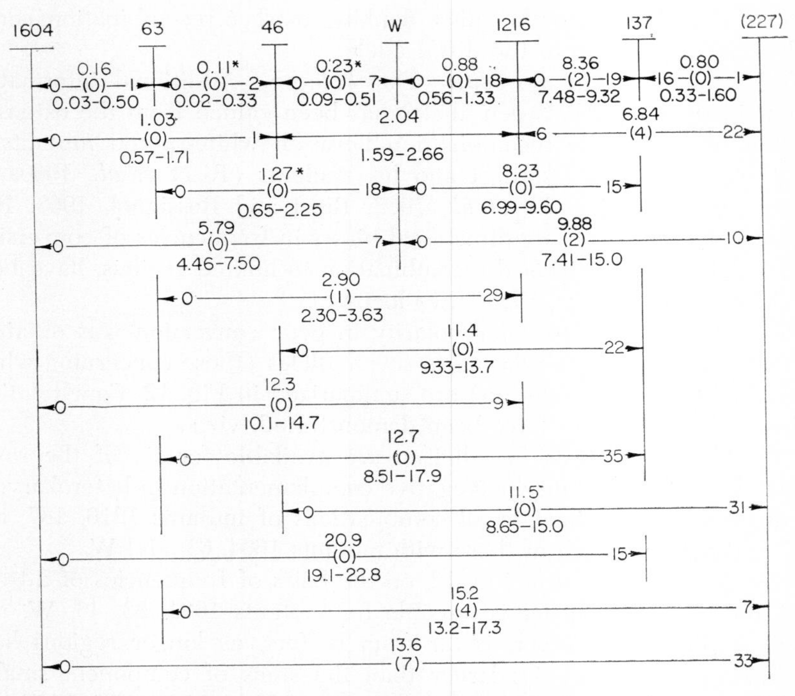

FIG. 12. Frequencies and kinds of wild-type recombinants between alleles of series 46 in *Ascobolus immersus*. The horizontal lines (double-headed arrows) designate the alleles between which recombination was sttudied. Observed frequencies of 2+:6m asci per thousand are shown above each such line, with the 95% confidence limits below the line. Numerals in the lines refer to the observed numbers of each of three kinds of recombination determined by genetic tests of a sample of 2+:6m asci from each cross: those bracketed in the center are reciprocal interallelic recombinants; those near the ends of the lines are convertants (segregating 6+:2m) of alleles at the corresponding ends. Data summarized from Lissouba (1960) and Rossignol (1964), the latter including much of that published in the former. Shown are frequencies from Rossignol; those from Lissouba are 63 to 46, 0.53 (0.20–1.14); 46 to W, 0.75 (0.43–1.41); 63 to W, 2.84 (1.70–4.51).

2+:6m asci resulting from nonreciprocal recombination have invariably had conversion (6+:2m segregation) at the mutant site to the right—never, in 198 asci, at that to the left. In crosses with mutant 277, on the other hand, 22 of 126 tested asci with nonreciprocal recombination resulted from conversion at the site on the left. Furthermore, all of these occurred at the two mutant sites presumed to be nearest it: 16 of 17 in the cross to 137, and 6 of 28 in the cross to 1216.

(4) There is the additional probability that, in crosses of individual

mutants to wild type, the frequency of conversion from mutant to wild type increases from left to right on the map. Lissouba (1960, Table IX) reported the following frequencies of 6+:2m asci in crosses to wild type: 4.59 per thousand for mutant 63, 1.90 for 48, 3.87 for W, and 7.17 for 137. He also reported (Lissouba, 1960, Table VIII) that phenotypic scoring often failed to give accurate frequencies of 6+:2m genotypes in this series—only 7 of 11 tested asci from the cross of 137 to wild type were correctly determined, the other four being phenocopies, so that only 63% (30–90%) of the observed should be considered as true conversions. Rossignol (1964) observed a similar relationship in a larger sample of asci from the same cross—only 69% (58–77%) of 108 asci had been correctly determined by phenotype. Rossignol observed a larger error in phenotypic determinations in the cross 63 X wild type in which only 9% (1–28%) of 23 tested asci had been correctly determined—and a still greater error in the cross 1604 X wild type, in which none (95% confidence interval 0–9%) were genotypically 6+:2m. Corrected frequencies of conversion to wild type thus range from zero (0–0.01) x 10^{-3} for 1604 at the left end of the map, through 0.22 (0.01–0.55) for 63, to 4.56 (2.10–6.39) at 137, from Lissouba's data, or 10.2 (8.15–13.6), from Rossignol's data. Judged solely from phenotypic counts, intermediate frequencies probably occur at mutant sites 46, W, and 1216, but there is no published record of genotypic determinations at these sites.

A possible fifth relationship in series 46 is a local restriction to the production of reciprocal intragenic recombination. Frequencies of such reciprocals in crosses between mutant 277 and other alleles was 6.0% (2.6–11%), and in other intercrosses only 1.5% (0.3–3.7%). The probability of chance variations as great or greater than observed between these two groups is about 0.04. Furthermore, all reciprocals which did not involve mutant 277, did involve either mutants 137 or 1216, both near the right end of the map.

Rossignol (1964) has studied a number of other mutants belonging to series 46. Although data concerning some of these are not too extensive, it is possible to place one between 1604 and 63, four between 46 and W, and two to the right of W, all in an order giving complete polarity in conversion, and reasonably good additivity in map units (making allowance for map expansion, and for the small numbers involved in some instances).

c. Locally Restricted Reciprocal Recombination and Reversal of Polarity. Somewhat different results have been observed in another series of closely linked white-spored mutants of *Ascobolus immersus* (Lissouba *et al.*, 1962). In series 19, mutants are arranged in three tightly linked groups—maximum interallelic recombination in each of groups A and B is 0.25×10^{-3}, and in group C only 0.02×10^{-3}. Because of the low frequencies, exact positioning of alleles within groups is difficult; and conversion polarity has not been determined for all combinations of

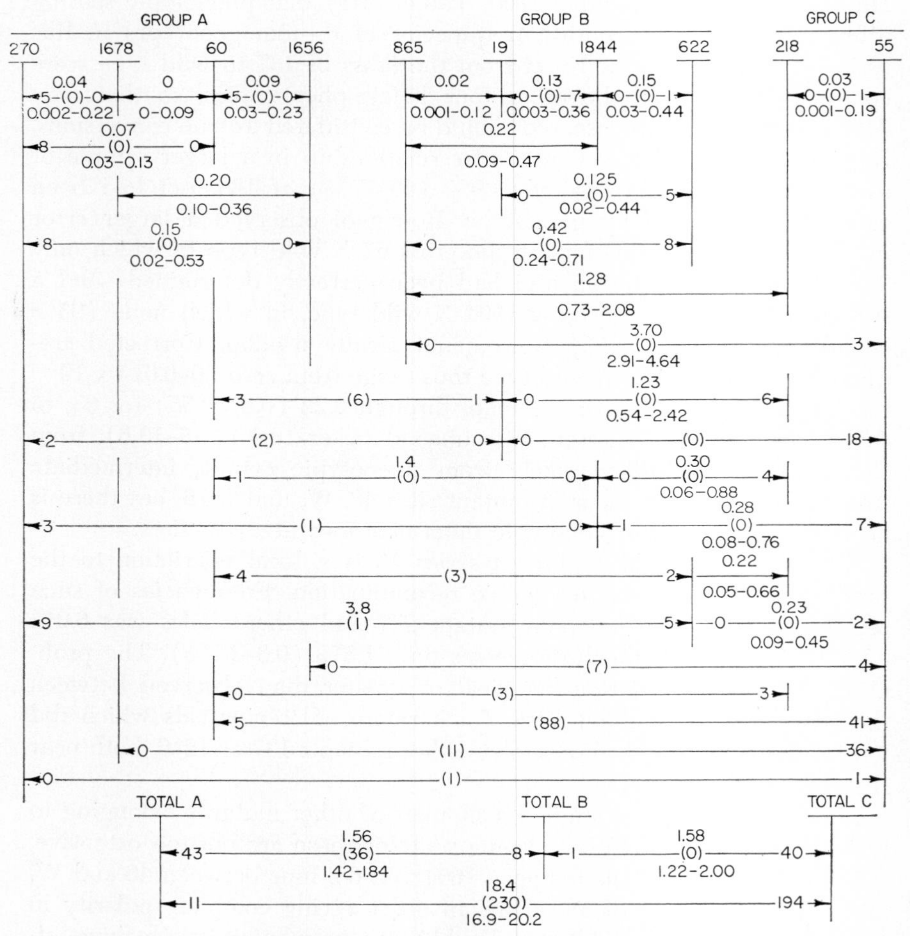

FIG. 13. Frequencies and kinds of wild-type recombinants in series 19 of *Ascobolus immersus*. Conventions as in Fig. 12. Data from Lissouba *et al.* (1962).

alleles. Nevertheless, it is possible to arrange 10 alleles in a reasonably consistent order, as shown in Fig. 13. On this arrangement, there is complete polarity within each group: of 26 tested asci from intercrosses between group A alleles, all have converted at the site to the left; whereas in 21 asci from group B intercrosses, and one ascus from those of group C, all have converted at the site to the right.

Data for intercrosses between groups are more extensive. In intercrosses between alleles of group A and group B, 84% (72–93%) of tested converted asci had conversion occurring at the site to the left; in intercrosses between alleles of groups A and C, this frequency was only 5.4% (2.7–9.4%); and between alleles of groups B and C, only 2.5% (0.06–12.5%). There is a definite reversal of polarity in A x B crosses from that occurring in A x C and B x C crosses.

No reciprocal recombination was observed in crosses between alleles of a single group (asci tested, 26, 21 and 1, in groups A, B and C, respectively), nor in intercrosses between groups B and C (41 asci tested). In crosses between members of group A and members of groups B or C, 41.4% (31.0–52.3%) and 44.6% (40.0–49.4%), respectively, of tested asci resulted from reciprocal recombination, all of which may be assumed to have occurred in the region between the rightmost allele of A and the leftmost of B.

3. *Intragenic Recombination in Unselected Octads*

a. Complete Genetic Analysis of Unselected Asci. In most known examples of intragenic recombination, very few of the total asci in an unselected population can be expected to show the effects of such recombination. To detect these rare examples it is necessary to isolate the eight spores from each of a very large number of asci, to derive cultures from each isolate for the detection of morphological or biochemical characteristics associated with commonly used flanking markers, and to intercross each isolate with tester stocks for the identification of the alleles involved in recombination. As a consequence, very few such analyses have been attempted; but very important information, unattainable by the use of the usual selective techniques, has resulted from those few.

The earliest study of this kind (Case and Giles, 1958a) involved a cross between alleles B5 and B3 at the *pan-2* locus of *Neurospora crassa*. In this instance, 1399 asci were dissected and analyzed in order to recover seven critical asci in which all eight products could be tested—four others also had obviously resulted from intragenic recombination, but one or more spores in each had failed to germinate. Of those completely analyzable, one resulted from reciprocal recombination between alleles, and four others from conversion from mutant to wild type at one mutant site—all of these could have been detected by selective techniques. Two others resulted from simultaneous conversion at both mutant sites: from mutant to wild type in one, and from wild type to mutant in the other, resulting in no wild-type recombinants, hence undetectable by selective methods. These two are important because they indicated that both heterozygous sites in a single chromatid had undergone simultaneous conversion to the allelic states present in the two sister chromatids of the homologous chromosome. (One further ascus, with only seven analyzable products, almost certainly involved postmeiotic segregation—an occurrence unexpected at that time.)

The double mutant (B5 B3), produced by reciprocal recombination in the above cross, was then crossed to wild type (Case and Giles, 1958b) to study recombination between genes in coupling (*cis* configuration), instead of in repulsion (*trans*). Here 8 of 1003 dissected asci were of the desired kind. There was conversion from mutant to wild type at the B5 site in 3, and at the B3 site in 2. Again, 2 asci showed simultaneous conversion at both sites, segregating 6+:2m at both sites. Among the 4 meiotic chromatids segregating at the second meiotic division, just as in the previous cross, 3 were of one parental constitution at both sites, and 1 of the other parental constitution, indicating a change covering both sites. The remaining ascus segregated 4+:4m for B3, but 8+:0m for B5. Such segregation could have been the result of spontaneous mutation—if due to conversion, more than two chromatids must have been involved at the same mutant site (see Section III,A,4,b).

More recently Case and Giles (1964) have studied recombination in a triallelic cross: (B23 + B72) x (+ B36 +), alleles in the *pan-2* cistron, again with appropriate flanking markers. Among 1457 analyzed asci there were 11 completely analyzed which showed intragenic recombination, and 2 others incompletely analyzed—none with wild-type recombinants. Of the 6 asci involving changes at only one mutant site, 1 had reciprocal recombination between B23 and B72; 3 had conversion from mutant to wild type and one from wild type to mutant at B23; 1 had conversion from wild type to mutant at B36; but site B72, that in the middle, was not involved without involvement of either B23 or B36, or both. Of the six asci in which more than a single site was altered: one had 2+:6m segregation at B23 and 6+:2m at B72 (i.e., three chromatids of one and one of the other parental constitutions over this region, but not involving site B36); two had 6+:2m segregation at B23, 2+:6m at B72, and 6+:2m at B23 [three chromatids of one and one of the other parental combinations over the entire region covered by the alleles studied (Fig. 14)]; one had 6+:2m segregation at B23, 5+:3m segregation at B72, and 3+:5m at B36 [simultaneous meiotic segregation, at B23, and postmeiotic, at the other two sites, the postmeiotic segregation involving the same chromatid at both B72 and B36 (Fig. 15)]. All three sites were again presumably involved in the remaining ascus (ascus 43), in which segregation was 3+:5m at B72, and 4+4m^p at B36 (i.e., postmeiotic segregation involving one chromatid at the mid site, involving both chromatids at the distal site), and apparently involving reciprocal recombination between B23 and B72, but occurring between one chromatid which was also involved in postmeiotic segregation at sites

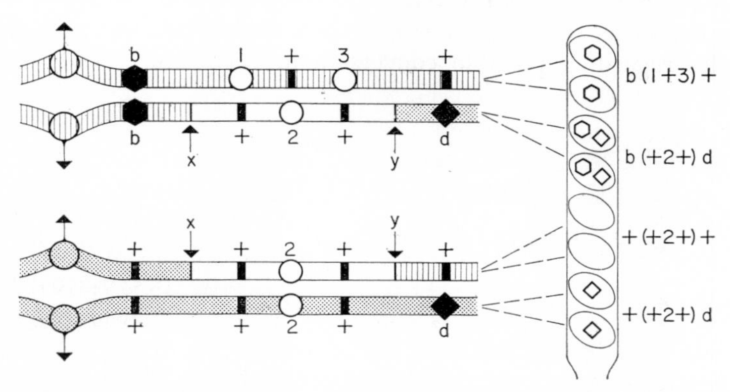

FIG. 14. Compositions of meiotic chromatids inferred from the spore pattern in ascus 565 of Case and Giles (1964). The proximal flanking marker, b, is *ad-1,* the distal flanking marker, d, is *tryp-2.* Alleles undergoing recombination, 1, 2 and 3, are B23, B72 and B36, respectively, at the *pan-2* locus of *Neurospora crassa.* Chromatids derived from one parent have vertical-line shading, those from the other stippled. Recombination between flanking markers could have occurred by reciprocal crossing over either at x or at y.

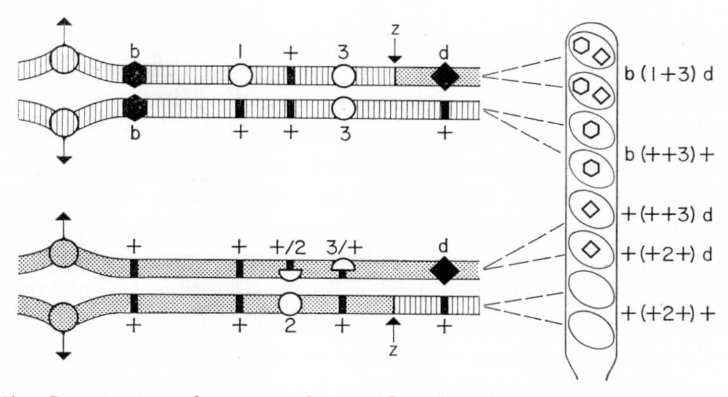

FIG. 15. Constitution of meiotic chromatids inferred from the spore pattern in ascus 529 of Case and Giles (1964). Recombination between flanking markers occurred between two chromatids not involved in interallelic recombination, z. Conventions as in Fig. 14.

B72 and B36, and a third chromatid which was not involved in postmeiotic segregation at site B36 (Fig. 16).

As can be seen from the examples just reviewed, genotypic analysis of octads in which no wild-type recombinants are present is essential to the detection of coincident conversion at different sites in a locus. The relative frequencies of such coincident occurrences are important to estimates of the lengths of regions involved in conversion; and the types of conversion occurring at adjacent sites, when coincident, are important to postulates regarding possible conversion mechanisms.

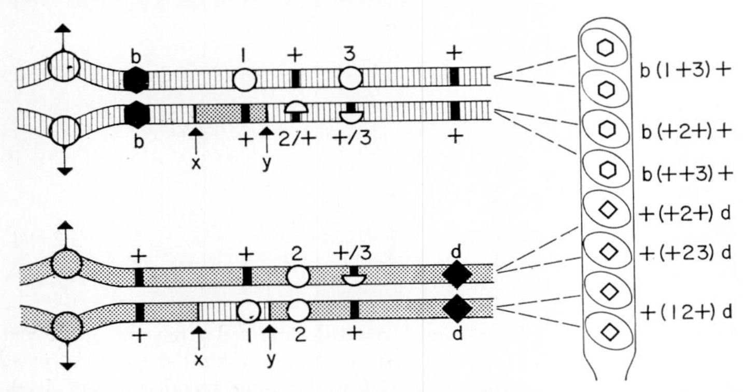

Fig. 16. Constitution of meiotic chromatids inferred from the spore pattern in ascus 43 of Case and Giles (1964). Of the two chromatids involved in conversion at sites 2 and 3, one (at y) has been involved in reciprocal interallelic recombination with a third chromatid; a second exchange, at x, is required to restore parental combinations of flanking markers.

b. Conversion Patterns Expected in Diallelic Crosses. Too few intragenic-recombinant asci have been analyzed either to tell whether or not all possible patterns can occur or to give much indication about their relative frequencies in unselected samples. Both kinds of information are needed for a thorough genetic characterization of intragenic recombination. If two chromatids are involved in intragenic recombination, and if the five segregation ratios characteristic of gene conversion (Fig. 5) occur at both mutant sites in diallelic crosses, nonreciprocal intragenic recombination should result in 54 different identifiable patterns. These will be discussed later in connection with an evaluation of one of the current models proposed to account for the phenomenon.

4. *Efficient Screening Techniques for Selecting Recombinant Asci*

Stadler and Towe (1963), in studying intragenic recombination between cysteine mutants of *Neurospora crassa,* and Fogel and Hurst

(1967) that between histidine mutants of *Saccharomyces cerevisiae*, identified asci with wild-type recombinants by the germination of prototrophic recombinant spores under conditions preventing germination of auxotrophic mutant spores. The spores of such asci were isolated before appreciable growth of the prototrophs had occurred, germinated, and tested for genetic constitutions. A similar, and somewhat simpler, method of selecting asci with wild-type recombinants had been used in connection with spore-color mutants of *Ascobolus immersus* by Rizet and associates (e.g., Lissouba *et al.*, 1962) and those of *Sordaria fimicola* by Kitani and Olive (1967a).

a. *Alleles with Different Phenotypes.* Kitani and Olive now have mutants with hyaline spores at the gray spore locus in *Sordaria fimicola*. That the hyaline mutants are functional alleles of gray is shown by the absence of complementation in heterokaryotic spores (Kitani and Olive, 1967b). The hyaline phenotype is epistatic to that of gray so that in crosses between gray, *g*, and hyaline, *h*, three phenotypes are produced: black spores, ++, gray spores, *g* +, and transparent spores, + *h* or *g h*. The numbers of spores of each phenotype and their positions in the linearly ordered asci permit the identification of 21 spore patterns indicative of intragenic recombination in crosses in repulsion (*trans* configuration) (Kitani and Olive, 1967a). These are illustrated in Fig. 17, with the letter designations used by Kitani, but arranged in an order to show segregation ratios at the *h* mutant site, these being directly observable. On the supposition (a reasonably well-established rule) that only two chromatids take part in interallelic recombination, each ascus is expected to have one pair of spores of each parental constitution, descended from the two noninteracting chromatids. (In Fig. 17, the pair with the constitution of the *g* parent is represented by the two top spores, that of the *h* parent by the two bottom spores, in each ascus.) It is apparent that in asci with 6+:2*h* segregation the six nonhyaline spores must then be either gray or wild type, so that segregation at the *g* site is evident. In asci with more than two hyaline spores, however, any or all hyaline spores in excess of two may be genotypically either + *h* or *g h*, and can be distinguished only by genetic test.

b. *Conditional Mutants.* The efficient screening method developed by Kitani and Olive (1967a) immediately suggests an extension to mutants which do not visibly affect the spores. At a number of loci there are known mutants which express a mutant phenotype under one set of conditions, but the wild-type phenotype under a different set of conditions—temperature-sensitive alleles and pH-sensitive alleles. (I shall adopt the terminology of bacteriophage geneticists, calling such mutants "conditional," the conditions engendering the mutant phenotype "restrictive," and those leading to the wild-type phenotype "permissive.") Screening would then be possible under either permissive or restrictive conditions. In interallelic crosses in which one mutant is conditional, parental ditype asci will contain four spores which react as wild type under permissive conditions, four as mutant; conversion at the site of the

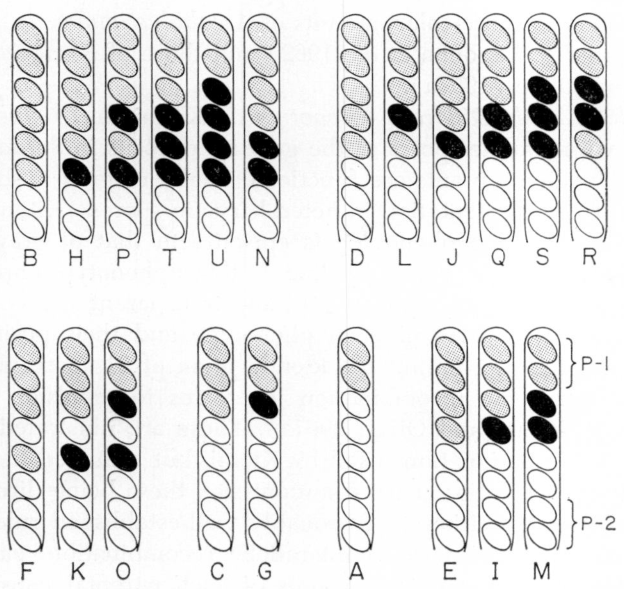

FIG. 17. Visibly detectable ascus patterns resulting from interallelic recombination between a mutant allele producing gray spores and a mutant allele producing transparent spores in which the transparent phenotype is epistatic to the gray. (After Kitani and Olive, 1967a). Groups: upper left 6+:2h, right 5+:3h, lower, left to right, 4+:4h^p, 3+:5h, 2+:6h, and 4+:4h^m.

conditional mutant can be recognized by the wild-type reaction of more or fewer than four spores. If it were feasible to retain the order of spores from linearly ordered asci under permissive conditions, five spore patterns indicating conversion at this site should be recognizable (6+:2m, 5+:3m, 4+:4m^p, 3+:5m, and 2+:6m); in tests of unordered asci 4+:4m^p segregation would not be detected, being indistinguishable from meiotic 4+:4m^m segregation, which is characteristic of parental ditype segregation. In ordered asci, phenotypes I and M (of Kitani and Olive) as well as E would be undetected, and in unordered asci so would phenotypes F, K, and O. However, phenotypes I, M, K, and O could be detected if a parallel screening were carried out under restrictive conditions, whereby all asci with wild-type recombinants could be selected. The situations occurring under the two sets of conditions can be visualized from Fig. 17: The stippled spores should appear black under permissive conditions, white under restrictive conditions.

C. Models of Gene Conversion

One well-established and generally accepted method of testing systems which are complicated beyond comprehension is to construct simple models and see whether they fit the systems in question. If they do, you will immediately become suspicious, and so will your colleagues most certainly, with the result that a blooming literature springs up (or breaks out) dealing with the problem of how you have managed to make all your errors cancel one another. If they do not, the beauty of the models themselves may shine for years untainted by the squalid awareness of reality.

Fizz-Loony and Linderstrøm-Lang (1956)

Models of genetic recombination have appropriately been discussed in Chapter IV on "Linkage and Recombination at the Molecular Level" by Bodmer and Darlington, and the reasonableness of the chromosome structures implied by the models in Chapter III on "Structure and Replication of Chromosomes" by Taylor. Nevertheless, some further discussion of the reciprocal bearing of models and genetic data upon one another, and the possible usefulness of models in the planning of pertinent experiments, may be warranted.

1. *Genetic Restrictions to Model Building*

An acceptable model embracing both intergenic crossing-over and intragenic recombination must satisfactorily account for known characteristics of each and for their frequent correlated occurrence. The kinds of information about intergenic exchange which are most important to restrained model building can be assumed to be already available; those relating to correlations between intergenic crossing-over and intragenic recombination are possibly, but not altogether convincingly, sufficient for the purpose; those relating specifically to intragenic recombination are still too incomplete to place much restraint on speculative interpretation.

Important characteristics of crossing over between loci include: (1) Two and only two chromatids are involved in any single recombinational event (data from tetrad and half-tetrad analysis). (2) The exchange between chromatids is reciprocal (data from tetrad and half-tetrad analysis). (3) The occurrence of crossing-over in one interval decreases the probability that crossing-over will occur in an adjacent, or nearby, interval (interference).

Known characteristics of correlation between intergenic and intragenic recombination include: (1) Among random products from crosses between alleles, recombination between flanking markers is much greater in the fraction consisting of wild-type recombinants than in the entire sample; to the best of my knowledge, this is true, without exception, in every example studied. (2) When such wild-type recombinants result from reciprocal recombination between alleles (detected by tetrad analysis), the correlation between flanking-marker and interallelic recombination may be absolute—this conclusion is substantiated by only one example (Fogel and Hurst, 1967), the only example adequately studied; but there is no strong evidence suggesting that complete correlation is not universal in equivalent circumstances. [In this connection it should be noted that all recombination between pseudoalleles of *Drosophila* (Lewis, 1963) is reciprocal for both pseudoalleles and flanking markers.] (3) As noted earlier (Section III,B,1 and 2,a), when wild-type recombinants result from nonreciprocal intragenic recombination, biased distributions of both recombinant and parental combinations of flanking markers usually occur. That this relationship is general, but probably different at least in degree in different examples, can be inferred from studies of random wild-type recombinants (e.g., Murray, 1963); it has again been established by tetrad analysis in only one example (Fogel and Hurst, 1967), in which (*a*) conversion to wild type at one site was nearly always associated with one parental combination of flanking markers, that at the other with the other; and (*b*) a larger fraction of convertants to wild type at one site was associated with one recombinant combination of outside markers than was the case when conversion had occured at the other site (Table IV). (4) Perhaps equivalent to the relationship just mentioned, conversion at a single heterozygous mutant site is associated with an increased frequency of recombination between flanking markers; and the extent of such association depends to some degree upon the particular type of conversion occurring (Kitani *et al.*, 1962; Kitani and Olive, 1967b). It is unfortunate that there are not more examples of this sort because, if general, models should account for the different relations to flanking-marker recombination of different types of conversion at a single site.

Additional characteristics of gene conversion and intragenic recombination include: (1) Just as in recombination between loci, intragenic recombination usually involves two and only two chromatids, but there are a few examples known (e.g., that illustrated in Fig. 16) in which at least three chromatids have apparently taken part in intragenic recombination within one cistron. (2) Both meiotic and postmeiotic segregation patterns of nonreciprocal intragenic recombination probably result from a common process—both kinds occur simultaneously at different allelic sites within a gene (as in the recombinants illustrated in Figs. 15 and 16) much more frequently than they should if separate events were involved. (3) The patterns of conversion occurring at one mutant site, or their relative frequencies, are often not independent of the kinds or frequencies of conversion occurring at a neighboring allelic site. The relative frequencies of meiotic and postmeiotic conversion patterns at allelic site *h* in *Sordaria fimicola* were observed by Kitani and Olive (1967a) to accompany a change from homozygosity to heterozygosity at allelic site *g*. Simultaneous conversion at nearby allelic sites has also been observed to involve changes to the allelic states introduced from a single parent more often than should be expected by chance (Emerson and Yu-Sun, 1967). Too few examples of either of these two kinds have been studied to warrant generalizing from them—or to indicate whether or not they may be similar to the "marker effects" observed in bacterial transformation (Ravin and Iyer, 1962; Sicard and Ephrussi-Taylor, 1965). (4) Polarity is sometimes observed in the frequency of conversion at different allelic sites within a locus (Lissouba *et al.*, 1962). (5) Polarity with respect to the flanking-marker combinations accompanying intragenic recombination seem to be the rule (Murray, 1963; Jessop and Catcheside, 1965).

2. *The Models*

A number of models have been proposed with the intention of accounting in molecular terms for both crossing-over between loci and recombination between mutant sites within loci. There are now two such models favorably considered by many fungal geneticists and discussed in most current papers dealing with conversion: that of Holliday (1964) and that of Whitehouse and Hastings (1965). It is of some interest to see how these two models account for various genetic observations, to note the differences between them, and to speculate on possible tests of their validity.

The two models have many points in common. (1) Both postulate that the meiotic chromatids consist of linear arrays of DNA molecules (single double-helices) attached end-to-end—each molecule perhaps corresponding to a gene or cistron. The end-to-end attachment of molecules may be by special structures, or by a special mode of bonding. (2) Both envisage the formation of hybrid DNA made up of one base chain derived from one chromatid and one from a homologous chromatid.

Whenever a heterozygous mutant site occurs in such a hybrid region, the bases at that site will be noncomplementary (hence mispaired) and will constitute a heteroduplex. (3) Both postulate that conversion results from repair of heteroduplexes by the replacement of one mispaired base by that one which is complementary to the other. (4) Both models involve a relationship between the formation of regions of hybrid DNA (a prerequisite to gene conversion) and reciprocal exchange between chromatids (which results in crossing-over between flanking markers). The two models differ in the way in which regions of hybrid DNA originate, and in the relation between hybrid DNA formation and an exchange between chromatids.

a. *The Whitehouse-Hastings Model.* The model of Whitehouse and Hastings (1965) requires several steps for the production of hybrid DNA. (1) Dissociation of one base chain of the DNA double helix of one chromatid of each parental type is initiated at a predetermined point which is identically located in both participating chromatids (Fig. 18a, disregarding base chains represented by broken lines). The two freed base chains are of opposite polarity. They must remain free from their original partners, and from each other, for a period of some duration. The detached chains are replaced (at least in part) in the original chromatids by resynthesis of chains complementary to the chains which had remained attached—newly synthesized base chains are represented by broken lines in Fig. 18a. Synthesis starts at the point of original separation, hence is opposite in polarity in the two chromatids. (3) The newly synthesized base chains then separate from their templates, starting at the end most recently synthesized (Fig. 18c). (4) The newly synthesized chains reassociate with the chains first dissociated from the homologous chromatids to reconstitute double helices of hybrid origin (Fig. 18e).

The reestablishment of the continuity (integrity) of chromatids is accomplished by the degradation of unpaired regions remaining in the base chains which so far had retained their original positions and end attachments, shown by the stippled regions in the left half of Fig. 18e, followed by reunion of free ends of base chains to make intact chromatids—shown in the right half of the same figure. An exchange between two chromatids necessarily accompanies such restoration of chromatid continuity.

b. *The Holliday Model.* The model of Holliday (1964) involves (1) dissociation of one base chain from the DNA double helix in each of two homologous chromatids (Fig. 18b). The base chains so freed are identical in polarity. (2) There follows an immediate reassociation between each freed chain and the base chain which had remained attached in the homologous chromatid (Fig. 18d). There results a half-chromatid chiasma between the hybrid DNA of the reassociated regions and the parental DNAs in the adjoining regions.

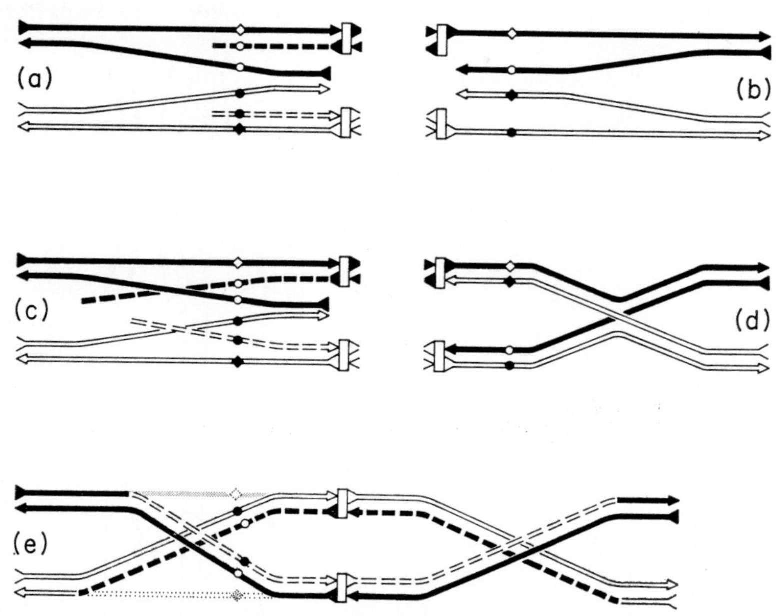

FIG. 18. Establishement of hybrid DNA and heteroduplexes according to the model of Whitehouse and Hastings (1965):—(a), (c), and (e)—and to that of Holliday (1964):—(b) and (d). The two base chains in one chromatid are represented by solid black lines; they have a mutant base pair represented by an open circle and an open square; the two base chains of the homologous chromatid are shown in outline, with the wild-type base pair at the mutant site represented by a solid black circle and solid black square; polarities of base chains are indicated by the arrows in which one end of each terminates; newly synthesized sections of base chains are represented by broken lines; sections of base chains undergoing breakdown are stippled; the open rectangles at which base chains end represent the points of which the separation of base chains is initiated—they may, or may not, be special structures. Further explanations in text.

The restoration of chromatid integrity takes place by resolution of the half-chromatid chiasma through breakage and reunion of two of the four base chains at the chiasma. Should the two chains illustrated as crossing in Fig. 18d be broken, and recombine to form two continuous chromatids, there would result no recombination between flanking markers, whereas breakage and recombination of the other two chains would result in flanking-marker recombination.

Because of the difficulty in visualizing a three-dimensional structure from a two-dimensional projection, a half-chromatid chiasma is diagrammed in different ways in Fig. 19. If the two base chains of a DNA molecule were truly represented by two parallel lines, then by viewing the structure from different points, different chains would appear to

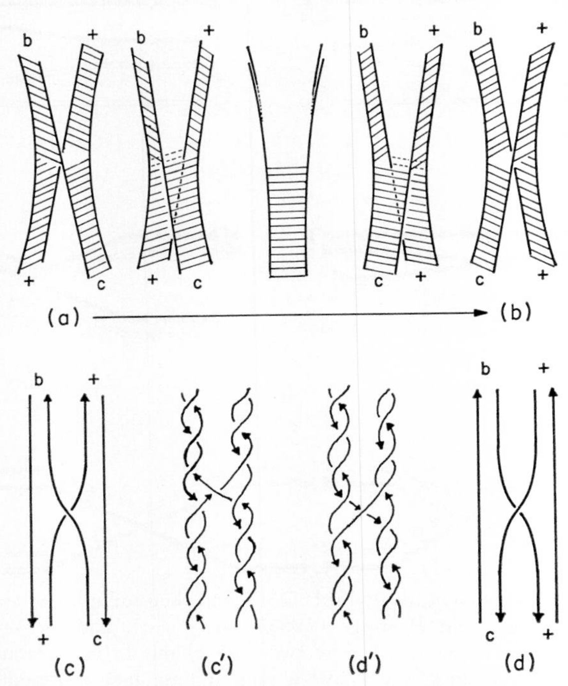

FIG. 19. Half-chromatid chiasmata. (a) and (b) Three-dimensional representations in which the heavier, nearly vertical lines represent base chains, the cross lines the hydrogen bonding between base pairs. Rotation of the figure causes different base chains to appear to cross one another. (c) and (d) Two-dimensional projections of figures shown at (a) and (b), respectively. (c′) and (d′) Equivalent to (c) and (d) but with base chains helically coiled.

cross, as shown by rotating the structure through 90°, from (a) to (b) in Fig. 19. The projection from position (a) is shown in (c), that from (b) in (d). In one, breakage and reunion of the crossed chains would result in flanking marker recombination, in the other it would not. If these diagrams represent a true picture we should expect recombination to occur in approximately half of all instances. If a model of a half-chromatid chiasma involves the helical coiling of base chains about one another, and the coiling is as tight as possible, the two chains which cross will be determined to a considerable extent by the coiling process itself, and not by the angle of observation (Fig. 19c′ and d′).

3. *Genetic Implications of the Models*

There are sufficient points of difference between the models of Holliday and of Whitehouse and Hastings to make it seem that genetic tests could distinguish between them. Up to the present time this has not proved to be possible in any significant way. There are, however, a number of ways in which a more thorough characterization of intragenic recombination would be useful in evaluating the models.

a. *Kinds and Frequencies of Heteroduplex Repair.* Whenever a heterozygous mutant site is included in a region of hybrid DNA in both of the chromatids participating in conversion (always, according to the model of Holliday, or usually, according to that of Whitehouse and Hastings), each heteroduplex may have three fates: It may be repaired to the wild-type homoduplex, repaired to the mutant homoduplex, or remain heteroduplex because of repair failure. The 6+:2m conversion ratio requires repair to wild type in both chromatids and the 2+:6m repair to mutant in both. Two ratios require repair failure in one hybrid chromatid and repair in the other—repair to wild type in 5+:3m, to mutant in 3+:5m. The postmeiotic 4+:4m^{p} ratio requires repair failure in both chromatids.

If conversion does occur in this way there must be still one more consequence. Repair must sometimes be to wild type in one chromatid and to mutant in the other, resulting in a 4+:4m^{m} ratio in which segregation was completed during the two meiotic divisions. Such 4+:4m segregations are indistinguishable by experimental means from normal segregations involving no exchanges of any kind (parental ditypes). The inability to detect this class of convertant asci makes it impossible to determine the total frequency of conversion at a mutant site.

Algebraic formulations for estimating frequencies of different kinds of conversion are simpler when based on the Whitehouse-Hastings model (Whitehouse, 1965) than on the Holliday model (Emerson, 1966b) because the heteroduplexes at one site in the two participating chromatids are identical in the former and not in the latter. Let the two bases of a mutant allele at a particular site be designated m_a and m_b, and the wild-type bases at that site as $+_a$ and $+_b$, with subscript *a* reflecting the polarity of one base chain, *b* that of the other. By the Whitehouse-Hastings model heteroduplexes in the two chromatids would be identical—sometimes both $+_a/m_b$, at other times both $m_a/+_b$, whereas by the Holliday model, one heteroduplex would be $+_a/m_b$ and the other $m_a/+_b$ in all instances. Identical base constitutions in the heteroduplexes of the two chromatids should result in equal frequencies of any one type of repair in both. Such equalities cannot be assumed if the heteroduplexes have different base compositions.

The frequencies of each type of repair occurring at a mutant site in crosses between a single mutant and wild type must be known in order to calculate frequencies of different genotypes that would occur in crosses

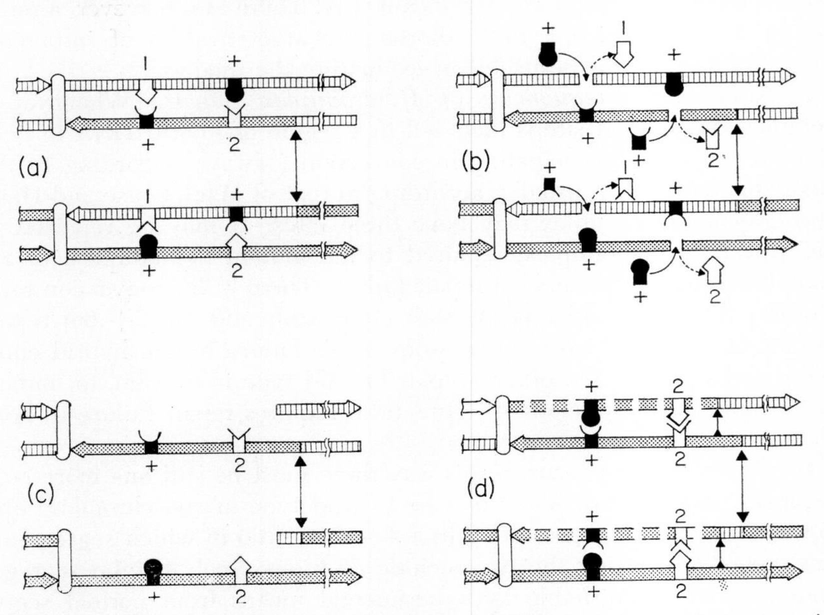

FIG. 20. Different modes of repair of heteroduplexes at two mutant sites. (a) Both alleles in common regions of hybrid DNA. (b) Independent excision of bases at the two allelic sites. (c) Extensive excision of single base chains. (d) Repair by resynthesis using the intact base chain as a template and resulting in coincident repair.

between alleles if repair at the mutant sites is completely independent. Algebraic formulations for estimating the frequencies of different kinds of repair at a single site are summarized in Addendum II,E. Formulations based on the Whitehouse-Hastings model assume that if a mutant site is in the hybrid region of one chromatid it is also in the hybrid site of the other, under which conditions the following relationship must hold:

$$\frac{\text{frequency of } 6+:2m}{\text{frequency of } 2+:6m} = \frac{\text{square of the frequency of } 5+:3m}{\text{square of the frequency of } 3+:5m}$$

Significant departure from this relationship does not necessarily contradict the Whitehouse-Hastings model—the discrepancies might be due to the occurrence of a heteroduplex in only one of the two participating chromatids provided the absence of a heteroduplex is more likely to be in the chromatid of one parental constitution than in that of the other.

b. *Extent of Hybrid Regions and of Repair.** There are essentially three possible structures of hybrid DNA of importance to the analysis of diallelic crosses. Heteroduplexes 1/+ and 2/+ may both be included in

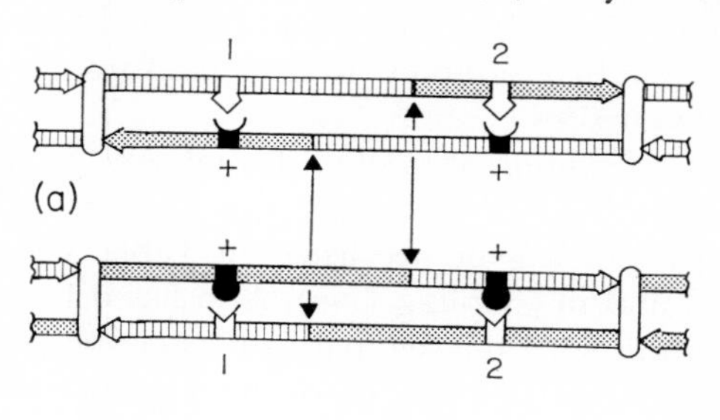

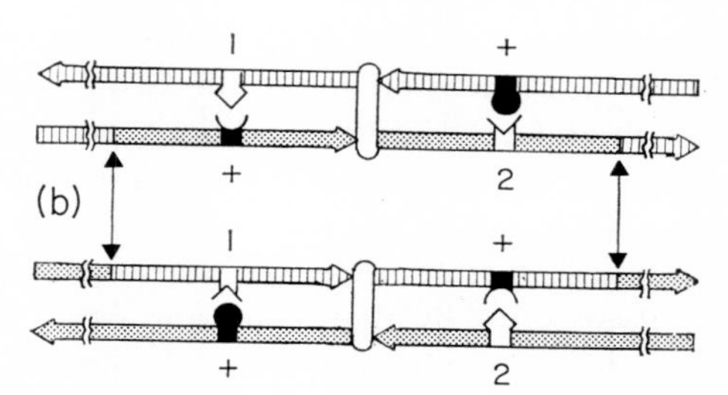

Fig. 21. Allelic sites in separate regions of hybrid DNA. (a) Hybrid regions at opposite ends of a gene. The hybrid regions are opposite in polarity; they would have been of common polarity had the same chains become dissociated at both ends. (b) Hybrid regions adjacent, with a common origin of dissociation.

a single continuous region of hybrid DNA, as illustrated in Fig. 20a; heteroduplex 1/+ may occur in one region of DNA hybridity and 2/+ in another, as illustrated in Fig. 21a and b; or only one of the two allelic sites may be included in a hybrid region, e.g., in only one of the hybrid regions in Fig. 21.

* In this and the following sections it will be assumed that when a mutant site is present in the hybrid region of one chromatid it is also included in the hybrid region of the other chromatid participating in recombination. This condition is not necessarily true for the Whitehouse-Hastings model, but should occur in the majority of instances.

Two possible variations in the mode of repair are important in instances in which both mutant sites appear as heteroduplexes in a common hybrid region. The replacement of one base of a heteroduplex by that complementary to the other may involve only, or little more than, a single mispaired base, as illustrated in Fig. 20b. In that case, the base removed from one heteroduplex may have no definite relationship to the base which is replaced at the other hybrid site. Alternatively, as illustrated in Fig. 20c and d, the correction of base mispairing may be accompanied by erosion of the base chain involved to an extent sufficient to include the second heteroduplex, so that bases from a single base chain are removed at both sites.

c. *Derivations of Different Octad Genotypes in Diallelic Crosses.* The 55 different octad genotypes which can result from crosses heterozygous for two alleles of a gene are listed in Table V for crosses both in repulsion (*trans*) and in coupling (*cis*). Also listed in this table are: (1) The allelic constitutions of the two pairs of sister spores derived from the sister half-chromatids of the two chromatids participating in intragenic recombination. (2) The segregation ratios occurring at each mutant site. (3) The phenotypic ratios observed in three different situations: (*a*) When the double mutant, 1 2, and both single mutants, 1 + and + 2, have identical phenotypes (or when one of the two mutants is conditional and tests are made under restrictive conditions, see Section III,B,4,b); (*b*) when the mutant at one site is conditional and tests are made under permissive conditions; (*c*) when the two mutants have different phenotypes but one is epistatic to the other, see Section III,B,4,a. Phenotypic ratios which are identical with that resulting from normal (parental ditype) segregation, hence not identifiable by selective screening techniques, are enclosed in boxes. (4) The possible derivations of each genotype from different hybrid structures by different modes of repair.

The symbols used to designate specific derivations are the following:

C. Coincident repair at the two mutant sites. Both sites had necessarily been present as heteroduplexes, followed by repair involving base substitution in the same chain at both sites to give either 6+:2m or 2+:6m at each site, or at one site and either 5+:3m or 3+:5m at the other. Examples:

$$\text{In repulsion } (\textit{trans}) \ \frac{1\,+}{+\,2} \text{ repaired to } \frac{1\,+}{(1\,+)} \text{ or } \frac{(+\,2)^{*}}{+\,2};$$

* Alleles in one base chain are shown above the line, those in the other below. Parentheses enclose bases which have been substituted for the mispaired bases which have been replaced by repair.

In coupling (*cis*) $\frac{++}{1\,2}$ repaired to $\frac{++}{(++)}$ or $\frac{(1\,2)}{1\,2}$.

(C). Coincident repair if both mutant sites were present in hybrid DNA, but inclusion of a single site in a hybrid region, S, is sufficient to produce the genotype in question.

P. The bases in a single base chain were of common parental origin at both hybrid sites—the two heteroduplexes may both be in one hybrid region, as in Fig. 20a, or in separate hybrid regions of the same polarity, as in Fig. 21b—and repair involved the replacement of a base from one chain at one site and from the complementary chain at the other (i.e., repair at one site was necessarily independent of that occurring at the other). Examples:

In repulsion (*trans*) $\frac{1\,+}{+\,2}$ repaired to $\frac{(+)+}{+(+)}$ or $\frac{1(2)}{(1)2}$;

In coupling (*cis*) $\frac{++}{1\,2}$ repaired to $\frac{(1)+}{1(+)}$ or $\frac{+(2)}{(+)2}$.

(P). Either coincident repair, C, is not excluded, or a heteroduplex at a single site, S, is sufficient.

N. Bases at the two hybrid sites were necessarily in nonparental combination in each base chain, requiring separate hybrid regions of opposite polarity, as in Fig. 21a, and no repair has occurred at either site.

In repulsion (*trans*) $\frac{1\,+}{+\,2}$ becomes and remains $\frac{++}{1\,2}$;

In coupling (*cis*) $\frac{++}{1\,2}$ becomes and remains $\frac{1\,+}{+\,2}$.

(N). Genotypes which could also have been derived from hybrid regions in which bases at the two mutant sites were in parental combinations in each base chain, P.

I. Genotypes which do not reflect any particular derivation except that both sites must have been present as heteroduplexes.

S. Only one site necessarily present in hybrid DNA—segregation must be $4+:4m^m$ at one site; S_1, site 1; S_2, site 2; $S_{1/2}$, either 1 or 2.

X. A genotype unexpected from either repair model. Heteroduplexes have remained unrepaired at both sites in both chromatids, but with bases in parental order in the chains of one chromatid, in

TABLE V

POSSIBLE KINDS AND ORIGINS OF SEGREGATION PATTERNS IN DIALLELIC CROSSES

Pattern	Converted spore pairs		Ratios at each site[a]			Cross in repulsion (*trans*)					Cross in coupling (*cis*)				
	One	Other	+:1	+:2	(F)	Possible origins[b]	Genotype ++:1+:+2:12	Phenotypes[c] (A)	(B)	(C)	Possible origins[b]	Genotype ++:1+:+2:12	Phenotypes[c] (A)	(B)	(C)
1	++ ++	++ ++	6:2	6:2	1	P(N)	4:2:2:0	4:4	6:2	4:2:2	C(P)(N)	6:0:0:2	6:2	6:2	6:0:2
2	++ ++	++ 1 +	5:3	6:2	1	P(N)	3:3:2:0	3:5	6:2	3:3:2	C(P)(N)	5:1:0:2	5:3	6:2	5:1:2
3	++ ++	++ + 2	6:2	5:3	1	P(N)	3:2:3:0	3:5	5:3	3:2:3	C(P)(N)	5:0:1:2	5:3	5:3	5:0:3
4	++ ++	++ 1 2	5:3	5:3	1/4	N	3:2:2:1	3:5	5:3	3:2:3	C(P)	5:0:0:3	5:3	5:3	5:0:3
5	++ 1 +	++ 1 +	4:4[p]	6:2	1	I	2:4:2:0	2:6	6:2	2:4:2	I	4:2:0:2	4:4	6:2	4:2:2
6	++ ++	1 + 1 +	4:4[m]	6:2	1	S_2(P)(N)	2:4:2:0	2:6	6:2	2:4:2	S_2(P)(N)	4:2:0:2	4:4	6:2	4:2:2
7	++ 1 +	++ + 2	5:3	5:3	1/2	I	2:3:3:0	2:6	5:3	2:3:3	I	4:1:1:2	4:4	5:3	4:1:3
8	++ ++	1 + + 2	5:3	5:3	1/4	P	2:3:3:0	2:6	5:3	2:3:3	N	4:1:1:4	4:4	5:3	4:1:3
9	++ 1 +	++ 1 2	4:4[p]	5:3	1/2	N	2:3:2:1	2:6	5:3	2:3:3	I (not N)	4:1:0:3	4:4	5:3	4:1:3
10	++ ++	1 + 1 2	4:4[m]	5:3	1/2	S_2(P)(N)	2:3:2:1	2:6	5:3	2:3:3	S_2(C)(P)(N)	4:1:0:3	4:4	5:3	4:1:3
11	++ + 2	++ + 2	6:2	4:4[p]	1	I	2:2:4:0	2:6	4:4	2:2:4	I	4:0:2:2	4:4	4:4	4:0:4
12	++ ++	+ 2 + 2	6:2	4:4[m]	1	S_1(P)(N)	2:2:4:0	2:6	4:4	2:2:4	S_1(P)(N)	4:0:2:2	4:4	4:4	4:0:4
13	++ + 2	++ 1 2	5:3	4:4[p]	1/2	N	2:2:3:1	2:6	4:4	2:2:4	I (not N)	4:0:1:3	4:4	4:4	4:0:4
14	++ ++	+ 2 1 2	5:3	4:4[m]	1/2	S_1(P)(N)	2:2:3:1	2:6	4:4	2:2:4	S_1(C)(P)(N)	4:0:1:3	4:4	4:4	4:0:4
15	++ 1 2	++ 1 2	4:4[p]	4:4[p]	1/4	N	2:2:2:2	2:6	4:4	2:2:4	I (not N)	4:0:0:4	4:4	4:4	4:0:4
16	++ ++	1 2 1 2	4:4[m]	4:4[m]	1/2	{Reciprocal $S_{1/2}$(P)(N)}	2:2:2:2	2:6	4:4	2:2:4	{P.D.T. $S_{1/2}$(C)(P)(N)}	4:0:0:4	4:4	4:4	4:0:4
17	++ 1 +	1 + 1 +	3:5	6:2	1	C(P)(N)	1:5:2:0	1:7	6:2	1:5:2	P(N)	3:3:0:2	3:5	6:2	3:3:2
18	++ 1 +	1 + + 2	4:4[p]	5:3	1/2	I (not N)	1:4:3:0	1:7	5:3	1:4:3	N	3:2:1:2	3:5	5:3	3:2:3
19	++ + 2	1 + 1 +	4:4[m]	5:3	1/2	S_2(C)(P)(N)	1:4:3:0	1:7	5:3	1:4:3	S_2(P)(N)	3:2:1:2	3:5	5:3	3:2:3
20	++ 1 +	1 + 1 2	3:5	5:3	1/2	I	1:4:2:1	1:7	5:3	1:4:3	I	3:2:0:3	3:5	5:3	3:2:3
21	++ 1 2	1 + 1 +	3:5	5:3	1/4	N	1:4:2:1	1:7	5:3	1:4:3	P	3:2:0:3	3:5	5:3	3:2:3
22	++ + 2	1 + + 2	5:3	4:4[p]	1/2	I (not N)	1:3:4:0	1:7	4:4	1:3:4	N	3:1:2:2	3:5	4:4	3:1:4
23	++ 1 +	+ 2 + 2	5:3	4:4[m]	1/2	S_1(C)(P)(N)	1:3:4:0	1:7	4:4	1:3:4	S_1(P)(N)	3:1:2:2	3:5	4:4	3:1:4
24	++ 1 +	+ 2 1 2	4:4[p]	4:4[m]	1	S_1(I)	1:3:3:1	1:7	4:4	1:3:4	S_1(I)	3:1:1:3	3:5	4:4	3:1:4
25	++ + 2	1 + 1 2	4:4[m]	4:4[p]	1	S_2(I)	1:3:3:1	1:7	4:4	1:3:4	S_2(I)	3:1:1:3	3:5	4:4	3:1:4
26	++ 1 2	1 + + 2	4:4[p]	4:4[p]	1/2	X	1:3:3:1	1:7	4:4	1:3:4	X	3:1:1:3	3:5	4:4	3:1:4
27	++ 1 2	1 + 1 2	3:5	4:4[p]	1/2	N	1:3:2:2	1:7	4:4	1:3:4	I (not N)	3:1:0:4	3:5	4:4	3:1:4
28	++ 1 +	1 2 1 2	3:5	4:4[m]	1/2	S_1(P)(N)	1:3:2:2	1:7	4:4	1:3:4	S_1(C)(P)(N)	3:1:0:4	3:5	4:4	3:1:4
29	++ + 2	+ 2 + 2	6:2	3:5	1	C(P)(N)	1:2:5:0	1:7	3:5	1:2:5	P(N)	3:0:3:2	3:5	3:5	3:0:5
30	++ + 2	+ 2 1 2	5:3	3:5	1/2	I	1:2:4:1	1:7	3:5	1:2:5	I	3:0:2:3	3:5	3:5	3:0:5

31	+ + 1 2	+ 2 + 2	5:3	3:5	1/4	N	1:2:4:1	1:7	3:5	1:2:5	P	3:0:2:3	3:5	3:5	3:0:5
32	+ + 1 2	+ 2 1 2	$4:4^{p}$	3:5	1/2	N	1:2:3:2	1:7	3:5	1:2:5	I (not N)	3:0:1:4	3:5	3:5	3:0:5
33	+ + + 2	1 2 1 2	$4:4^{m}$	3:5	1/2	S_2(P)(N)	1:2:3:2	1:7	3:5	1:2:5	S_2(C)(P)(N)	3:0:1:4	3:5	3:5	3:0:5
34	+ + 1 2	1 2 1 2	3:5	3:5	1/4	N	1:2:2:3	1:7	3:5	1:2:5	C(P)	3:0:0:5	3:5	3:5	3:0:5
35	1 + 1 +	1 + 1 +	2:6	6:2	1	C(P)(N)	0:6:2:0	0:8	6:2	0:6:2	P(N)	2:4:0:2	2:6	6:2	2:4:2
36	1 + 1 +	1 + + 2	3:5	5:3	1/4	C(P)	0:5:3:0	0:8	5:3	0:5:3	N	2:3:1:2	2:6	5:3	2:3:3
37	1 + 1 +	1 + 1 2	2:6	5:3	1	C(P)(N)	0:5:2:1	0:8	5:3	0:5:3	P(N)	2:3:0:3	2:6	5:3	2:3:3
38	1 + + 2	1 + + 2	$4:4^{p}$	$4:4^{p}$	1/4	I (not N)	0:4:4:0	0:8	4:4	0:4:4	N	2:2:2:2	2:6	4:4	2:2:4
39	1 + 1 +	+ 2 + 2	$4:4^{m}$	$4:4^{m}$	1/2	{ P.D.T. / $S_{1/2}$(C)(P)(N) }	0:4:4:0	0:8	4:4	0:4:4	{ Reciprocal / $S_{1/2}$(P)(N) }	2:2:2:2	2:6	4:4	2:2:4
40	1 + + 2	1 + 1 2	3:5	$4:4^{p}$	1/2	I (not N)	0:4:3:1	0:8	4:4	0:4:4	N	2:2:1:3	2:6	4:4	2:2:4
41	1 + 1 +	+ 2 1 2	3:5	$4:4^{m}$	1/2	S_1(C)(P)(N)	0:4:3:1	0:8	4:4	0:4:4	S_1(P)(N)	2:2:1:3	2:6	4:4	2:2:4
42	1 + 1 2	1 + 1 2	2:6	$4:4^{p}$	1	I	0:4:2:2	0:8	4:4	0:4:4	I	2:2:0:4	2:6	4:4	2:2:4
43	1 + 1 +	1 2 1 2	2:6	$4:4^{m}$	1	S_1(P)(N)	0:4:2:2	0:8	4:4	0:4:4	S_1(P)(N)	2:2:0:4	2:6	4:4	2:2:4
44	1 + + 2	+ 2 + 2	5:3	3:5	1/4	C(P)	0:3:5:0	0:8	3:5	0:3:5	N	2:1:3:2	2:6	3:5	2:1:5
45	1 + + 2	+ 2 1 2	$4:4^{p}$	3:5	1/2	I (not N)	0:3:4:1	0:8	3:5	0:3:5	N	2:1:2:3	2:6	3:5	2:1:5
46	1 + 1 2	+ 2 + 2	$4:4^{m}$	3:5	1/2	S_2(C)(P)(N)	0:3:4:1	0:8	3:5	0:3:5	S_2(P)(N)	2:1:2:3	2:6	3:5	2:1:5
47	1 + 1 2	+ 2 1 2	3:5	3:5	1/2	I	0:3:3:2	0:8	3:5	0:3:5	I	2:1:1:4	2:6	3:5	2:1:5
48	1 + + 2	1 2 1 2	3:5	3:5	1/4	P	0:3:3:2	0:8	3:5	0:3:5	N	2:1:1:4	2:6	3:5	2:1:5
49	1 + 1 2	1 2 1 2	2:6	3:5	1	P(N)	0:3:2:3	0:8	3:5	0:3:5	C(P)(N)	2:1:0:5	2:6	3:5	2:1:5
50	+ 2 + 2	+ 2 + 2	6:2	2:6	1	C(P)(N)	0:2:6:0	0:8	2:6	0:2:6	P(N)	2:0:4:2	2:6	2:6	2:0:6
51	+ 2 + 2	+ 2 1 2	5:3	2:6	1	C(P)(N)	0:2:5:1	0:8	2:6	0:2:6	P(N)	2:0:3:3	2:6	2:6	2:0:6
52	+ 2 1 2	+ 2 1 2	$4:4^{p}$	2:6	1	I	0:2:4:2	0:8	2:6	0:2:6	I	2:0:2:4	2:6	2:6	2:0:6
53	+ 2 + 2	1 2 1 2	$4:4^{m}$	2:6	1	S_2(P)(N)	0:2:4:2	0:8	2:6	0:2:6	S_2(P)(N)	2:0:2:4	2:6	2:6	2:0:6
54	+ 2 1 2	1 2 1 2	3:5	2:6	1	P(N)	0:2:3:3	0:8	2:6	0:2:6	C(P)(N)	2:0:1:5	2:6	2:6	2:0:6
55	1 2 1 2	1 2 1 2	2:6	2:6	1	P(N)	0:2:2:4	0:8	2:6	0:2:6	C(P)(N)	2:0:0:6	2:6	2:6	2:0:6

[a] $4:4^{m}$, completely meiotic segregation; $4:4^{p}$, postmeiotic segregation in two spore pairs; (F), fraction of the combination of ratios at two sites leading to a particular pattern.

[b] The designations for possible origins are explained in the text. Subscripts to S designate the allele which is in hybrid DNA: S_1, allele 1 only; S_2, allele 2 only; $S_{1/2}$, either allele 1 or allele 2.

[c] (A), phenotypic ratios when mutants 1 and 2 and double mutant 1 2 have identical phenotypes, that is, + +:(1 +, + 2, and 1 2); (B), phenotypic ratios when one allele, 1, is a conditional mutant grown under permissive conditions (+ + and 1 +):(+ 2 and 1 2); (C) phenotypic ratios when alleles 1 and 2 have different phenotypes, and mutant 2 is epistatic to mutant 1, + +:1 +:(+ 2 and 1 2).

non-parental order in the other chromatid: $\frac{1\,+}{+\,2}$ and $\frac{+\,+}{1\,2}$. To derive this genotype from either original *cis* or *trans* configurations would require either (*a*) recombination between base chains of opposite polarity, (*b*) the removal of one pair of mispaired bases and their reinsertion in inverted order, or (*c*) the removal of one pair of mispaired bases and their replacement by the pair of mispaired bases complementary to those removed.

One striking feature in the summary of genotype derivations (Table V) is the small number of genotypes which have only one probable derivation. Two genotypes, the P's, have arisen by independent repair at two hybrid sites which had been present either in a common region of DNA hybridity or in separate hybrid regions of common polarity. Nine genotypes, the N's, have arisen from hybrid structures in which the two sites were in separate regions of opposite polarities. One genotype, the X, is hardly expected to have resulted from any of the postulated modes of heteroduplex formation. Information about the hybrid structures and modes of repair would be more readily obtainable if a large fraction of the resultant genotypes could arise in only one manner, but a great deal can be learned anyway.

d. *Concurrent Occurrences of Heteroduplex Formation.* As can be seen from the derivations listed in Table V, there are no octad genotypes which must have arisen from a hybrid DNA structure including only one mutant site—there are, however, many requiring inclusion of both sites. Even if heteroduplex formation at one mutant site occurred completely independently of that at another site, some coincident occurrences are usually to be expected.

It is probable that complete independence in heteroduplex formation at two mutant sites within a gene is unlikely to be encountered. There is very likely some condition prerequisite to heteroduplex formation which would affect both mutant sites similarly, though not necessarily identically. Such conditions might be the intimacy of chromosome pairing in the general region of the gene concerned, the coincidence of breaks in base chains essential to the formation of hybrid DNA, etc.

e. *Polarity in Conversion Frequencies.* If there are definite sites along the chromosome at which the base-chain separation is initiated (a prerequisite to the formation of hybrid DNA in both models), and if such separation and subsequent reassociation progresses for varying distances in different instances, mutant sites closer to such points of separation initiation should be included in regions of hybrid DNA (and undergo conversion) more often than more distantly located mutant sites. There are some known examples (Lissouba, 1960, see Section III,B,2,b; Yu-Sun, 1969) in which there is an orderly increase in conversion frequencies (in crosses of single mutants to wild type) from one end of a locus to the other. In these examples it was also noted that in intercrosses between two alleles (crosses in repulsion—*trans* configurations) great discrepancies occurred in the relative frequencies of octad pattern 6, in which only site 2 need have been heteroduplex (derivation S_2), and pattern 12, in which only site 1 need have been heteroduplex (derivation S_1), with sites undergoing more frequent conversion again in the same orderly array. In such examples other octad genotypes, most with phenotypic ratios other than 2+:6m, would be expected to show similar differences in frequencies of S_1 and S_2 derivations even though all could have been derived from instances in which both sites were heteroduplex.

f. *Coincident Repair at Two Mutant Sites.* More precise information about the frequency in which repair has occurred by replacement of bases at two sites in the same base chain can be obtained from reciprocal crosses than from either the cross in repulsion (*trans*) or the cross in coupling (*cis*) alone. This situation is summarized in Table VI. The comparisons involved are between segregation patterns of P(N) and of C(P)(N) derivation, both of which require repair at both mutant sites, at least in one chromatid.

The patterns of both derivations could have arisen by independent repair at the two sites in a hybrid chromatid whether both sites were in the same region of hybrid DNA, or in separate regions of either polarity. Whenever repair at one site is independent of that occurring at the other, conversion of heteroduplexes at each site to wild type and to mutant can be expected to occur in the same frequencies as they do in crosses of each individual mutant to wild type. Thus, with independent repair, the frequencies of octads of pattern 1 (Table V) should be the product of the frequencies of 6+:2m segregation at both sites, of those of pattern 2 the product of the frequency 5+:3m segregation at site m_1 and that of 6+:2m segregation at site m_2, and so on. The expected frequencies listed in the table are based on observed segregation frequencies of mutant alleles w-10 and w-78 of *Ascobolus* (Emerson and Yu-Sun, 1967) when crossed individually to wild type—12.7% and 11.0%, respectively, 6+:2m; 3.7 and 4.8%, 2+:6m; 0.6 and 1.4%, 5+:3m; and 0.9%, 3+:5m in each. As this example illustrates, when 6+:2m segregations at each site occur more frequently than 2+:6m, and when 5+:3m and 3+:5m segregations are infrequent, the expected frequencies of patterns of derivation P(N) will be greater than those of derivation C(P)(N) in

TABLE VI

COINCIDENT VERSUS INDEPENDENT REPAIR IN RECIPROCAL DIALLELIC CROSSES

Pattern numbers	Segregation at: Site 1	Segregation at: Site 2	Cross in repulsion *trans* configuration: Octad genotypes ++:1+:+2:12	Cross in repulsion *trans* configuration: Hybrid chromatids (1 + / + 2) repaired to	Hybrid chromatids (1 + / + 2) repaired to	Expected fraction[a] (%)	Cross in coupling *cis* configuration: Octad genotypes ++:1+:+2:12	Cross in coupling *cis* configuration: Hybrid chromatids (+ + / 1 2) repaired to	Hybrid chromatids (+ + / 1 2) repaired to
				Derivation P(N)			Derivation C(P)(N)		
1	6+:2m	6+:2m	4:2:2:0	(+)+ / +(+)	(+)+ / +(+)	1.40	6:0:0:2	+ + / (+ +)	+ + / (+ +)
55	2+:6m	2+:6m	0:2:2:4	1(2) / (1)2	1(2) / (1)2	0.18	2:0:0:6	(1 2) / 1 2	(1 2) / 1 2
				Subtotal		1.58			
2	5+:3m	6+:2m	3:3:2:0	(+)+ / +(+)	1 + / +(+)	0.07	5:1:0:2	+ + / (+ +)	+ + / 1(+)
3	6+:2m	5+:3m	3:2:3:0	(+)+ / +(+)	(+)+ / + 2	0.18	5:0:1:2	+ + / (+ +)	+ + / (+)2
49	2+:6m	3+:5m	0:3:2:3	1(+) / (1)2	1 2 / (1)2	0.03	2:1:0:5	(1)+ / 1 2	(1 2) / 1 2
54	3+:5m	2+:6m	0:2:3:3	1(2) / + 1	1(2) / (1)2	0.04	2:0:1:5	+(2) / 1 2	(1 2) / 1 2
				Subtotal		0.32			
				Total		1.90			

				Derivation C(P)(N)		Derivation P(N)	
35	2+:6m	6+:2m	0:6:2:0	$\frac{1\ +}{(1\ +)}\ \frac{1\ +}{(1\ +)}$	0.61	2:4:0:2	$\frac{(1)+}{1(+)}\ \frac{(1)+}{1(+)}$
50	6+:2m	2+:6m	0:2:6:0	$\frac{(+\ 2)}{+\ 2}\ \frac{(+\ 2)}{+\ 2}$	0.41	2:0:4:2	$\frac{+(2)}{(+)2}\ \frac{+(2)}{(+)2}$
				Subtotal	1.02		
17	3+:5m	6+:2m	1:5:2:0	$\frac{1\ +}{+(+)}\ \frac{1\ +}{1(+)}$	0.11	3:3:0:2	$\frac{+\ +}{1(+)}\ \frac{(1)+}{1(+)}$
29	6+:2m	3+:5m	1:2:5:0	$\frac{(+)+}{+\ 2}\ \frac{(+\ 2)}{+\ 2}$	0.10	3:0:3:2	$\frac{+\ +}{(+)2}\ \frac{+(2)}{(+)2}$
37	2+:6m	5+:3m	0:5:2:1	$\frac{1\ +}{(1\ +)}\ \frac{1\ +}{(1)2}$	0.03	2:3:0:3	$\frac{(1)+}{1(+)}\ \frac{(1)+}{1\ 2}$
51	5+:3m	2+:6m	0:2:5:1	$\frac{(+\ 2)}{+\ 2}\ \frac{1(2)}{+\ 2}$	0.05	2:0:3:3	$\frac{+(2)}{(+)2}\ \frac{+(2)}{1\ 2}$
				Subtotal	0.29		
				Total	1.31		

[a] Based on frequencies among total converted asci of the different conversion ratios in crosses $m_1 \times +$ and $m_2 \times +$, respectively: 6+:2m, 72 and 47%; 2+:6m, 25 and 19%; 5+:3m, 1 and 16%; and 3+:5m, 1 and 17%. Data from lines 17 and 18, respectively, in Table II, Section III,A,2).

crosses in repulsion, and just the reverse in crosses in coupling. [Each pattern has the same expected frequency in the two reciprocal crosses, but its derivation is P(N) in one, C(P)(N) in the other.]

Whenever both mutant sites are in the same region of hybrid DNA and base replacement occurs in a single base chain in each hybrid chromatid, on the other hand, the type of repair occurring at one site cannot be independent of that occurring at the other site, and only C(P)(N) derivations are possible. Hence, if patterns with C(P)(N) derivations occur more frequently than expected by independent repair, the excess over that expected must have arisen from coincident repair in a hybrid region which includes both sites. In the example on which the expected frequencies in Table VI are based, a little over 95% of all octads produced in the cross in repulsion, w-10 + × + w-78, were of phenotype 0+:8m; among which 3 of 86 genotypically tested (about 3.7% of total octads) were pattern 35 (0.61% expected), and 8 (about 9.8%) were pattern 50 (0.41% expected)—a considerable excess of patterns of C(P)(N) derivation. Moreover, in the same cross there were fewer than 0.03% of octads of phenotype 4+:4m (i.e., pattern 1—1.4% expected), indicating a deficient frequency of patterns of P(N) derivation. Thus, in this example at least, it has been possible to infer from a cross in only one direction that both sites must frequently be included in a single hybrid DNA region in which repair is often coincident.

Even in examples in which coincident repair is frequent, but especially when infrequent, more efficient tests are possible when both reciprocal crosses can be made; the critical patterns are of contrasting derivation in the two crosses but, if repair is independent, each pattern should occur in equal frequencies in both crosses in repulsion and in coupling, eliminating the need for estimates of expected frequencies. Crosses in coupling have an additional advantage in that all 12 patterns which are important in tests for coincident conversion have phenotypic ratios different from that of parental ditypes, permitting the use of selective screening methods; whereas half of them (those with 0+:8m ratios) in crosses in repulsion are indistinguishable by phenotypic ratios from octads in which no recombination had occurred. (It is often quite difficult to obtain the double mutant needed for the cross in coupling, especially for alleles between which reciprocal recombination is rare.) The double mutant w-10:w-78 has only recently been obtained, but the substantial frequency in the cross in coupling of 6+:2m phenotypic ratios (9.5%), which should all be pattern 1 (see Table V), and of C(P)(N) derivation, already shows a great increase over the frequency of this pattern in crosses in repulsion. In examples in which 6+:2m conversion ratios have high relative frequencies at both sites when alleles are separately crossed to wild type, a greater frequency of 6+:2m phenotypic ratios in diallelic crosses in coupling (*cis*) than 4+:4m phenotypic ratios in repulsion (*trans*) is of itself an indication that there is an appreciable amount of coincident repair.

Estimates obtained as outlined above relate to the frequency of only a fraction of all instances in which both allelic sites were present in a single hybrid structure (Fig. 20a, Section III,B,3,b)—that fraction in which coincident repair (Fig. 20c and d) has occurred in both hybrid chromatids, or in one chromatid accompanied by repair at one site, repair failure at the other, in the second chromatid. In either reciprocal cross there are only six octad patterns of these two sorts. Hence, only minimal estimates of occurrences of this particular hybrid structure are obtained.

It is even more difficult to obtain minimal estimates of the presence of two allelic sites in separate regions of hybrid DNA. Among the 54 octad patterns that can arise by repair of heteroduplexes at one or both allelic sites (Table V) only those with derivations designated as N, of which there are eight in each reciprocal cross, require the presence of the two allelic sites in separate hybrid regions. In this instance the hybrid regions must be opposite in polarity (Fig. 21a)—other patterns requiring independent repair at two sites, P and P(N), could have arisen from separate hybrid regions of the same polarity (Fig. 20b), or from independent repair at two sites in a single hybrid region (Fig. 21b). Furthermore, those with only N derivations require postmeiotic segregation (5+:3m, 3+:5m, or 4+:4m^{p}) at both sites in both chromatids, hence infrequent in most cases. The estimated maximum expected frequency (if repair at two sites were always independent) of all octads with N derivations in the example used above (alleles w-10 and w-78 in *Ascobolus*) is less than 0.7×10^{-5}.

g. *Polarity in Coincident Repair.* When there is coincident repair at two mutant sites in both hybrid chromosomes in which mutant alleles were in *trans* positions, the resultant genotypic ratio is either $2m_1+$: $6+m_2$ (6+:2m segregation at site 1, 2+:6m at site 2) or $6m_1+$:$2+m_2$ (2+:6m segregation at site 1, 6+:2m at site 2). In most known examples there is a greater frequency of 6+:2m segregation than 2+:6m in crosses of individual mutants to wild type. When two mutants with this characteristic are intercrosses (in repulsion) the relative frequencies of 6+:2m and 2+:6m segregations occurring in single mutant crosses cannot be maintained at both sites in the progeny from coincident repair. If, however, base replacement is initiated at (or near) one site, the relative frequencies of segregation ratios normally occurring could be maintained at that site during coincident repair—frequencies of segregation ratios at the other site would be dependent upon those established at the initiating site.

Such a polar situation is mildly suggested by the data already cited (from crosses between alleles w-10 and w-78 in *Ascobolus*), and is summarized in Table VII. The data rule out the possibility that w-10 is at a

TABLE VII
TESTS FOR POLARITY IN REPAIR INITIATION[a]

Genotypes ++:1+:+2:12	Observed	Expected when control is at Site 1	Site 2	Random
0:6:2:0	3	8.52	3.34	5.5
0:2:6:0	8	2.48	7.66	5.5
		$P < 0.001$	$P \approx 0.8$	$P \approx 0.2$

[a] Data from cross in repulsion (*trans*), with w-10 at site 1, w-78 at site 2.

site controlling base replacement; they are in extremely good accord with the supposition that w-78 is at such a site; but they do not exclude the possibility that neither site exerts an influence on the choice of base chains in which repair is to occur.

4. *Biochemical Implications of the Models*

Postulated repair of heteroduplexes in hybrid DNA has been invoked in models of gene conversion because of apparent similarities between conversion and the repair of radiation damage in bacteria. The discovery in fungi of enzymes affecting genetic recombination of any kind would certainly be useful for learning more about recombination mechanisms. Without awaiting such favorable discoveries, however, it may be possible to learn something by the use of biochemical genetic methods.

During the last three years we have been testing chemical agents known to interfere with dark repair of radiation damage in bacteria—caffein, acridines, various basic dyes, etc. The negative results we have obtained thus far are possibly without meaning because of unsuitability of the material (*Ascobolus*) and methods used. (It is possible that agents interfering with heteroduplex repair also interfere with processes necessary to growth and normal fruiting; and the use of pulses of higher concentration may have failed to produce significant results because of the lack of synchrony in early stages of meiosis. A positive result in those asci which were at a critical stage at the time of treatment might also be obscured to a considerable extent by the variation in relative frequencies of different conversion patterns over different intervals of the spore-shedding period.)

The successful inhibition of steps in the repair process would, of course, support the repair interpretation. Inhibition of excision of mispaired bases should result in the production of postmeiotic 4+:4m^p segregation to the exclusion of all other conversion ratios. Inhibition of extensive erosion of a base chain after base excision should alter the relative frequencies of independent and coincident repair in diallelic crosses. Inhibition of resynthesis of base chains should affect an earlier step in recombination in the Whitehouse-Hastings model than in that of Holliday. In this author's opinion, the greatest chance of learning more about recombination at this time will come from experimentation somewhat along these lines.

5. *Genetic Control of Intragenic Recombination*

Another line of investigation likely to lead to a better understanding of intragenic recombination involves its genetic control. For example, Jessop and Catcheside (1965) have reported an approximately 20-fold increase in recombinations between at the *his-1* locus of *Neurospora crassa* in crosses homozygous for *rec-1*, a recessive allele at an independently inherited locus. The *rec-1* gene specifically affects recombination within the *his-1* locus, and has no effect on intragenic recombination at any other tested locus. Inasmuch as such genetic control is limited to a single locus, it is reasonable to conclude that the recombinational process normally occurring in one short segment of a chromosome is, at least in part, independent of that occurring in other intervals.

IV. In Retrospect

All, or nearly all, of the essential characteristics of crossing-over between genes have been known for a long time (see Section II) with one overriding exception: There is no definite knowledge of the mechanism by which it occurs.

The most promising clue to the mechanism of intergenic crossing-over is its correlation with intragenic recombination, or with gene conversion. There is a very strong possibility that there are common steps in the intergenic and intragenic processes and, of the two, the intragenic process offers the larger number of routes of investigation with genetic as well as with biochemical methods.

Although the known characteristics of intragenic recombination (Section III), including its correlation with intergenic crossing-over in an adjacent region, can be satisfactorily accounted for by postulated mechanisms involving heteroduplex formation and repair, there is no direct proof that genetic recombination in eucaryotes occurs in this way.

Necessary to the establishment of a satisfactory interpretation of recombination in eucaryotes at the molecular level are (1) the resolution of the molecular structure of DNA in meiotic chromatids as it relates to genetic and reproductive units; (2) more accurate determinations of the timing of DNA replication with respect to stages in meiosis in a species in which correlation with intragenic recombination can be determined.

To establish the occurrence and repair of heteroduplexes it would be desirable (1) to detect heteroduplexes by biochemical or biophysical methods and (2) to identify either the enzymes responsible for repair or the products of repair. The inhibition of repair by specific agents might be an adequate substitute for enzyme identification in showing that repair does occur. If it becomes established that meiotic chromatids effectively consist of single DNA double helices in regions undergoing intragenic recombination, and if normal DNA replication occurs at the times indicated in Fig. 3 (Section II,E,1,b), the occurrence of postmeiotic segregation is sufficient for inferring that heteroduplexes do occur.

While awaiting the essential and desired kinds of information just enumerated, it is possible that additional limitations to legitimate speculative interpretations of recombination may result from a more complete genetic characterization of the intragenic recombination process (Section III,C,3).

There are a number of observed characteristics of crossing-over which have not been referred to in this review. Examples of such are: changes in recombination frequencies as effected by temperature and other exogenous agents, and as effected by age of the female (in *Drosophila*) and time of maturing of meiotic products (in certain 8-spored Ascomycetes); negative correlations in frequencies of crossing over in nonhomologous chromosomes (in *Drosophila*). It seems to me that the relations of these phenomena to mechanisms of recombination can be fully understood only after the mechanisms have been established.

Addendum I: Historical Notes

By the mere citation of authority for a statement a reviewer may give the appearance of assigning undue historic importance to the work cited. The selection might be one of expediency: for example, a historical survey may seem out of place in the context of the review and any good reference with which the author is familiar may seem suitable. On the other hand, the historical implication may have been intended, but reasons for the choice omitted as not being of primary importance to the immediate discussion. Such a deliberate choice of suitable reference may have omitted the earliest one for any of a number of reasons: (1) The author may have been unaware of the earlier study; (2) the first paper published on the subject may be inadequate in that the problem was not thoroughly discussed, or not supported by sufficient evidence; (3) the early work may not have led to further extension of the subject, whereas a later one definitely did (in some instances the field of science was at first not prepared to follow the new lead); or (4) the early reference may appear to fall outside the scope of the review in question—as when the review is limited to work on a particular organism, or group of organisms.

Certain choices of references cited in the body of this chapter may need to be accounted for. In order to keep diversionary topics out of the main text, explanations have been reserved to this addendum. There will be no attempt to make a thorough statement of the history of the study of genetic recombination, but there are a few points to which I believe I may contribute.

A. Discovery of Chromatid Crossing-Over

The first suggestion that genetic tests could distinguish between the involvement of whole chromosomes in crossing-over, an interpretation then commonly held by geneticists (see T. H. Morgan *et al.*, 1915, pp. 59–69), and the involvement of only two nonsister chromatids, as required by the chiasma-type interpretation (Janssens, 1909), was made by Bridges (1916, pp. 120–136). He pointed out that among daughters receiving both X chromosomes from the mother, as a result of nondisjunction, a particular genetic constitution could result only if crossing over took place at the 4-strand stage. The specific example cited was equivalent to the recovery in a single egg nucleus of chromatids 2 and 3, or 3 and 4, in Fig. 1d (Section II,B,1), one of which is a double crossover and the other a single crossover which is the reciprocal of one of the crossovers in the double crossover chromatid. In an addendum included while the paper was in press, Bridges (1916, footnote, p. 136) reported the occurrence of one such nondisjunctional daughter.

Surprisingly enough, Bridges' discussion of the problem, and his limited verification, had little effect on genetic thought: All standard texts of genetics discussed crossing-over as occurring between undivided chromosomes until that of Sturtevant and Beadle (1939). Direct studies of the problem were not undertaken until the fall of 1921 when Anderson proposed to Bridges that an intensive study be made of products of nondisjunction (half-tetrads) in an attempt to learn more about the mechanism of crossing-over. Anderson's proposal involved making use of Bridges' high nondisjunction stock, which had unfortunately been lost. Bridges and Anderson then worked out a plan to accomplish the same purpose by a study of triploid daughters (arising from diploid eggs) of triploid mothers. Bridges spent considerably more than a year in building up suitable triploid stocks, in which multiple mutants were viable and fertile. (I spent the summer of 1922 in Anderson's laboratory at Woods Hole and audited almost daily discussions between Bridges and Anderson relative to the progress, or lack thereof, in stock building.) In carrying out their experiments, Bridges made the primary crosses, and classified the progenies from them. Triploid daughters from these were then tested by Anderson, in a further generation in which alleles carried by the two maternally derived chromosomes could be identified. The results of the triploid experiment did show that crossing-over was between chromatids, but complete analysis of their data was difficult, and was not completed until after Anderson had analyzed crossover data from attached-X flies, and both experiments were reported at the same time (Anderson, 1925; Bridges and Anderson, 1925).

The data from attached-X chromosomes were not only more readily analyzable than those from triploids, but also reflected more directly the recombinational processes occurring in normal diploid meiosis. As a consequence, Anderson's 1925 paper is generally considered the real beginning of analysis of recombination at the chromatid level. Bridges and Anderson's triploid study was also of considerable additional importance to the area of cytogenetics. Newton and Darlington (1929) obtained cytological evidence showing that intimate pairing in triploids was between only two of the three homologs, with occasional switches of pairing partners along the length of the trivalent, correlating with Bridges and Anderson's observation of crossing-over between only two homologs at any one level, but sometimes involving different homologs at different levels.

B. Early References to Gene Conversion

A theory of gene conversion was proposed by Winkler (1930) as a substitute for the chiasma-type interpretation of crossing-over between genes. Gene conversion, in this sense, was never accepted by geneticists because of the great amount of evidence (reviewed in Section II) supporting the standard interpretation.

In the modern usage, gene conversion is the process responsible for ratios other than 4+:4m (in octads; 2+:2m in tetrads) resulting from segregation at a single heterozygous site. I believe that Zickler (1934) was the first to use the term in this sense. He was one of the first fungal geneticists in Europe to recognize that second-division segregation patterns resulted from crossing-over between the centromere and the segregating locus. Included in the discussion of 6+:2m and 2+:6m segregation in his material there is a comparison made between crossing-over and conversion (Zickler, l.c., pp. 606–616). It seems surprising now that practically no attention was paid this aspect of his study—especially as another part of the same report, that dealing with unequal frequencies of symmetrical and alternate second-division segregation patterns, did attract considerable attention (Catcheside, 1944; Whitehouse and Haldane, 1946; Emerson, 1963).

A paper by Lindegren (1953), dealing with aberrant tetrad ratios in Saccharomyces, and using the term gene conversion in the same sense as Zickler did, had a very different reception—his interpretation was quite generally disbelieved. The manifest reason for not believing his interpretation, as judged by published criticisms and remembered informal discussions, was that his experiments did not rule out other established causes of aberrant segregation in yeast (review in Emerson, 1956); this is true enough, but I now think that there was also a stronger, though tacit, reason: Other established causes of aberrant segregation did no violence to the concept that genes were never altered through contact with their alleles, whereas conversion did.

General acceptance of the concept of gene conversion, and of its importance in intragenic recombinations, started with a study by Mitchell (1955). Hers was the first study of conversion in a diallelic cross, in which the effects of conversion are manifold. In well-executed experiments she observed that (*a*) wild-type recombinants between alleles were accompanied by a greatly increased frequency in recombination between closely linked flanking markers; (*b*) that wild-type intragenic recombinants were associated in nearly equal frequencies with both parental and both nonparental combinations of flanking markers; (*c*) by test crosses with both mutant parents of all mutant segregants in asci containing wild-type recombinants, that no double mutant was present, i.e., that recombination was nonreciprocal; and (*d*) that nonreciprocal intragenic recombination involved aberrant 6:2 segregation, at least at one mutant site. In sum, the deviations from the results to be expected from intergenic crossing over demonstrated the occurrence of a new phenomenon. Perhaps also, the new knowledge that recombination was probably not reciprocal in bacteria or in bacteriophages may have made acceptance easier in 1955 than earlier.

Addendum II: Critique of Some Data, Their Acquisition, and Their Handling

A. Preachment on Statistics: Their Purpose and Interpretation

Two questions need to be answered by statistical treatments. One is, do the data, at a selected level of significance, exclude a given hypothesis: Is there a significant difference between different sets of data, or between observed and theoretically expected frequencies? The other arises when the answer to the first is negative—when observed differences are not significant at a reasonable level of confidence. It asks just how large an actual difference would not be excluded at the level of significance employed. The importance of answering this second question is too often overlooked.

One example of a possible, or even probable, error in interpretation arising from failure to ask the second question is to be found in the handling of data concerning recombination in attached-X chromosomes of *Drosophila,* with which I was directly concerned. These data were the basis for the interpretation that the choice of chromatids taking part in any exchange was completely random and not influenced by the particular chromatids taking part in an adjacent exchange. The available data are not in serious disagreement with this standard interpretation but, as I shall attempt to show in a following section, they can equally well be construed to support a different interpretation.

There are a number of reasons why it is easy to be satisfied with a negative answer to the first question. If the fit to expectation is fairly good there is a good probability that the interpretation is correct, and the chance of establishing the reality of a small discrepancy may not seem to justify the sometimes enormous effort necessary to accumulate numbers sufficient to make so small a difference significant. Besides, it is usually impossible to prove two observed frequencies are identical, or that one is identical with a theoretical expectation—there is always some probability that the observed frequency deviates from the true frequency occurring in an infinitely large population. It is helpful, when possible, to calculate confidence intervals which give an indication of how far off the observed values might be in either direction.

In instances in which a large number of samples are available, plus and minus deviations are expected to be equally frequent, and to have a normal distribution. When this is true one has considerable assurance that the mean is approximately the true value. When not true, with an important exception, a further look at the situation is advisable. The exception just referred to involves situations in which frequencies are so low that zero is not significantly different from the true, or expected, mean. In such instances, or in examples in which such is true for some classes, only positive, or an excess of positive, deviations are expected.

Due to complexities in many biological problems it is often very difficult (impossible?) to apply simple and direct methods to determine the range of divergence from expectation which is not statistically excluded by the data. In such cases one may turn to brute force or Robin Hood's barn methods, such as some to be used in later sections. Leads to methods to be adopted may sometimes become apparent from the appearance of patterns among deviations when the data are summarized in different ways. While it is possible that sufficiently adept mathematicians could develop elegant solutions to some of these statistical problems, it has been my experience with professional statisticians that the elegant solutions they develop do not give answers to the questions I am interested in.

B. Chromatid Interference in Attached-X Chromosomes

The interpretation that the chromatids involved in two adjacent exchanges represent a random choice—that each of the two homologous chromatids involved in one exchange has no more and no less than a random chance of involvement in the second—was developed from recombination data obtained from the attached-X chromosomes of *Drosophila melanogaster* (Anderson, 1925; Sturtevant, 1931; Emerson and Beadle, 1933; Beadle and Emerson, 1935; Bonnier and Nordenskjöld, 1937; Welshons, 1955; Weltman, 1956). If this interpretation is correct, the relative frequencies of 2-strand, 3-strand, and 4-strand double exchanges should occur in the proportions 1:2:1. These relative frequencies cannot be directly determined for the attached-X chromosomes (as they can be by tetrad analysis), but they can be estimated from two relationships which can be observed. (1) Following a nonreciprocal proximal exchange, homozygosis occurs at loci between that exchange and the nearest second exchange distal to it. All second exchanges are detected because of the restoration of heterozygosity at loci distal to them. Two types of such doubles occur: (*a*) Those in which one attached chromatid is a double crossover, the other a noncrossover, and which could have arisen from either 2-strand or 3-strand double exchanges; (*b*) those in which one chromatid is a single crossover at the level of one exchange, the other a single crossover at the level of the other exchange, and which could have arisen from either 3-strand or 4-strand double exchanges. (2) Following a reciprocal proximal exchange two types of double exchanges are again recognizable: (*a*) Those in which both attached chromatids are double crossovers, and heterozygous throughout, and which could have arisen only from 2-strand double exchanges; (*b*) those in which one chromatid is a double crossover, the other a single crossover at the level of the proximal exchange, which is heterozygous at loci between exchanges and homozygous at loci distal to the more distal exchange, and which could have arisen only from 3-strand double exchanges. Different types of analyses are necessary for these two relationships.

1. *Relative Frequencies of 2-Strand and 4-Strand Double Exchanges*

The four kinds of second exchanges following a nonreciprocal proximal exchange are diagrammed, together with their products, in Fig. 22. Points to be noted are: (1) All products are homozygous for alleles from one or the other homolog in the interval between the two exchanges. (2) Each product of a 2-strand double exchange has one double crossover chromatid and one noncrossover, one product being the reciprocal of the other. (3) Each product of a 4-strand double exchange has one chromatid which is a single crossover at the level of the proximal exchange, the other a single crossover at the level of the more distal exchange—again one product is the reciprocal of the other. (4) The products of a 3-strand double exchange are dissimilar and not the reciprocals of one another, one being identical with a product of a 2-strand double exchange, the other with a product of a 4-strand double exchange. (5) There are two different 3-strand double exchanges, the products of one being the reciprocals of those of the other. (6) All products of these double exchanges are detectable and can be identified as being derived from adjacent double exchanges—any intervening exchange (in appropriately marked chromosomes) would have interrupted the homozygosity in the interval between the exchanges being studied.

The contribution from 3-strand double exchanges to the class of "apparent 2-strand" doubles and "apparent 4-strand" doubles should be equal, but the fraction of each class originating from 3-strand double exchanges cannot be determined from double crossovers of this general type. Unless there is some mechanism whereby one kind of 3-strand double exchange occurs more often than the other, and a mechanism whereby one product is more likely than the other to be shunted into the polar body, there is no reason to suspect that contributions to apparent 2-strand and 4-strand double exchange classes would be unequal. Hence, any divergence in equality between the two classes can reasonably be assumed to reflect an unequal proportion of actual 2-strand and 4-strand double exchanges.

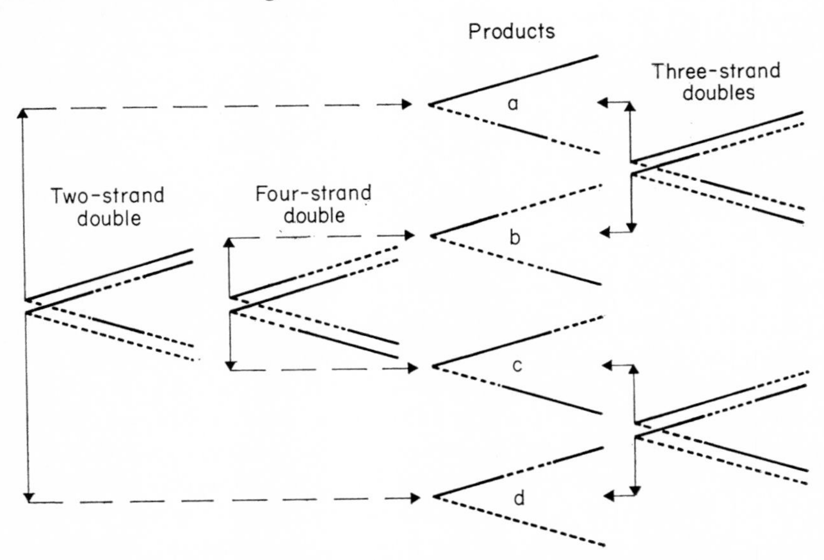

FIG. 22. Products of the four kinds of double exchanges in attached-X chromosomes of *Drosophila* which result in homozygosis at loci bracketed by the two exchanges.

Largely because of the ease of detection of double crossovers which are homozygous for alleles flanked by the two exchanges, there is a fairly large body of data concerning them. These have been summarized, and evaluated to some extent (Emerson, 1963). This summary shows a slight deficiency in apparent 4-strand doubles—47.7% of a total of 1105. The summary excludes the data reported by Bonnier and Nordenskjöld (1937) which, in disagreement with those from all other sources, showed an excess of apparent 4-strand doubles—93.5% of a total of 108—and was interpreted to be the result of misclassification (fully discussed in the review cited). The probability of a deficiency of apparent 4-strand doubles as great or greater than that reported in the summary is about 40%, and the heterogeneity between the 60 samples included has about the same probability of being due to chance fluctuations—nothing to be very worried about.

A closer look at the deviations from equality of the individual samples can be had from the plot (Fig. 23) of observed numbers of apparent 4-strand doubles as a function of the numbers expected. There are a

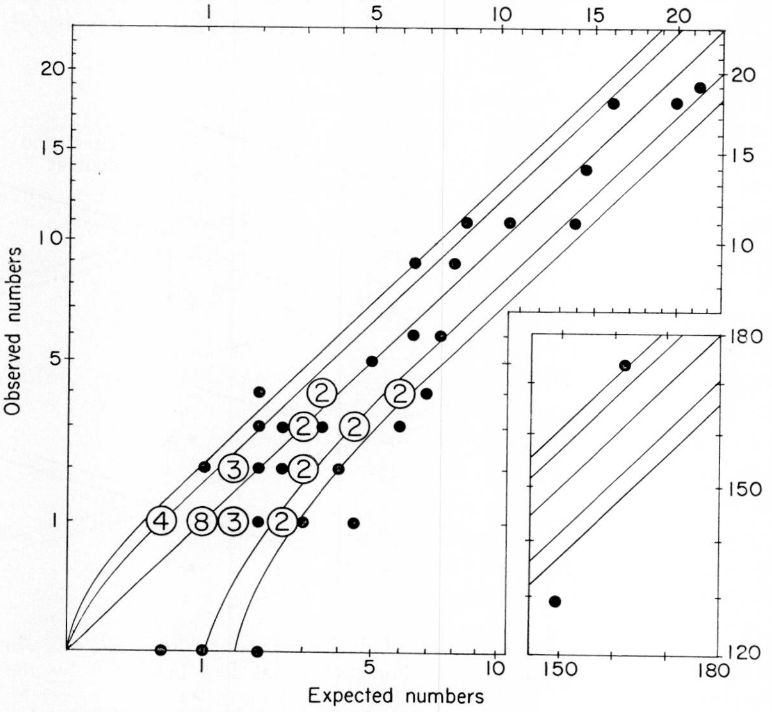

FIG. 23. A plot comparing observed and expected numbers of apparent 4-strand double exchanges in 60 samples of daughters homozygous at loci between the two exchanges. Solid dots represent single samples, larger open circles represent the numbers of samples indicated by the enclosed numerals. The central diagonal line represents perfect fit between observed and expected, the curved lines becoming parallel to that diagonal represent upper and lower 50 and 80% confidence limits.

larger number of significant minus than plus deviations, which becomes more evident when one recalls that zero is not a significant minus deviation at the 80% confidence limit in samples of size three or less. Among the 39 samples with four or more individuals, three exceeded the upper 80% confidence limits and seven the lower, where the expected in each direction is 3.9 (10%).

a. *Relation to Distance between Exchanges.* Welshons (1955) was the first to call attention to a possible correlation between the relative frequency of apparent 4-strand double exchanges and the distance between the two exchanges. The total frequencies of apparent 2-strand and 4-strand doubles in his material were 93 and 84, respectively, whereas among those in which exchanges occurred in adjacent intervals, they were 22 and 10, respectively.

Before looking into a similar relationship in the total available data it is necessary to select some reasonable estimate of the actual distance between exchanges. In some samples in which exchanges occurred in adjacent intervals the maximum possible distances between exchanges are larger than those in other samples with an intervening interval. Means of the distances within which exchanges could have occurred have been used (Emerson, 1963), but these also are inaccurate, they may give a fair value in instances in which distances are long, but when short it is likely that most double exchanges occurred at nearer the maximum possible distance. A number of mapping functions have been developed (e.g., Barratt *et al.*, 1954), but these relate recombination frequency to distance between loci flanking the exchanges, whereas the need here is for a measure of the distance between the exchanges themselves. Further, to the best of my knowledge, such mapping functions neglect the decreasing effect of interference with increasing map distance. Being unable to derive an appropriate mathematical formulation I have made use of the following empirical procedure:

(1) The data used are those summarized by Weinstein (1936), comprising recombination frequencies (single-strand data) covering the major portion of the X-chromosome of *Drosophila,* and thus closely corresponding to the data from attached-X chromosomes. Summaries of two crosses are used: one encompassing 16,136 sons of mothers heterozygous for *sc ec cv ct v s f car bb,* the other 28,239 sons of mothers heterozygous for *se ec cv ct v g f.* The intervals between loci are sufficiently short that the possible occurrence of double exchanges within them can be neglected—intervals range from 5.1 to 13.1 map units in the first cross, from 7.1 to 11.1 in the second.

(2) Coincidence values, standardly derived, showed that interference did not extend beyond five intervals; consequently, the distributions of double crossovers in segments of five intervals were determined—from *sc* to *s* or *g,* from *ec* to *f, cv* to *car,* and *ct* to *bb.*

(3) In each of these segments of five intervals, the total number of double crossovers having one crossover in a particular terminal interval of the segment was ascertained; the fraction of these which had the sec-

ond crossover in each succeeding interval was then determined. Such determinations were made from each end of each segment.

(4) In Fig. 24 these fractions are represented by the areas of the fine-line rectangles, each of whose height is the fraction of doubles divided

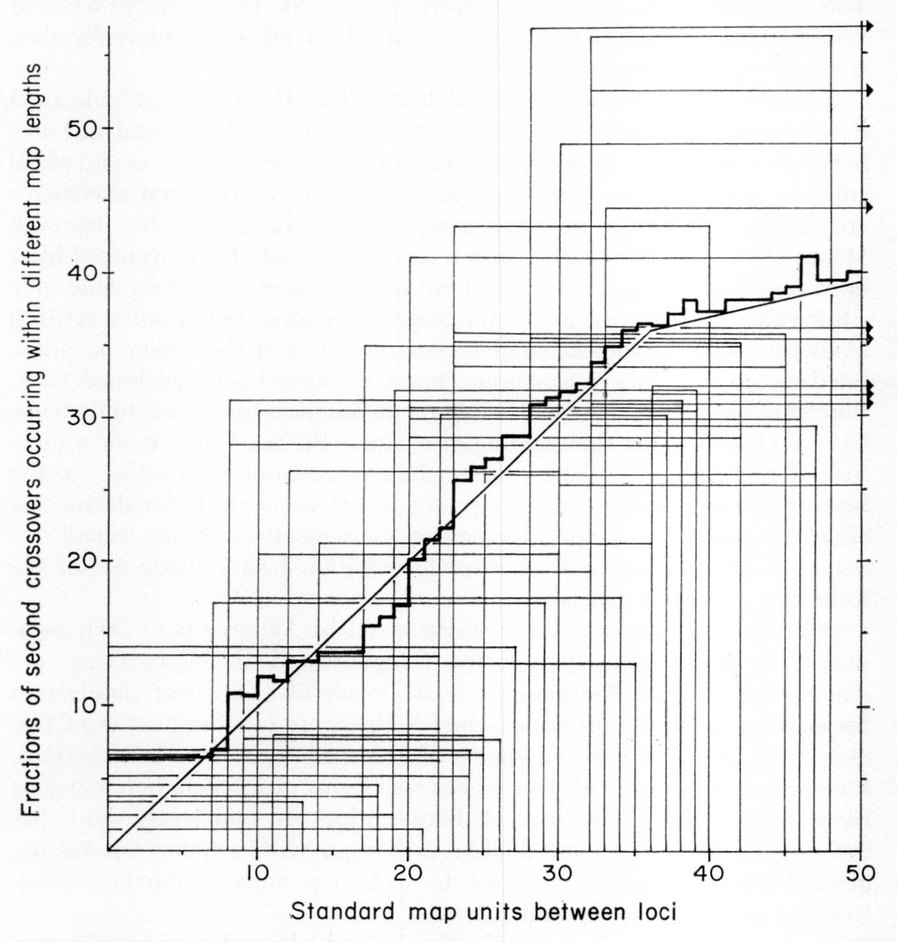

FIG. 24. Basis for estimating adjusted mean map distances between exchanges as explained in text.

by the sum of the map units in the two intervals, whose length is the sum of the two intervals in map units, and whose position along the base line is determined by the number of map units intervening between the two intervals in which crossovers occurred.

(5) The heights of all rectangles which include each succeeding map unit along the base line were then averaged; the mean fraction of doubles arising from exchanges separated by different numbers of map units is thus obtained. The relationship of these means to increasing map distance is shown by the stepped heavy line in the plot (Fig. 24).

(6) An approximate fit to this stepped line is represented by the two medium-weight straight lines (Fig. 24) with slopes 1.0, from the origin to 36 units of map length, and 0.25 from 37 to 51 units of map length, after which the slope is supposed to be zero. These lines represent the fraction of double crossovers per map unit, expressed in percent, as a function of the number of map units separating the two exchanges.

(7) In practice, the area under this line (with slopes 1.0, 0.25, and 0.0, over different stretches), starting at a point on the base line corresponding to the minimum map distance between exchanges (the sum of intervening intervals), and extending to a point on the base line corresponding to the maximum map distance between exchanges (the total distance between loci flanking the double exchange), is ascertained. To half of the area so obtained is added the area covering the base line from the origin to a point corresponding with the minimal map distance between exchanges. The length, in map units, of the base line, starting from the origin, which would just be covered by the area so derived is then taken as the adjusted mean map distance between exchanges.

As one possible test of the validity of the method just described, coincidence coefficients for all pairs of intervals in the data just used are plotted as a function of adjusted mean map distances between exchanges (Fig. 25). At less than 35 units of map length the calculated relationship is $y = 0.0367\ x - 0.419$, where y is the coefficient of correlation and x the number of map units between exchange. (At greater than 35 units of map distance, the calculated relationship is $y = 0.00026\ x + 0.935$.) The fit is not too good, probably reflecting differences in interference intensities in different regions of the chromosome. Such differences must be averaged out in any generalized treatment, and the one here proposed is probably not much worse than any other.

The attached-X data relating to the relative frequencies of apparent 2-strand and 4-strand double exchanges can be examined for correlation with map distance between exchanges in a number of ways. In Fig. 26a the fraction of apparent 4-strand doubles in each of the 60 samples is plotted as a function of adjusted mean map distance between exchanges. There is at least a suggestion of a correlation; the deficiency of apparent 4-strand doubles is somewhat greater when distances between exchanges are shorter. The same trend is evident in the plot (Fig. 26b) of absolute probabilities (one-tailed) of deviations from equality as great or greater than observed: deviations to excess apparent 2-strand doubles with probabilities less than 40% are clustered in the half of the plot having lower map distances between exchanges; and deviations to excess apparent 4-strand doubles with probabilities of less than 50% do not occur at less than 22 map units between exchanges.

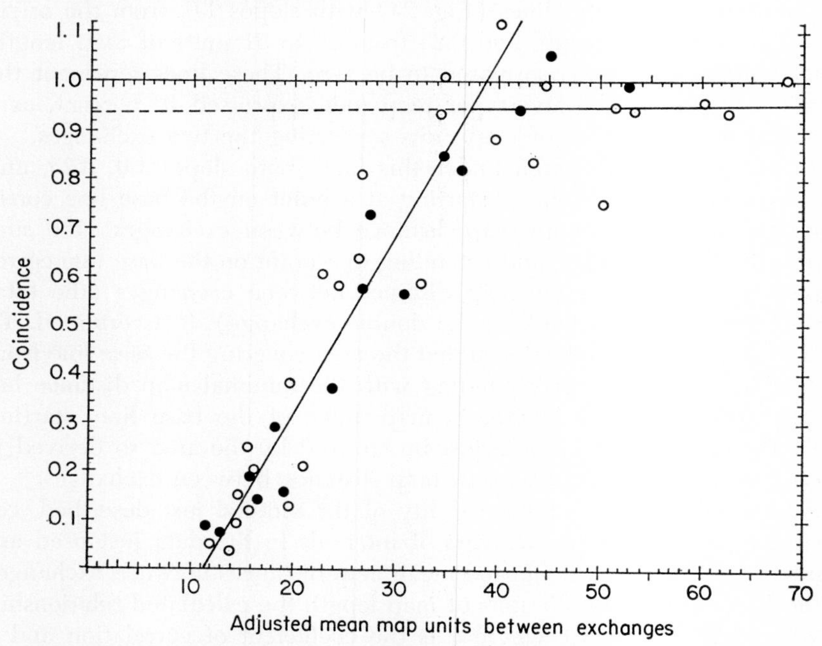

FIG. 25. Coincidence values in *Drosophila* as a function of adjusted mean map distances between exchanges. Open circles represent data from crosses heterozygous at nine loci, solid dots data from crosses heterozygous at seven loci, from the single-strand data from *Drosophila* summarized by Weinstein (1936).

To obtain a quantitative measure of the trend just noted it is convenient to group the samples into lots for which confidence intervals will not be excessively large. To do this, the two largest samples, with totals of 285 and 324, at 31.9 and 41.6 map units between exchanges, were handled separately, and the remaining 58 samples, totaling 496, were divided as equally as possible into four groups according to map length between exchanges—a total of 125 had map distances ranging from 7.0 to 13.6 units, 122 from 14.0 to 20.8, 131 from 21.5 to 25.9, and 118 from 26.4 to 56.9. The weighted means, for both map units and frequencies, of these are plotted in Fig. 27. The regression line calculated from this distribution has the relationship $y = 35.86 + 0.3963\ x$, where y is the fraction of apparent 4-strand doubles ($\times$ 100), and x is the number of adjusted mean map units between exchanges. Whereas this relationship gives the best fit to the data as here summarized, it must be accepted as only a rough estimate; for example, grouping samples into lots of greater or less size would most probably result in a different regression constant, as use of a different estimate of mean map distance is known to do.

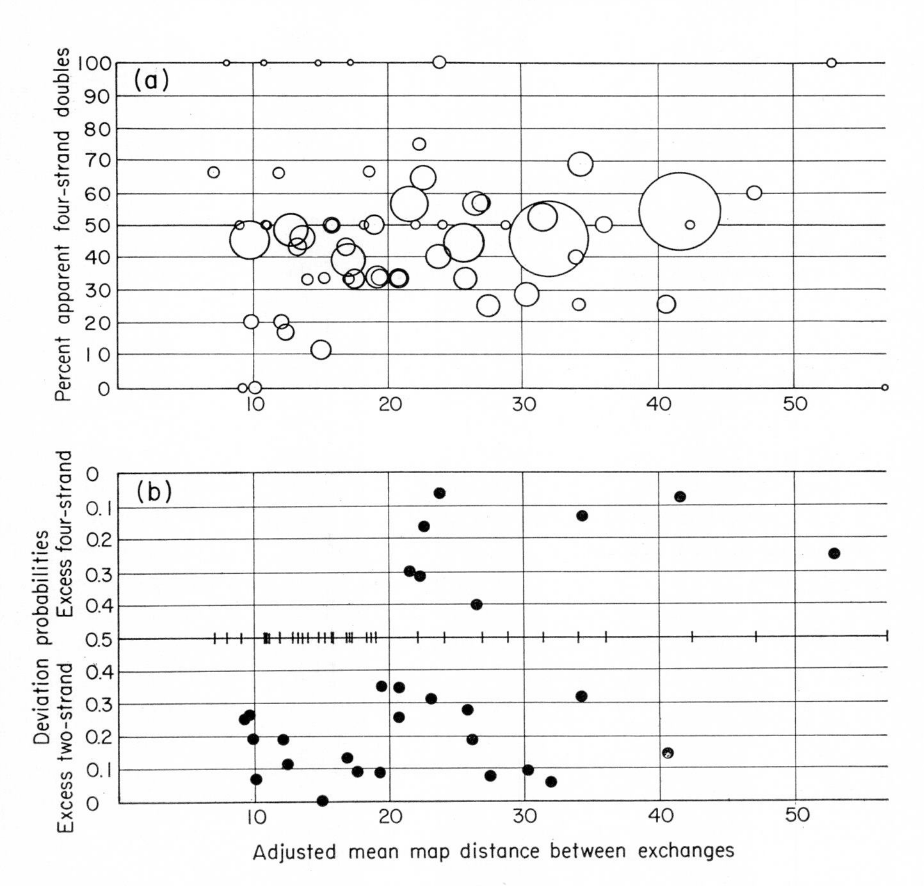

FIG. 26. (a) Fractions of apparent 4-strand double exchanges plotted as a function of adjusted mean map distance between exchanges. Areas of circles are proportional to sample sizes. Data from attached-X chromosomes as summarized by Emerson (1963). (b) One-tailed probabilities of deviations from equality as great or greater than observed in the samples plotted in (a). Probabilities of 50% are indicated by vertical hatches on the 0.5 line, others by dots; probabilities from tables of Warwick (1932).

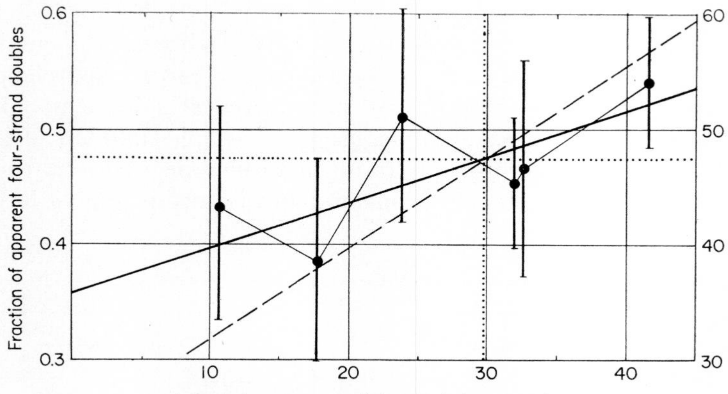

FIG. 27. Regression plot showing the relation between frequencies of apparent 4-strand doubles and map distances between exchanges. Vertical lines represent 95% confidence intervals of the groups of samples plotted; the heavy solid line is the calculated regression; the dashed line represents regression with twice the calculated slope; and dotted lines show mean values of the entire population.

2. *Relative Frequencies of 2-Strand and 3-Strand Double Exchanges*

The four kinds of second exchanges following a reciprocal proximal exchange are diagrammed in Fig. 28. Points to be noted are: (1) Products of 2-strand double exchanges are of two kinds: one product is identifiable as a derivative of a 2-strand double exchange, consisting of two chromatids which are reciprocal double crossovers; the other is not a recognizable product of double exchange, consisting of two noncrossover chromatids. (2) Products of both types of 3-strand double exchanges are also of two kinds: one can be identified as a product of a 3-strand double exchange, consisting of one double crossover chromatid and one chromatid which is a single crossover at the proximal level; the other product again cannot be recognized as resulting from a double exchange, having one chromatid which is a single crossover at the distal level and one which is noncrossover. (3) Neither product of a 4-strand double exchange is diagnostic of double crossing-over—each consists of two chromatids which are reciprocal single crossovers at the levels of one or the other of the two exchanges. (4) Whereas only three-eighths of the products of these double exchanges can be identified as to origin, those which can be are completely identifiable with respect to the kind of double exchange involved. (5) Products of 2-strand and 3-strand double exchanges differ genotypically, and also phenotypically if recessive alleles are present beyond the level of the distal exchange in the homolog be-

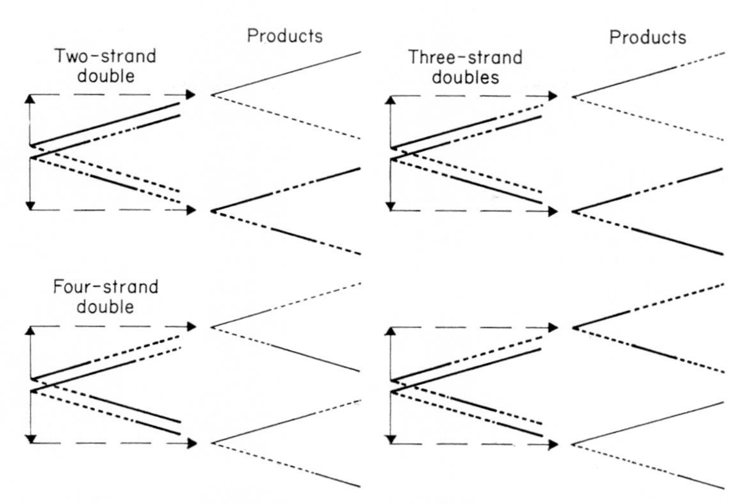

FIG. 28. Products of the four kinds of double exchanges in attached-X chromosomes of *Drosophila* which result in reciprocal crossovers at the first exchange.

coming homozygous at that point; hence there is no basis for selective screening for products of double exchanges of this class.

Data concerning the relative frequencies of 2-strand and 3-strand double exchanges in which the proximal exchange was reciprocal are not extensive and, to the best of my knowledge, have never been summarized. In addition, in examples in which recessive alleles are present in both homologs it is necessary to make use of proportional frequencies —the products of frequencies of phenotypes in the total population and frequencies of the products of double exchange within the phenotypes in which they occur. This method is illustrated in Section D of Table VIII, which summarizes the relatively more extensive data in published accounts. The 95% confidence limits given in the table are based solely on expected fluctuations of double crossovers within phenotypes, which is sufficient in those examples (A through C) in which recessive alleles were carried in only one homolog, and in which only wild phenotypes were subjected to genotypic testing. Where recessive alleles were carried in both homologs (D), expected fluctuations in phenotypic frequencies should properly also have been taken into account. On a percentage basis, however, such fluctuations are expected to be small relative to those of genotypes within phenotypes tested, and their neglect may not be too serious.

In each of the four experiments summarized in the table, total frequencies of 2-strand double exchanges and of both types of 3-strand double

TABLE VIII

Relative Proportional Frequencies of 2-Strand and 3-Strand Double Exchanges in Attached-X Chromosomes of *Drosophila*

Source[a]	Phenotype	Frequency (%)	No. tested	Type of double exchange	Frequency in phenotype		Proportional frequency (%) in total sample[b]	Mean map units
					No.	Percent		
A	+	81.6	516	2-strand	11	2.13	1.47 (0.75–2.58)	30.5
				3-strand	16	3.10	2.14 (1.23–3.45)	31.4
B	+	79.3	1638	2-strand	7	0.43	0.35 (0.14–0.70)	21.6
				3-strand	6	0.37	0.30 (0.11–0.63)	21.0
C	+	69.0	690	2-strand	6	0.69	0.55 (0.21–1.15)	22.9
				3-strand	5	0.58	0.46 (0.15–1.04)	21.0
D	+	51.1	705	2-strand	19	2.70	1.36 (0.86–2.12)	30.2
				3-strand	5	0.71	0.36 (0.12–0.83)	
	ec	6.20	109	3-strand	13	11.93	0.74 (0.40–1.32)	
	ec ct	8.06	123	3-strand	6	4.88	0.39 (0.15–0.83)	
			Total	3-strand	24		1.49 (1.06–2.35)	30.2
	sc	5.39	94	3-strand	18	19.15	1.03 (0.63–1.61)	
	sc cv	8.16	90	3-strand	6	6.67	0.54 (0.20–1.14)	
	sc cv v	5.69	54	3-strand	1	1.85	0.11 (0.003–0.56)	
			Total	3-strand	25		1.68 (1.14–2.63)	30.5
Combined totals[c]				2-strand	43		3.73 (2.76–5.24)	27.6
				3-strand	51.5		4.48 (3.39–6.21)	28.7

[a] Sources: A, Bonnier and Nordenskjöld (1937); B and C, Welshons (1955); autosomal inversions in C; D, Beadle and Emerson (1935).

[b] 95% confidence limits based solely on frequencies within phenotypes.

[c] Includes averages of two kinds of 3-strand doubles in source D.

exchanges [when distal homozygosity for alleles in both homologs was ascertained (D)] are approximately equal, as expected in the absence of chromatid interference; and the mean map distance between exchanges is approximately the same for 2-strand and 3-strand doubles, as expected if there were no bias with respect to map distances. Consequently, there is little in the data themselves to suggest any deviation from randomness in choice of chromatids taking part in adjacent exchanges. On the other hand, inasmuch as the data relating to 2-strand and 4-strand double exchanges rather strongly suggest that 2-strand doubles are in excess in short intervals, a similar excess relative to 3-strand doubles would not be unexpected. The total data relative to 3-strand double exchanges here summarized involves only 149 genotypically tested double exchanges distributed among 31 samples of different mean map distances between exchanges. By combining the samples, on the basis of map distances, into seven groups with approximately equal frequencies, a slight correlation of 3-strand exchanges with map distance is observed, as shown by the expression $y = 67.56 + 0.196\, x$ (where y is the percent of 3-strand doubles among the total of 2-strand and 3-strand doubles, and x is the number of adjusted mean map units between exchanges). It should be noted that the data are really insufficient for estimating regression; had the sample at map distance 13 had a 1:2 frequency of 2-strand to 3-strand doubles, instead of the 0:3 observed, regression with a very much greater slope would have been estimated. All that can be truly said is that the data do not distinguish between a constant ratio between 2-strand and 3-strand doubles at all map distances, on the one hand, and a very considerable increase in the fraction of 3-strand doubles with increasing map distance, on the other.

3. *Relative Frequencies of All Kinds of Double Exchanges*

In those double exchanges in attached-X chromosomes which result in homozygosis between the two exchanges (Fig. 22) all double exchanges are detected; but half of the products of 3-strand doubles are of the same constitution as all products of 2-strand doubles, and half have the same constitution as all products of 4-strand doubles. Hence

$$p_4 + 0.5p_3 = f_{4(x)} \qquad (1)$$
$$p_2 + 0.5p_3 = 1 - f_{4(x)} \qquad (2)$$

when p_2, p_3 and p_4 are, respectively, 2-strand, 3-strand and 4-strand double exchanges actually occurring, and $f_{4(x)}$ is the fraction of 4-strand doubles observed when the two exchanges are separated by x map units.

In those double exchanges not resulting in homozygosis between exchanges (Fig. 28) no 4-strand doubles are detected, and 2-strand and 3-strand doubles are detected in only half of the products of each. Those products that are detected, however, definitely indicate the type of double exchange from which they arose. Hence

$$\frac{p_3}{p_2 + p_3} = f_{3(x)} \tag{3}$$

in which p_2 and p_3 are the frequencies in which 2-strand and 3-strand double exchanges actually occur, $f_{3(x)}$ the fraction of observed products of 3-strand double exchanges among the total observable products of double exchanges.

Equations which relate the frequency of occurrence of each kind of double exchange to the relative frequencies of their products are derived from Eqs. (1), (2), and (3):

$$p_2 = \frac{2\,[1 - 3\,f_{3(x)}][1 - f_{4(x)}]}{2 - f_{3(x)}} \tag{4}$$

$$p_3 = \frac{2\,f_{3(x)}[1 - f_{4(x)}]}{2 - f_{3(x)}} \tag{5}$$

$$p_4 = \frac{2\,f_{4(x)} - f_{3(x)}}{2 - f_{3(x)}} \tag{6}$$

The numerical expression already derived (Addendum II,B,1,a) which relates the observed fraction (in percent) of apparent 4-strand doubles to the number of map units, x, between exchanges is

$$f_{4(x)} = 35.86 + 0.396\,x \tag{7}$$

and that (Addendum II,B,2) which relates the observed fraction of 3-strand doubles to map distance between exchanges is

$$f_{3(x)} = 67.56 + 0.196\,x \tag{8}$$

By substituting the numerical values from Eqs. (7) and (8) into Eqs. (4), (5), and (6), the frequencies of occurrences of each kind of double exchange, as related to map distance between exchanges, are obtained.

The approximate best fit to the data obtained by the method just described is plotted in Fig. 29, curves a-2, a-3, and a-4. As noted at the top of that figure, the data used actually cover only the midrange of map distances between exchanges, and the extrapolation of the curves to greater and, more especially, to lesser distances between exchanges is probably not justifiable. In any case, the extrapolations do not lead to an obvious interpretation based on chromatid interference.

It was noted earlier that the interpretation of no chromatid interference (curves b-2, b-3, and b-4 in Fig. 29) is not definitely excluded

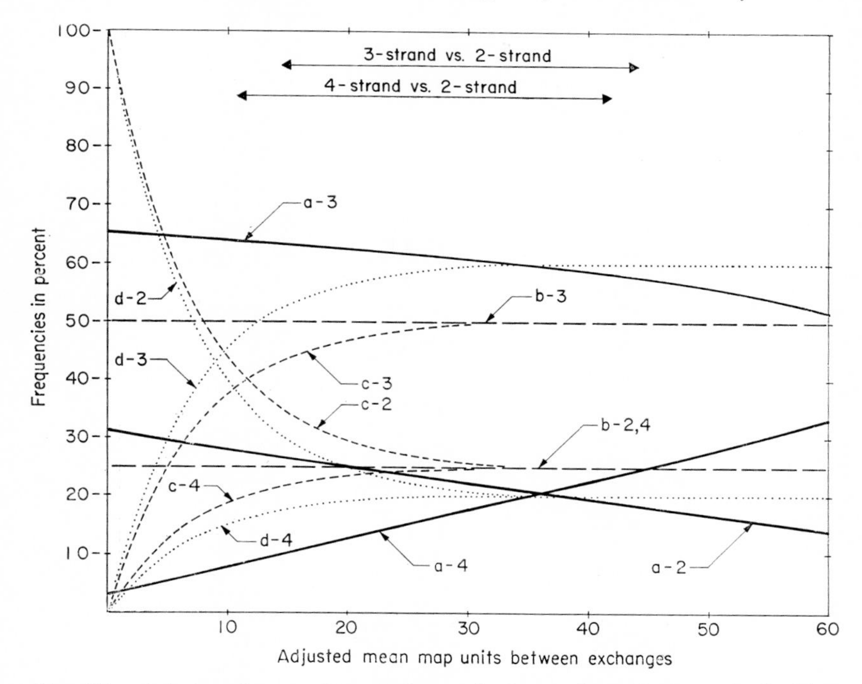

FIG. 29. Relative frequencies of 2-strand, 3-strand, and 4-strand double exchanges as a function of map distance between exchanges. The heavy solid lines show values calculated from the two regression formulas discussed in the text (a-2, for 2-strand; a-3, for 3-strand; and a-4, for 4-strand double exchanges. The long-dashed lines (b-2,4 for 2-strand and 4-strand; b-3 for 3-strand doubles) represent values expected with no chromatid interference and no effect of map distance. The short-dashed lines (the c-curves) and the dotted lines (the d-curves) show values expected on two other interpretations discussed in the text. The lines at the top show the range of map distances covered by data used in estimating linear regressions.

on statistical grounds by the data. The purpose of this analysis, however, is to see if other null hypotheses are also not excluded. One such alternative interpretation is that exchanges occurring in the closest possible proximity to one another invariably involve the same two chromatids in both exchanges (i.e., complete negative interference), with a gradual relaxation of chromatid interference accompanying increasing distances between exchanges until with, say, 30 or more map units between exchanges there is a random participation of homologous chromatids in the two exchanges. Such a situation is illustrated by curves c-2, c-3, and c-4 in Fig. 29.

For a statistical comparison of the goodness of fit of the data to alternative interpretations it seems appropriate to use the observed mean fraction of apparent 4-strand doubles, 47.7%, instead of 50% (expected if there is no chromatid interference), and the observed mean fraction of

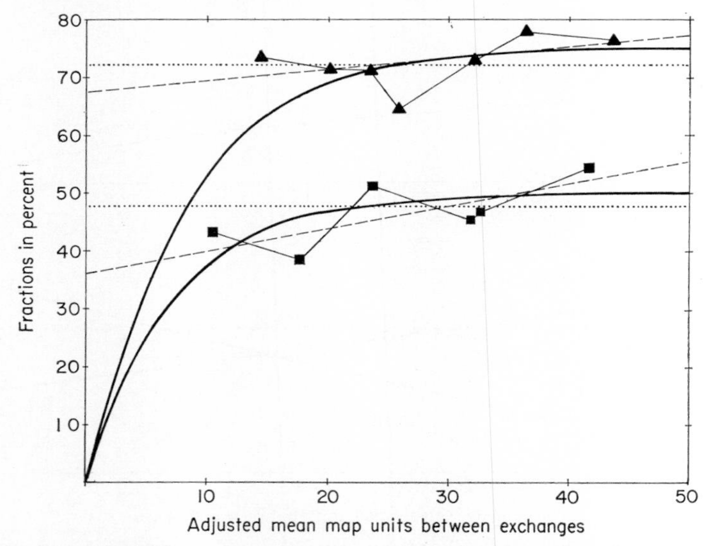

FIG. 30. Relation of observed frequencies of 3-strand (triangular points) and of apparent 4-strand (quadrangular points) double exchanges to map distance between exchanges. Heavy solid-line curves, theoretical expectation based on d-2, d-3, and d-4 in Fig. 29; light dashed lines, calculated linear regressions; dotted lines, mean frequencies at all map distances.

3-strand doubles, 72.4%, instead of 66.7% (see Fig. 30). A modification of the alternative interpretation (high negative chromatid interference over short intervals) to agree with these mean fractions results in curves d-2, d-3, and d-4 (Fig. 29). [The shape of the curve for 2-strand doubles is based on the arbitrary formulation $p_2 = 0.8 \times 0.9^{Kx} + 0.2$, in which p_2 is the frequency of 2-strand doubles, x is the number of map units between exchanges, and K is a constant, 4/3 in this instance.] Expected fractions of 3-strand doubles and of apparent 4-strand doubles as functions of map distances between exchanges are shown by the heavy-line curves in Fig. 30. The points plotted in this figure are the observed values of the grouped data used in estimating regressions. Observed deviations from expected values are a little less when expected values are based on the interpretation of decreasing (negative) chromatid interference with increasing distance between exchanges than on the interpretation of no change with map distances: standard deviations 2.71 and 2.81%, respectively, for the 60 samples relating to apparent 4-strand doubles; 0.72 and 0.75%, respectively, for the 30 samples relating to 3-strand doubles. Consequently, the available attached-X data are insufficient to distinguish between the complete absence of chromatid interference, on the one hand, and strong negative chromatid interference over short map distances, on the other.

C. Mechanism of Mitotic Recombination

1. *Incipient Meiosis*

In the yeast *Saccharomyces cerevisiae*, unlike the situation in a number of other species in which there is commonly only one somatic exchange in only one chromosome arm in any nucleus (Roper, 1966), somatic exchanges have been observed to occur in multiples in those nuclei in which it occurs at all. This observation has led to the interpretation that there has been a generalized pairing between homologous chromosomes in some cells of the population. Direct support for this interpretation was obtained in an ingenious experiment of Wilkie and Lewis (1963):

In yeast there are no specialized sex organs, but vegetative diploid cells are transformed directly into sporocytes in which meiosis occurs; four ascospores resulting from the two meiotic divisions. This transformation can be induced in a large fraction of the cells in a diploid clone by culture on so-called sporulating media. By a fairly short (2-hour) exposure to sporulating medium followed by return to a medium favoring vegetative growth, Wilkie and Lewis obtained diploid progeny in which the frequency of somatic recombination (judged by homozygosity for recessive alleles) was about 20 times greater than in the progeny of cells continuously maintained in vegetative medium (line 1 in Table IX). It has long been known (or at least surmised [Emerson, 1955]), that sporadic sporulation can occur in cultures which are largely in the vegetative state—it is possible that, in normal vegetative cultures, some cells start to enter the meiotic phase and then revert to the vegetative phase, as a larger fraction were induced to do in the experiment of Wilkie and Lewis, and thus account for all somatic recombination on a common basis.

a. *Effect of Irradiation.* It was noted above (Section II,A,3,b) that the frequency of somatic recombination in *Saccharomyces cerevisiae* and *Ustilago maydis* can be considerably augmented by exposure to ultraviolet (UV) irradiation. An example of this effect is included in Table IX. It is interesting that the radiation effect upon cells induced to enter early stages of meiosis is less than one-tenth that on cells kept in the vegetative phase, more or less irrespective of dosage. I am not sure what this result means: (*a*) whether or not the effect of radiation is to induce vegetative cells to enter meiotic stages; or (*b*) whether or not the mechanism of *UV*-induced recombination differs from that of spontaneous recombination.

TABLE IX
EFFECT OF CELL STATE AND OF UV IRRADIATION ON SOMATIC RECOMBINATION[a]

Minutes' exposure to ultraviolet	Cells in log phase			Cells in sporulating medium[b]		
	Total colonies	Recombinants per 10^3	Radiation effect	Total colonies	Recombinants per 10^3	Radiation effect
0	3920	0.25		2180	5	
1	3533	21	84 ×	4650	40	8 ×
2	1629	93	372 ×	2005	98	20 ×
3	7355	85	340 ×	6607	115	23 ×

[a] Data from Wilkie and Lewis (1963).
[b] Two hours before irradiation and returned to vegetative medium.

2. *Mitotic vs. Meiotic Disjunction*

The idea that, in somatic recombination in yeast, chromosome disjunction as well as chromosome pairing might be of the meiotic type was, I believe, first suggested by James and Lee-Whiting (1955). Meiotic disjunction results in homozygosis (at the end of the first meiotic division) from the centromere to the first exchange on either side of it, whereas mitotic disjunction maintains heterozygosis from the centromere to the nearest exchanges. The data obtained by James and Lee-Whiting support the mitotic disjunction interpretation, as do the more recent data of Hurst and Fogel (1964), in that homozygosis increases with distance from the centromere (also true in *Drosophila* and *Aspergillus*). The data of Wilkie and Lewis (1963), on the other hand, show homozygosis greatest near the centromere and decreasing with map distance, to account for which they revived the interpretation of meiotic segregation followed by restoration of diploidy. Hurst and Fogel suggest that this discrepancy in observation results from the differences in methods employed, that growth in liquid medium prior to UV irradiation offered the opportunity for undetected sporulation. I wish to show the data of Wilkie and Lewis do not critically test the alternative interpretations.

The distinction between the two interpretations in the relation of map distance to homozygosis is sharp as long as exchanges, especially multiple exchanges, are infrequent, but does not hold if there is too great a frequency of multiple exchanges. To account for the data of Wilkie and Lewis on either interpretation it is necessary to suppose that there actually was an excessive frequency of multiple exchanges, as is shown in

TABLE X

DISTRIBUTIONS OF PHENOTYPES RESULTING FROM THE MULTIPLE EXCHANGES NECESSARY FOR THE PRODUCTION OF THREE SEGREGANTS OF PHENOTYPE abc

	Numbers homozygous for recessive alleles							
	abc	ab	ac	bc	a	b	c	—
Observed	3	8	6	8	47	61	34	?
Mitotic disjunction								
Requiring 192 products of quadruple exchanges								
Double in region (1), and single in regions (2) and (3)	3	9	6	9	18	27	30	90
Meiotic disjunction								
Requiring 96 products of quadruple exchanges								
Doubles in regions (1) and (2)	3	3	3	9	6	6	30	36
Doubles in regions (1) and (3)	3	9	0	9	9	24	9	33
Requiring 192 products of sextuple exchanges								
Doubles in regions (1), (2) and (3)	3	9	6	9	18	27	30	90

Table X. The data summarized are from their Table 4, and involve three loci on chromosome VII (*tryptophan-5*, designated a in the table, *leucine-1*, designated b, and *adenine-6*, designated c). The cross may be diagrammed a (1) + (2) centromere (3) $c \times +$ (1) b (2) centromere (3) +, in which the numerals in parentheses indicate the regions in which exchanges are detected.

According to the mitotic disjunction interpretation it is possible to obtain cells homozygous for all three mutant alleles, abc, only following a quadruple exchange which is a 4-strand double in region (1), and a single exchange in each of regions (2) and (3), with the (2)-(3) exchange also being a 4-strand double. Only one quarter of the products of such quadruple exchanges which are 4-strand doubles in the two critical regions would be homozygous at the three loci (say the product consisting of chromatids 1 and 3, numbering chromatids as in Fig. 2), the other three (then consisting of chromatids 2 and 4, 1 and 4, and 2 and 3) would be homozygous for no recessive alleles. (Apparently only one of the two sister products of somatic recombination was likely to survive in this experiment ". . . at high UV doses, largely whole-colony auxotrophs arise on plating out cells directly, due to sister cells at the first division carrying a lethal factor": Wilkie, 1963.) Further, assuming no chromatid interference (as do the authors in their calculations), only one quarter of all doubles within region (1), or between region (2) and region (3), should be 4-strand doubles. Hence, if the recombinations occurred in exactly the expected frequencies, 192 quadruple exchanges [double in region (1) and single in regions (2) and (3)] would be required to produce three cells homozygous for abc, and quadruples involving other strand relationships would produce the numbers of cells homozygous for other combinations of recessives as indicated in the table. It will be noted that all observed multiple recessives are accounted for. The remaining observed single recessives could be accounted for by about 60 single exchanges in region (2), producing a and b homozygotes each in one quarter of the products, and 16 singles in region (3), producing c in one quarter of the products. This would have to be the total of the exchanges—a number of possible double exchanges would result in double homozygotes which have already been completely accounted for by the quadruples.

Estimations based on the meiotic disjunction interpretation are a little more complicated because the triple homozygote abc can arise in two ways: either from a quadruple exchange which is a 4-strand double in both regions (1) and (2), or one in which there is a 4-strand double in both regions (1) and (3). Ninety-six quadruple exchanges of either type should produce three abc individuals, with the number of other homozygous types indicated in the table. Inspection suggests that both types of quadruples have contributed. It is interesting (to this author) that 192 sextuple exchanges, with 4-strand doubles in each of regions (1), (2), and (3), would give exactly the same expected numbers as indicated following the quadruple exchange used with the mitotic disjunction interpretation above. It should be remembered that, by the meiotic interpretation, nonexchange tetrads (in the regions considered) would result in homozygosis for ab in half of the products, for c in the other half. There could not be many of these without producing larger numbers than observed in some classes.

In conclusion it may be said that neither interpretation of these data is very satisfactory. On any interpretation one must concede a great frequency of multiple exchanges—almost to the exclusion of all others. This excess of multiples can well be due to the levels of ultraviolet irradiation used. Augmentation of somatic recombination in this way may not be very helpful to an understanding of the mechanisms of spontaneous recombination.

D. Tetrad Analyses

1. *Characteristics of Independent Inheritance*

Whereas linkage between two heterozygous loci can often be most efficiently detected in random meiotic products (Perkins, 1953), a number of characteristics by which linked and independent assortment differ, and which are not evident in studies of random products, are apparent when tetrads are analyzed. This is especially true when analyses are made of linearly ordered tetrads or octads. The six different spore patterns observable in linearly ordered asci are indicated in Table XI, and are identified by the Arabic numerals 1 to 6 which will be used to designate each specific pattern in the following discussions.

TABLE XI
Designations for Spore Patterns[a]

Segregation division	I		II			
			Alt.		Sym.	
Spore pattern number	1	2	3	4	5	6
Apical spore pair	m m	+ +	m m	+ +	m m	+ +
Second spore pair	m m	+ +	+ +	m m	+ +	m m
Third spore pair	+ +	m m	m m	+ +	+ +	m m
Basal spore pair	+ +	m m	+ +	m m	m m	+ +

[a] Linear distributions of spores carrying any mutant allele, m, and its wild-type allele, +, following segregation at the first (I) or second (II) meiotic division. Alt., alternating division II pattern; Sym., symmetrical division II pattern (compare with Fig. 4).

As illustrative data we shall use those of Mitchell (1964) involving two loci in the same arm of one chromosome and one locus in an independent chromosome of *Neurospora crassa* (data summarized in her Table 5). We shall designate the mutants at the three loci as a, b, and c, with a independent of b and c, which are linked. The cross may be represented

TABLE XII

NUMBERS OF ASCI WITH DIFFERENT PATTERNS IN CROSSES BETWEEN INDEPENDENT GENES[a,b]

Locus a/+		Locus +/b I 1	Locus +/b I 2	Locus +/b II Alt. 3	Locus +/b II Alt. 4	Locus +/b II Sym. 5	Locus +/b II Sym. 6	Locus c/+ I 1	Locus c/+ I 2	Locus c/+ II Alt. 3	Locus c/+ II Alt. 4	Locus c/+ II Sym. 5	Locus c/+ II Sym. 6	Totals		
I	1	92	95	13	18	17	24	62	51	32	31	39	44	259	522	
		(93.7)		(18.8)				(57.7)		(36.4)						
	2	102	91	16	13	22	19	58	66	40	23	34	42	263		
II Alt.	3	7	8	3	2	4	1	7	3	4	2	4	5	25	52	91
	4	8	7	3	5	1	3	3	5	6	3	6	4	27		
		(8.2)		(1.6)				(5.0)		(3.2)						
II Sym.	5	6	11	1	0	1	1	3	5	2	5	4	1	20	39	
	6	8	5	1	3	0	2	3	5	4	2	1	4	19		
				37	41	45	50			88	66	88	100			
Totals		223	217					136	135							
				78		95				154		188				
		440		173				271		342				613		

[a] From Mitchell (1964).

[b] Spore patterns designated as in Table XI. Numbers in parentheses are those expected for each pattern in that block of patterns. Cross, a + c × + b +.

a(+ c)× + (b +), with the linked genes in parentheses. As spore patterns at each of the three loci were identified, a total of 216 different combinations is expected, but for purposes of the present discussion it is simpler to consider only two loci at a time, with only 36 different combinations each. Such summaries for the two independent combinations a + × + b, and a c × ++ (the first being in repulsion, the second in coupling) are given in Table XII.

a. *Polarity in Meiotic Divisions.* If heterozygous alleles, mutant and wild-type, pass at random to opposite poles at the first meiotic division, spore patterns 1 and 2 should occur equally often. Similarly, if the alleles pass at random to the two poles of both second-division spindles, alternating and symmetrical second-division segregations (Fig. 4, Table XI) should be equally frequent, as should the two patterns comprising each of them. The observed frequencies in each of these categories for each of the three loci are listed in Table XIII, together with the probabilities of obtaining deviations from equality as great or greater than observed, and the 95% confidence limits for observed frequencies of one member of each pair of patterns being compared. Deviations from equality among

TABLE XIII
TESTS FOR POLARITY IN MEIOTIC DIVISIONS

Patterns	Locus	Observed	P^a	Frequency[b] (%)	Confidence interval[c] (%)
		In first meiotic division:			
(1):(2)	a/+	259:263	0.85	49.6	45.3–53.9
	+/b	223:217	0.77	50.9	46.0–55.8
	c/+	136:135	0.94	50.2	44.3–56.1
		In second meiotic division:			
Alt:Sym	a/+	52:39	0.17	57.1	46.8–66.8
	+/b	78:95	0.20	45.1	37.9–52.5
	c/+	154:188	0.05	45.0	39.9–50.3
(3):(4)	a/+	25:27	0.77	48.1	35.2–61.2
	+/b	37:41	0.64	47.4	36.8–58.3
	c/+	88:66	0.08	57.1	49.3–64.7
(5):(6)	a/+	20:19	0.84	51.3	36.2–66.2
	+/b	45:50	0.54	47.4	37.7–57.3
	c/+	88:100	0.38	46.8	39.8–53.9

[a] Probabilities of deviations from equality as great or greater than observed, based on χ^2 with one degree of freedom.

[b] Fraction of first pattern in total of both.

[c] 95% confidence levels based on Table VIII_1 of Fisher and Yates (1957).

second-division spore patterns are perhaps somewhat large (the mean probability is 38%), but those among first-division patterns are unexpectedly small (mean probability 85%), and the mean probability of all 12 tests is 54% where 50% is expected. The data can be said to show no significant indication of polarity or nonrandomness in the distribution of alleles at the three loci during meiosis. On the other hand, they also fail to exclude, at the 95% level of confidence, rather large possible deviations from randomness, especially among second-division spore patterns in which the numbers involved are relatively small.

There does seem to be a trend in the data suggesting that the more crossing-over there is between centromere and heterozygous locus the greater is the probability of nonrandom distributions. A large fraction of all first-division patterns results from instances in which no exchange occurred proximal to the segregating locus, and deviations from randomness are less among first-division patterns than among second-division patterns, all of which resulted from instances in which there was at least one exchange proximal to the locus in question. There is also a correlation between the frequency of second-division segregation at a locus, itself a function of the amount of crossing-over between it and the centromere, and the probability of nonrandom distribution. In the combined tests of deviations from randomness at both meiotic divisions, locus a/+, with 14.8% second-division segregation, had a mean probability (of deviations as great or greater than observed) of 66%; +/b, with 28.2% second-division segregation, a mean probability of 54%; and c/+, with 55.8% second-division segregation, a mean probability of 36%. However, inasmuch as the data fail to establish any nonrandomness, this trend must also be considered to be unestablished. If one were to accept the validity of a correlation between nonrandomness and centromere, it would mean that the data just reviewed offer still stronger support for the interpretation that the centromeres themselves are assorted completely at random during meiosis.

b. *Correlated Distributions of Nonhomologous Centromeres.* A phenomenon, known as "affinity," whereby the disjunction of homologous centromeres in one pair of chromosomes is related to that occurring in another pair of chromosomes in a manner causing a particular homolog of each pair to segregate together, has been reported to occur in the house mouse (Wallace, 1958) and in yeast (Lindegren *et al.*, 1962). [In passing, it should be noted that the yeast data used to support this interpretation have been shown (Emerson, 1963) to have the distribution of spore patterns expected in examples of reciprocal translocation; I know of no direct evidence which definitely distinguishes between these alternative interpretations.] Tests for affinity between the two chromosomes involved in Mitchell's data are given in Table XIV. No significant indication of affinity is evident, and a difference greater than 56:44 is excluded at the 95% level of confidence. Again, the locus closer to the centromere gives the better fit to the interpretation of complete randomness.

TABLE XIV
TESTS FOR AFFINITY

Patterns[a]	Loci	Observed	P	Frequency	Confidence interval
PDT:NPDT	a/+ +/b	197:183	0.47	51.9	44.2–56.1
	a/+ c/+	128:109	0.21	54.0	44.7–62.0

[a] PDT, parental ditype ascus, when mutant alleles are in repulsion, as in a/+ +/b, pattern (1) at a/+ with pattern (2) at +/b, and pattern (2) at a/+ with (1) at +/b; when in coupling, as in a/+ c/+, pattern (1) at both loci, and pattern (2) at both loci. NPDT, nonparental ditype asci, combinations of patterns just the reverse of those in PDT's, i.e., with mutants in repulsion, pattern (1) at both loci, etc. Other designations as in Table XIII.

c. *Tests of All Patterns for Randomness of Combination.* As shown in Table XV, combinations of first- and second-division segregation at one locus (a/+) with those at either locus on the independent chromosome (+/b or c/+) are in very close agreement with complete random-

TABLE XV
TESTS OF COMBINATIONS OF FIRST- AND SECOND-DIVISION PATTERNS

Loci			Pattern combinations[a]				
			I+I	I+II	II+I	II+II	P^b
a/+ +/b	Observed numbers		380	142	60	31	0.78
	Expected numbers		374.7	147.3	65.3	25.7	
	Pattern deviations within combination	P^c	0.83	0.53	0.87	0.49	0.91
a/+ c/+	Observed numbers		237	285	34	57	0.91
	Expected numbers		230.8	291.2	40.2	50.8	
	Pattern deviations within combination	P^c	0.51	0.22	0.77	0.81	0.87

[a] Sums of all patterns with first- or second-division segregation; the first Roman numeral refers to patterns of the first listed locus.

[b] Probabilities based on χ^2 with 3 degrees of freedom for deviations from expected numbers in combinations; with 35 degrees of freedom for deviations of all patterns from theoretical expectations.

[c] Deviations from theoretically expected numbers; degrees of freedom: 3 in I+I, 7 in I+II and II+I, 15 in II+II.

ness—probabilities of as great or greater deviations are 78 and 91%. Further, there is no significant deviation from randomness among the 4, 8, or 16 pattern combinations comprising groups with common first and second-division combinations. The fits of all 36 patterns listed in Table XII are extremely good: probabilities 91% for a/+ and +/b, 87% for a/+ and c/+.

E. Estimation of Frequencies of Different Kinds of Heteroduplex Repair

If regions of hybrid DNA arise from an exchange of base chains of one polarity, as in the Holliday model (Fig. 18d), the two heteroduplexes at the same mutant site in the two hybrid chromatids must have dissimilar base compositions. Inasmuch as the bases themselves may determine to some extent the type of repair occurring, the relative frequencies of repair failure, repair to wild type, and repair to mutant cannot be assumed to be identical in the two hybrid chromatids. An algebraic formulation, in which identity in repair in the two hybrid chromatids is not assumed, is summarized in Table XVI.

TABLE XVI
Expected Fractions of Different Conversion Ratios[a]

		Dissimilar heteroduplexes		Identical heteroduplexes
		One	Other	
Repair failure		$1-p$	$1-q$	$1-p$
Repair to wild type		pr	qs	pr
Repair to mutant		$p(1-r)$	$q(1-s)$	$p(1-r)$
Ratio	Observed fraction	Expected fraction		Expected fraction
6+:2m	A	$pqrs$		p^2r^2
2+:6m	B	$pq(1-r)(1-s)$		$p^2(1-r)^2$
5+:3m	C	$p(1-q)r+(1-p)qs$		$2p(1-p)r$
3+:5m	D	$p(1-q)(1-r)+(1-p)q(1-s)$		$2p(1-p)(1-r)$
4+:4m^{p}	(E)[b]	$(1-p)(1-q)$		$(1-p)^2$
4+:4m^{m}	(F)[c]	$pqr(1-s)+pq(1-r)s$		$2p^2r(1-r)$

[a] After Emerson (1966b).
[b] Requires detection of postmeiotic 4:4 segregation.
[c] Undetected because of identity with parental ditype segregation.

From the observed relative frequencies of asci with 6+:2m, 5+:3m, and 3+:5m ratios, it is possible to derive expressions for the parameters p and q in terms of parameters r and s:

$$p = \frac{A(r - s)}{A(r - s) - Crs + (C + D)r^2s} \qquad (9)$$

$$q = \frac{A(s - r)}{A(r - s) - Crs + (C + D)rs^2} \qquad (10)$$

From the observed relative frequencies of 6+:2m and 2+:6m segregations it is possible to derive an expression for parameter s in terms of parameter r:

$$s = \frac{A(1 - r)}{A + (B - A)r} \qquad (11)$$

Hence, parameters p, q, and s can be determined for all valid values of r (i.e., when p, q, r, and s all have positive values not exceeding 1.0). Then, by substituting corresponding values for all parameters in the formula for expected frequencies of ratios (Table XVI), one obtains expected frequencies for 6+:2m, 2+:6m, 5+:3m and 3+:5m ratios in exactly the same relative proportions as their observed fractions (A, B, C, and D). Unique values for expected frequencies of the other two ratios are not obtained by this method (Emerson, 1966b). Expected 4+:4m^{p} ratios range from zero to a definite maximum and, if observed values for this ratio fall within this range, there will usually be two sets of parameter values (or four if there is an independent means of differentiating p and r from q and s) in perfect accord with the observed fractions of all detectable ratios, but with no way of determining which is correct. For each set of parameter values which is in agreement with observed fractions of all five detectable ratios there will be a predicted frequency for 4+:4m^{m} ratios (which are completely undetectable as products of heteroduplex repair).

When r is exactly equal to s, as required by the Whitehouse-Hastings models, Eqs. (9) and (10) do not give valid solutions, p and q both becoming zero. Formulas based on this model (setting q equal to p and s equal to r) are listed in the last column of Table XVI. From the relationships

$$\frac{A}{B} = \frac{p^2r^2}{p^2(1 - r)^2}$$

$$= \frac{r^2}{(1 - r)^2} \qquad (12)$$

and

$$\frac{C}{D} = \frac{2p(1-p)r}{2p(1-p)(1-r)}$$

$$= \frac{r}{1-r} \qquad (13)$$

it follows that

$$\frac{A}{B} = \frac{C^2}{D^2} \qquad (14)$$

Hence, to accord with the Whitehouse-Hastings model, the ratio of observed frequencies of 6+:2m and 2+:6m segregations must not differ significantly from the ratio of the squares of the observed frequencies of 5+:3m and 3+:5m segregations.

REFERENCES

Anderson, E. G. (1925). *Genetics* **10**, 403–417.

Barratt, R. W., Newmeyer, D., Perkins, D. D., and Garnjobst, L. (1954). *Advan. Genet.* **6**, 1–93.

Beadle, G. W., and Emerson, S. (1935). *Genetics* **20**, 192–206.

Bonnier, G., and Nordenskjöld, M. (1937). *Hereditas* **23**, 257–278.

Bridges, C. B. (1916). *Genetics* **1**, 1–52 and 107–163.

Bridges, C. B. (1935). *J. Heredity* **26**, 60–64.

Bridges, C. B., and Anderson, E. G. (1925). *Genetics* **10**, 418–441.

Case, M. E., and Giles, N. H. (1958a). *Proc. Natl. Acad. Sci. U.S.* **44**, 378–390.

Case, M. E., and Giles, N. H. (1958b). *Cold Spring Harbor Symp. Quant. Biol.* **23**, 119–135.

Case, M. E., and Giles, N. H. (1964). *Genetics* **49**, 529–540.

Catcheside, D. G. (1944). *Ann. Botany* [N.S.] **8**, 119–130.

Catcheside, D. G. (1966). *Australian J. Biol. Sci.* **19**, 1047–1059.

Catcheside, D. G., and Overton, A. (1958). *Cold Spring Harbor Symp. Quant. Biol.* **23**, 137–140.

Catcheside, D. G., Jessop, A. P., and Smith, B. R. (1964). *Nature* **202**, 1242–1243.

Chiang, K. S., Kates, J. R., and Sueoka, N. (1965). *Genetics* **52**, 434–435 (abstr.).

Creighton, H. B., and McClintock, B. (1931). *Proc. Natl. Acad. Sci. U.S.* **17**, 492–497.

Darlington, C. D. (1932). "Recent Advances in Cytology." McGraw-Hill (Blakiston), New York.

de Serres, F. J. (1963). *Genetics* **48**, 351–360.

Dobzhansky, T. (1930). *Biol. Zentr.* **50**, 671–685.

Emerson, S. (1955). *In* "Hoppe-Seyler/Thierfelder Handbuch der physiologisch- und pathologisch-chemischen Analyse" (K. Lang and E. Lehnartz, eds.), 10th ed., Vol. 2, Part 2, pp. 443-537. Springer, Berlin.

Emerson, S. (1956). *Compt. Rend. Trav. Lab. Carlsberg, Ser. Physiol.* **26**, 71–86.
Emerson, S. (1963). *In* "Methodology in Basic Genetics" (W. J. Burdette, ed.), pp. 167-208. Holden Day, San Francisco, California.
Emerson, S. (1966a). *In* "The Fungi" (G. C. Ainsworth and A. S. Sussman, eds.), Vol. 2, pp. 513–566. Academic Press, New York.
Emerson, S. (1966b). *Genetics* **53**, 475–485.
Emerson, S. (1968). Unpublished observations.
Emerson, S., and Beadle, G. W. (1933). *Z. Induktive Abstammungs- Vererbungslehre* **65**, 129–140.
Emerson, S., and Yu-Sun, C. C. C. (1967). *Genetics* **55**, 39–47.
Emerson, S., and Yu-Sun, C. C. C. (1968). Unpublished data.
Fincham, J. R. S. (1966). "Genetic Complementation." Benjamin, New York.
Fisher, R. A., and Yates, F. (1957). "Statistical Tables for Biological, Agricultural and Medical Research," 5th ed. Oliver & Boyd, Edinburgh and London.
Fizz-Loony, F., and Linderstrøm-Lang, K. (1956). Reprinted in "Selected Papers of Kaj Linderstrøm-Lang" (H. Holter, N. Neurath, and M. Ottesen, eds.), p. 581. Academic Press, New York, 1962.
Fogel, S., and Hurst, D. D. (1967). *Genetics* **57**, 455–481.
Freese, E. (1957a). *Genetics* **42**, 671–684.
Freese, E. (1957b). *Z. Induktive Abstammungs- Vererbungslehre* **88**, 388–406.
Holliday, R. (1961). *Genet. Res.* **2**, 231–248.
Holliday, R. (1964). *Genet. Res.* **5**, 282–304.
Hurst, D. D., and Fogel, S. (1964). *Genetics* **50**, 435–458.
James, A. P., and Lee-Whiting, B. (1955). *Genetics* **40**, 826–831.
Janssens, F. A. (1909). *Cellule* **25**, 389–411.
Jessop, A. P., and Catcheside, D. G. (1965). *Heredity* **20**, 237–256.
Kitani, Y., and Olive, L. S. (1967a). *Genetics* **56**, 571 (abstr.).
Kitani, Y., and Olive, L. S. (1967b). *Genetics* **57**, 767–782.
Kitani, Y., Olive, L. S., and El-Ani, A. S. (1962). *Am. J. Botany* **49**, 697-706.
Kruszewska, A., and Gajewski, W. (1967). *Genet. Res.* **9**, 159–177.
Leupold, U. (1961). *Arch. Julius Klaus-Stift. Vererbungsforsch. Sozialanthropol. Rassenhyg.* **36**, 89–117.
Lewis, E. B. (1963). *Am. Zoologist* **3**, 33–56.
Lindegren, C. C. (1953). *J. Genet.* **51**, 625–637.
Lindegren, C. C., and Lindegren, G. (1939). *Genetics* **24**, 1–7.
Lindegren, C. C., and Lindegren, G. (1942). *Genetics* **27**, 1–24.
Lindegren, C. C., Lindegren, G., Shult, E., and Hwang, Y. L. (1962). *Nature* **194**, 260–265.
Lissouba, P. (1960). *Ann. Sci. Nat.: Botan. Biol. Vegetale* [12] 641-720.
Lissouba, P., Mousseau, J., Rizet, G., and Rossignol, J. L. (1962). *Advan. Genet.* **11**, 343–380.
Luzzati, M. (1965). Doctoral Thesis, University of Paris.
McClintock, B. (1945). *Am. J. Botany* **32**, 671–678.
Mitchell, M. B. (1955). *Proc. Natl. Acad. Sci. U.S.* **41**, 215–220.
Mitchell, M. B. (1964). *Am. J. Botany* **51**, 88–96.
Morgan, L. V. (1922). *Biol. Bull.* **42**, 267–274.
Morgan, T. H., Sturtevant, A. H., Muller, H. J., and Bridges, C. B. (1915). "The Mechanism of Mendelian Heredity." Holt, New York.
Murray, N. E. (1963). *Genetics* **48**, 1163–1183.

Newton, W. F. C., and Darlington, C. D. (1929). *J. Genet.* **21**, 1–16.
Painter, T. S. (1933). *Science* **78**, 585–586.
Perkins, D. D. (1953). *Genetics* **38**, 187–197.
Perkins, D. D. (1955). *J. Cellular Comp. Physiol.* **45**, Suppl. 2, 119–149.
Prakash, V. (1964). *Genetics* **50**, 297-321.
Ravin, A. W., and Iyer, V. N. (1962). *Genetics* **47**, 1369–1384.
Rizet, G., and Rossignol, J. L. (1963). *Rev. Biol. (Libson)* **3**, 261–268.
Rizet, G., Lissouba, P., and Mousseau, J. (1960a). *Bull. Soc. Franc. Physiol. Vegetale* **6**, 175–193.
Rizet, G., Lissouba, P., and Mousseau, J. (1960b). *Compt. Rend. Soc. Biol.* **154**, 1967–1970.
Roman, H., and Jacob, F. (1958). *Cold Spring Harbor Symp. Quant. Biol.* **23**, 155–160.
Roper, J. A. (1966). *In* "The Fungi" (G. C. Ainsworth and A. S. Sussman, eds.), Vol. 2, pp. 589–617. Academic Press, New York.
Rossen, J. M., and Westergaard, M. (1966). *Compt. Rend. Trav. Lab. Carlsberg* **35**, 233–260.
Rossignol, J. L. (1964). Doctoral Thesis, University of Paris.
Sicard, A. M., and Ephrussi-Taylor, H. (1965). *Genetics* **52**, 1207–1227.
Singleton, J. R. (1953). *Am. J. Botany* **40**, 124–144.
Stadler, D. R., and Towe, A. M. (1963). *Genetics* **48**, 1323–1344.
Stern, C. (1931). *Biol. Zentr.* **51**, 547–587.
Stern, C. (1936). *Genetics* **21**, 625–730.
Sturtevant, A. H. (1915). *Z. Induktive Abstammungs- Vererbungslehre* **13**, 234–287.
Sturtevant, A. H. (1931). *In* Sturtevant, A. H., and Dobzhansky, T. (1931). *Carnegie Inst. Wash. Publ.* **421**, 61–68.
Sturtevant, A. H., and Beadle, G. W. (1939). "An Introduction to Genetics." Saunders, Philadelphia, Pennsylvania.
Sueoka, N., Chiang, K. S., and Kates, J. R. (1967). *J. Mol. Biol.* **25**, 47–66.
Wallace, M. E. (1958). *Phil. Trans. Roy. Soc. London* **B241**, 211–254.
Warwick, B. L. (1932). *Texas Agr. Expt. Sta., Bull.* **463**, 1–28.
Weinstein, A. (1936). *Genetics* **21**, 155–199.
Welshons, W. J. (1955). *Genetics* **40**, 918–936.
Weltman, A. S. (1956). Doctoral Thesis, University of Missouri.
Westergaard, M., and von Wettstein, D. (1966). *Compt. Rend. Trav. Lab. Carlsberg* **35**, 261–286.
Whitehouse, H. L. K. (1965). *Proc. 11th Intern. Congr. Genet., The Hagne, 1963* Vol. **2**, 87–88. Pergamon Press, Oxford.
Whitehouse, H. L. K., and Haldane, J. B. S. (1946). *J. Genet.* **47**, 208–212.
Whitehouse, H. L. K., and Hastings, P. J. (1965). *Genet. Res.* **6**, 27–92.
Wilkie, D. (1963). Personal communication.
Wilkie, D., and Lewis, D. (1963). *Genetics* **48**, 1701–1716.
Winkler, H. (1930). "Konversion der Gene" Fischer, Jena.
Woodward, D. O., Partridge, C. W. H., and Giles, N. H. (1958). *Proc. Natl. Acad. Sci. U.S.* **44**, 1237–1244.
Yu-Sun, C. C. C. (1966). *Genetica* **37**, 569–580.
Yu-Sun, C. C. C. (1969). Unpublished data.
Zickler, H. (1934). *Planta* **22**, 573–613.

VI MEIOTIC AND SOMATIC PAIRING

RHODA F. GRELL

I. Introduction

Pairing is a well recognized and tacitly accepted fact of chromosome behavior. At the same time, it constitutes one of the least understood categories of chromosome activity. In order for chromosomes to pair, they must first recognize one another; yet virtually nothing is known of the physicochemical basis of recognition, and the various hypotheses which have been advanced, borrowing from established physical or chemical principles, all suffer from the common malady of experimental inaccessibility. Recent findings of a new type of recognition, based on chromosome size or mass, add a further complication to the little understood picture. Perhaps in the rapidly emerging field relating molecular structure and biological specificity, fruitful approaches to the problems of chromosome pairing will be found. At this particular point in time, however, a review of the pairing phenomenon must of necessity be limited to its grosser aspects.

Initially it seems desirable to attempt to fit the various manifestations of pairing into some meaningful relationship. Variations in chromosome pairing are manifold and puzzling. Chromosomes may pair in a condensed state, but according to most cytologists they also pair when they are greatly elongated. They may pair side-by-side (parasynapsis) or end-to-end (telosynapsis); chromosome pairing occurs most frequently between homologs, but nonhomologous pairing is a well-recognized phenomenon; in some instances the union is restricted to two chromosomes at any particular level, whereas in other cases no such restriction exists.

Unfortunately, classification based on any of the above cytological criteria soon breaks down. For instance, a dichotomy into homologous and nonhomologous pairing conceivably might be a fundamental one if the boundaries of homology were well defined. Frequently, however, it is not clear from cytological inspection where homology begins and ends. The danger inherent in such assumptions is exemplified in interspecies studies in plants where chromosome pairing at the late prophase or metaphase of meiosis I has been interpreted as an indication of homology between the participating chromosomes, and such presumptive homology has been used as the basis for determining evolutionary pathways. It is now clear from genetic studies that pairing at this time may not be restricted to homologs* or homeologs but may also occur between nonhomologs. In a similar vein, whether chromosomes pair as condensed bodies or in a greatly elongated condition undoubtedly reflects differences in the specificity, the mechanism, and the functions involved; yet the numerically unrestricted homologous pairing in polytene nuclei and the restricted two-by-two pairing preceding exchange are both considered to take place between extended chromosomes.

The present chapter will be primarily concerned with neither the vast and detailed cytological data on pairing nor the vague and untestable physicochemical models that have been invoked. Instead, its orientation will be largely a genetic one, for it is from this source that much of our knowledge concerning the pairing processes derives. The adoption of a genetic point of view in the present treatment means that the analysis of pairing will proceed chiefly from its genetically detectable results. As a consequence, the appropriate criterion for classification is a functional one.

Recent studies utilizing autoradiographic techniques have provided clear demonstrations that chromosomes function in different ways at different times in the cell cycle and in different kinds of cells; the evidence

---

* Homologs are the two members of a chromosome pair; homeologs are homologs that have undergone evolutionary divergence in different species but still retain a large degree of homology for one another.

indicates that the functions they perform are reflected in their physical state; it is clear that for certain of these functions there is a pairing requirement; and it seems justifiable to assume that the kind of pairing necessary for different functions varies. Based on these considerations we propose to group the pairing processes into the two functional categories: (1) genetic and (2) metabolic. Genetic pairing will be further subdivided into exchange and distributive.

Although much information is available on the analysis of the genetic categories, little is known about the pairing requirements for metabolic activities. Indeed, it may be justifiably asked whether such requirements exist. The premise that they do rests largely on the phenomenon of somatic pairing, which has been observed in a variety of organisms. In this regard, a knowledge of the spatial relationship of homologs during interphase is sorely needed. If there is a pairing requirement for metabolic function, then this requirement might be applicable to meiotic as well as somatic cells. In this case the different kinds of pairing observed during the maturation divisions need not all be concerned with genetic events. In a similar vein, somatic crossing-over is presumably preceded by an exchange pairing. Yet, despite the fact that genetic pairing need not be restricted to meiotic cells nor metabolic pairing to somatic cells, our knowledge of each has come almost entirely from such sources. In practice, a primary division based on cell type proves more workable than one based on function; therefore, pairing will be considered under two main headings—meiotic and somatic.

II. Meiotic Pairing

A. Exchange Pairing

1. *Characteristics*

Although a vast amount of information has accumulated on the exchange process, the actual mechanism of exchange has thus far eluded analysis. In one sense the extent and diversity of the data have undoubtedly contributed to the difficulty of the problem; for genetic recombination, in its secondary aspects at least, has probably evolved in a variety of ways, thus precluding the possibility of incorporating all of the information into one uniform and yet coherent picture.

Nevertheless, certain basic features of exchange pairing appear to be true for most systems. First, and despite the absence of direct cytological confirmation, it is generally conceded that an intimate pairing between homologs must precede the interchange of their genetic information. Second, for exchange to occur, virtually without error, the intimacy at the site of exchange must be so great as to require that the pairing specificity reside ultimately at the nucleotide level. The molecular basis for specific nucleotide pairing is most easily conceived as hydrogen bonding between single, complementary, and nonsister strands of deoxyribonucleic acid (DNA). Third, genetic evidence demonstrates that four units are present at the time of exchange pairing. Whether a unit represents a single strand of DNA, a Watson-Crick duplex, or even some higher multiple is not known. Should the unit turn out to be a single strand, the sites for exchange pairing would probably be established before DNA replication of that region. Fourth, for most eucaryotes, exchange pairing or the exchange event itself is of such a nature as to place a spatial limitation upon the occurrence of a second event. This limitation is given the name of positive interference. Fifth, if the homologous chromosomal segments available for exchange pairing are greater than the diploid number, pairing at any particular level appears to be both two-by-two and competitive. Sixth, exchange pairing is apparently restricted to nonsister strands.

2. *Initiation*

a. *Traditional View.* It is assumed that chromosomes become cytologically visible as fine, unpaired threads present in the diploid number (leptonema). Point-by-point association between morphologically distinct homologous chromomeres is initiated at zygonema, and complete pairing of homologs achieves a reduction to the haploid number of chromosome bodies. This stably paired state is called pachynema, for by now the chromosomes have become shorter and thicker (*pachy,* "thick"; *nema,* "thread"). During pachynema it becomes evident that each of the homologs comprising the bivalent is divisible into two parts, and until recently it was assumed that chromosome duplication occurred at this time. A tendency for the homologs to dissociate from one another marks the beginning of diplonema. Now a repulsion force comes into play that causes the homologs to separate longitudinally, and unless crossing-over has produced a chiasma to hold the homologs together, this force eventually dissolves the bivalent into its component univalents. The onset of repulsion may be considered to represent the termination of exchange pairing.

In his precocity theory of pairing (1937), Darlington interpreted this sequence to mean that chromosomes, which normally exist as paired chromatids, enter meiotic prophase prematurely in an unreplicated condition and that it is this unsaturated state which is responsible for homologous pairing at zygonema. Upon replication at pachynema, a saturated state between sister strands is achieved and a repulsion force characteristic of diplonema replaces the previous attraction between homologs. This model assumes that crossing-over occurs after replication when torsion incident to the repulsion of relationally coiled homologs causes breakage and reunion of nonsister chromatids. A variant of Darlington's model, proposed by Belling (1933), uses copy-choice as the mechanism for recombination of genetic material at pachynema.

Cytophotometric and autoradiographic studies of the chromosomes at meiosis, beginning with the work of Swift (1950) and including the intensive investigations of Taylor (1953) and his co-workers (Taylor and McMaster, 1954; Taylor *et al.*, 1957), have shown that in nearly all cases, the last DNA replication is completed during interphase or leptonema and well before pachynema. Claims have been made by Wimber and Prensky (1963) for *Triturus viridescens* and by Hotta *et al.* (1966) for *Lilium longiflorum* that a small amount of DNA synthesis occurs as late as pachynema. Attempts by Taylor and Callan to confirm these results in another species of salamander, *Triturus vulgaris*, have been unsuccessful (Taylor, 1967). In the case of Hotta *et al.*, meiotic stage determination was made in some cases by bud length. The reliability of this criterion is open to question, since not only a single bud but even a single anther may include several stages, as shown in Fig. 1. In any case it is abundantly clear that the bulk of DNA synthesis is accomplished prior to pachynema. This finding clearly invalidates Darlington's precocity theory of pairing. More important, if the stage sequence described above is correct, DNA replication and crossing-over are discrete events, well separated in time. How convincing, then, is the evidence that pairing of homologs is initiated at zygonema?

b. *Cytological Evidence.* i. *Historical.* Originally outlined by Winiwarter (1901) and set forth in a well-integrated fashion by E. B. Wilson (1928), the formulation of the traditional sequence represented the culmination of many years of controversy among cytologists as to the correct interpretation of their observations. According to Wilson the theory of synapsis in the earliest form was visualized as an end-to-end pairing to form a single spireme, similar to that assumed to occur in a somatic cell but one where reduction was accomplished by transverse divisions into the haploid number of chromosomes, in contrast to the mitotic division in which the transverse divisions were assumed to give the diploid number. The discovery of crossing-over clearly required a side-by-side association of homologs at some phase of meiosis and this undoubtedly

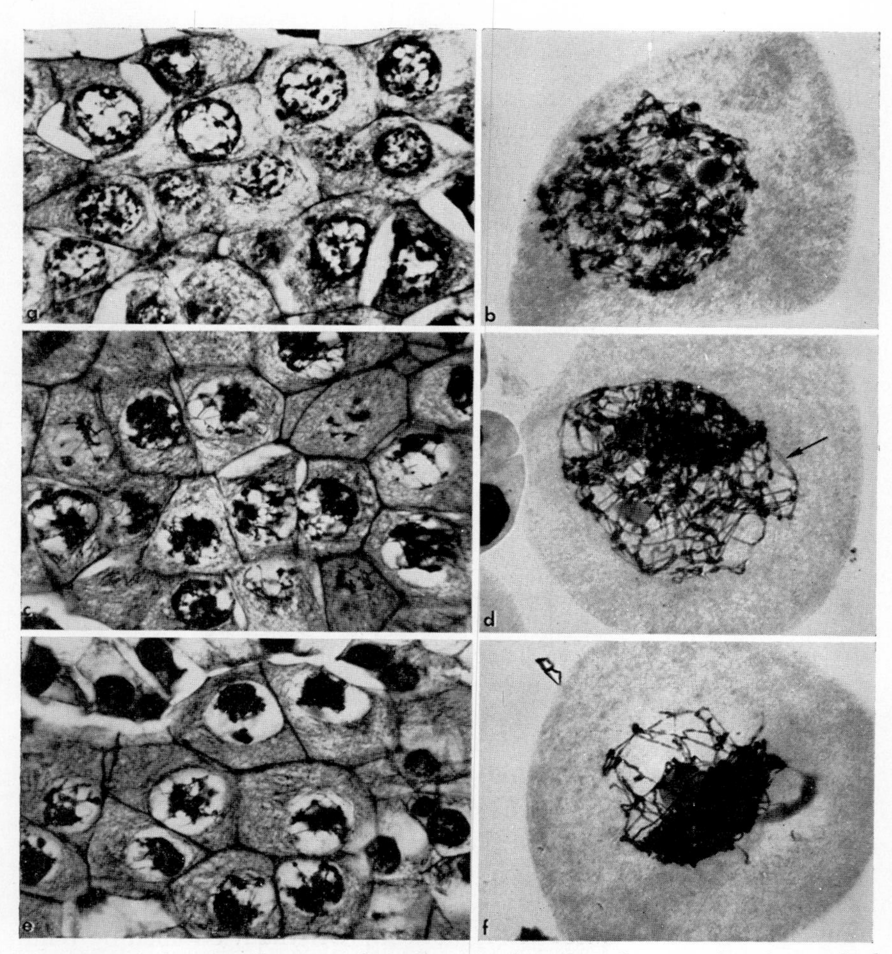

FIG. 1. Stages found in a single anther of *Lilium longiflorum* (taken from a bud 10.7 mm long) which illustrate the degree of asynchrony in development between the base and the tip of the anther; a,b = base, c,d, = center, and e,f = tip. Magnification: ×620. (Courtesy of P. Moens.)

proved a powerful deterrent to the telosynaptic hypothesis. Nevertheless, the adherence of many competent cytologists to the telosynaptic view over a long period of time attests to the difficulties inherent in the problem of determining the limits and the associations of chromosomes at prepachytene stages.

E. B. Wilson (1928) recognized two distinct types of synapsis bridged by various intermediate conditions. In the first type, which is present in higher plants, some arthropods, and in some other invertebrates, the leptotene threads show no definite polarization but become concentrated into a densely staining chromatic knot, often enclosing the nucleolus and situated at one side of the nucleus. This is the contraction figure or synizesis (synizesis, a collapse). Wilson adopted the view, first set forth by Henking (1891), that the leptotene threads as they draw together to form the knot become associated, two-by-two. It is admittedly impossible to count the number of leptotene threads before synizesis, even when this stage can be found, so the evidence that they are diploid rests on counts of interphase prochromosomes from which they arise. As will be discussed below there is not universal agreement that prochromosomes are diploid in number (see Section II,A,2,b,iii).

In the second type of synapsis, which is frequently found in animals but rarely in plants, the contraction figure is usually lacking, and the chromosome threads take the form of loops that show a polarization with their distal ends turned toward that pole of the nucleus where the centriole lies. The polarized threads are described as beginning to thicken at their polar ends by a process Wilson interpreted as parasynapsis (i.e., a side-by-side pairing) of homologs proceeding from the pole to the antipole, so that in the middle part of the period, known as the bouquet or amphitene stage, thick threads (pachytene) are present in the polar region and thin ones (leptotene) at the antipolar region. At this time, Wilson stated, it is possible to determine the number of loops in the bouquet in certain favorable material, such as Gelei's preparations of *Dendrocoelum* (1921), when the diploid number of fourteen is considered to be present. The Y-shaped figures—which are often visible at this time and which presumably represent the thick, polar pachytene portion and the thin, antipolar leptotene threads—are, in Wilson's opinion, very convincing evidence of parasynapsis. He noted, however, that in some cases uncertainty exists as to the stage represented and that some of the figures showing Y's may actually be early diakinesis with the threads in the process of separation rather than zygonema with parasynapsis in progress (E. B. Wilson, 1928, p. 554).

Thus, the cytological evidence for parasynapsis at zygonema rests primarily on the number of loops in the polarized leptotene bouquet or on the number of prochromosomes preceding synizesis. Were it possible to determine unambiguously the number of loops for a stage that could with certainty be identified as leptonema (and there is some doubt that this is ever the case), then the presence of more than the haploid number might even then be interpreted as interrupted pairing in certain regions of otherwise paired homologs. Loops in excess of the haploid number would also be expected if the centromeric region of metacentric chromosomes represents a region of association with the nuclear membrane, since each arm would form a loop (Wenrich, 1917). The recent studies of Woolam *et al.* (1966) on the attachment of the synaptinemal complex to the nuclear membrane at pachynema in the golden hamster suggests, however, that metacentrics attach only at their ends and not medially. As for the number of prochromosomes, an initial diploid number followed by a haploid number has been reported in a variety of well-studied cases (see Section II,A,2,b,iii). Thus, the cytological evidence for initial synapsis at zygonema is neither entirely convincing nor conclusive.

ii. *Premeiotic anaphase or telophase association.* What cytological evidence, if any, is there for prezygotene synapsis? For descriptions of this critical period, it is useful to refer to the work of early cytologists, who studied the final gonial mitosis and early premeiotic interphase in great detail in an attempt to demonstrate the individuality and permanence of the chromosomes from one cell generation to the next. Smith (1942) has tabulated the information from many investigations which support the concept of a premeiotic association of homologs at the final gonial division in many invertebrate phyla and in certain vertebrates.

Pairing at this time is often confined to the proximal tips of the chromosomes so that V-shaped figures are formed from two acrocentric homologs. Sutton (1902) described pairing of this kind between each of the eleven pairs of homologs in *Brachystola magna* (diploid number = 23). Although the chromosomes of the haploid set are nearly all different in length, the eleven double chromosomes are each made up of two limbs of equal size, which Sutton interpreted as members of a pair joined together at their polar ends. Montgomery (1900, 1901) in studies of *Peripatus* and certain Hemiptera found the association in pairs to occur during telophase of the last gonial division, and a similar conclusion was reached by Nichols (1902) for *Oniscus*, and by Downing (1905) for *Hydra*. Dublin (1905) found that in *Pedicellina*, union into pairs occurs during the anaphase of the last spermatogonial division. Overton's studies (1909) of dicotyledons—including *Helleborus*, *Podophyllum*, and *Cam-*

panula—showed a paired condition in the earliest interphase nuclei and convinced him that homologs are paired when they enter the mother cell.

In *Culex*, Stevens (1908, 1910) and S. M. Grell (1946b) agree that parasynapsis occurs at telophase and continues without a typical resting stage into the synizesis stage when the nucleolus appears to receive material from the chromosomes wound about it. Stevens states that synizesis has no relation to synapsis since the latter has already occurred. When the synizetic knot loosens, the paired pachytene chromosomes become evident.

A similar early description is given by Blackman (1903) for the myriopod *Scolopendra*. He wrote that it "can be stated with the greatest certainty that pseudoreduction [synapsis] occurs during the telophase of the last spermatogonium and is completed before the reconstruction of the nuclear membrane." The nuclear space is then occupied by sixteen elongated segments of chromatin (diploid number = 33) and the accessory chromosome which takes no part in the synaptic process.

Illert (1956) has described meiosis in *Aphrophora salicina*, a homopteran, in which there is a tendency for homologs to associate at metaphase of the gonial divisions, which is most strongly expressed as a "somatic" pairing in the final premeiotic division. The bouquet stages follow directly after the premeiotic telophase with no intervening interphase. At first only the heteropycnotic X is visible, oriented as a loop, with both ends attached close to one another at the nuclear membrane. After this the autosomes also orient into a bouquet. No pairing stage is observed, and the chromosomes first become clearly visible as a fully paired pachytene bouquet!

iii. *Prochromosomes.* The presence of strongly condensed chromatic bodies, or "prochromosomes," at preleptotene stages of meiosis has been reported in a wide variety of organisms by many cytologists. Wilson refers to this period as stage "c" of meiosis where stage "a" corresponds to the final gonial telophase and stage "b" to a resting or net-like stage of short duration in which the individual chromosomes are lost to view. In stage c the chromatin again draws together into dense chromosome-like bodies. According to Wilson, the number of these bodies is nearly or exactly equal to the diploid number. By contrast, many cytologists who have described prochromosomes agree that although initially appearing in the diploid number, they shortly arrange themselves into pairs conforming to the haploid number.

Two of the earliest reports of prochromosomes come from the studies on *Triton* by Moore and Embleton (1905) and on *Periplaneta* by Farmer and Moore (1905), who, incidentally, introduced the term "maiosis," meaning to lessen. In the premeiotic resting cells of *Triton,* Moore and Embleton describe chromatic structures, which had been regarded as nucleoli (Flemming) but which correspond in number to the diploid set of chromosomes. They infer that these Anlagen represent the structures alluded to as "prochromosomes" by Overton (1905), Miyake (1905), and Strasburger (1905). They further state that the prochromosomes do not remain single but undergo a diminution in number "produced by a pairing of the bodies while the nuclei themselves remain at rest," leading to the formation in *Triton* of 12 "gemini" in place of the 24 single Anlagen (prochromosomes). Subsequently, "by growth and elongation the gemini constitute the polarized loops of the first meiotic prophase," and still later these loops become longitudinally split.

Guyénot and Naville (1933–1934) have reported for *Drosophila melanogaster* females that shortly after the last gonial division three chromatic bodies are distinguishable in the oocyte which correspond without any doubt to the three pairs of large chromosomes. These bodies represent the prochromosomes, intimately united two-by-two by parasynapsis, and correspond to the same figures they find in *Calliphora.* The prochromosomes then pass into a diffuse stage during which only the sex chromosomes remain contracted. The diffuse chromatin network, resulting from the transformation of the prochromosomes, gives rise to the very fine, partially paired leptotene threads.

Spermatogenesis in the tenebrionid beetle *Blaps* has been described in detail by Nonidez (1920). The last spermatogonial division is followed by a resting stage. The diploid number of prochromosomes (36) appears; once formed the prochromosomes begin to shift to a peripheral position at that pole of the nucleus near the centriole. They become crowded, sometimes into a dense mass of clumped chromosomes, corresponding to a synizesis stage. The prochromosomes then unravel to form the thin threads of the leptotene stage. Nonidez could not recognize a definite synaptic stage, although he speculated that it might take place in the prochromosome stage but that because of the crowded condition it is difficult to observe actual pairing.

Descriptions of prochromosomes are not restricted to the germ cells of animals. Overton (1909), one of the first cytologists to draw attention to them, wrote of the dicot *Thalictrum purpurascens* that

> after the formation of the nuclear membrane, following the final pre-meiotic division, the chromatin material becomes rather regularly distributed in the nuclear cavity, the greater portion of the stainable substance lying in the prochromosomes, each suggesting by its form and size that it is derived from a chromosome of the preceding telophase. In all nuclei at this stage, the prochromosomes show as distinct,

rather uniform bodies arranged in pairs. With ordinary magnification the prochromosomes often appear as 24 single distinct bodies while with higher magnification each prochromosome appears double so that each of the apparently single bodies is composed of two, or 48 in all.

This pairing of the prochromosomes may possibly occur in the telophase of the last premeiotic division, so that they are already paired when the mother cell nucleus is formed. The paired prochromosomes pass into the synizetic contraction and presumably remain individual bodies associated in parallel pairs from presynizesis to pachynema. Overton observed a similar sequence for *Calycanthus floridus,* where in the presynizesis stage the homologous prochromosomes are so closely associated as to appear fused. Hiraoka (1941)—in studies of the preleptotene stages of meiosis in *Trillium, Tradescantia,* and *Fritillaria*—also noted the condensation of coiled structures from the homogeneous interphase nucleus, which at maximum condensation are clearly distinguishable from one another and from which the fine leptotene threads emerge. Linnert (1955) reports that premeiotic interphase nuclei of *Salvia nemorosa* show the haploid number of prochromosomes, and a similar report has been published by Abel (1965) for the liverwort *Sphaerocarpus* and by Stack and Brown (1968) for *Haplopappus* and *Rhoeo.*

iv. *The "diatene" and "peritene" stages (McClung).* The most detailed cytological evidence of prezygotene pairing comes from the remarkable studies of McClung (1927) on *Mecostethus* and *Leptysma.* This work represents the product of some twenty-five years of research on maturation in Orthoptera species in collaboration with Eleanor Carothers. The pertinence of McClung's findings coupled with the clarity, comprehension, and judgment displayed in this work merits its presentation here in some detail. It should be emphasized that McClung did not consider this sequence representative for all forms but rather stressed the diversity of the process.

The chromosomes of the *Mecostethus lineatus* male are all acrocentric and are 23 in number, consisting of 2 small pairs, 3 large pairs, 6 intermediate pairs, and an odd one, the accessory chromosome. In order to establish the chronological seriation of events it is necessary to employ objective criteria for the age of the germ cell. McClung uses four: first, the position of the cyst in the follicle; second, the size of the cells in the cyst (since the earliest spermatocytes possess the smallest cells): third, the nucleus:cytoplasm ratio (which again seems to identify the earliest spermatocytes, for while the nucleus remains constant in size, the cytoplasm is progressively reduced during the spermatogonial divisions so that it forms a mere shell about the nucleus at the final gonial division); fourth, the precocious and characteristic behavior of the accessory chromosome (which epitomizes the behavior of the autosomes in many respects).

McClung used the term "diatene stage" to describe the first period following the last spermatogonial telophase. At the end of telophase the irregularly spiral chromosomes extend across the nucleus, with each end resting against opposite sides of the nuclear membrane. The accessory chromosome is seen at the surface of the nucleus as a slender heteropycnotic rod, extending like the other chromosomes across the nucleus. Now, with one end attached to the nuclear membrane, the accessory elongates, showing a compensatory reduction in thickness, and bends upon itself so that the two ends come to lie on the same side of the nucleus, producing a pear-shaped structure. During the looping process irregularities appear at different points along its length, so that at times it appears bipartite or even divided.

While the accessory chromosome is behaving in this way, the euchromosomes (autosomes) become loose in structure, so that their outline is indistinct, and run close to the nuclear membrane. During this time the cell doubles in size, presumably reflecting the metabolic activities of the autosomes. Then the autosomes withdraw their processes, become more compact and definite in outline, and at the same time distribute themselves in a regular fashion through the nucleus. At first they are present in the diploid number, but when they begin to contract they arrange themselves in a paired condition. Following rapidly upon this contracted state is another period of extension and chromomere movement, during which it appears they are going into the synaptic state since the approximation of homologous chromomeres is often very marked. McClung wrote: "For *Mecostethus* I have no doubt that thus occurs the beginning of the close association between the most minute portions of the chromosomes which is so characteristic and significant a feature of synapsis."

After diatene the prochromosomes are present in the haploid number and are composed of two longitudinal moieties. One end of each body is attached to the nuclear membrane, and the other end is free. At the attached end the chromosomes now begin to spin out a fine thread which McClung described as approaching the invisible, so that there is no possibility of recognizing any differentiation. In the case of the shorter chromosomes, the main body has drawn away, leaving the fine thread stretched well across the nucleus; while in that of the longer chromosomes, the free distal end winds toward the attached proximal end and reaches near to it. The result is a chromosome—the proximal half of which is a slender thread, and the distal part a heavy, double, twisted body. From the polar view, the slender threads all converge on the nuclear membrane within a small area, leaving the massive ends to appear upon the periphery of a circle below them. Eventually all of the material of the chromosomes becomes drawn out into a thin thread, so that the whole body consists of a thread of regular diameter in the form of a loop. By the end of the peritene stage these loops become uniform in structure and arrangement and what corresponds to the conventional leptotene stage has been attained. At this time one finds the extreme linear extension of the chromomeres so that the nucleus sometimes seems full of unconnected granules. After this extreme elongation each thread becomes gradually shorter and thicker, and the nuclear stage might be called pachynema.

McClung emphasized that the fine thread is spun out from a block which represents a pair of homologs, and it is from this thread, at a later stage, that the four chromatids become evident. It is when these four threads are in the most intimate contact, drawn out to an extreme length, that "the condition for physical reaction approaches molecular dimensions," and we can readily see how the four elements might be "altered by mutual exchange as in the manner demanded by genetical 'crossovers'." Could the fine thread anchored to the nuclear membrane represent two paired homologs undergoing replication and perhaps the initial stages of recombination as suggested by McClung? Relevant to this question is the increasing recognition being given to the role of membranes in the regulation of DNA replication. For instance, Comings and Kakefuda (1968), using synchronized human amnion cells exposed to thymidine-^{3}H so as to incorporate label for the first 10 minutes of the S period, have demonstrated with electron microscope autoradiography that label is restricted to the periphery of the nucleus and to the nucleolus (Fig. 2). The recent studies of Mückenthaler (1964) on *Melanoplus differentialis* using thymidine-^{3}H short-term labeling have shown that in this species of grasshopper, at least, the last replication occurs at premeiotic interphase. A determination of the precise time of replication in the spermatocyte of *Mecostethus gracilis* would clearly be of great value.

McClung's version for *Mecostethus* stands in marked contrast to the traditional sequence where leptotene threads are considered to unite in pairs to from the zygotene and pachytene threads.

v. *A reinterpretation of zygonema.* Further uncertainty is added to the cytological picture by Moens' finding (1964) that the traditional sequence of meiotic stages does not hold for *Lycopersicon esculentum* (tomato). Longitudinal sections of the anther of the tomato show a

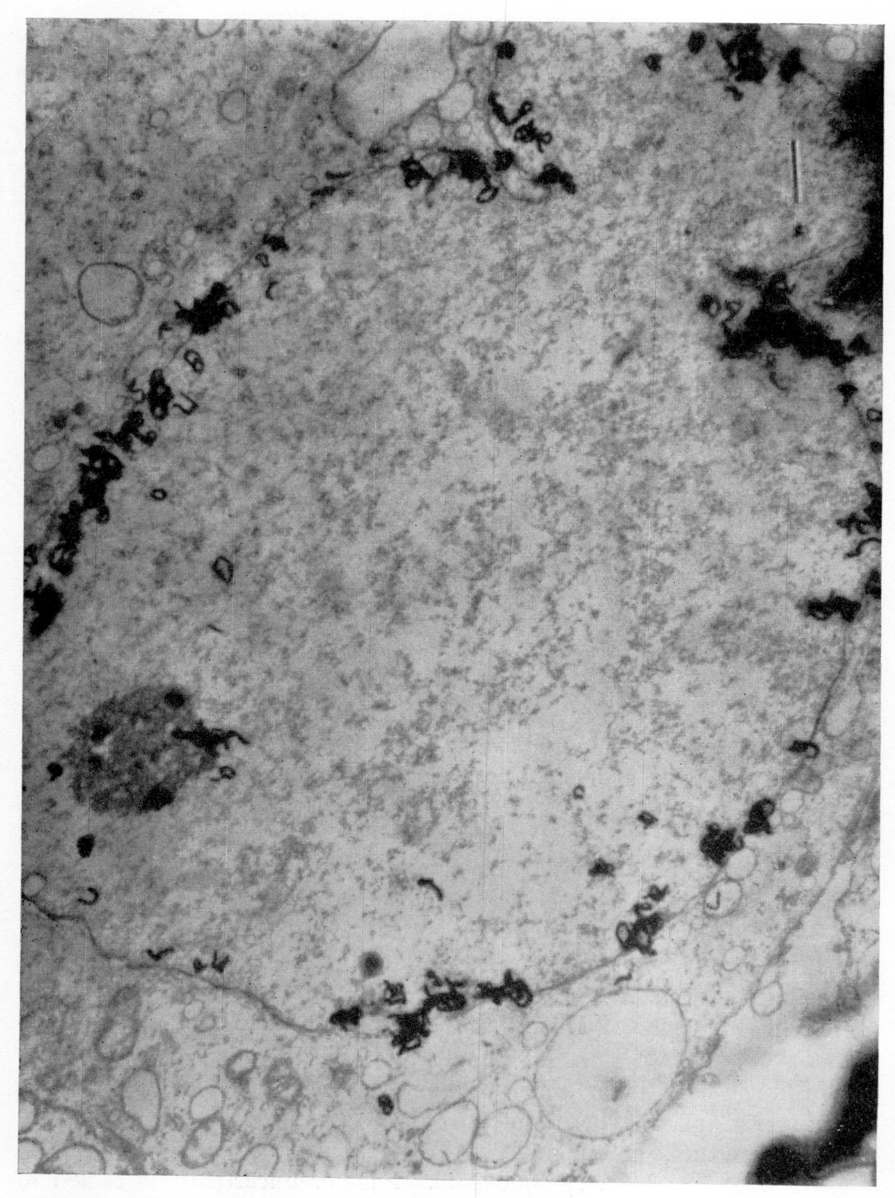

FIG. 2. Electron microscope autoradiography of human amnion cells synchronized to incorporate thymidine-^{3}H at the beginning of the S period and labeled for 10 min. The grains are predominantly restricted to the nuclear membrane and the periphery of the nucleus. The marker represents 1 μ. Magnification: × 7600. (Courtesy Comings and Kakefuda.)

maturity gradient with pollen mother cells at the apex in a much earlier meiotic stage than those at the base, as well as a sequence of chronologically ordered intermediate stages between. Using the location along the gradient as an independent criterion for the sequence of stages, Moens has carefully studied pollen mother cells at different positions. He finds that following premeiotic interphase the chromosomes first appear as fully paired, tightly clumped pachytene chromosomes. The chromosomes then loosen up and correspond to what is generally called pachynema. After this, homologs separate except at the tips and centromere regions, and Moens suggests that this stage, which has traditionally been called zygonema, be renamed schizonema. The nucleus now becomes filled with a network of thin strands resembling the descriptions given for leptonema. During this diffuse stage the centromeres become separated and the chromosomes reappear as diplotene chromosomes, held together only at the points of contact. As determined by Moens, the sequence is interphase-pachynema-schizonema-diffuse-stage-diplonema-diakinesis rather than interphase-leptonema-zygonema-pachynema-diplonema-diakinesis. Moens' interpretation has, in the main, been confirmed by Menzel (1965), who had previously expressed doubts concerning the interpretation of the zygotene stage (1962) and had indicated that cells containing this stage are found in anthers in which the range of stages lies between pachynema and diakinesis.

There is some evidence that the reinterpretation is applicable in forms other than the tomato. Maguire (1960) observed that pairing failure in maize at pachynema could have been misinterpreted as zygonema by one unacquainted with the length of corn chromosomes at this time. As pairing failure appears to increase rather than to decrease with chromosome shortening, a zygonema interpretation would seem incorrect. Maguire suggested that a similar misidentification may have been made for most or all of the material that has been identified as zygonema in maize.

The fully paired condition of the chromosomes at the beginning of cytologically visible meiosis in *Lycopersicon* again provokes questions concerning the time of onset of exchange pairing. Examination of the premeiotic interphase cells indicated to Moens that pairing does not occur in the transition to the condensed stage, for chromatic spots in the interphase nucleus (prochromosomes?) correspond to the heterochromatic areas of the paired pachytene chromosomes. Moens suggested that pairing in the tomato may occur during or before premeiotic interphase. Menzel and Price (1967) report that chromosomes in *Lycopersicon* are unpaired prior to synizesis, but the basis for this conclusion is not apparent.

c. *Electron Microscopy.* Electron microscopy offers the opportunity to explore the intimate details of chromosome structure and behavior at a resolution not previously possible with light microscopy. Initial studies were disappointing in that they revealed only homogeneous masses of granules within the chromosomes. But soon preparations of the nuclei of germ cells of the crayfish (Moses, 1956) and of the pigeon, cat, and man (Fawcett, 1956) in early meiotic stages disclosed characteristic structures or cores. As described by Fawcett (1956) (Fig. 3), the cores were ribbon-like and consisted of a pair of parallel dense fibrils each 450 Å in width and equidistant from a delicate, linear midregion. Recently, Brinkley and Bryan (1964), using silver aldehyde staining specific for DNA, reported that in the primary spermatocyte of the European corn borer, each of the lateral 450 Å fibrils is bipartite and contains two parallel 100-Å DNA fibrils separated by a distance of 200 Å and embedded in a material which may be protein (Coleman and Moses, 1964). Silver staining is not evident in the midaxial region, but lateral loops or microfibrils of 100 Å dimension, which make up the bulk of the chromatin and which are apparently continuous with the lateral component, were found to contain DNA (Brinkley and Byran, 1964).

The tripartite core structure was given the name "synaptinemal complex" by Moses (1958) and considered to represent a pair of synapsed homologs involved in the precise point-to-point pairing presumed to be a prerequisite for exchange. Maillet and Folliot (1965), who have studied these structures during spermatogenesis of *Philaenus spumarius,* a homopteran, question Moses' interpretation on several grounds. Their electron micrographs, which are of remarkable clarity (Fig. 4), indicate that the midaxial region sometimes shows a very distinct morphological structure which, coupled with the absence of DNA, would suggest that it does not represent a single pairing line of fibrils coming from the two lateral regions. Moreover, they find the complex existing sometimes in multiples whose configuration becomes difficult to interpret as congregations of paired bivalents. Sotelo and Trujillo-Cenoz (1960) had earlier observed that the medial component was complex in structure but variable with stage and with the organism, and sometimes absent altogether. This component has been considered by some workers to have a bridging nature between the two lateral components (Nebel and Coulon, 1962).

Sotelo and Wettstein (1965) report that in *Gryllus argentinus,* a multilayered composite body—apparently formed by the nucleolus, the sex chromosome, and tripartite structures of the autosomes—is still retained at the spermatid stage. Schin (1965) also observed multiple complexes of the axial core structure in both the spermatocyte and the spermatid nuclei of *Gryllus domesticus* which, though having dimensions and fine structure identical to those of axial complexes regularly found in paired pachytene chromosomes, sometimes assemble as sheets rather than ribbons and often in a system associated with the nucleolus. Schin believes that the core complexes (synaptinemal complexes) represent protein molecules which typically assemble under the control of, and in close

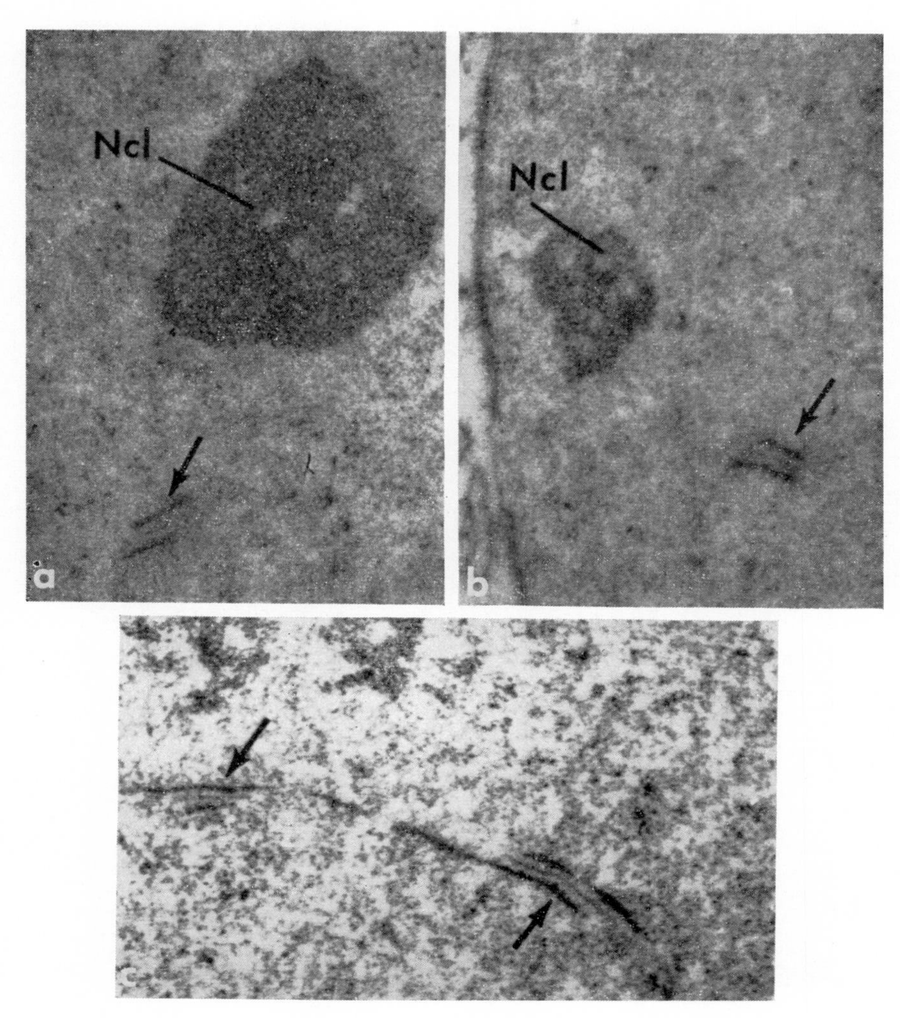

FIG. 3. Filamentous cores in spermatocytes at a very early stage preceding the formation of recognizable chromosomes; (a) and (b) from human spermatocytes, (c) from cat spermatocyte. Magnification: × 29,000, 30,000, and 28,000, respectively. (From D. W. Fawcett, 1956.)

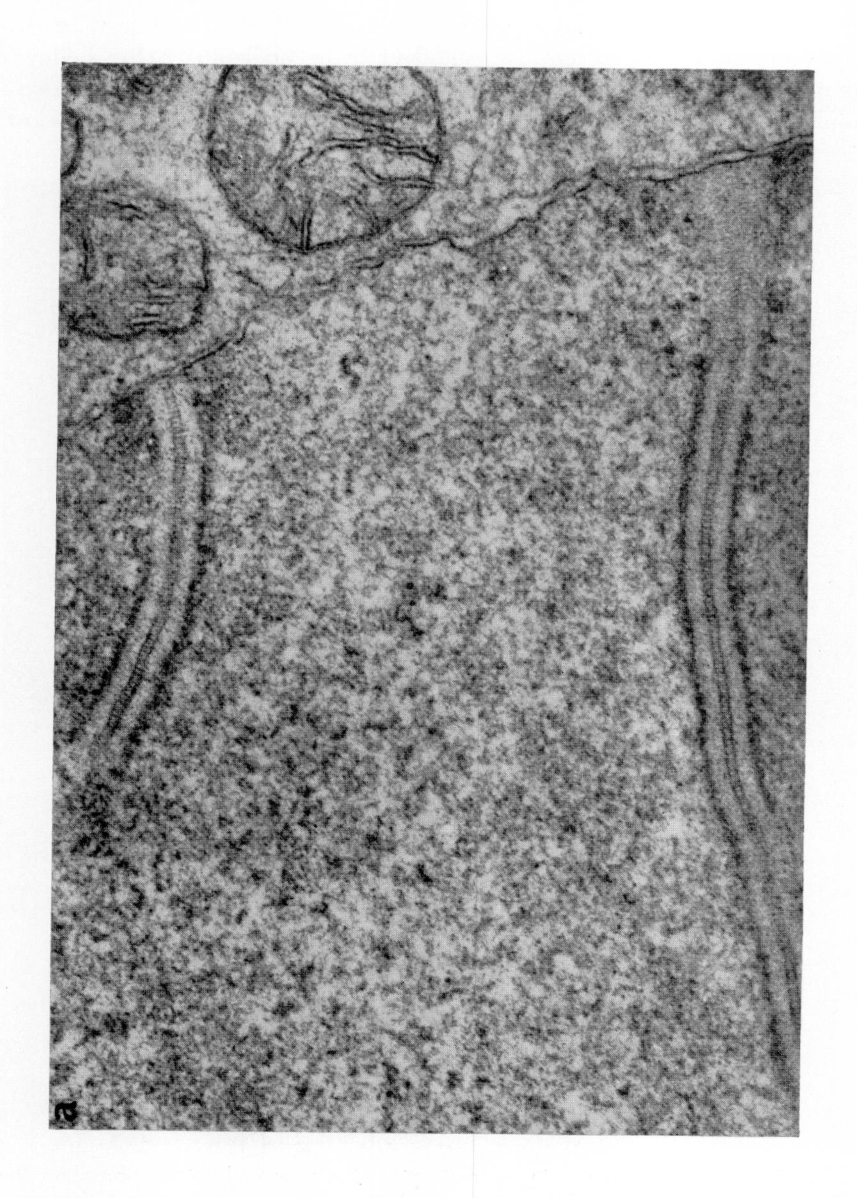

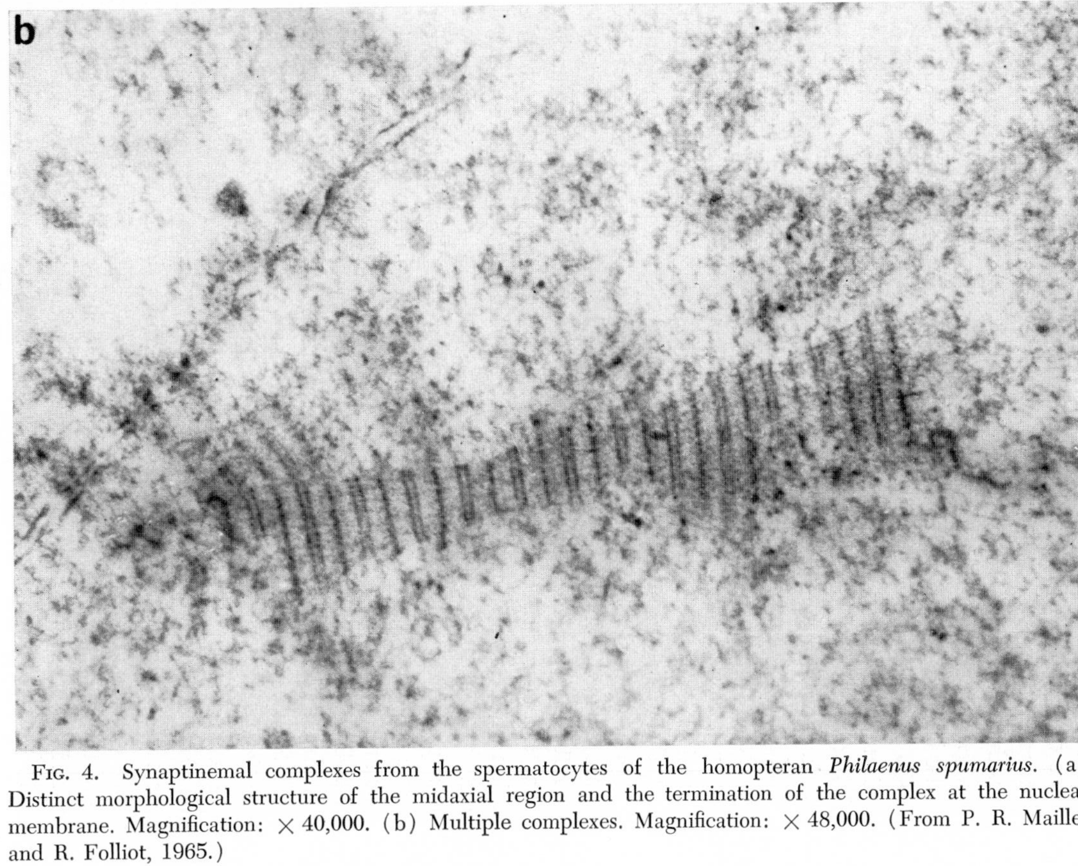

FIG. 4. Synaptinemal complexes from the spermatocytes of the homopteran *Philaenus spumarius*. (a) Distinct morphological structure of the midaxial region and the termination of the complex at the nuclear membrane. Magnification: × 40,000. (b) Multiple complexes. Magnification: × 48,000. (From P. R. Maillet and R. Folliot, 1965.)

association with, the pachytene chromosomes and that their appearance in spermatids indicates that this material may continue in the cell, dissociated from the chromosome, after the meiotic divisions. The reported presence of DNA in the lateral components is difficult to reconcile with this interpretation.

Recently, Wolstenholme and Meyer (1966) have examined spermatids of *Gryllus domesticus* and report that the axial core structure and the round body, present in both X and O spermatids, are composed of the same 100 Å fibers found in the X chromosome. Their cytochemical examination shows that the outer ribbons of the axial core as well as the round body contain ribonucleic acid (RNA), DNA, and basic protein which they consider evidence against Sotelo and Wettstein's identification (1965) of the round body as a nucleolus whose protein would be predominantly acidic. Moreover, this evidence would seem to rule out Schin's hypothesis (1965) that the axial core structure of *Gryllus* spermatids consists of protein molecules dissociated from the chromosomes. Instead, they conclude that the axial core structures and the round body are condensed portions of the chromosomes; that the presence of the axial core structure in postmeiotic haploid cells indicates that the condition is not restricted exclusively to homologs about to undergo exchange; and that the core structure may be related to the relative state of condensation of chromatin and the spatial relationship between condensed regions.

The supposition that the complex may represent homologs paired for exchange is supported by its apparent restriction to meiotic or postmeiotic cells and generally to those meiotic cells where crossing-over is known to occur. Thus, the primary spermatocytes of *Drosophila,* where crossing-over is absent, lack such structures (Meyer, 1961), as do the spermatocytes of *Steatococcus,* which are haploid, and of *Tipula oleraceae* and *Phyrne fenestralis,* which are achiasmate; however, the complexes are found in a species of *Tipula* with chiasmate spermatocytes (Moses, 1964). In line with these observations, female *Drosophila* homozygous for the mutant *c3G,* which eliminates meiotic exchange, do not show the complex, while normal females do (Meyer, 1964). The fairly usual occurrence of the complex at early meiotic prophase, as well as the presence of DNA in the lateral elements, adds further credibility to the assumption that these elements could sometimes represent synapsed homologs. An apparent exception to the "meiotic line" restriction is the discovery that polycomplexes may be present in nurse cell nuclei of mosquito ovaries (Roth, 1966).

The failure to find the complex in *Drosophila* males, where pairing of homologs is known to occur but where crossing-over is absent, has led Moses (1964) to propose that the complex is more closely allied with the process of exchange than with synapsis. On the other hand, Menzel and Price (1966) find complexes in the meiotic prophase of the diploid hybrids of *Lycopersicon exculentum* and *Solanum lycopersicoides* that are typical except for a narrower midaxial region. Light microscopy shows the chromosomes of these hybrids to be completely synapsed, but they often fail to form chiasmata and subsequently fall apart, indicating a marked reduction in exchange. Occasionally, Menzel and Price find completely normal complexes in the haploid of the tomato in which chiasmata rarely, if ever, occur. These authors suggest, as an equally tenable hypothesis, that synaptinemal complexes are a characteristic feature of synapsis and only indirectly related to crossing-over. Apparently the "synapsis" that occurs between homologs in *Drosophila* males and that found in the diploid hybrid differ, perhaps in their degree of intimacy, and thus in their potential to permit exchange.

The view that the complex is a reflection of synapsis, and only indirectly associated with exchange, receives direct support from the studies of Gassner (1967) with *Panorpa nuptialis,* a North American mecopteran. Light microscopy shows that male meiosis in this species is achiasmate, which is in agreement with the report of Ullerich (1961) for three European *Panorpa* species. Electron micrographs reveal, however, that the synaptinemal complex is present in the primary meiotic nuclei of both sexes. Gassner has proposed the very interesting hypothesis that the bipartite feature of the lateral elements represents homologous chromatids rather than sisters, as is commonly supposed. In Gassner's model, the plane of homologous separation corresponds to the plane which bisects each lateral element. This interpretation is based on the observation that the lateral elements become increasingly more bipartite as meiosis progresses, until complete separation occurs, whereas the distance between lateral elements remains relatively constant.

If the "synaptinemal complex" reflects exchange pairing, then the time of its occurrence must coincide or follow, but not precede, initial pairing of homologs. Unfortunately, the correlation between the meiotic stage, as identified by light microscopy, and the initial appearance of the synaptinemal complex is somewhat ambiguous. Fawcett (1956) reports that the cores seem to be present very early in prophase before chromosomes are visible in the electron micrograph and when the nuclei appear homogeneous except for a prominent nucleolus. Furthermore, according to Fawcett's description (1956) only the tripartite core is initially present, while later masses of delicate granules accumulate around the parallel

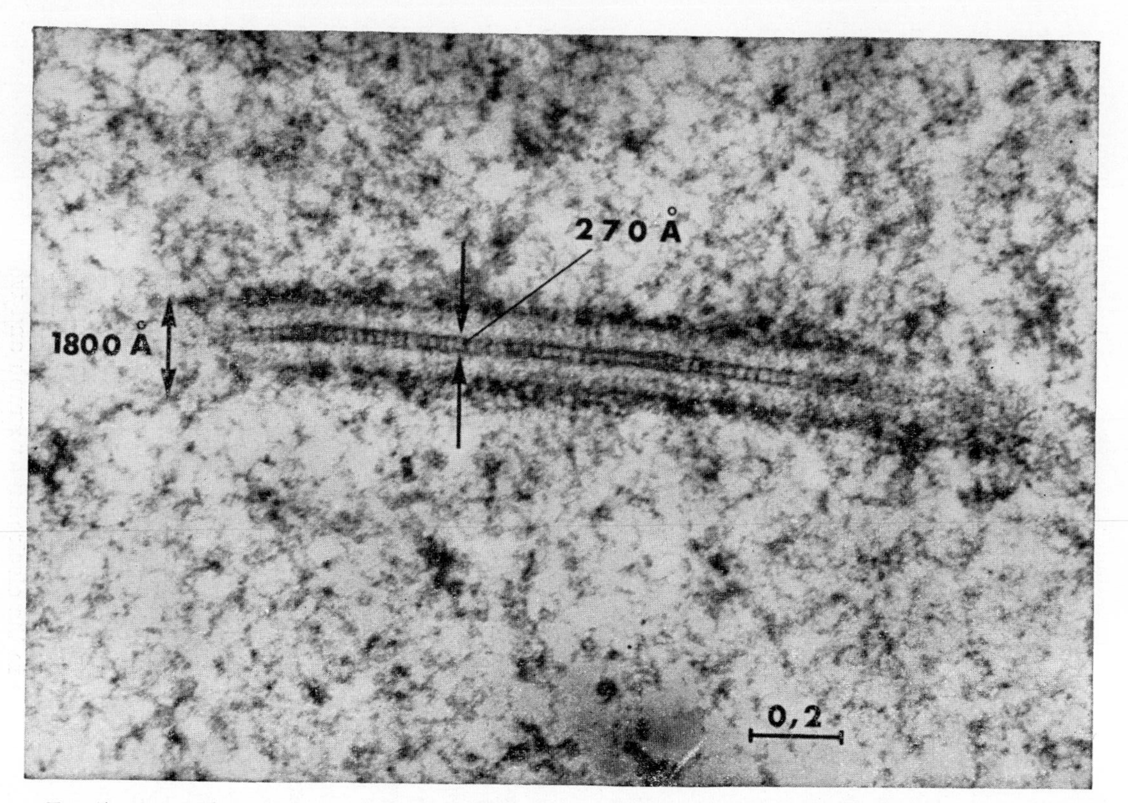

FIG. 5. Synaptinemal complex from the spermatocyte of *Philaenus spumarius* showing the dimensions of the complex and the absence of chromatin material at either side. Magnification: $\times$ 78,000. (From P. R. Maillet and R. Folliot, 1965.)

fibers to form recognizable chromosomes. Similarly, Sotelo and Trujillo-Cenoz (1958) find the complex to be present when the nucleus does not yet show identifiable chromosomes with electron or light microscopy. Moses (1958), on the other hand, states that synapsis has been positively identified as the stage of meiotic prophase at which the complex occurs, although he notes that no leptotene stages were found. Presumably by synapsis Moses means zygonema and pachynema. His electron micrograph of a zygotene or zygotene-pachytene stage shows a considerable region of chromosomes to each side of the core (Moses, 1958). Electron micrographs of pachynema from pigeon spermatocytes (Nebel and Coulon, 1962) show a still wider expanse of chromatin material. The core without chromatin material is shown in Fig. 5. Its width is ~ 1800 Å, which places it at the limit of visibility in the light microscope. If it first appears in this form at zygonema, the zygotene chromosomes as seen with light microscopy must represent more than the core, whereas it is apparently the only manifestation of the chromosome in the electron microscope. This assumption, however, is in conflict with Moses' electron micrographs of zygonema showing associated chromatin. The dilemma is resolvable if it is assumed that the appearance of the core precedes zygonema, as is suggested by the observations of Fawcett and of Sotelo and Trujillo-Cenoz, by the report of electron-dense threads in the rat spermatocyte at leptonema (Franchi and Mandl, 1962), and of cores in the *Drosophila* oocyte at interphase or leptonema (Koch *et al.*, 1967), as shown in Fig. 6.

An interesting feature of the complex is its frequent termination at the nuclear membrane (Moses, 1958), where dense thickenings often occur at the end of each axial component (Fig. 3). As pointed out by Menzel and Price (1966), this may reflect the polarization of the bouquet stage found in many animal spermatocytes in which the ends of the paired or pairing chromosomes are attracted to the nuclear membrane at the centriole pole. In *Solanum* and *Lycopersicon*, where the bouquet stage and the centriole are missing but where a synizetic knot occurs, terminations of the complex at the nuclear membrane are never found. Instead, the chromosomes are attracted to the nucleolus, although no tendency to terminate there was observed (Menzel and Price, 1966).

d. *Cytogenetic Evidence*. In contrast to the cytological studies at the level of the light or electron microscope, many investigators have combined cytological and genetic techniques. Moreover, some of these studies have raised questions about the traditional view of pairing initiation at zygonema. One such case has been that of asynaptic maize. The asynaptic (*as*) gene in maize is recessive and in the homozygous

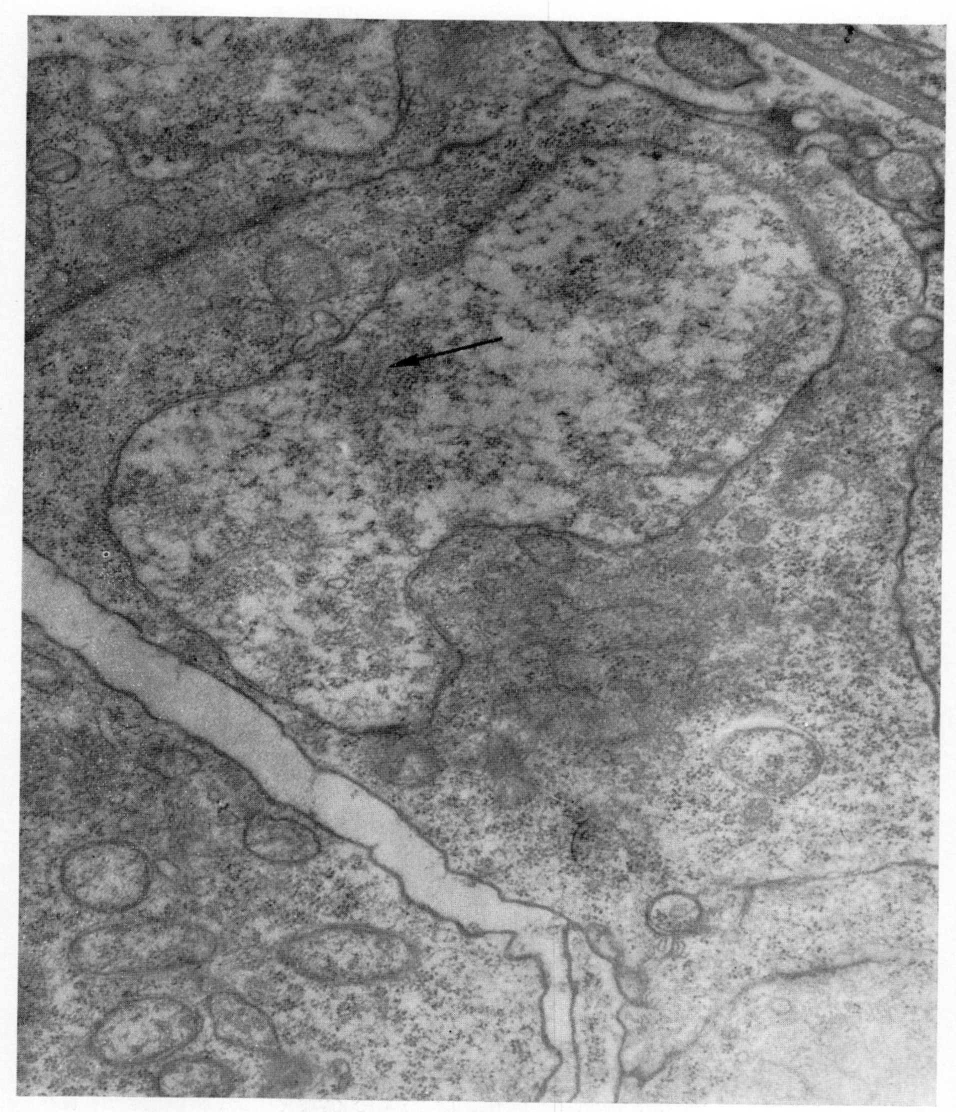

FIG. 6. Synaptinemal complex in an oocyte of *Drosophila melanogaster*. Oocyte taken from region 1 of the germarium and at a stage corresponding to premeiotic interphase or possibly leptonema. Magnification: × 46,500. (Courtesy of E. A. Koch, R. C. King, and P. A. Smith.)

condition frequently causes the chromosomes to reach metaphase in an unpaired state. The frequency of bivalents at metaphase is variable, and Beadle's original studiees (1933) of microsporogenesis showed a range of bivalents in different plants of from 0–10, although in only one of the eight plants studied were 10 bivalents encountered. Cytological studies indicated the term "asynaptic" to be incorrect, for synapsis apparently occurs. Synizesis, preceding pachynema, finds the chromosomes drawn into a tight knot, but Beadle, by careful study of occasional segments drawn out at the periphery as well as of identifiable regions of specific chromosomes, found them to be double. The doubleness does not represent the equational split in unpaired chromosomes, because in the later stages an increases in the proportion of single threads is observed (Beadle, 1933). Pachynema shows a high frequency of asynapsis, the amount varying with the particular plant, but correlating roughly with the amount at metaphase. Desynapsis presumably occurs at early pachynema.

If crossing-over takes place at pachynema and if chiasmata are the cytological equivalents of crossovers, then the reduction or absence of pachytene pairing and of chiasmata should find its genetic expression in a reduction in exchange. Beadle examined the frequency of exchange in the *sh-wx* region of chromosome 9 and failed to find the anticipated decrease. He assumed that the sample tested, which came from the most fertile plants, represented haploid gametes which were more nearly normal in terms of crossing-over and did not reflect the average exchange frequency. While haploid gametes may come from plants with normal metaphase pairing, diploid gametes, by contrast, should come from cells with a very high frequency of univalents at metaphase, and an absence of chiasmata should reflect an absence of exchanges. Beadle had in fact reported greatly elongated, abnormal spindles in cells with many univalents, which might be indicative of a division failure responsible for diploid gametes. In an attempt to verify this prediction, Rhoades (1947) and Rhoades and Dempsey (1949) analyzed crossing-over in the diploid gametes and, surprisingly, found a higher than normal frequency of single crossovers. Moreover, their reinvestigation of crossing-over in haploid gametes showed a higher frequency than was observed for their normal sibs and an especially high frequency of double exchanges. Dempsey (1959), speculating that the increase in doubles might be a selection artifact, corrected for the total ovules rather than viable seed set and found the absolute number of double cross-overs from *as* ears exceeded that for normal ears. A similar correction for single crossovers gave a reduction in crossovers in *as* plants of $\sim 1/2$; but this, of course, represents a maximal reduction, since it assumes that all inviable zygotes carry noncrossover chromosomes.

O. L. Miller (1963) reexamined the cytology of asynaptic plants and reported, contrary to Beadle's finding, a total absence of pairing from leptonema to metaphase I in completely asynaptic plants. He concluded that chromosomes which were unpaired at late pachynema had not previously been paired. Miller, however, made no mention of the synizetic stage (zygonema), which is precisely where Beadle found completely paired strands. A search with the electron microscope for the synaptinemal complex at the synizetic stage in *as* plants might be informative in this regard.

To determine if asynapsed chromosomes might have undergone precocious chiasma resolution, Miller looked for equational separation of heterozygous knobs at diakinesis. Among 102 univalents he failed to find a single case of equational separation of knobs. The two plants that he used had a mean of 9.4 and 9.2 bivalents. Since the synaptic behavior of these plants was almost normal, the univalents observed might well represent chromosomes that also behaved normally and had merely failed to undergo exchange. A study of heteromorphic bivalents in completely asynaptic plants would be more cogent.

A number of points are still unclear. The high frequency of crossovers in diploid gametes could be attributed to failure of first division cytokinesis in meiocytes with a high frequency of bivalents (O. L. Miller, 1963). The division failure, in this interpretation, would not be a consequence of failure of metaphase pairing leading to abnormal spindles, as suggested by Beadle (1933), but would merely be another expression of the pleiotropic asynaptic gene. Miller cites the case of one low asynaptic plant with regularly occurring elongated spindles to substantiate this possibility although he notes that spindle abnormality is generally correlated with the frequency of univalents. The absolute increase in double crossovers found by Dempsey could be attributed to an alteration in the pattern of exchange such that interstitial regions would remain unsynapsed, while a compensatory increase in crossing-over would occur in distal and proximal regions where exchange has been measured (O. L. Miller, 1963).

The cause of the paradoxical situation found in asynaptic maize (i.e., a reduction or absence of both pachytene pairing and chiasmata coupled with the observed increase in exchange) has not yet been resolved. It may arise from true asynapsis with a variety of factors, including selective recovery of crossover gametes, failure of first division cytokinesis, and an alteration in the pattern of exchange, all acting together to produce the results; or it may arise from a precocious resolution of chiasmata produced at a prepachytene exchange event, with a selective recovery of higher exchange tetrads; or other undeciphered factors may be responsible.

The discovery that the final DNA replication of the meiotic cell occurs well before pachynema provided a different means of access to the problem. By using the final replication as a fixed point in the meiotic cycle, the chronological relationship between exchange (and hence exchange-pairing, which precedes exchange) and DNA replication can, hopefully, be ascertained. One approach to the problem has been to make use of an organism with a zygotic type of meiosis. In such forms the meiotic divisions occur just after instead of just before the union of the gametes. Hence, the initial divisions of the zygote give rise to haploid cells, the diploid phase of the life cycle exists only as a transitory result of karyogamy during the meiotic stages, and crossing-over in such forms must take place during the diploid period. Rossen and Westergaard (1966) have attempted to determine whether the meiotic DNA replication occurs in the prefusion nuclei or, alternatively, whether it follows karyogamy. In the former case, crossing-over would necessarily be independent of the main DNA synthesis.

Their choice of an organism, *Neotiella rutilans*, was dictated by the requirement of Feulgen spectrophotometry for large nuclei, a condition rare among Ascomycetes. The results of their studies indicated to them that there is a replication in the prefusion nuclei, from which they infer that crossing-over cannot take place at the time of replication. A careful examination of the subject reveals a number of ambiguities surrounding their work which should be resolved before any final conclusion can be reached. First, *Neotiella rutilans* cannot be grown in the laboratory, so that the studies were conducted on specimens collected and fixed in the field over an undisclosed period of time. Earlier cytological studies of this species have placed the haploid number at 16 (Guillermond, 1904; Fraser, 1907; I. M. Wilson, 1937); yet Rossen and Westergaard report that their material has an $n = \sim 23 + 3 - 4B$ chromosomes. Should they be working with a mixture of species with different chromosomal numbers, their DNA determinations would be uninterpretable. Second, although *Neotiella rutilans* has been the subject of several careful cytological investigations, practically nothing is known of its genetics. Both Fraser (1907, 1908) and I. M. Wilson (1937) have concluded that this species reproduces apogomously (i.e., without sexual fusion), a conclusion substantiated to some extent by Rossen and Westergaard's suggestion (1966, p. 238) of what may happen in crozier formation, namely, "the terminal cell nucleus divides so that the crozier is formed from a terminal binucleate cell." If the terminal cell nucleus of the ascogenous hypha does divide twice to produce the four nuclei found in the crozier, the subsequent nuclear fusion between two of them to produce the zygote nucleus must involve identical nuclei. If apogamy is indeed the rule, the question of whether crossing-over occurs in this form becomes a pertinent one.

A third problem arises from Rossen and Westergaard's identification of the prefusion nuclei in the crozier. I. M. Wilson (1937) identifies the prefusion nuclei in *Neotiella rutilans* as the two nuclei in the penultimate cell of the crozier, and this is, in fact, the same identification generally given for *Neurospora* and Ascomycetes. The present authors, however, identify the nuclei of the stalk and terminal cells of the crozier as the prefusion nuclei, and it is on the DNA determination of these nuclei that their conclusion has been reached. This identification appears to be in direct conflict with the observations and figures of Wilson, who in living material of a related form, *Sepultaria sepulta,* has watched a migration of the terminal cell nucleus into the stalk cell to be followed by the growth of this binucleate cell into a new crozier after its two nuclei undergo a simultaneous mitosis. Fourth, perhaps the most serious question concerning this study stems from the failure of Rossen and Westergaard to demonstrate convincingly that no DNA replication occurs in the zygote nucleus following karyogamy. Chiang and Sueoka (1967a,b), who have studied DNA replication in *Chlamydomonas reinhardi,* a form which also has a zygotic type of meiosis, have found that despite the strong implication that a replication must take place in the gamete preceding karyogamy, a second chromosomal DNA replication occurs durnig the germination of the zygote and precedes the first meiotic division in both 4 and 8 zoospore lines. For these reasons, any unqualified interpretation of Rossen and Westergaard's results is clearly premature.

Another study, one that utilizes the effect of elevated temperature on chiasma frequency to mark the time of exchange, and label, to mark the final DNA replication, has been carried out by Henderson (1966) on *Schistocerca gregaria.* He has interpreted his results to mean that temperature affects crossing-over at late zygonema or pachynema. The experiment was not a critical one because changes in chiasma frequency, in this situation, are not necessarily equatable to changes in exchange frequency. A chiasma may be subject to resolution by elevated temperature for some time after the actual exchange event, in which case crossovers may be present in the absence of their cytological equivalent. Or, put another way, univalents may be crossovers. Furthermore, the study was apparently carried out in the absence of controls. Instead, Henderson relied on values obtained three years before (Henderson, 1963), despite a statement to the effect that the pattern of chiasma response to temperature in the present studies differed from the earlier one—a difference Henderson attributes to genetic variations between the stocks. Interestingly, in his 1963 paper he concluded that heat treatment had little or no effect on metaphase chiasma frequency until day 6, which coincides precisely with the appearance of label at metaphase on day 6 in his present study. Finally, an examination of his data indicates that the temperature effect on chiasmata cannot be construed to occur at late zygonema or pachynema as Henderson concludes, but must actually take place earlier, since the interval between a significant decrease in chiasmata and the appearance of label at metaphase is 2 days (Fig. 12), whereas the interval between synthesis and pachynema at the same temperature is given as $\sim$ 10 days.

The proper subject for the study of the relation of crossing-over to DNA replication, is, of course, crossing-over itself. Investigations utilizing this approach have been carried out in *Sphaerocarpus*, in *Drosophila*, and in maize. Abel (1965) found that in *Sphaerocarpus* alterations in temperature increased the frequency of crossing-over. The latest time that temperature changes, initiated at different points during the meiotic cycle from premeiotic interphase to late prophase, and continued through tetrad formation, are capable of increasing recombination is taken to indicate the earliest time that recombination can occur. Abel observed an effect up to day 15, at which time the majority of spore mother cells were found from cytological studies to be in leptonema. The time of chromosome duplication had been localized previously (Abel, 1963) by irradiating spore mother cells at daily intervals during interphase and comparing the frequencies of tetrads with a single mutant to those with two identical mutants. With this method, the first evidence for doubling occurred on day 15. Abel, however, does not assume a coincidence of the two events, because dissimilar growth conditions in the two experiments altered the timetable for maturation such that day 15 corresponds predominantly to leptonema in the recombination experiment but to interphase in the replication experiment. Abel concludes that replication precedes intergenic recombination, although he notes that since in *Sphaerocarpus* the homologs are already paired in premeiotic interphase, intragenic recombination could be occurring during DNA synthesis. Abel's conclusions would carry greater weight if the two experiments had been done simultaneously and had employed a labeling technique to localize the synthetic period. On the basis of cytological studies of *Impatiens balsamina* ($n = 7$) and *Salvia nemerosa* ($n = 7$) whose chromosomes at premeiotic interphase associate in groups of two's, so as in some cases to show only the haploid number, Chauhan and Abel (1968) are now inclined to believe that crossing-over occurs prior to zygonema.

A similar type of study has been carried out by R. F. Grell and Chandley (1965) in *Drosophila melanogaster* females. They injected thymidine-^{3}H into newly eclosed females and by sacrificing groups at daily intervals, were able to follow cytologically the progression of labeled oocytes through the ovary. The final replication was found to occur in a very young oocyte, located in the anterior portion of the germarium and in a stage most probably corresponding to premeiotic interphase. The interval between synthesis and the appearance of the first labeled mature oocyte occupied 6 days, while for the majority of labeled oocytes, maturation required 8 days. Parallel genetic experiments, utilizing an elevated temperature of 35°C for 24 hours to increase exchange, demonstrated that the first significant increase in exchange occurred among those progeny originating from eggs laid 6½ days after treatment, while maximal exchange values occurred among progeny from eggs laid 8–9 days after treatment. The times of DNA replication and the heat-sensitive period for crossing-over in the *Drosophila* oocyte are therefore roughly coincident. It has been subsequently demonstrated by genetic means (R. F. Grell, 1966) that the temperature-induced crossovers are exclusively of meiotic origin, eliminating the possibility that some gonial crossovers contribute to the observed rise in exchange values. Chandley's labeling studies (1966) have likewise shown that thymidine-^{3}H does enter the oocyte nucleus so that the first labeled oocytes that were followed cytologically had actually incorporated their label as oocytes and not as oogonia. This was demonstrated by sacrificing females 15 minutes after the injection of thymidine-^{3}H and identifying labeled 16-cell cysts representing the labeled oocyte and nurse cells in the anterior third of the germarium. The supposition that temperature affects exchange indirectly by conditioning the germ cell sometime prior to exchange (Whittinghill, 1955) cannot presently be excluded. It appears unlikely, however, that a preoocyte stage is the sensitive target, since heat treatment given to developing females does not significantly affect recombination until day 6, which coincides closely with the formation of the oocyte (R. F. Grell, 1967). Subsequent studies utilizing a pupal system which provides a well-synchronized oocyte population have repeatedly corroborated the coincidence between DNA replication and the temperature-sensitive period for exchange in the *Drosophila* oocyte (R. F. Grell, 1967).

A coincidence between the temperature-sensitive period for recombination and the meiotic replication period has been reported in maize as well. Cronenwett and Maguire (1967) established that DNA synthesis in the pollen mother cell occurs at premeiotic interphase, possibly extending into synizesis; Maguire's studies (1967a) have shown that elevated temperature increases exchange when given during premeiotic interphase and possibly during synizesis (Maguire, 1968). Should future work demonstrate that temperature acts directly at the time of exchange, as suggested by McNelly-Ingle *et al.* (1966), the foregoing results would mean that exchange pairing, and at least the initiation of the exchange process, occur close to the time of synthesis.

The cytogenetic investigations of Maguire (1965) are highly pertinent to the question of initiation of exchange pairing. From backcross progenies of maize-*Tripsacum* hybrids Maguire obtained a derivative stock carrying a pair of maize 2-*Tripsacum* interchange chromosomes, 2^T and T^2, in addition to a standard 2. The *Tripsacum* chromosome possesses a terminal knob which permits cytological identification of this segment. The interchange is considered to have been homologous in nature, for the *Tripsacum* segment contains the dominant alleles for three loci found in the corresponding region of the short arm of 2. The exchange point was estimated cytologically to have been located at $\sim$ 25% of the total length from the distal end of the short arm of 2 and genetically at $\sim$ 54 crossover units from this end. The total translocated segment is apparently not completely homologous to the corresponding segment in 2, since exchange occurs only rarely and since two maize segments pair preferentially with almost complete exclusion of the *Tripsacum* segment when all three are present. Secondary derivative crossovers have been obtained which give 21-chromosome plants carrying duplicated (or triplicated) segments of varying size and position.

The original interchange and two of the secondary derivative chromosomes, whose behavior will be described, are shown in Fig. 7. Derivative I, which was obtained by a secondary crossover between 2^T and T^2 at a point $\sim$ 11% of the total cytological length of 2 from the end of the short arm, carried the distal half of the *Tripsacum* segment, including the terminal knob, back to T^2. Derivative II was obtained by a crossover between a normal 2 and 2^T which also moved the distal segment of 2^T, including the knob, this time to one of the second chromosomes. The first 21-chromosome plant carries two standard maize 2 chromosomes and derivative I; the second plant carries two derivative II chromosomes and the original T^2. In both cases a triplication for a small segment of the short arm of 2 is carried in the extra chromosome so that pairing is competitive in this region. The frequencies of pachytene and metaphase trivalents were studied in both types of 21-chromosome plants. Trivalent formation at pachynema depends in plant 1 on the pairing of derivative I and in plant 2 on the pairing of T^2, since the two normal second chromosomes in plant 1 and the two derivative II chromosomes in plant 2 virtually always pair in the long arm region and in the proximal short arm region. Trivalent formation at metaphase depends on crossing-over

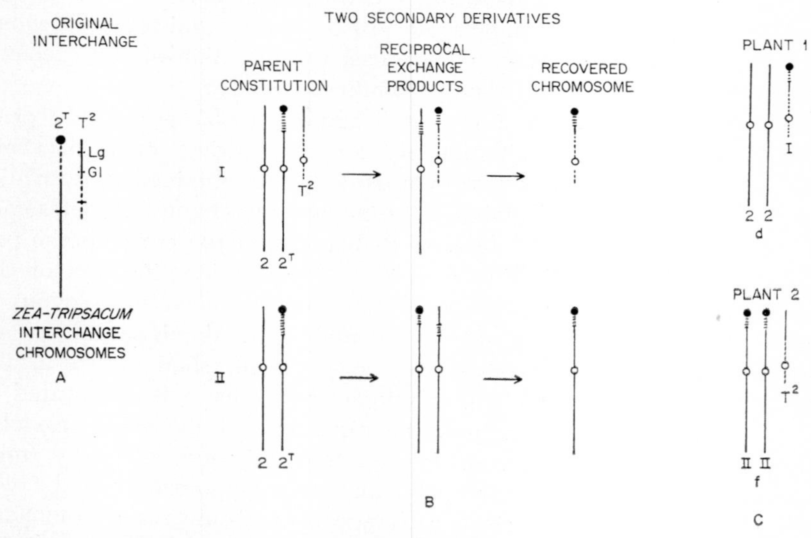

FIG. 7. A. Original interchange products 2^T and T^2 from a crossover between a second chromosome of maize and a *Tripsacum* chromosome. B. The origin of the two secondary derivatives, I and II. C. The composition of the two 21 chromosome plants studied. ——, *Zea* chromosome 2; . . . . , *Tripsacum* chromosome. (Modified from M. P. Maguire, 1965.)

of derivative I in plant 1 and on crossing-over of T^2 in plant 2. The extent of homologous pairing available for exchange in the triplicated region is estimated to have included a maximum of 29 genetic units. In the case of plant 1, the frequencies of pachytene and metaphase trivalents were found to be 17 and 15%, respectively; in the case of plant 2 the frequencies were 18 and 17%, respectively. The close similarity in the frequency of pachytene and metaphase trivalents can be interpreted in two alternative but unorthodox ways. Either (1) crossing-over has occurred before pachytene pairing and the requirement for trivalent association at both pachytene and metaphase is the presence of a preestablished crossover, or (2) crossing-over invariably occurs in the pertinent arm of the trivalent regardless of the genetic length of the synapsed segment which in these cases represents a maximum of 29 units.

Maguire (1966) has recently approached the problem in another way; she has employed an inverted region rather than a triplication. Plants heterozygous for a paracentric inversion with breakpoints at 0.70 and 0.87 of the cytological distance from the centromere to the distal end of the long arm of chromosome 1 were used. The long arm of chromosome 1 has a genetic map length of ~ 112 units so that if genetic length were to correspond to cytological length, the inverted region should contain ~ 19 crossover units and well under 50 units in any case. Pachytene configurations were scored as containing reverse synapsis in the inverted region, pairing throughout the arm, or pairing failure in the region. Only reverse pairing indicates homologous synapsis. Anaphase I cells were scored for (a) presence of bridge and fragment, (b) presence of fragment only or (c) absence of bridge and fragment; (a) and (b) indicated exchange within the inverted region and (c) no such exchange. The data are summarized in Table I.

TABLE I

FREQUENCIES OF HOMOLOGOUS SYNAPSIS AND BRIDGE AND FRAGMENT FORMATION IN MICROSPOROCYTES OF MAIZE, HETEROZYGOUS FOR A SHORT PARACENTRIC INVERSION[a]

	Frequency of reverse synapsis at pachynema		Combined anaphase I bridge and fragment and fragment only frequency	
Plant	No.	Percent	No.	Percent
1	182/505	36.0	466/1303	35.8
2	149/495	30.1	303/1023	29.6
3	190/544	34.9	426/1244	34.2

[a] From Maguire (1966).

There is a very close correspondence between the frequencies of homologous pachytene pairing in the inverted region, as estimated by reverse pairing, and the frequency of crossing-over within the inversion, as estimated by the frequency of bridge and fragment formation. The results are again interpreted to mean that either a preexisting crossover determines homologous pairing in the inverted region at pachytene or that homologous pachytene pairing invariably leads to a crossover even when the paired segment represents considerably less than 50 crossover units, a rather unlikely possibility.

e. *Models of Prezygotene Exchange Pairing.* i. *Discontinuous pairing.* Genetic analysis of minute regions in a variety of organisms has disclosed that an enhancement of recombination appears to occur within very short distances of the genetic material. Pritchard (1955), for instance, in the case of *Aspergillus nidulans,* found that for heteroallele crossovers in the *ad* 8 cistron, the frequency of exchange in the closely adjoining regions was much higher than expected. Pritchard also showed that the closer together the two heteroalleles lie, the more recombination in the adjoining regions exceeded expectation. In *Aspergillus* this effect is restricted to a segment which Pritchard (1960a) calculates to be $\sim$ 0.4 map units. Localized negative interference has also been reported for *Escherichia coli* (Rothfels, 1952), for bacteriophage (Streisinger and Franklin, 1956; Chase and Doermann, 1958), for *Neurospora* (Giles, 1955; Mitchell, 1956), and for other organisms.

Rothfels (1952) favored the view that in *E. coli* the apparent excess of multiple crossovers is due to an excess of noncrossovers through irregular and incomplete pairing. To account for this clustering of exchanges in eucaryotes Pritchard (1955) postulated, as an alternative to the traditional meiotic sequence, that exchange may occur before zygotene pairing. Homologs are assuming to be in contact for only a small part of their length, so that pairing is discontinuous, and the region of contact has been called an effective pairing segment (Pritchard, 1955). Pairing of this kind at premeiotic interphase or early leptonema, which would not be cytologically visible, could well coincide with the time of DNA replication in the meiotic cell.

With the Pritchard model the probability of establishing an exchange site is considered to be low, but once established the probability of exchange within it is considered to be high. For two intervals distant from one another on the genetic map, an exchange in both requires two regions of effective pairing; for intervals close enough to lie within a single region of effective pairing, the probability of two exchange events in a paired region is expected to be the product of the frequencies of single exchanges in the region. When selective techniques permit examination of the segment of the population in which effective pairing for a particular region is presumed to have occurred, the frequency of multiple exchanges is found to be in good accord with the expectation based upon singles in the region (Doermann, 1965). Pritchard's model thus provides a plausible solution to the phenomenon of negative interference.

For most organisms that have been studied genetically, with the possible exception of *Aspergillus* (Strickland, 1958), one crossover tends to suppress the occurrence of a second crossover in the vicinity (i.e., positive interference is present). In terms of the Pritchard model, this would mean that the establishment of one effective pairing site should greatly decrease or eliminate the possibility of establishing a second one close by. Positive interference has been interpreted as a consequence of the steric configuration of the paired homologs such that one crossover produces a structural hindrance to another occurring nearby. The Pritchard model provides no ready explanation for positive interference. In fact, a model in which exchange events are wholly dependent upon the chance meeting of very small regions of homology, superficially, at least, excludes the concept of events exerting effects over long distances.

Another difficulty inherent in the Pritchard model is its failure to explain why bivalents do not become interlocked during the exchange process. If homologs are greatly extended and randomly arranged in the nucleus, such that exchange depends on chance pairing of minute regions, the breakage and reunion of different strands which must accompany exchange whether a copying-choice mechanism is invoked or not, in order to account for 3- and 4-strand double crossovers, should frequently lead to entanglements which would become evident at late prophase as interlocked bivalents or multivalents. Interlocking seems to be a rare event. For instance, Suomalainen (1947) and Therman (1953, 1956) report that in the tribe Polygonatae not a single case of a true interlocking has been observed. In liliaceous plants as well, interlocking appears to be extremely infrequent. Upcott (1936) records interlocking in 1 or 2% of the cells of *Eremurus spectaculis*, but Oksala and Therman (1958), in studies of *Eremurus himalaecus*, found no interlocking, although the number of cells examined is not given.

To test the model in a more direct way, Pritchard (1960b) utilized a rearrangement which moved a large segment of one chromosome terminally (or subterminally) onto another chromosome so that the translocated segment was present as a triplication. Pairing, according to the classic model, is complete along the chromosome, and at any particular region it is restricted to two strands, a and b. The introduction of a third strand, c, should add no increased opportunity for pairing or exchange. Pritchard's model, which assumes pairing is discontinuous and infrequent, predicts that the addition of a third region of homology, c, should increase the opportunity for homologous contact and hence pairing threefold from ab to ab + bc + ac. With the probability of exchange primarily dependent on the probability of the establishment of an effectively paired region, the frequency of exchange with the triplication should approach three times that found in the diploid.

Pritchard's results, which showed an increase from 16% in the diploid to 22–38% with the triplication, are difficult to evaluate since no numbers nor standard errors are given. The possibility that *Aspergillus* displays some measure of positive interference in this situation is suggested by the 50% reduction in crossing-over in two regions neighboring the triplication.

ii. *Rough pairing preceding intimate exchange pairing.* Cytologically visible, continuous pairing and prezygotene discontinuous pairing do not represent the only alternatives. The frequent reports (reviewed above) of pairing at premeiotic anaphase or telophase, as well as the many observations of the haploid number of prochromosomes at premeiotic interphase, suggest a third possibility (Maguire, 1960; R. F. Grell, 1965). A rough alignment of homologs may be accomplished just prior to or during early premeiotic interphase. McClintock (1945) and Singleton (1953) have described pairing between condensed chromosomes shortly after fusion in *Neurospora crassa.* Parasynapsis of this type would then be followed by the more intimate exchange pairing established discontinously along the chromosome. The additional feature of a rough alignment of homologs, preceding a more intimate association, provides a way for positive interference to operate along the chromosomes as well as a method for positioning homologos together prior to their elongation so as to avoid interlocking. Recombination might then occur close to the time of synthesis, if not coincidentally with it, by whatever mechanism one chooses to invoke. While the initiation of exchange points may occur at this time, their resolution might be either concurrent or delayed. With this scheme, pachynema, instead of representing the time of initiation of exchange, would represent the final expression of exchange pairing, following which homologs generally fall apart except at exchange points and perhaps at centromeres. The chief advantage of this alternative is that homologous contact at a molecular level could be established at the time of maximal extension of the chromosomes. Whether a similar intimacy can be achieved at pachynema when the chromosomes have undergone considerable spiralization, particularly at those points where differentiation into distinct chromomere patterns have occurred, remains open to question.

f. *Conclusions.* This section has mustered an array of circumstantial evidence which raises some doubts concerning the classic interpretation of the onset of exchange pairing. Admittedly, the presentation has been one-sided. The traditional view is so well known and so widely accepted as to require no further elaboration or defense. On the other hand, legitimate questions do exist. For certain plants that include a synizesis stage in their meiosis, Moens' studies indicate that no true leptotene stage is present and that pairing probably occurs prior to the cytologically visible stages. For organisms with a bouquet stage, McClung's very careful observations suggest that what would be called a leptotene strand in *Mecostethus* represents paired homologs. In fact, many cytologists have demonstrated that chromosome threads at late leptonema are clearly double (Guyénot and Naville, 1933–1934; Oehlkers and Eberle, 1957), and it has not been unequivocally established that the doubleness represents sister chromatids rather than homologs.

The dimension of the synaptinemal complex, before lateral loops are formed, is ~ 2000 Å, which places it at the threshold of visibility with light microscopy. Its attachment to the nuclear membrane, in forms with a polarized bouquet stage, bears a certain resemblance—with regard to dimensions, nuclear location, and perhaps time of appearance—to the fine threads McClung observes in *Mecostethus* at the beginning of the peritene stage. Significantly, in the tomato, which lacks both centrioles and a bouquet stage, but where the paired chromosomes first appear in a clumped synizetic knot, Menzel reports that the synaptinemal complex is not attached to the nuclear membrane but is found intimately associated with the nucleolus.

Cytogenetic studies of rearranged or triplicated regions well under 50 crossover units in length, where pachytene pairing is often not expected to be followed by an exchange, show an excellent correlation between the frequencies of pachytene pairing and of exchange. The latest heat-sensitive period for increasing exchange frequency in the organisms where this test has been carried out (namely, in *Sphaerocarpus,* maize, and *Drosophila*) is prepachynema. Although these observations may all turn out to have explanations which are consistent with the classic picture, at this juncture several interpretations are tenable and unreserved acceptance of any version appears premature.

3. *Competition in Exchange Pairing*

a. *Early Studies.* The introduction of extra homologs or homeologs into the genome provides a method whereby information about the process of exchange pairing may be obtained. Such studies have been pursued in polyploids, in trisomics, and in diploids which carry an extra segment of variable length so that a particular region is present in triplicate.

In early genetic studies of this kind, Bridges and Anderson (1925) investigated crossing-over in the free X chromosomes of triploid *Drosophila.* Total exchange was found to be equivalent to that in the diploid, or if calculated on a chromosomal basis, reduced from that in the diploid. This result suggested that pairing in the presence of an extra homolog is competitive. Redfield (1930, 1932) carried out similar experiments for chromosomes 2 and 3 in triploid *Drosophila,* and in these cases the amount of crossing-over per chromosome was approximately the same in the diploid and triploid. The discrepancy in results between the X and the autosomes may be merely a reflection of the classes of progeny that were studied. In calculating crossing-over frequencies Bridges and Anderson used female progeny carrying two maternal X chromosomes, whereas Redfield used females carrying a single maternal second or third chromosome. A variety of experiments have shown that in the progeny from triploids, noncrossover chromosomes are recovered more frequently in a gamete with a homolog ($2n$), and crossover chromosomes in a gamete without a homolog (n) (see below). As a consequence, Bridges and Anderson's results probably underestimate crossing-over for the X, and Redfield's results overestimate it for the autosomes. Thus, while it seems clear that pairing and exchange per chromosome are not increased in a triploid, it is not certain whether competition for pairing reduces exchange. The results in all cases are complicated by marked regional differences, such that the distal and proximal regions of the X (Beadle, 1934) and the proximal regions of the autosomes show a conspicuous increase while other regions remain unaltered or show a decrease.

Studies of crossing-over in the X chromosomes of triploids are difficult because the manner of assortment of the two extra autosomes is not independent of the X's. To circumvent this complication, Beadle and Ephrussi (1937) performed an ingenious experiment. Females which carry three X chromosomes and two sets of autosomes are known as superfemales and, despite the implications of their title, are always sterile. Ovaries from larvae of superfemales were transplanted into diploid larvae homozygous for a female sterile gene. In this situation all eggs that are produced originate in the transplanted ovary. Although only ~ 1% of the eggs laid gave rise to adults, it was clear that crossing-over in the trisomic was much reduced, since only 1 crossover X and 26 noncrossover X chromosomes were recovered. The reduction may well be a consequence of this particular chromosomal imbalance on pairing and/or exchange rather than an inherent property of trisomics.

b. *Structural Heterozygosity and Competition.* The advantage of strict sequential homology for exchange pairing becomes evident from the studies that have been carried out in allopolyploids where it is assumed that some structural divergence has occurred and where the degree of divergence is reflected in the pairing, exchange, and segregation relationships. Operationally, when two chromosomes from genome A are marked with the dominant R allele and two from genome B with the recessive r allele, strictly homologous pairings will lead exclusively to homogenetic bivalents and to gametes that are Rr in genotype. Consequently, progeny showing a recessive phenotype will be absent and structural differentiation of the chromosomes should be considerable. Homeologous pairing, on the other hand, will lead to some frequency of heterogenetic bivalents or multivalents and to some rr gametes. The frequency of progeny with the recessive phenotype then becomes a measure of the degree of homeologous pairing, which in turn is presumed to reflect the degree of structural differentiation that the chromosome has undergone.

Gerstel and Phillips (1958) have studied gene segregation in synthetic allopolyploids of *Gossypium* and *Nicotiana* and have found ratios of dominant to recessive phenotype of 4.3:1 to 1:0 in *Gossypium* and from 3:1 to 1:0 in *Nicotiana*. In subsequent studies (Gerstel, 1963) it has been possible to show that different chromosomes in the genomes of *Nicotiana tabacum* and *Nicotiana sylvestris* vary in their degree of preferential affinity. Segregation ratios for different characters in the amphipolyploid were found to vary from 12.2:1 to 3.2:1, indicating that while some chromosomes of the two species have remained completely homologous, others have become differentiated. In cytological studies of synthetic allopolyploids of *Gossypium*, Sarvella (1958) found a good correlation between the frequency of multivalents and the percentage of recessive segregants. Hybrids of perennial teosinte (4N) and tetraploid *Zea mays* also display a high frequency of preferential pairing (Shaver, 1962).

Maguire (1961) used three types of hyperploid 21-chromosome maize plants to study the position of an extra chromosome at pachynema in one strongly competitive and two weekly competitive situations. The extra chromosome was in all three cases a *Zea-Tripsacum* interchange chromosome, Z*, which carried the distal half of the short arm of *Zea* chromosome 2, the remainder being of *Tripsacum* origin. In one of the three plants (ZTZ*)—the reciprocal exchange chromosome, which bears a *Tripsacum* knob—a normal chromosome 2 and the extra chromosome were present; the second plant (ZZZ*) carried two normal *Zea* chromosomes 2 and the extra chromosome; the third plant (TTZ*) carried two reciprocal exchange chromosomes 2 and the extra chromosome (Fig. 8). The knob-bearing segment from *Tripsacum* is only partially homologous to the distal half of the short arm of *Zea* chromosome 2 with which it has interchanged and some structural divergence may be assumed to have occurred.

When the extra chromosome was not involved in trivalent formation with the other second chromosomes at pachynema, significant tendencies were noted for it to lie near its homologous region in chromosome 2. The tendency was strongest in the ZTZ* complement (weakly competitive), in which the only exact homolog for the distal half of the short arm of 2 is found in the extra chromosome. A lesser tendency is observed in ZZZ* plants, where half of the short arm of 2 is present in triplicate (strongly competitive). No measurable tendency is found in TTZ* plants (weakly competitive), where the extra chromosome has no region of exact homology to either exchange chromosome 2. As one possibility for the proximity, Maguire suggests that the nearness found in ZTZ* plants may reflect earlier pairing between the normal 2 and the extra chromosome, since some pairing failure was observed at the knob end of the chromosome 2 bivalent. If the extra chromosome had been synapsed with the normal 2 at an earlier stage and exchange had failed to occur, the prox-

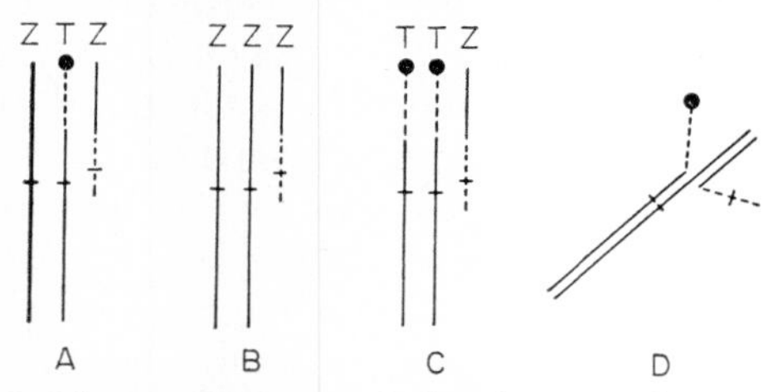

FIG. 8. A, B, and C. The compositions of the three 21 chromosome plants used for studies of pachytene pairing. D. The usual trivalent configuration formed in ZTZ plants. ——, *Zea* chromosome; - - - -, *Tripsacum chromosome.* (From M. P. Maguire, 1961.)

imity could reflect desynapsis. Synaptic dissociation is known in this material (Maguire, 1960), particularly between corn and *Tripsacum* segments in ZTZ* plants. Of particular interest are the frequencies of trivalents at pachynema in ZTZ* and ZZZ* plants, which are 90 and 69%, respectively. A frequency of 69% corresponds very well to the 67% which is expected if random mating occurs between the extra chromosome and the two normal second chromosomes in their regions of shared homology, despite synapsis of the remainder of chromosome 2, which is present only in duplicate.

Instead of employing undetermined structural heterozygosity that has originated between related but diverging forms, well-defined rearrangements may be introduced into the genome. Kozhevnikov (1940) made the important observation that when rearrangements of the second or third chromosome were known to be present in triploids of *Drosophila*, the rearranged chromosome tended to be recovered in the 2 N gametes. This is, of course, the expected result if the two normal autosomes pair preferentially, crossover, and segregate regularly while the rearranged noncrossover chromosome assorts randomly with respect to its two homologs.

Rhoades (1957) synthesized triploids of *Zea mays* in which one chromosome 3 was inverted with respect to the other two. The structurally different chromosome was found in the progeny from diploid gametes more frequently than a similarly marked chromosome from control matings with structurally homozygous triploids. Doyle (1959) also studied preferential pairing in trisomes, triploids, and tetraploids of maize which carried inverted and normal chromosomes. Alterations in gene segregation ratios, in multivalent frequencies, and in anaphase bridge frequencies in the inversion-bearing genomes (as compared to the controls) all indicated that pairing had occurred preferentially between structurally similar chromosomes.

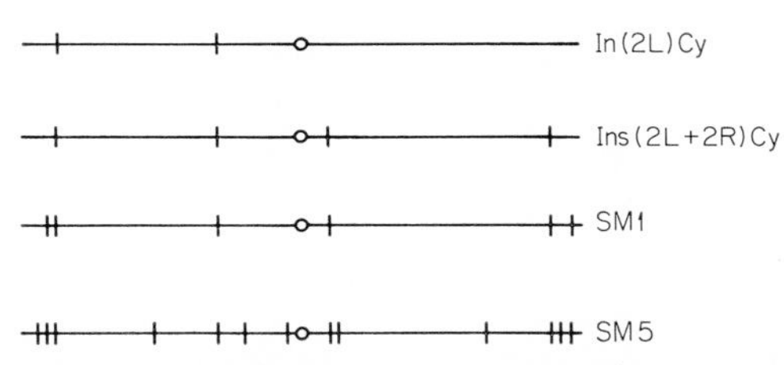

FIG. 9. Diagram of the breakpoints of a series of rearranged second chromosomes that reflect increasing inversion complexity. These chromosomes were used to determine degrees of preferential segregation. (From E. H. Grell, 1963).

The most definitive work of this kind was carried out in *Drosophila* by E. H. Grell (1961). A series of sequentially derived, inverted second chromosomes, each representing a greater degree of complexity than the preceding one and each carrying the inseparable dominant marker *Curly*, was employed for the study. The structure of the four rearranged chromosomes is shown in Fig. 9. Four types of triploid females were synthesized, each of which possessed two normal second chromosomes but differed by the particular rearranged two that they carried. The degree of preferential pairing of the two normal chromosomes (i.e., the degree of exclusion from pairing of the inverted chromosome) was evaluated by comparing the frequencies with which the *Curly* phenotype was recovered in progeny from haploid and diploid gametes.

The results (Table II, column 3) clearly demonstrate that the frequency with which the inverted chromosome is found at the haplo-2 pole is negatively correlated with the complexity of the inversion system. Disregarding trivalents, if pairing were nonpreferential among the three homologs, each should be paired in 67% of the oocytes and each should be found at the haplo-2 pole 33%. If pairing were completely preferential between two of them, the third should never be recovered at the haplo-2 pole. Grell observed that a single inversion in one arm reduces the frequency that the inverted two is found at the haplo-2 pole to 27%; one inversion in each arm causes a further reduction to 21%; a pericentric inversion superimposed on the two paracentric inversions brings the value down to 10%; and with the complexly rearranged SM5 chromosome, which has 13 breakpoints, the frequency is only 3%.

TABLE II
SEGREGATION FROM INVERSION-BEARING TRIPLOID FEMALES[a]

Chromosomes of triploid	Frequencies of segregations producing the following gametes	
	In/+ and +	+/+ and In
Random (+*/+/+)[b]	0.67 (+*/+ and +)[b]	0.33 (+/+ and +*)
In (2 L)Cy/+/+	0.73	0.27
Ins (2 L + 2 R)Cy/+/+	0.79	0.21
SM1/+/+	0.90	0.10
SM5/+/+	0.97	0.03

[a] From E. H. Grell (1963).
[b] Asterisk (*) designates any one of the three normal second chromosomes.

Pairing in a diploid organism may be achieved frequently despite complex structural heterozygosity, because long matching sequences of nucleotides still remain and because inverted regions may achieve pairing through loop configurations. In a competitive situation, homologs that are isosequential possess a distinct pairing advantage. The greater the degree of structural departure of one chromosome from its two homologs, the more infrequently will it engage in exchange pairing.

c. *Location of a Triplicated Region.* The ability of a homolog or a segment of of a homolog to compete for exchange pairing hinges to some extent on its position in the genome. Beadle (1934) studied crossing-over in triploids that carried two attached-X chromosomes and one free X. He found that in the distal portions of the three X chromosomes, crossing-over occurred at random, but that near the region of the spindle attachment the two attached-X chromosomes were preferentially involved in exchange. Crossover values in this experiment were based upon progeny from XX, 1A gametes. Subsequent studies (Beadle, 1935) disclosed that this preference did not hold for all classes of gametes, although the total crossover values, calculated on the basis of the four types of gametes (1X, 1A; 1X, 2A; 2X, 1A; 2X, 2A) indicated that the two X chromosomes attached to one centromere did undergo more pairing and exchange in the region of closer proximity. The difference in crossing-over that Beadle observed for the various classes of gametes was one of the first indications of an interchromosomal effect on chromosome assortment which selectively involved noncrossover chromosomes (see Section II,B,5,b,i).

Dobzhansky (1931, 1932, 1933) investigated extensively the effects of rearrangements on crossing-over in *Drosophila* and correctly interpreted the decreases observed, particularly in the neighborhood of breakpoints, as due to a competition for the establishment of pairing regions between the rearranged segments and their homologous regions in the normal chromosomes. For instance, if, as the result of a translocation, a chromosome is produced which carries segments homologous to two different chromosomes, both chromosomes will attempt to pair with their respective homologous segments in the translocation. Both may be successful in part, but close to the break point a conflict will ensue and pairing may be achieved by neither homolog.

Dobzhansky's interest in competitive pairing led to an investigation of the behavior of triplicated segments. In *Drosophila,* segments present in triplicate are known as duplications or duplicating segments. They may be present as a fragment on their own centromere, in which case they are known as free duplications; they may be present in their normal location in tandem or reverse tandem sequence; or they may be inserted or attached at a foreign location in the genome. Rhoades (1931) was the first to study the effect of a duplication on crossing-over. A short segment of a second chromosome attached to a Y chromosome in *Drosophila* females was found to depress crossing-over between the normal second chromosomes, especially in the region homologous to the duplication and to a lesser extent in an adjacent region.

Dobzhansky's studies (1934) with X duplications disclosed an unusual aspect of their behavior. While they greatly decrease exchange in homologous regions of normal chromosomes, they themselves are rarely involved in exchange. Consequently, the measurement of the effect of duplications on exchange pairing is assessed indirectly through the reduction in exchange pairing and exchange that they impose on their homologs. As evidenced by the large effects they produce, the ability of very small duplicating fragments to find and recognize homologous regions attests to the extraordinary specificity involved in exchange pairing.

Among the duplications studied by Dobzhansky, three carried approximately the same extent of distal X euchromatin [*yellow* (0.0) to *echinus* (5.5)], but while one was recovered as a free duplication, the other two arose as translocations attached to the distal tip of the second chromosome in one case and of the third chromosome in the other. The amount of crossing-over between the two normal X chromosomes, for the region between yellow and echinus (normally 5.5%), was 5.0% for the free X duplication and 1.6% and 1.3% for the attached duplications. The failure of the free duplication to reduce exchange was pointed out by Morgan to be inconsistent with other results, especially with those of Dobzhansky in which other free X duplications, carrying only a section of the region between *y* and *ec*, effected significant reductions. Morgan (1938) studied the behavior of a duplication for the tip of X, which was attached to a sizable proximal duplication of the X, including fused and all loci to the right of it. She observed a large reduction in crossing-over caused by the duplication and noted that such segments were less effective in reducing exchange in their distal region of homology than were distal segments of comparable length attached either to proximal regions lacking euchromatic loci or to the ends of long autosomes.

E. H. Grell (1964) selected duplications which carried the X euchromatin situated proximal to the locus of Bar, but which were placed at different positions in the genome (Fig. 10). Dp(1;f)B^S exists as a free

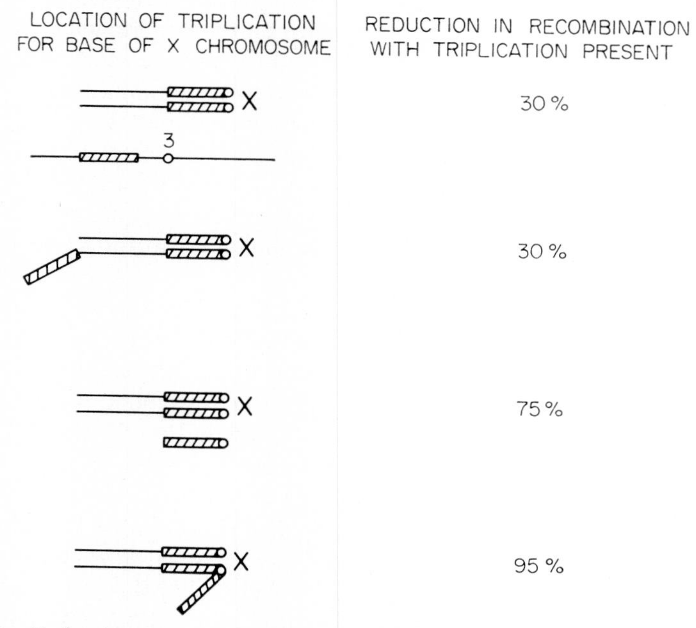

FIG. 10. Reduction in crossing-over in the proximal portion of the X chromosome, correlated with the position in the genome of a duplication for this region. (From data of E. H. Grell, 1964.)

fragment; the TAG chromosome of Lindsley carries the duplication at the distal tip of X; the Doubler chromosome carries it as a proximal appendage; and Dp(1;3)B^{S3} has the duplicated region inserted into the left arm of chromosome three. Crossing-over was measured in the normal X chromosomes between forked and the centromere. In all cases a marked reduction in exchange in the triplicated region is observed, but the extent of reduction depends on the proximity of the duplication to its homologous region. If it is attached to the distal tip of the X chromosome or inserted into the third chromosome, the reduction is only 30%; if it is free in the genome, the reduction is 75%; but if it is attached to the base at X, very close to its normal location, it induces a 95% reduction. Clearly the location of a duplicated segment plays a prime role in its ability to establish a pairing relationship with a homologous segment.

An extra piece of a chromosome, instead of being displaced or free on its own centromere, may occupy a normal location in its chromosome either in tandem or in reverse tandem sequence. The first duplication of this kind to be recognized and studied was one that produces the Bar phenotype (Sturtevant and Morgan, 1923; Sturtevant, 1925). As described by Bridges (1936), the duplicated region, present in tandem, includes six bands between 16A1–16A6. Unequal crossing-over occurs infrequently within the duplicated regions in *B/B* females, leading to individuals that carry one copy of the region and are normal in phenotype, and to individuals that carry the region in triplicate and are more extreme in phenotype. Bridges (1935) has suggested that "repeats," as tandem duplications have been called, may be the raw material from which new genes arise. Other repeats in *Drosophila* are responsible for the Hairy Wing phenotype, the Beadex phenotype and the Confluens phenotype, while the t locus in the mouse, the β and δ hemoglobin loci in man, and the Q locus in *Neurospora* are considered to also represent examples of tandem duplications.

Tandem duplications are of particular interest in competitive studies, for the extra segments are located in chromosomes of approximately normal length and at their normal position. Green (1962) studied the effects of three tandem duplications of the X chromosome, Bar, Beadex, and Dp(1)*z-w* for their effects on crossing-over in their environs. Each duplication was investigated in the heterozygous and homozygous condition and Beadex was studied as a homozygous triplication and quadruplication as well. Heterozygotes for a duplication of the region between zeste and white gave the same frequency of crossing-over between yellow and white as a control lacking the duplication, although both values were somewhat lower than standard map values; the heterozygotes for

Bx and for *B,* for which there were no controls, gave values slightly lower than standard. In all cases the values for the homozygous duplications were greater than those for their respective heterozygotes and in excess of the additional genetic length introduced by the duplications. The homozygous triplication and quadruplication for *Bx* gave values which were alike, and which were greater than that of the homozygous duplication. Green interprets the increase in terms of the effective pairing hypothesis of Pritchard. If it is assumed that small, duplicated regions adjacent to one another increase the probability for establishment of effective pairing of the segment (and this may not be a valid extension of the hypothesis), then it becomes difficult to account for the absence of an increase in exchange with the heterozygous duplications or with a quadruplication as compared with a triplication. In any case, it would seem that a small duplication in its normal position is capable of both exchange pairing and exchange. The great depression in crossing-over induced by free or displaced duplications may be attributable to their relatively unimpaired ability to establish a pairing relationship with a homologous segment but to their inability to undergo exchange.

The largest tandem duplication studied by Green represented a genetic length of 0.7 crossover units. Roberts (1966a) has described the behavior of a tandem duplication, located in the left arm of chromosome 2, which extends from 26A to 28E of the salivary chromosome map and probably corresponds to a genetic length of 5–10 crossover units (Roberts, 1966b). In the heterozygous and homogygous conditions this duplication reduces crossing-over in 2L to ~ 1/6 and 2/5 of normal, respectively. This drastic reduction with a large tandem duplication stands in marked contrast to the increase in recombination with small, homozygous, tandem duplications. Roberts believes that the greater length permits intrachromosomal pairing between duplicate regions which, in turn, hinders pairing between homologs.

d. *Size of a Triplicated Region.* Dobzhansky (1934) and Morgan (1938) recognized that the ability of a triplicated segment to reduce exchange is related to its size. For the case in which the entire genome is present in triplicate (i.e., a triploid), the generalization does not hold, since, as indicated above (Section II,A,3,a), the total amount of crossing-over in triploids may actually exceed that in the corresponding diploid. On the other hand, Beadle and Ephrussi's meager data (1937) for a trisomic of the X chromosome in *Drosophila* suggest that the presence of an extra homolog drastically reduces exchange, although this result may be a consequence of chromosomal imbalance.

A study has been made of crossing-over between two closely linked X chromosome markers, yellow (0.0) and white (1.5), in the presence of free X duplications which carry various lengths of the yellow to white region as well as with one carrying the entire region and a portion of the adjacent region (Fig. 11) (R. F. Grell, 1967). If the distance between *y* and *w* is arbitrarily given a value of 1, the cytological length of euchromatin carried by each duplication (based on salivary chromosome length) is shown in Table III, column 3, and the genetic length in Table III, column 4. The percentage of reduction between *y* and *w* effected by each duplication is shown in Table III, column 6. It is evident that the

TABLE III

CROSSING-OVER *y*-*w* WITH DIFFERENT LENGTHS OF TRIPLICATIONS FOR THE REGION[a]

Duplication	Total length of Duplication	Cytological length of euchromatin[b] (y-w = 1)	Genetic length of euchromatin [y-w(1.5) = 1]	Observed: Crossing-over [y-w(%)]	Observed: Crossover reduction [y-w(%)]
None	—	—	—	1.57 ± 0.12	—
1144	1.1	0.06	< 0.01	1.41 ± 0.18	11
856	3.0	0.21	0.07	1.21 ± 0.13	23
1337	1.4	0.44	0.20	0.85 ± 0.10	46
w^{m5}	1.9	1.00	1.00	0.54 ± 0.14	66
z^{9}	3.8	1.23	4.6	0.51 ± 0.10	68

[a] From R. F. Grell (1967).

[b] Based on salivary gland X-chromosome map of Bridges.

greater the amount of euchromatin carried by the duplication the greater the reduction in exchange. The data suggest that the reduction shows a better correspondence with cytological length of euchromatin than with the genetic length (column 3 versus column 4). The total length of the duplications varies from 1.1 to 3.8 times the length of chromosome 4 (Table III, column 2). The difference between total length and euchromatic length represents the amount of heterochromatin of the X base or, in the case of w^{m5}, of the fourth chromosome base present in the duplication. A comparison of columns 2 and 6, Table III, shows that the reduction in exchange is not affected by the extent of heterochromatin present in the duplication but only by the length of euchromatin.

A maximal reduction of two thirds appears to be reached when the entire *y*-*w* segment is present in triplicate as with w^{m5}. The same reduction frequency is observed when adjacent loci are also present, since the z^{9} duplication extends from *yellow* to *diminutive* (at 4.6 crossover

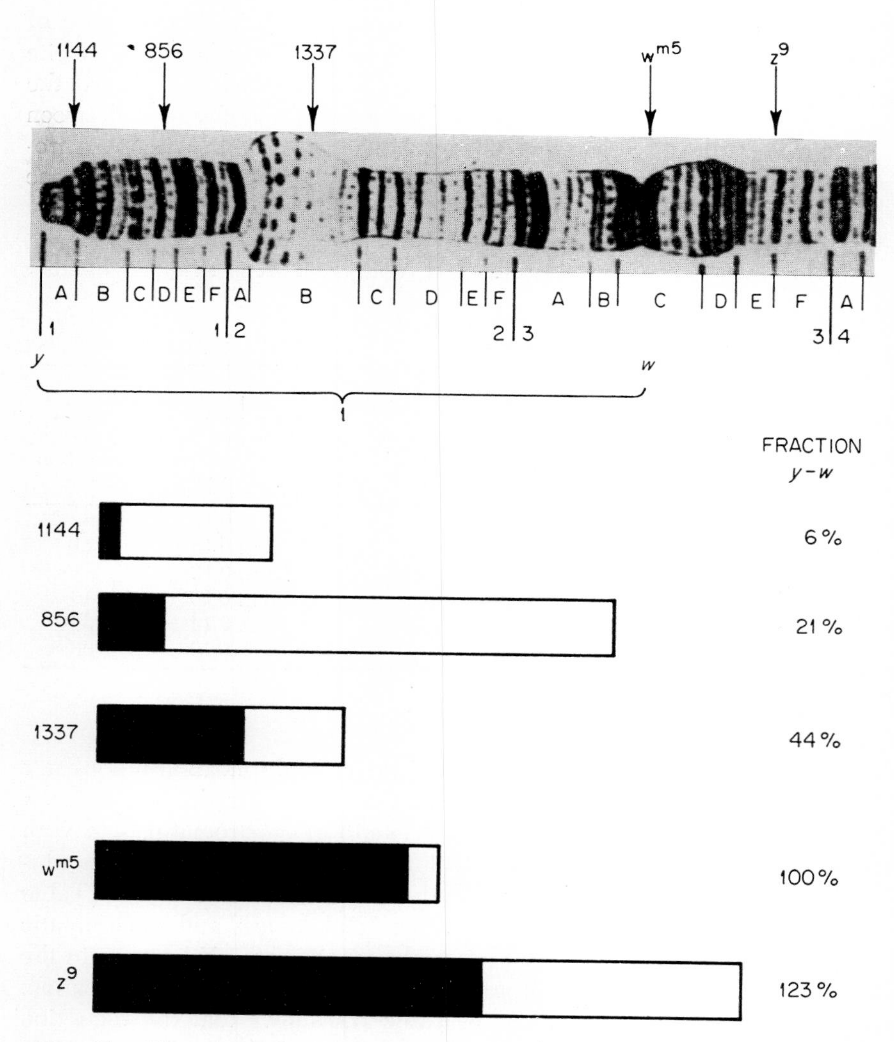

FIG. 11. The extent of X euchromatin in five free X duplications. Dark region, euchromatin; white region, heterochromatin. The diagram accurately portrays the differences in euchromatic content and in total length among the five duplications. The fraction of heterochromatin within each duplication has been decreased for illustrative purposes. (From R. F. Grell, 1967.)

units from yellow). If it is assumed that exchange pairing is two-by-two, that pairing occurs randomly among the three elements, and that the duplicating fragments are capable of pairing but not of exchange, a two-thirds reduction conforms to the theoretical expectation from a continuous pairing model; whereas the discontinuous model would predict unaltered crossover values for duplications that undergo exchange pairing but not exchange (see Section II,A,2,e,i). The achievement of the theoretical two-thirds reduction by w^{m5} and z^9 suggests that pairing does occur randomly between the duplication and the two normal homologs. This would mean that the extent of homology represented by the region from *y* to *w* is sufficient, in this system, for recognition as complete as that achieved by the same region when present in a normal chromosome; and that the minimal requirement in terms of euchromatic length for such recognition may, in fact, be less than the *y* to *w* region.

Random pairing, as determined cytologically by the frequency of trivalent formation at pachynema, was reported by Maguire (1961) for a maize segment present in triplicate. E. H. Grell (1964) observed, however, a reduction of 75% with Dp(1;f)B^S, and should this value be significantly different than 67%, two thirds may not represent a maximum. In cases in which the duplication has a position advantage, as it does with Doubler (Fig. 10), a reduction greatly in excess of two thirds may be observed.

4. *Asymmetrical Pairing*

a. *Origin of Duplications and Deficiencies from Asymmetrical Pairing.* Duplications of gene sequences, arising from some rare event, are thought to constitute the material from which new genes evolve. Initially the duplicated region would be identical to the original, but with time it would diverge. The divergence might be greater in some portions of the duplication than in others, so that a region of quasi-homology might become separated from the original segment by a region which no longer retains genetic or functional homology. Spatial separation might also occur through subsequent rearrangements. When a duplicated region exists that is no longer contiguous with the region from which it arose, and is still capable of undergoing exchange with it, asymmetrical pairing and exchange between the two will lead to a duplication for or to a loss of the region which separates them. Thus, in contrast to unequal crossing-over between contiguous tandem duplications, which can only restore the normal genome or increase the number of doses of the duplicated region, asymmetrical exchange between noncontiguous duplications will produce deficiencies that are frequently lethal and may also constitute that rare event by which a new duplication arises.

The detection of noncontiguous duplications depends on the recognition of the abnormal products that result from the asymmetrical exchange. In many cases the duplication does not alter the phenotype, and it is more often the deficiency that is recognized through its lethality. One of the first cases of reciprocal deficiency-duplication products was reported by Catcheside (1947) in *Oenothera blandina* for plants which showed a variegated-type position effect for certain genes involved in an interchange. Among the progeny of the interchange plants, a fraction possessed a new phenotype, and pollen carrying the abnormality was lethal. Genetic analysis revealed that the new phenotype was associated with the presence of a direct, contiguous duplication. A number of independent occurrences of both the duplication and the reciprocal deficiency product indicated that both must arise by asymmetrical interchange between two noncontiguous regions in the same chromosome arm.

Green's studies (1959b, 1961) with *Drosophila* have shown that a similar situation exists for the zeste locus, located at band 3A3, and the white locus, located at one or more bands in the 3C region. White is a pseudoallelic series consisting of a minimum of 5 thus far recombinationally indivisible loci numbered 1, 2, 3, 4, 5 from left to right and probably associated with salivary gland chromosome band 3C2. Gans' remarkable investigation (1953) of the zeste (z) locus had revealed that a functional relationship exists between zeste and white, although the genes are separated by about ten bands. Gans found that mutant alleles, subsequently localized at locus 4 of the white region, acted as dominant suppressors of zeste, and Green (1959a) has since shown that alleles at locus 5 have a similar property. In crosses of females carrying two different white-apricot (w^{a2}/w^{a}) alleles to white (w) males, Green regularly recovered exceptional white crossover products that were male lethal. Cytological examination of the salivary gland chromosomes showed a deficiency for most of 3A and all of 3B that presumably included bands 3A4–3C1. Subsequently, Green recovered a tandem duplication which apparently extended from bands 3A4 to 3C1 and which was reciprocal to the previously reported deficiency.

Green proposed that pairing and exchange between bands 3A3,4 (the z locus) and 3C1,2 (the w locus) of homologous chromosomes would yield both products, thus:

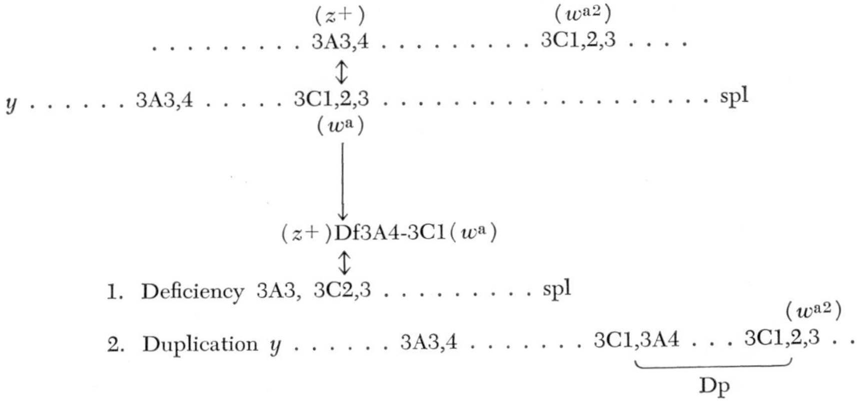

```
                    (z+)                 (wa2)
        . . . . . . . . . 3A3,4 . . . . . . . . . 3C1,2,3 . . . .
                      ↕
y . . . . . . 3A3,4 . . . . . 3C1,2,3 . . . . . . . . . . . . . . . . . . . spl
                    (wa)
                      |
                      ↓
               (z+)Df3A4-3C1(wa)
                      ↕
    1. Deficiency 3A3, 3C2,3 . . . . . . . . . spl
                                                                (wa2)
    2. Duplication y . . . . . . 3A3,4 . . . . . . . 3C1,3A4 . . . 3C1,2,3 . .
                                                     └──────────────┘
                                                           Dp
```

There is good evidence that the particular alleles present at the zeste and white loci in the cross which produced the duplication and deficiency may possess greater homology for each other than do other alleles. First, the pairing situation leading to the event appears to be a fixed one. On the basis of the outside markers recovered in association with the two products, pairing occurs specifically between bands 3C1,2 of one chromosome and 3A3,4 of the other. Substitution of the *z* mutant or the z^+ allele from either a Canton-S or Oregon strain for the original z^+ allele reduces the frequency of the event and results in a switch in the marker association. Similarly, substitution of a different w^a allele at the 3C1,3 locus, but not of a different allele at the w^{a2} locus of the homolog, decreases the frequency.

In a similar study, Judd (1961a), using females carrying w^a in trans-configuration with w^a, $+$, w^{bf} and w^{Bwx}, recovered three different types of deficiency products and one duplication product. Genetic and cytological analyses indicated that a different kind of asymmetrical pairing and exchange was responsible for each type. The specific bands involved are postulated to be 3A2,3 with 3C1,2, 3C1,2 with 3D2,3, and 3A7,8 with 3C1,2. In confirmation of Green's results, only a single type of asymmetrical exchange was found to occur in any given cross, indicating that the asymmetrical pairing is specific and requires a considerable extent of shared homology. Green's original description of such events as "nonhomologous pairing and crossing-over" would seem to be a misnomer, as he points out (1961).

Reciprocal deficiency-duplication crossover products were also recovered from w^{a4}/w^a females (Green, 1959c, 1963); but in this case the proof of their nature rested upon genetic grounds, since cytological examination disclosed no aberration. Since w^+ recombinants recovered from females w^a/w^{ch} (but not w^{a4}/w^{ch}) were as effective as w^{a4}/w^a in producing duplication-deficiency products, Green concluded that the w^a locus itself is not causally associated with the event but rather that genetic factors lying immediately to the right of the w^a locus are responsible. The evidence indicates that a noncontiguous duplication is present within the white locus. The deficiency product, which results from asymmetrical exchange, is not lethal but males carrying it are white-eyed. The segment separating the duplicated region must be much shorter than that separating *z* and *w* and, unlike the latter, must not be necessary for survival. Judd (1961b) achieved the simultaneous recovery of both products of an asymmetrical exchange within the white region by utilizing an attached-X chromosome carrying *z* w^a in one arm and w^{bf} in the other. The recognition of the reciprocal exchange product apparently depends on the interaction of a tandem duplication in one chromosome with the *z* mutant in the reciprocal deficiency chromosome so that a white eye is produced in place of the "light buff" eye which would be present if the fly were homozygous for z^+. The ability of the duplication to act as an enhancer of zeste indicates that it too belongs to locus 4 or 5 in the white series.

There seems to be little doubt that asymmetrical pairing and exchange could provide the source of much variability in the germ plasm and thus be of profound importance in evolution. It has further been suggested that it might play a role in the mutation of an established gene. Sax (1931), noting a high correlation between crossover frequency and mutations in different parts of the *Drosophila* genome, suggested that one mechanism for mutation might be unequal crossing-over. Since euchromatic regions, in which most genes lie, normally undergo crossing-over, while heterochromatic regions, which possess few detectable genes, cross over infrequently (Brown, 1940; Baker, 1958; Roberts, 1965), the correlation between high crossover frequency and mutation is undoubtedly the consequence of the clustering of genes in certain regions. Nevertheless, the possibility of spontaneous mutation arising through exchange is an attractive one and one for which the following experimental evidence exists.

b. *The "Selfing" Phenomenon.* The phenomenon of "selfing" in *Salmonella* was discovered by Demerec (1962). Auxotrophic strains transduced with homologous phage (i.e., phage grown in the same strain) were found to give rise to significantly larger numbers of wild-type variants than occur by spontaneous reversion in uninfected cells. A "selfer" may thus be defined as a mutant which gives rise to wild types in mating with itself. Originally, Demerec supposed that the excess wild types were the result of unequal crossing-over between adjacent nucleotides during transduction. Subsequent tests (Demerec, 1963) utilizing phage grown on bacteria carrying a deletion for the mutant region also produced the excess wild-type colonies and led Demerec to relinquish his original explanation. In its place, he proposed that synapsis of the transducing fragment with the homologous region of the bacterial chromosome in the neighborhood of the selfer gene stimulated mutation of the gene.

Clark (1964) suggested as still another alternative that recurrent nucleotide sequences in the bacterial genome provide the basis for replacing the mutant sequence with the wild type. Recurrent sequences could represent either convergent evolution of certain sequences which were selected because of special functional utility, such as the active site of certain enzymes, or they could represent vestiges of similar sequences remaining after the divergence of two originally identical genes. While both of Demerec's explanations assumed that a fragment must be introduced which is derived from a region at or close to the selfer mutation, Clark's hypothesis permits the fragment to be derived from any portion of the genome that carries the recurrent sequence which might represent as few as 12 or 15 nucleotides.

Clark's suggestion that selfing occurs through a recombinational event which replaces a mutant sequence with a wild-type sequence, taken from elsewhere in the genome, is an appealing one. *A priori*, however, the possibility of short sequences of nucleotides, such as those which might be held in common by the active sites of certain enzymes, undergoing pairing and double exchange in a foreign region would seem inimical to the preservation of the integrity of the DNA molecule. The number of nucleotides required for recognition at exchange pairing is unknown, but transformation studies suggest that it may be fairly large (see Section II,A,5,a). Moreover, the selfing phenomenon appears to be limited to certain loci and to be a function of the location of the site within the locus (Demerec, 1963). An extensive genetic analysis by Hartman and his associates (reported in Demerec, 1963) of 8 loci in the *his* operon including more than 100 mutants studied with homologous phage and 32 mutants studied with deletion donors, uncovered only one selfer site. In addition, Demerec's studies using deletion strains as donors with a number of *arg* and *cys* selfer mutants have demonstrated that not only does selfing always occur with the deletion but that in some cases deletions produce an excess of selfers. It is highly significant that in the case of 1 *arg*, 5 *cysH* and the single *his* site, mutants which failed lf with themselves acted as selfers with deletions. Clark proposes this latter result is accounted for if it is assumed that a new nucleotide sequence is formed through the restoration of the continuity of the DNA after deletion has occurred. The new order would then provide the wild-type sequence for the nonselfer mutant. This explanation seems implausible, particularly in view of the frequency with which this situation has been observed.

Another explanation (R. F. Grell, 1964c) is afforded from the results of a study by Demerec *et al.* (1963) of five *cys* genes (*C,D,H,I,J*) which are present as a cluster on the *Salmonella* chromosome. A long "silent section" occurs between *D* and *H* and a short one between *H* and *I*. Eighty-two long deletions, called "ditto deletions" of similar extent and covering *H*, *I*, and *J*, have been recovered for the region. All of the ditto deletions have one end in the longer of the two silent regions. Demerec *et al.* postulated that the long silent region is a duplication of a section beyond *cysJ* and that during replication the duplicated regions synapse to form a loop, with the material within the loop being excluded so as to produce the recurring ditto deletions. If a duplicated region with quasihomology for *cysH* lies in the vicinity of the *cys* cluster, the higher frequency of selfers with the deletion donors may be explained by competitive exchange pairing. When homologous phage is used, synapsis during transduction might generally proceed in a strictly homologous fashion and only rarely would the quasihomologous region pair with the *cysH* region to produce selfing; with phage carrying a deletion for *cysH*, the opportunity for such pairing would be increased. Similarly, for mutants which fail to self with themselves but do so with a deletion, it might be postulated that in these cases the opportunity for quasihomologous pairing only exists when the strictly homologous region is absent. The difference between the two types could depend on the extent of homology retained in the duplication. Loci which fail to self, such as the *his* loci, would represent those for which duplicated regions, retaining sufficient homology to permit pairing and exchange, are absent from the genome. The explanation offered here differs from that of Clark's in the greater specificity (i.e., number of nucleotides in common) that is presumed to be a prerequisite for synapsis which can lead to exchange and in the infrequency that such critical recognition lengths are assumed to be duplicated in the genome.

c. *Mutation in Eucaryotes.* Mutation associated with exchange has been reported in a number of eucaryotes. Diploid strains of *Saccharomyces cerevisiae,* homozygous for certain biochemical mutants, have reversion rates in meiosis that are higher than those in mitosis (Magni and von Borstel, 1962), and the majority of meiotic revertants are associated with crossing-over in the region of the reverted gene. Kiritani (1962) has presented evidence that mitotically stable isoleucine-valine mutants in *Neurospora* undergo reversions at higher rates in selfing crosses and that the revertants show a much higher frequency of recombination of outside markers than do controls. New alleles at the T locus in the mouse appear spontaneously with a frequency of $\sim$ 1/500, and the mutational events are usually associated with exchange of a neighboring outside marker (Bennett and Dunn, 1964; Dunn, 1964). Asymmetrical exchange between neighboring and quasihomologous regions seems a distinct possibility in all of these cases. A selfing mechanism during meiosis presents one complication that is not present with transduction where it is possible to imagine a simple replacement of a segment by a double exchange between the transducing fragment and the bacterial chromosome. In meiosis, as noted earlier, asymmetrical exchange between noncontiguous, quasihomologous genes results in an increase or decrease of the genetic material. When such products are not at a disadvantage, a selfing mechanism could well operate.

In man, the genes specifying the β- and δ-chains of hemoglobin are linked, and amino acid sequence analysis has shown that they possess $\sim$ 95% of their residues in common. Individuals homozygous for an abnormal hemoglobin, Hb-Lepore, lack both the normal Hb-A ($\alpha + \beta$) and the minor component Hb-A_2 ($\alpha + \delta$) but possess a new type of hemoglobin with an unaltered α-chain and a second chain which has characteristics of both the β- and δ-chains. Illegitimate crossing-over between corresponding points of the β- and δ-genes is postulated by Baglioni (1962) to account for the formation of a hybrid β δ-gene with some accompanying loss of genetic material. A reciprocal kind of event (i.e., unequal crossing-over with a gain of genetic material) has been suggested by Smithies *et al.* (1962) as the mechanism responsible for the duplication in length of one peptide chain of the human haptoglobins.

5. *Control of Exchange Pairing*

a. *Sequential Homology.* Control of exchange pairing is mediated in two distinct ways. First, the primary requirement for the establishment of intimate synapsis must be sequences of nucleotides held in common by chromosomes. The most feasible mechanism for ensuring precision in alignment would seem to be complementary base pairing between single strands of DNA. This may be considered to occur after DNA replication is completed, so that only repair DNA synthesis is involved in exchange

(Whitehouse, 1963; Whitehouse and Hasting, 1965; Holliday, 1964), or it might precede replication, in which cases the four strands participating in exchange could be the single strands comprising the two parental double helices. The recent finding that, after mating, the Hfr chromosome is recovered in minicells as a single strand is consistent with the latter interpretation (Cohen *et al.*, 1968). Bodmer (1965), extrapolating from a model that he has proposed for transformation, suggests that homologous chromosomes, replicating synchronously through the same replicating point during meiosis, would provide a natural structure at which crossing-over could occur by breakage and reunion with subsequent DNA synthesis. Complementary base pairing between homologous single strands would then be analogous to a region of effective pairing.

One difficulty with this model is that it fails to provide a way in which 3-strand double exchanges can occur. This shortcoming could be overcome in the following way. Let us assume that a single strand is composed of units which may reverse their polarity at some unusual linkage point. Then, for a particular effectively paired segment, multiple crossovers could include the 2-strand double or 4-strand double varieties, whereas a 3-strand double would necessarily involve two regions of effective pairing. Switches in polarity could conceivably constitute obligatory termination points of effectively paired segments, and extrapolating from transcription evidence, end points would be permissible between but not within operons. A single-stranded recombination model provides a possible molecular explanation for the restriction of exchange to nonsister strands, a restriction difficult to explain if crossing-over occurs between four Watson-Crick double helices.

The minimal number of nucleotides required for homologous recognition is not known. Possibly the requirement may vary from one organism to another. If the number for both recognition and exchange were to be very small, constancy in the composition of the genetic material probably could not be maintained. On the other hand, the phenomena of transduction and transformation, as well as the ability of a small duplicated segment drastically to reduce exchange between normal homologs in *Drosophila* (Section II,A,3,c) and even, though rarely, to engage in exchange, demonstrate that recognition and exchange do not require parasynapsis of the entire chromosome. In the case of the duplication, it seems possible to draw a distinction between the recognition and the exchange process, for the discrepancy in the frequencies between the two introduces the possibility that recognition requires fewer nucleotides than exchange, although other explanations are possible.

Evidence from transformation experiments with *Bacillus subtilis* (Bodmer, 1966), with pneumococcus (Fox, 1966), and with *Hemophilus influenzae* (Notani and Goodgal, 1966) gives provisional estimates for the size of the integrated region of ~ 1500, 2000, and 20,000 nucleotides, respectively. In no case has it been shown for certain that the amount of integrated DNA, as measured biochemically, is in one continuous piece. Single-strand breaks produced in transforming DNA by deoxyribonuclease (DNase) were found to inhibit transformation to a greater extent than integration, indicating that not all integrated DNA produces transformants (Bodmer, 1966). In the T4 DNA transformation system, the critical recognition length was estimated to be ~ 25 nucleotides (Bautz and Bautz, 1967).

Large, heterozygously inverted chromosomal segments are known to inhibit pairing and exchange and, as such, have been successfully exploited as balancers to maintain certain mutations in a heterozygous condition. The effect of an accumulation of small differences in nucleotide sequence (such as is presumed to exist between closely related bacterial genera) on genetic recognition and exchange, has been the subject of a series of elegant investigations by Ravin and his co-workers. A comparison of interspecific and intraspecific transformation, utilizing certain strains of *Pneumococcus* and *Streptococcus*, has shown that homospecific transformations occur with greater efficiency than the heterospecific ones, and that the difference in efficiency is entirely attributable to difficulties following uptake of DNA into the host of foreign species (Chen and Ravin, 1966a). The integration efficiency of a marker (defined as the ratio of the yield of transformants bearing a specified marker to that of transformants bearing some standard reference marker) varies between allelic markers but remains a characteristic feature of the marker and one which is independent of the concentration of the DNA used for transformation (Chen and Ravin, 1966b). When reciprocal heterospecific transformations were compared with the corresponding homospecific transformations, it was found that linkage was reduced in the former case. The reduction occurred for weakly linked loci as well as for closely linked markers within a single locus, although the reduction was greater in the former instance (Ravin and Chen, 1967). Both the inefficiency of heterospecific transformation and the reduction in linkage associated with such transformation have been interpreted by Ravin and Chen (1967) as a consequence of incomplete homology in the regions undergoing recombination, leading in turn to a reduction in the average amount of DNA integrated into the recipient genome.

In eucaryotes, one way that incomplete homology of this kind can be studied is through the synthesis of genomes carrying homologous regions of identical origin, in one of which new marker mutants have been induced. A comparison of recombination frequencies in this situation with one in which the two homologs are of diverse origin should provide information about the effectiveness of isosequentiality versus heterozygosity in recombination. Work relevant to this problem has been carried out in *Neurospora crassa* by de Serres (1958, 1962, 1967). He has studied the effect of differences in the genetic composition of homologous chromosomes in ~ uniform background, on map distances between close markers, as well as on the patterns of chromosome interference, by comparing intergenic recombination in crosses where markers had originated in the same strain with crosses having markers that arose in different strains. Isosequentiality was found to be associated with low map distances and negative interference for double exchange events in immediately adjacent regions (I, II and II, III doubles), but positive interference was found for double exchanges in nonadjacent regions (I, III doubles). By contrast, heterozygosity gave high map distances and positive chromosome interference for all double exchange events. The manner by which isosequentiality leads concomitantly to reduced exchange and to negative interference, and heterozygosity to increased exchange remains a matter for conjecture.

b. *Genic Control.* Second, the control of pairing may be mediated through specific genes. When the frequency of recombination per nucleotide pair is compared for phage, bacteria, *Aspergillus, Drosophila,* and the mouse (Hayes, 1964, p. 343), it becomes evident that recombination has been greatly curtailed during evolution. The frequency per nucleotide pair is ~ 1000 times greater for phage and ~ 100 times greater for bacteria than for *Drosophila.* Curiously, it has been found that alterations in temperature in either direction from a certain point will increase exchange in *Drosophila* (Plough, 1917), in *Neurospora* (Towe and Stadler, 1964; McNelly-Ingle *et al.,* 1966), and in *Sphaerocarpus* (Abel, 1964). This result is not characteristic of any simple enzyme system where a $Q_{10} = 2$ might be expected for every 10°C increase in temperature. The control is perhaps most easily visualized as acting through gene products which suppress recombination rather than through gene products participating in the process.

Experimentally, it is difficult to distinguish between control exerted via pairing and control via exchange. As one possibility, a reduction in exchange may be considered to be accomplished by a genetic restriction in pairing. If exchange occurs at discontinuously paired sites, the amount of exchange could be reduced through a reduction in the number or in the length of such sites.

That control may be exerted genetically is demonstrated by mutant genes which affect pairing or exchange. An extreme case is known in *Drosophila* in which the mutant *c3G*, in homozygous condition, virtually eliminates meiotic exchange. Meyer (1964) reports that oocytes from females homozygous for *c3G* show no synaptinemal complex, suggesting that the effect of the mutant may be upon exchange pairing.

Interestingly, C. W. Hinton (1966) has found that recombination in females heterozygous for the mutant *c3G* is significantly increased over normal. The difference in behavior of this gene in heterozygous and homozygous condition could provide a model to explain the reduction in chiasma frequency which has been observed when certain normally outbreeding species are inbred. For instance, *Secale cereale* shows reduced chiasmata upon inbreeding (Lamm, 1936), and intercrossing of inbred lines produces progeny which show higher chiasma frequencies than either of the parents (Rees and Thompson, 1956). Apparently the genetic system of this outbreeding species operates most efficiently with heterozygosity of genes, presumably less drastic in their effects than c3G, but perhaps of a similar nature. By contrast, *Neurospora crassa*, which also represents an outbreeding species, has been found by Stadler and Towe (1962) and by Cameron *et al.* (1966) to show increased crossing-over and decreased interference with successive backcrossing. Cameron *et al.* suggest that homozygosity may lead to an increased general effectiveness of pairing on a chromosomal level, and that the results of de Serres (see above) might be expected if local heterozygosity within a particular region that is paired stimulates crossing-over there.

A number of mutant genes have been classified as "asynaptic," but whether they eliminate pairing as the name implies, whether they prevent exchange between paired homologs so that bivalents fall apart, or whether they act precociously to resolve chiasmata, is not known. Any interpretation of their behavior is, of course, intimately related to the unresolved question of when pairing is initiated and when crossing-over occurs. With other genes termed "desynaptic," less ambiguity exists since homologous pairing is observed up to a particular point in prophase, after which homologs fall apart. Genetic analyses would be required to determine if the cause of desynapsis is failure of chiasma formation or early resolution of chiasmata.

Studies on common wheat, *Triticum aestivum* (Sears and Okamoto, 1959; Riley and Chapman, 1958), have shown in a very convincing way that in this species pairing is under genic control. Wheat carries the full diploid complement of three closely related species, *Triticum monococcum* (A genome), *Aegilops speltoides* (B genome), and *Aegilops squarrosa* (D genome). Since each species contributes 7 pairs, wheat possesses 42 chromosomes comprising 7 groups of 6 related chromosomes. Among a group of 6, each chromosome has 1 homolog and 4 homeologs. Despite the close relationship of all 6 chromosomes, pairing is restricted entirely to homologs. Furthermore, in the 21 chromosome haploid, in which homologs are absent, most chromosomes remain univalents (Okamoto and Sears, 1962). This behavior is in striking contrast to the high frequency of homeologous pairing that is observed when the three ancestors are crossed in any combination to form 14 chromosome hybrids (Riley and Chapman, 1958). The failure of pairing in the haploid is not due to structural divergences from the ancestral type, since hybrids between *T. aestivum* and either *T. monococcum* or *A. squarrosa* show regular homologous pairing (Riley, 1965). The restriction of pairing to homologs was found to be under the control of the long arm of chromosome 5B (Sears and Okamoto, 1959; Riley and Chapman, 1958). In the absence of the $5B^L$ chromosome, homeologous pairing occurs. Since the ancestral types do not possess a system to suppress pairing between homeologs, the gene in 5B is assumed to have arisen after polyploidization. Dosage studies (Feldman, 1966) have shown that while pairing at first meiotic prophase remains normal with four doses of $5B^L$, six doses permit pairing between homeologs and, in addition, induce asynapses of homologs and a high frequency of interlocking bivalents. In the hexaploid species of oats, *Avena byzantina*, genetic control of homologous pairing has been localized to specific chromosomes (Singh and Wallace, 1967).

Gerstel (1961) has looked for similar control in *Nicotiana tabacum*. Using as one parent a stock of *N. tabacum* which carried one substituted pair of chromosomes from the taxonomically distant species *N. ghitinosa* and as the other parent, *N. tomentosiformis*, a closely related species to *N. tabacum*, Gerstel was able to determine whether in the amphiploid, preferential pairing is under genic control as it is in wheat. Segregants for different duplex loci in the amphiploid gave a characteristic gametic output of 3:1, indicating absence of differential affinity, but for a factor on the substituted chromosome the output was found to be 59:1. Apparently, in this case preferential pairing is determined by chromosome homologies and is not under genic control. Gerstel postulates that in tobacco there was no need for a genetic control to have arisen, since amphiploids between species of different ancestral types already exhibit fairly regular bivalent pairing (Greenleaf, 1942).

One can only speculate about the mechanism responsible for isolating homeologs during exchange pairing. Menzel and Price (1966) have made the very attractive suggestion that, since the entire genome is affected simultaneously, the control might be exerted through asynchrony of the pairing process in the different genomes. If the chromosomes of each set became available for pairing at a time when the chromosomes of the other sets are not, homeologous pairing would be averted.

B. Distributive Pairing

1. *Definition*

In its present context, distributive pairing is defined as any achiasmate association or coorientation of chromosomes that is responsible for their segregation pattern. In its original context, based on studies of the *D. melanogaster* female, the term distributive pairing had a more restricted usage, namely, that of a postexchange pairing between noncrossover chromosomes which determined their manner of segregation. Subsequent studies with the *melanogaster* female have shown that the specificity for this pairing is based on chromosome size and is independent of of homology, thus permitting nonhomologs to pair and segregate regularly. Whether the *melanogaster* female type of distributive pairing occurs in other organisms remains problematical, since too little is known of the genetic details in most forms. The question arose, therefore, as to whether cytological descriptions of pairing events, which might conform to the *melanogaster* female type but in which the cogent details were missing, should be included under distributive pairing.

Also, it seemed desirable, in view of the functional approach that has been adopted, that all kinds of pairing serving primarily a segregational purpose be included within a single category. Among these, however, one encounters forms of segregational pairing clearly not of the *melanogaster* female type. For instance, in the female of *Bombyx*, the specificity is not one of size but rather of homology; the term postexchange pairing is inappropriate because crossing-over is absent throughout the genome; and the telosynaptic pairing found at metaphase I does not constitute an independent second pairing but rather appears to be a continuation of a parasynaptic pairing established prior to pachynema.

The choice lay, then, between devising a new term to cover all achiasmate, segregational, pairing other than that found in the female of *D. melanogaster* or broadening the original definition. The latter choice seemed to permit a more unified approach and has been adopted. The category of distributive pairing includes all cases in which the absence or precocious resolution of chiasmata is the rule for certain or for all of the chromosomes of the genome and in which some device has evolved to ensure the regular segregation of such chromosomes. The postexchange pairing which has been characterized for the female of *D. melanogaster* will be designated the *melanogaster* female type of distributive pairing.

2. *Distributive Pairing of the Entire Genome*

a. *Following Desynapsis.* Agar (1911) described spermatogenesis in *Lepidosiren paradoxa,* a dipnoan restricted to the Amazon and La Plata Rivers of South America. The diploid chromosome number of 38 re-

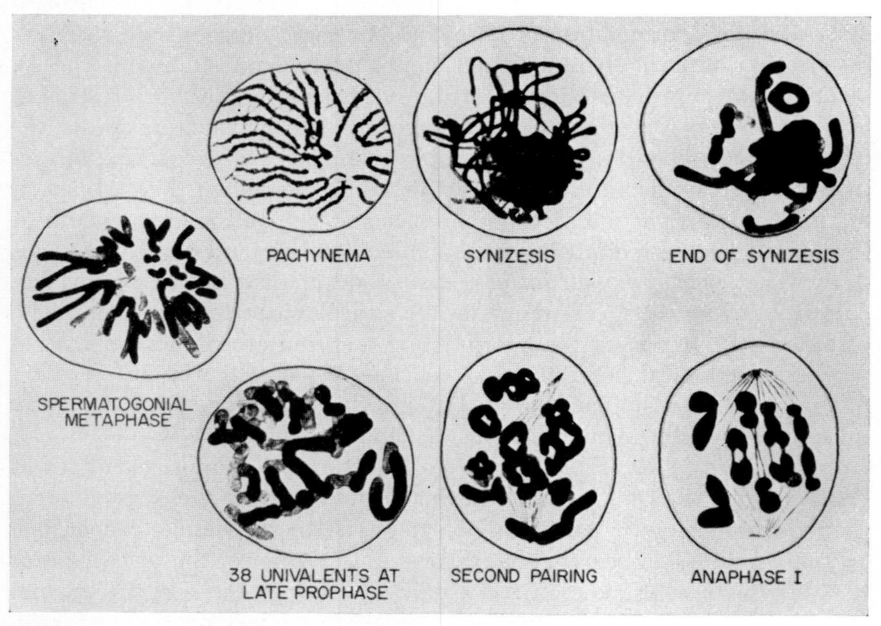

FIG. 12. Spermatogenesis in the dipnoan *Lepidosiren paradoxa,* illustrating an early and a late pairing. (After Agar, 1911.)

ported by Agar has recently been confirmed by Ohno and Atkins (1966). The course of spermatogenesis is shown in Fig. 12. At pachynema, 19 loops are present, each representing a pair of closely synapsed homologs; next, the chromosomes enter a contraction stage or modified diplonema from which there eventually emerge 38 univalent chromosomes. After spindle formation, and as the homologs move toward the equational plate, they pair for a second time and assume their places on the plate as bivalents. At anaphase I the members of each bivalent move to opposite poles. Similarly, in the elasmobranch *Pristiurus,* Rückert (1892, 1893) described a typical parasynapsis in the early meiotic stages, a subsequent separation into the diploid number in the middle period, and a return to the haploid number at a later period. Henking (1891) found a diakinetic conjugation in the hemipteran *Pyrrhocoris* after the second contraction figure, and Gross (1904) has described a similar situation in the closely related genus *Syromastis*.

There are two possible explanations for the phenomenon described above. Either crossing-over is absent in these forms, or chiasmata are prematurely resolved. In either case, the second or distributive pairing is responsible for the regular segregation of homologs.

b. *Without Desynapsis.* In certain animals, but not in plants as far as is known, an achiasmate type of meiosis occurs in one or the other of the sexes. Spermatogenesis without visible chiasmata has been reported for some of the mantids (White, 1938), certain grasshoppers (White, 1965a,b), roaches (Suomalainen, 1946), Mecoptera (Ullerich, 1961), scorpions (Piza, 1943), mites (Keyl, 1957), the Nematocera among the lower Diptera, and in nearly all of the higher Diptera with the exception of the family Phoridae. Oogenesis without chiasmata has been observed in certain Lepidoptera (Maeda, 1939) and in some of the copepods (W. Beermann, 1954; S. Beermann, 1959). In addition, a few of the protozoans (Le Calvez, 1950) are achiasmate, and in certain of the hermaphroditic enchytraeid worms, chiasmata are absent in both the oocyte and the spermatocyte (Christensen, 1961). Only in *Drosophila* males (with the exception of *ananassae*) and in *Bombyx* females, where genetic studies have confirmed the absence of meiotic exchange, can it be stated unequivocally that the absence of visible chiasmata reflects an absence of exchange. Paradoxically, Cooper (1949) has reported instances of chiasmata in the *Drosophila* spermatocyte, although it is clear that meiotic exchange does not occur here.

In the achiasmate orthopteran and dipteran males, at least, no second pairing is involved, yet the elimination of chiasmata does not interfere with the orderly segregation of homologs. The meiotic mechanism which permits regular segregation in these cases consists of the telescoping or complete elimination of the diplotene and diakinesis stages (White, 1965b). This device assures that the repulsion force, which is responsible for the separation of homologs and, incidentally, about which even less is known than of the pairing forces which bring chromosomes together, does not come into play, so that homologs remain paired until metaphase or anaphase despite the absence of chiasmata.

The method by which elimination of those stages, inimical to the retention of homologous pairing, is accomplished differs in the Orthoptera and in the Diptera. In the former case, opening out of the bivalents in the reductional plane, which normally occurs at the beginning of diplonema is postponed until first metaphase or until the homologs are pulled apart at anaphase. Accompanying the delay in separation of homologs is a very rapid and direct increase in the degree of condensation from pachynema to prometaphase, which probably involves a "switching off" of biosynthetic activity (White, 1965b).

The achiasmate Diptera have adopted a more drastic solution to the problem; for, unlike the Orthoptera, the prophase stages do not remain normal through pachynema. According to Metz and Nonidez (1921) for *Asilus* and to Metz (1926) for five *Drosophila* species, synapsis of homologs is effected in the telophase preceding meiosis by an intimate association of chromosomes that were already loosely paired in anaphase. Homologs apparently remain associated during the early growth stages following the last division (Stevens, 1908; Metz, 1926). Typical leptotene, zygotene, pachytene, and diplotene stages are missing (Guyénot and Naville, 1933–1934). Whether homologs are held together by a force analogous to somatic pairing or by a more intimate type of pairing, perhaps comparable to exchange pairing, is not clear. The pairing segments of the X and Y, which according to Cooper (1944, 1964) consist of localized, cohesive elements or "collochores," appear to be much more closely associated than the chromatids of the autosomal bivalents (White, 1954), which show a side-by-side association along their entire length, and this may reflect a different origin for the achiasmate pairing of the sex chromosomes and of the autosomes. Elimination of exchange between the sex chromosomes probably preceded elimination of exchange for the autosomes, and the mechanisms evolved for ensuring regular segregation in the two cases may well have developed along different lines.

In the female of *Bombyx mori,* where genetic investigations have shown that crossing-over is absent, there is neither a loss of early prophase stages nor a telescoping of midprophase stages. Instead, cytological studies by Maeda (1939) have disclosed that oogenesis proceeds normally except for the absence of chiasmata. At diplonema, the homologs, which had been tightly twisted about one another in the preceding stage, begin to untwist. The process of untwisting is completed in all bivalents by late diplonema, and the two homologs are found paired end-to-end and stretched out to form a single rod. They remain associated in this way until segregation occurs at anaphase.

3. *Distributive Pairing of Small Chromosomes*

The "m" chromosomes (microchromosomes) are a pair of minute to small chromosomes peculiar to the coreid Hemiptera where they were first described in *Anasa tristes* (Paulmier, 1899) and have since been found in all other Coreidae studied. They vary widely in size in different species. In *Pachylis* they are about the size of a centriole; in *Anasa, Alydus,* and *Syromastis* they are markedly larger; in *Leptoglossus* they are only slightly smaller than an autosome; and in *Protenor* equivalent in size to the small autosomes. The unusual feature of the m chromosomes is their tendency for delayed synapsis (E. B. Wilson, 1905a,b). They often remain separate until the final prophase stages and unite to form a bivalent after the disappearance of the nuclear membrane when the chromosomes are moving on to the spindle. Segregation, in this case, is achieved by a secondary pairing mechanism.

Small chromosomes are also found in a number of plant genera (e.g., *Uvularia, Yucca, Agave,* and *Fritillaria,* to mention a few). Darlington (1930) assumed that the metaphase pairing of fragments in *Fritillaria* as well as in other forms depends entirely on chiasma formation. Thus, he writes, "In the fragments on the other hand the chiasmata are inferred from the observation of pairing. I interpret them as 'terminal chiasmata' merely because the chromosomes are paired. The observations are directly a verification of this interpretation and therefore indirectly a verification of the theory that chromosome pairing depends on chiasma formation." Darlington's reasoning at this point is clearly of the circular variety. It seems likely that the metaphase pairing of small chromosomes in some plants, as in some animals, is an achiasmate, secondary pairing for the purpose of segregation.

4. *Distributive Pairing of Sex Chromosomes*

Many forms possess an XY system of sex determination in which crossing-over does not occur between the heteromorphs, and their regular segregation depends on a secondary pairing mechanism. Sex chromosomes of this type have been demonstrated in a large number of animals, including Hemiptera, Coleoptera, Diptera, Orthoptera, nematodes, and vertebrates (E. B. Wilson, 1928). In the spermatocyte of the dipteran *Olfersia,* for example, the X and Y are unpaired during early prophase, and may even lie on opposite sides of the nucleus, but by late diakinesis or prometaphase they have contracted and joined to form a bivalent, with regions close to the kinetochore of both chromosomes intimately united (Cooper, 1944). In most cases the X and Y are heteropycnotic during the prophase stages. Frequently their association is mediated by the nucleolus, as with the sex vesicle of mammals. The association may persist until metaphase or it may terminate earlier. In the latter case they may undergo a late end-to-end pairing or merely a coorientation during the formation of the metaphase plate. In Hemiptera, the X and

Y, although often united during the growth period, typically separate again during diakinesis, each dividing equationally at metaphase I; and in the late anaphase of this division they conjugate to form an XY pair that disjoins at anaphase II (E. B. Wilson, 1928). In certain Coleoptera physical contact between the sex chromosomes does not occur and regular segregation is achieved through correct orientation of the sex multivalent at metaphase I (Virkki, 1967, 1968).

5. *Distributive Pairing in Drosophila melanogaster Females*

a. *Early Evidence Suggesting Two Pairings.* Meiosis in *Drosophila melanogaster* follows markedly different pathways in the two sexes. The events that occur during oogenesis have been deduced from genetic analyses, since meiosis in the oocyte has been unamenable to cytological study. The limitation implicit in this method is that only processes with genetic consequences are detectable. For instance, a temporary separation of homologs, such as described cytologically in *Lepidosiren* at late prophase, would not be revealed by genetic means because the homologs subsequently undergo a second pairing at metaphase and segregate regularly. It is not unexpected, therefore, that genetic analysis of the *Drosophila* female for many years gave results that were for the most part compatible with expected ratios from the classic, single-pairing model.

In practice, genetic detection of a second pairing depends on the participation of nonhomologs. Only then will segregation results show a departure from the expected Mendelian ratios based on the assumption of independent assortment. Since the early work of Bridges, studies of secondary nondisjunction, in which pairing involves the quasihomologs X and Y, had given inexplicable results. Bridges (1916) showed that the presence of a Y chromosome in the female greatly increased the frequency of nondisjunction of the X's. Since secondary exceptions (i.e., individuals arising from XX $\leftrightarrow$ Y segregation) are noncrossovers (in contrast to most arising from regular XY $\leftrightarrow$ X segregation), the result was interpreted by Bridges to mean that the difference between the two types was initiated prior to exchange and that the exceptions arose from those cases in which XY synapsis occurred. A prediction from this model is for a decrease in crossing-over between the X's that parallels the frequency of preexchange XY pairing. Instead, Bridges found an average increase in crossing-over of $\sim$ 13% in the presence of a Y, and a similar result was obtained by Falk (1955). Sturtevant and Beadle (1936), in their classic paper on the relation of inversions in the X chromosome to crossing-over and disjunction, observed that while the Y had a strong effect on the frequency of nondisjunction, it had little effect on exchange, a result again not anticipated from the Bridges model.

Anderson (1929), in studies of a heterozygous X-autosomal translocation in XXY females, found that despite a marked reduction of crossovers in the exceptional classes, the percentage of X-chromosome crossovers among the combined regular and exceptional progeny was apparently the same as that found in the progeny from XX mothers. Again, this result is inconsistent with that expected from Bridges' model of competitive preexchange pairing among the two X's and the Y chromosome. Anderson concluded that synapsis and crossing-over between the X's is not affected by the presence of a Y, but that after crossing-over had taken place, the Y could cause "the more loosely paired X chromosomes to be distributed to the same pole." For the first time a postexchange event that affected segregation was postulated for the *Drosophila* female.

In retrospect, it becomes apparent why inconsistencies with the classic model were revealed initially by studies of secondary nondisjunction. As indicated above, the consequences of a second pairing are detectable only when associations at this time take place between nonhomologs. Although the X and Y possess at least one locus in common, they very rarely undergo meiotic exchange and fall more readily into the category of nonhomologs than homologs. Understandably, their segregation pattern, which is a consequence of a postexchange or distributive mechanism, is not consistent with the results expected on the premise that they pair only once, prior to exchange, in the manner of most homologs.

b. *Nonrandom Assortment of Nonhomologs.* The discovery that nonhomologous chromosomes, under proper conditions, are capable of extremely high frequencies of pairing, as inferred from their segregation behavior, offered a new and fruitful approach to the problem of the interrelationship between pairing, exchange, and segregation. Unlike earlier studies utilizing secondary nondisjunction, this method possessed the special virtue of being uncomplicated by homology between the participating chromosomes.

i. *Early indications of nonhomologous pairing.* Segregation frequencies incompatible with the assumption of independent assortment of nonhomologs were obtained for triploid *Drosophila* by Bridges and Anderson (1925) and by Beadle (1934, 1935). In this situation, the distribution of the third set of chromosomes differed markedly from randomness, showing instead a tendency for equal numbers of chromosomes to move to opposite poles. As a result, oocytes tended to possess one X chromosome and two sets of autosomes or two X chromosomes and one set of autosomes, so that the recovery of euploid progeny (either 2X:2A, XY:2A, or 3X:3A) was greatly reduced from expectation. This deviation from randomness, known as the "crowding effect," was assumed by Beadle to be a consequence of the size and mass of the chromosomes, causing them to orient in a particular way on the metaphase plate. Alternatively, Sandler and Novitski (1956) interpreted the crowding effect in triploids as being due to the formation of a bivalent and univalent for each set of homologs followed by nonhomologous association between the X and an autosomal univalent. The basis for association was considered to be regions of shared homology, located in the proximal heterochromatin, and presumed to be common to all of the chromosomes of the genome.

The inference that nonhomologs might be pairing could also be drawn from Sturtevant's studies (1944) of the effect of heterozygous autosomal inversions on X chromosome nondisjunction. Sturtevant found that the inversion, in addition to greatly increasing exchange between the X's also caused an increase in X nondisjunction, despite the fact that the exceptional progeny always carried noncrossover X chromosomes. The apparent paradox was explained by Cooper *et al.* (1955) as a consequence of interactions at meiosis between nonhomologous chromosomes, leading to increased X nondisjunction and concomitant autosomal nondisjunction. The latter effect was detectable as dominant lethals in excess of those anticipated from XXX and OY zygotes or from multiple strand crossing-over within inverted regions. Cooper *et al.* believed that the heterozygous inversions employed acted to cause pairing difficulties between homologs which they assumed facilitated pairing between nonhomologs.

ii. *Genetic demonstration of nonhomologous pairing.* Compelling genetic evidence for a nonhomologous pairing phenomenon came from the experiments of R. F. Grell (1957, 1959a) and Oksala (1958) in studies of genotypes in which it was possible to follow the segregation of the two participating nonhomologs among the progeny and to show that they were recovered separately in equal but significantly higher frequencies than expected from independent assortment. Thus R. F. Grell (1959a), employing females carrying a Y chromosome and a single free fourth chromosome, found that in over 90% of the progeny, if the Y was present, the free four was absent; and conversely if the free four was present, the Y was absent (Fig. 13). Furthermore, the frequency of eggs carrying the maternal free four and lacking the Y was equivalent to the frequency of eggs carrying the Y and lacking the four. In other words, the two kinds of eggs represented the two reciprocal products resulting from the segregation of the two nonhomologs, the Y and the fourth chromosome. This phenomenon clearly violated the law of independent assortment.

Additional studies by a number of workers disclosed that nonhomologous associations are not restricted to particular chromosomes but may

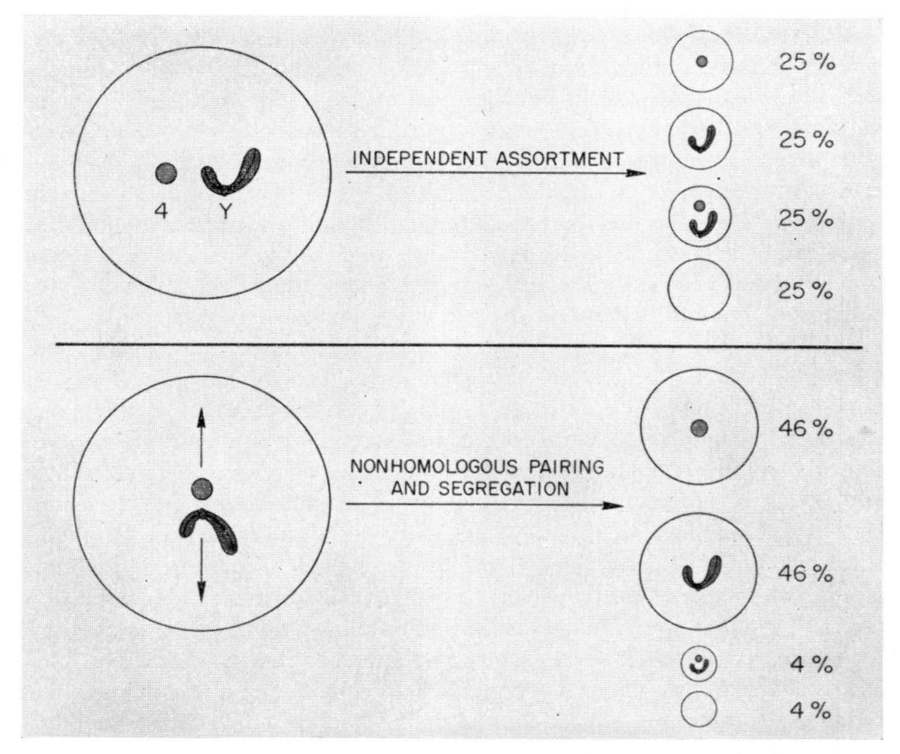

FIG. 13. The expected (top) and observed (bottom) distribution of two nonhomologous chromosomes in the female of *Drosophila melanogaster*. (From R. F. Grell, 1967.)

involve any of the chromosomes or their parts. In experiments similar to the one described above, evidence was obtained for meiotic associations between the Y and a second chromosome (Oksala, 1958); for an X and a fourth chromosome (R. F. Grell, 1959b; R. F. Grell and Grell, 1960); for a Y and a 3;4 translocation (R. F. Grell, 1957); for an X and a 3;4 translocation (R. F. Grell and Grell, 1959); for an X and a second chromosome (Forbes, 1960); for an X and a 2;3 translocation (Forbes, 1960); for a Y and a third chromosome (Miller and Grell, 1963); and for an X duplication and a fourth chromosome (R. F. Grell and Grell, 1960).

In all cases the segregation frequencies show a highly significant departure from the 50% expected with independent assortment. In many cases they exceed 90% and in certain cases range as high as 99.9%. The reasons for the variations in frequencies are described below (Section II,B,5,d).

iii. *Nonhomologous pairing and the interchromosomal effect on exchange.* Early studies (Sturtevant, 1919; Redfield, 1930, 1932) demonstrated the existence of an interbrachial and interchromosomal effect of heterozygous inversions on crossing-over, i.e., the presence of a heterozygous inversion in one arm or in one pair of chromosomes caused an increase in exchange in certain regions of the other arm or in certain regions of heterologous chromosomes. Although this phenomenon has been extensively studied, the mechanism responsible for the interaction has not been resolved. Hypotheses to explain the effect fall into two categories, namely, a structural or a physiological explanation.

After the discovery of an interchromosomal effect on segregation, attempts were made to relate this phenomenon to the interchromosomal effect on exchange. Cooper *et al.* (1955) proposed that the simultaneous presence of heterozygous inversions in more than one pair of chromosomes could lead to nonhomologous pairing between such heterologs, and hence to nondisjunction of homologs, to aneuploidy, and to dominant lethality. Since nonhomologous pairing should reduce or eliminate exchange between homologs, it might be expected that in such dominantly lethal eggs, noncrossover or single crossover strands would be selectively eliminated, and such elimination might, then, partially account for the interchromosomal effect on crossing-over. Shortly thereafter Redfield (1957) investigated the possibility of explaining the interchromosomal recombination effects in terms of major elimination at the egg stage of low ranking strands. Her results negated the possibility, first because egg mortality is much lower than that which would be necessary to produce the observed increases in exchange and, second, because no amount of elimination of low rank strands could lead to the changed proportion of higher rank strands nor to the appearance of crossover strands of extremely high rank that are observed with heterozygous inversions. Moreover, it should be emphasized that the interchromosomal effect is observed in structurally normal homologs which do not participate in nonhomologous associations and hence do not undergo selective elimination.

Oksala (1958) put forth an elaborate hypothesis to relate pairing, exchange, and segregation; it was based on the speculation that the

Drosophila chromosomes are polarized in a bouquet formation during early meiosis and that the alterations in the spatial configurations incident to rearrangements are responsible for the interchromosomal effects on both crossing-over and segregation. Unfortunately, the consequences of the model are, as Ramel (1962a) points out, inconsistent with the observed interchromosomal effects in triploids and in inversion heterozygotes as well as with the observed segregation pattern between compound sex chromosomes and heterozygously inverted autosomes.

Levine and Levine (1954) and C. W. Hinton (1965) have proposed that part of the interchromosomal effects of inversions on recombination might be attributed to their associated genic modifiers rather than to their structure. C. W. Hinton's discovery (1962, 1966) that the mutant *c3G*, in heterozygous condition, can elicit specific regional increases in exchange, resembling those produced by heterozygous inversions, suggested to him that this case might serve as a prototype for this category of genes.

Probably the strongest evidence against a structural interpretation comes from the cogent approach of Ramel (1962a), in which he compared the effects of heterozygous autosomal inversions on exchange in the X chromosome in standard and inverted sequence. Although the results are not entirely unambiguous, Ramel found the response of a particular region to be more or less characteristic of that region, regardless of its position along the chromosome. The specific and apparently autonomous reaction properties of the different regions of the X chromosomes to heterologous inversions led Ramel to conclude that the basic mechanism underlying the interchromosomal effect on exchange is chemical-physiological in nature rather than mechanical. Additional experiments making use of this approach would be desirable.

iv. *The time of nonhomologous pairing.* Initially, it was supposed that only one pairing, that which precedes exchange, takes place during meiosis. Thus, homologous and nonhomologous pairing were thought to be coincidental and competitive events. The models for nonhomologous pairing that were proposed by Cooper, by Sandler and Novitski, and by Oksala all proceeded from this assumption, as did virtually all thinking about meiosis up to this time. A consideration of the facts soon discloses a serious difficulty with such schemes. The requirement for precise, highly specific pairing at the time of exchange would seem to be incompatible with nonspecific associations occurring with a high frequency at the same time. Moreover, a number of anomalous results, particularly those coming from studies of secondary nondisjunction (see Section II, B,5,a), has shown that none of the models proposed to explain the relationship between pairing, exchange, and segregation was consonant with the known genetic facts. For instance, it was recognized that a decrease in X-chromosome exchange is accompanied by an increase in secondary nondisjunction, yet it was clear from the behavior of X-inversion heterozygotes or the fourth chromosomes that exchange is not necessary for regular disjunction. A reappraisal of the meiotic process appeared therefore to be in order.

c. *Elaboration of the Distributive-Pairing Model.* i. *Tests of the single-pairing hypothesis.* If homologous and nonhomologous pairing are concurrent, they should influence one another. Thus, the genetically detectable consequence of homologous pairing, namely, exchange, should be reduced when a successful nonhomologous competitor is present. To examine the validity of the single-pairing hypothesis, an experiment to test this prediction was carried out (R. F. Grell, 1962a). The effect of a Y chromosome on the amount of crossing-over between two second chromosomes was measured. In order to involve the Y with the second chromosome, a multiple inversion was present in the latter, which restricted exchange to the distal half of the right arms of the second chromosomes. The other second chromosome was part of a reciprocal translocation with chromosome 3 and despite greatly reduced crossing-over between the 2's, the attachment of one of them to chromosome 3 ensured that it was generally part of a crossover complex. When a Y chromosome was introduced into females of this genotype, the Y was found to segregate from the inverted 2 ~ 75% of the time. Translated into association frequency (R. F. Grell and Grell, 1960), this means that association had occurred between these nonhomologs about 50% of the time. Genetic analysis revealed that when the inverted 2 segregated from the Y, it was always a noncrossover, and that when it assorted independently with the Y, it was a crossover. Yet when crossing-over between the second chromosomes was measured in the presence and absence of a Y chromosome, the values were found to be almost identical. How could nonhomologous pairing, occuring ~ 50% of the time between exclusively noncrossover chromosomes and at the expense of homologous pairing, leave crossover values unaltered?

A similar paradox existed in the case of secondary nondisjunction; however, in this instance the relationship is not as sharply defined, because homology exists between the X and Y chromosomes at least at the bobbed and/or the nucleolar organizing region. As described above (Section II,B,5,a), when a Y is present in a female, X nondisjunction is greatly increased. Since such nondisjoining X chromosomes are always noncrossovers, Bridges (1916) postulated that the difference between the two types was initiated prior to exchange by competitive pairing between the two X's and the Y. Pairing as an XY bivalent and an X univalent would lead to X nondisjunction one half of the time (i.e., when the single X happened to pass to the same pole as the segregating X). This scheme is subject to the same criticism as that raised against the models for nonhomologous pairing. If pairing for exchange is based on homology, and there is every reason to believe it is, why should an X chromosome pair and segregate preferentially from a Y with which it shares only one known locus (as happens in the case of heterozygous X inversions which yield $> 50\%$ secondary nondisjunctionals) rather than the other X? On the other hand, if an XXY trivalent (Cooper, 1948a), occurring prior to exchange, is responsible for X nondisjunction, the effect of the Y on crossing-over should be localized to the proximal region where the X and Y share homology. Distal regions should show crossover values approaching normal both in the regular and the exceptional progeny. Yet it is well established that secondary exceptions are noncrossovers throughout their length.

The Bridges model is testable, for the formation of XY bivalents and X univalents before exchange should lead to a corresponding decrease in XX bivalents and a correlated decrease in X exchange. When such a test was performed (R. F. Grell, 1962b) using XX and XXY sisters carrying two normal X chromosomes, it was found that crossing-over, instead of showing a decrease compatible with the 40-fold increase in X nondisjunction produced by the Y, showed a slight increase. Similar studies employing a variety of heterozygously inverted X chromosomes in XX and XXY sisters again demonstrated that the Y produced large increases in nondisjunction but never the anticipated total decrease in exchange or the uniform reduction in exchange throughout the chromosome predicted by Bridges' model. Instead, localized effects of the Y were observed that depended on the position of the inversion. With proximal inversions which reduced or eliminated exchange in the proximal region, the depression presumably caused by competitive pairing of the Y was eliminated, a compensatory distal increase was retained, and the overall effect of the Y was to increase exchange. With distal inversions, where exchange was curtailed distally, a compensatory proximal increase was decreased, presumably by the competitive pairing of the Y, and here the overall effect of the Y was to reduce exchange, but never to the extent expected from the frequency of secondary nondisjunction. These findings failed to support Bridges' model for heterosynapsis preceding exchange as the source of the nonexchange X chromosomes that are recovered as exceptions. This was most clearly shown with normal X's where it was possible to perform a tetrad analysis to determine the number of noncrossover X bivalents. The results (Table IV) showed for two independent studies that the frequency of noncrossover X tetrads (E_0's) is not increased by the presence of a Y. Since the effect of the Y on X nondisjunction is independent of its effect on X exchange, it was concluded that the two effects cannot be exerted concurrently.

TABLE IV
TETRAD ANALYSIS OF CROSSING-OVER IN NORMAL X CHROMOSOMES[a]

Experiment		E_0[b]	E_1[c]	E_2[d]	E_3[e]	Distance measured	Percent total crossing-over
1	XX	4.73	57.62	36.00	1.64	Distal tip (*sc*) to centromere	67.24 ± 0.95
	XXY	4.98	51.77	41.34	1.91		70.09 ± 1.00
2	XX	5.23	62.91	30.94	0.91	Distal tip (*y*) to *car*	63.76 ± 1.14
	XXY	3.67	60.73	34.61	0.98		66.44 ± 0.95

[a] From R. F. Grell (1962b).
[b] No-exchange tetrad.
[c] Single-exchange tetrad.
[d] Double-exchange tetrad.
[e] Triple-exchange tetrad.

ii. *The distributive pairing model of meiosis.* The apparently conflicting genetic results can be reconciled by simply assuming that two pairing events, rather than one, occur during meiosis in the *Drosophila* female. This premise serves as the basis for the distributive pairing model of meiosis illustrated in Fig. 14. According to this interpretation, chromosomes that share homology with one another pair before exchange. This pairing is called exchange pairing. Following exchange pairing, exchange may or may not occur. If it does, a chiasma ties the participating chromosomes together, and this bond precludes their subsequent involvement with other chromosomes. Hence, they remain associated during the following stages and segregate regularly. If exchange fails to occur (either because a homolog is absent, or because the homologs are small and crossing-over occurs infrequently or not at all, or because rearrangements are present which reduce exchange), all such chromosomes become part of a group of nonexchange chromosomes called the distributive pool. Within this pool, a second type of pairing, called distributive pairing, takes place. Distributive pairing may occur between two homologs, in which case segregation is again regular. Alternatively, it may occur between two nonhomologs, and when this happens, the segregation

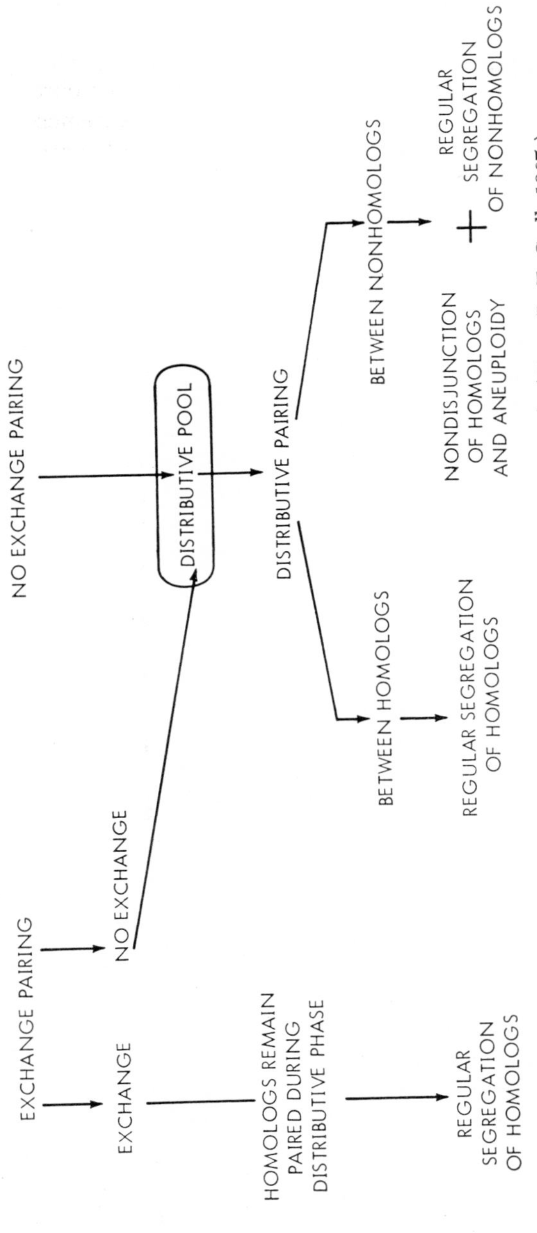

FIG. 14. The distributive pairing model of meiosis (= 2 pairings). (From R. F. Grell, 1967.)

of the nonhomologs will lead, part of the time, to nondisjunction of homologs and to aneuploid gametes.

With this sequence the seemingly irreconcilable genetic data fall neatly into place. For instance, the paradox that crossover chromosomes segregate regularly, that nondisjoining chromosomes are nearly always noncrossovers, but that exchange is not a prerequisite for regular disjunction, is easily resolved. First, as Darlington postulated, homologs which undergo exchange remain associated, do not become involved with nonhomologs, and segregate regularly at anaphase I. On the other hand, absence of exchange need not lead to nondisjunction as Darlington assumed, because homologs which fail to cross over are provided with a second opportunity to pair with each other at the distributive phase of meiosis. Intercalation of this step in the meiotic process should be of high selective value for genomes in which the probability of exchange between certain homologs is low, either because of small size, inversion heterozygosity, or reduced homology. For example, it is this kind of pairing that is responsible for the regular segregation of inversion heterozygotes when the majority of them are noncrossovers, or of the fourth chromosomes which are invariably noncrossovers in the diploid female (see Section II,B,5,d,iv).

Segregation of noncrossover homologs remains regular, however, only so long as a suitable nonhomolog is absent from the pool. Should a Y chromosome, for instance, be added to the pool, of which the X chromosomes are frequently members, associations between the X's and the Y will result in a high frequency of X nondisjunction (i.e., the process known as secondary nondisjunction). The more efficient the inversion system in producing noncrossover X tetrads, the more frequently will the X's become members of the pool and the more frequently will secondary nondisjunction occur.

Although the ability of the Y to increase X nondisjunction is well recognized, it is not so well known that autosomes may do likewise. R. F. Grell and Grell (1959), Forbes (1960), and Roberts (1962) succeeded in producing high frequencies of X nondisjunction with a variety of autosomes. Roberts employed an inversion translocation system in which one element of a reciprocal 2:3 translocation was a frequent member of the distributive pool to show that the exceptional females produced in this way, like secondary exceptions from XXY females are invariably noncrossovers for the X chromosome. In this system, Roberts found that a greater than 10-fold increase in X nondisjunction was accompanied by a significant increase in X exchange, a result hardly compatible with a single pairing scheme in which high frequencies of X-autosomal pairing, prior to exchange, would be expected to effect a reduction in X exchange. The term distributive nondisjunction is appropriate to describe nondisjunction of any homologous pair induced through distributive pairing.

The distributive pairing model provides a ready explanation, then, for the failure of nonhomologous pairing or XY pairing, although confined to noncrossover chromosomes, to produce corresponding increases in the frequency of such noncrossover chromosomes; to wit, associations of this kind occur only after exchange and exclusively between those chromosomes which have failed to undergo exchange. Hence, they do not alter exchange frequency. Furthermore, with this model, the requirement for precise, highly specific, preexchange pairing is maintained, since associations which precede exchange are necessarily homologous, and it is only those which follow exchange which may be nonhomologous.

iii. *The Novitski "alternative" and the Merriam model of meiosis.* Novitski (1964), while recognizing that the data are consistent with the distributive pairing model for *melanogaster* females, believes the data need not be restricted to this interpretation. He has offered an alternative, prompted partially by the contention that one pairing should suffice for segregation (discussed below) and partially by the understandable viewpoint that since the opportunity for nonhomologous pairing would, under natural conditions, give rise to a high frequency of lethal zygotes, there should be considerable selective pressure against the incorporation of such a mechanism. To validate the claim that lethality would ensue from the presence of this kind of mechanism, Novitski cites the experiments of Terzhagi and Knapp (1960) with *D. pseudoobscura* and of Cooper *et al.* (1955) with *D. melanogaster*, in which cases dominant lethality is unquestionably a consequence of nonhomologous pairing. It should be abundantly clear, however, that if lethality results from nonhomologous pairing in the cases Novitski cites, a mechanism which permits such pairing must be present in these species and therefore must have been incorporated and retained in these forms. The solution to this apparent paradox lies in the assumption that it is not the pairing mechanism which need be selectively eliminated, but only the unusual genetic conditions which permit nonhomologs to make use of this mechanism.

Despite his objection to the retention of a mechanism which permits nonhomologs to pair and segregate, Novitski proposes as his alternative a model which incorporates such a mechanism. In this case it is a chromocenter, established prior to exchange rather than a second pairing initiated subsequent to exchange. Novitski considers the lethality that is a concomitant and necessary adjunct of nonhomologous pairing to be the price the organism must pay for the facilitation of pairing provided by the chromocenter. The sequence Novitski proposes is as follows: (1) An association of all chromosomes, homologous or not, prior to exchange which places homologs in an advantageous position for the onset of synapsis; (2) a departure (either actual or otherwise) as independent bivalents from the chromocenter of those homologs achieving a synaptic configuration; (3) the pairing "after all synapsing pairs had removed themselves" between those chromosomes, homologs or not, which were not involved in bivalent formation; and (4) subsequent segregation of such paired chromosomes.

In Novitski's scheme, the critical factor which separates those chromosomes which may undergo a postexchange pairing and those which may not is synapsis. In the distributive model it is exchange. Actually little is known of the time or mode of synapsis in the *Drosophila* oocyte. However, Novitski indicates the conditions under which he considers synapsis would fail to occur. When we examine these, we find that in each case the absence of synapsis either is, or is equivalent to, the absence of exchange. The conditions Novitski enumerates which lead to failure of synapsis between homologs are (1) absence of exchange, (2) absence of a homolog, which is tantamount to absence of exchange, (3) the attachment of homologs to a common centromere [in which case the compound chromosome invariably enters the distributive pool, since the requirement for a crossover embraces two independent centromeres (E. H. Grell, 1963)], and (4) structural differences between homologs, and here again it has been shown (see Section II,B,5,c,i) that crossovers between heterozygously rearranged chromosomes are unavailable for distributive pairing and that only noncrossovers engage in such pairing.

When the operational genetic term *exchange* replaces the postulated cytological term *synapsis,* it becomes evident that the sequence Novitski envisions is no different than the distributive pairing sequence. First, a type of pairing for exchange, which we call exchange pairing, must precede exchange. This is followed by the exchange event. After the exchange period, those chromosomes which have failed to crossover pair with one another, either homologously or nonhomologously. We call pairing of this latter type, which will determine the segregation pattern of noncrossover chromosomes, distributive pairing. Whether or not chromocentral formation is invoked prior to exchange, has no bearing on the segregational events, since any postulated association prior to exchange is with respect to segregation, completely indeterminate. The determination of the segregation pattern must await the exchange process, for it is only with the realignment of chromosomes following exchange that the segregation pattern emerges.

A model whose genetic predictions differ markedly from those of Grell's or Novitski's has been proposed by Merriam (1967). The essential difference lies in the intercalation of the final and definitive pairing step at a point preceding exchange. Pairing at this time may occur between homologs, between nonhomologs, or a chromosome may be simultaneously involved with a homolog and a nonhomolog. In the latter case, segregational consequences can result from nonhomologous pairing only if the homologs fail to undergo exchange. Once preexchange pairing occurs, however, a chromosome is irrevocably committed to a permanent alignment with the chromosome or chromosomes of its choice. A nonhomolog, once paired, remains associated with the homologous bivalent until metaphase whether exchange occurs or not. Therefore the Merriam model places marked restrictions upon the segregational consequences of nonhomologous pairing since its potential is diminished to the extent that either participant subsequently becomes part of an exchange bivalent.

Merriam's test of his model depended, first, upon the observation that a Y chromosome, added to females carrying w^{m4} in homozygous condition, apparently decreased X exchange and increased noncrossover X tetrads; and, second, upon the demonstration that these alterations in X exchange by the Y chromosome (presumably marking the position of the Y at exchange) were eliminated through the introduction of a nonhomolog which competed successfully for pairing with the Y. Unfortunately, in all but one instance, Merriam's data were derived solely from an examination of the proximal two thirds of the genetic length of the w^{m4} chromosome.

A reexamination of w^{m4}, including the distal third of the chromosome, has failed to confirm any of Merriam's findings (Grell, 1968b). The w^{m4} chromosome, much like the normal X, displays both a proximal decrease and a compensatory distal increase in crossing-over in the presence of the Y, so that total exchange remains unaltered. Tetrad analysis reveals the frequency of noncrossover X tetrads is virtually identical whether the Y is present or absent. The introduction of a successful nonhomologous competitor for the Y, utilizing both the nonhomolog tested by Merriam as well as two alternative nonhomologs not previously tested, fails in each case to alter the depression in proximal X exchange induced by the Y. Finally, calculations of the sum of the frequencies that the Y segregates both from the X chromosomes and from the newly introduced nonhomolog, indicate that the frequency of Y pairing, as computed from the Merriam model, would have to significantly exceed the maximal limit of 100% for each competitor tested; whereas the frequency of Y pairing, as calculated from the distributive model, either falls within or reaches this limit but in no case significantly exceeds it. These results are incompatible with any model (e.g. Merriam model) which postulates that nonhomologous pairing precedes exchange.

d. *Size-Dependency at Distributive Pairing in* Drosophila melanogaster *Females.* i. *The problem of recognition between nonhomologs.* Although the forces which bring chromosomes together for exchange pairing are not understood, there seems little doubt that the basis for recognition at this time resides ultimately in the similarity of nucleotide sequences between participating chromosomes. By contrast, the phenomenon of nonhomologous pairing offers no immediate clue as to the nature of recognition in this case. Originally, Sandler and Novitski (1956) proposed that regions of shared homology, common to all of the *Drosophila* chromosomes and located in the proximal heterochromatin, were responsible for "nonhomologous" associations. Viewed in this way nonhomologous pairing becomes entirely a matter of proximal, heterochromatic pairing or a special kind of homologous pairing.

In an experiment to test this notion (R. F. Grell and Grell, 1960), three chromosomes were introduced into the distributive pool, two of which carried the same centromere and heterochromatic base but euchromatin of different origin and length, while the third chromosome possessed no known homology for either, but resembled the larger one in size. If nonhomologous pairing is in fact equivalent to proximal heterochromatic pairing, associations are expected to occur at least as frequently between the two chromosomes with shared homology as between either of these and the third chromosome. Instead, it was found that the two larger chromosomes segregated very regularly (98%), while the small chromosome, although sharing proximal homology with one, assorted randomly with both. Apparently factors other than identity of the centromere and proximal heterochromatin are involved in recognition between nonhomologs.

ii. *Size in noncompetitive pairing.* Proceeding from the assumption that size might play a central role in recognition at distributive pairing, a study was made of the relationship between the size of two chromosomes and the regularity of their association (R. F. Grell, 1964a). This was accomplished by measuring the frequency of nondisjunction between a single free fourth chromosome and each of a series of free X duplications. The sizes of the X duplications ranged from $\leqq 0.3$ to greater than

4 times the length of chromosome four at mitotic metaphase. Since regular segregation plus nondisjunction = 1, the higher the nondisjunction frequency, the lower the pairing and segregation frequencies. The results showed that the highest nondisjunction values occurred with the smallest and largest duplications [i.e., those most dissimilar to the fourth chromosome in size (Fig. 15)]. Then as the sizes of the duplications approached the size of the four, from both directions, nondisjunction values decreased. The lowest values were observed with those duplica-

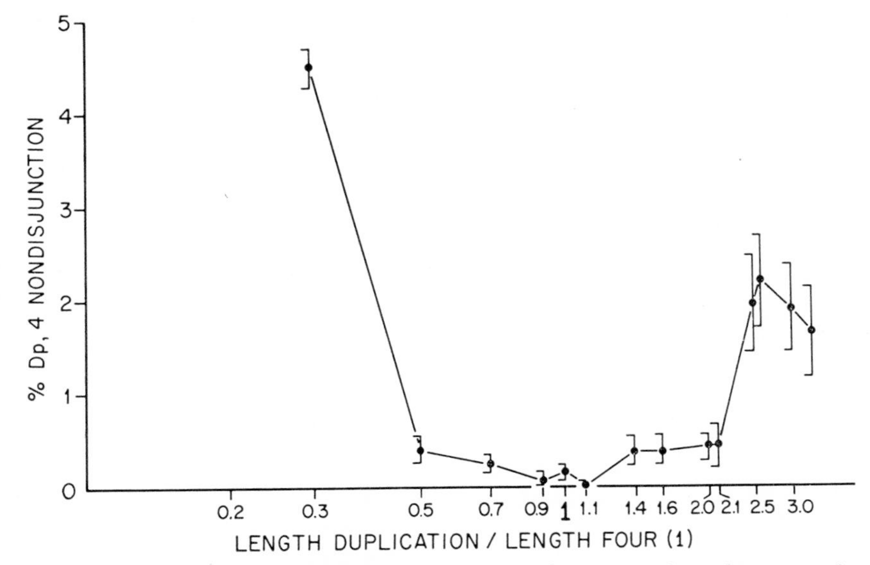

FIG. 15. The effect of X duplication size on the frequency of nondisjunction between the X duplications and a free fourth chromosome under noncompetitive conditions. (From R. F. Grell, 1967).

tions closest to the four in length (0.9 to 1.1). In this range, the average nondisjunction value between the duplication and the four was 0.1%, which in terms of segregation is equivalent to 99.9%. These results demonstrated that the critical factor in recognition at distributive pairing is either chromosome size or a property closely related to size such as chromosome mass.

iii. *Size in competitive pairing.* The finding that recognition is size-dependent suggested, as a reasonable extension, that preference in distributive pairing might also be size-dependent. To study preference, it becomes necessary to provide a competitive situation, which means that a minimum of three chromosomes are required in the distributive pool.

Experimentally, this was accomplished by adding another chromosome to the pool and studying the effect of the third chromosome on the frequency of nondisjunction of the X duplication and the four (R. F. Grell, 1964a). The chromosome that was added, T_4, was one part of a reciprocal 3;4 translocation. T_4 size at metaphase is ~ 5½ times chromosome four so that it is larger than any of the X duplications. As seen in Fig. 16, the effect of the competitor, T_4, depends on the similarity in size between

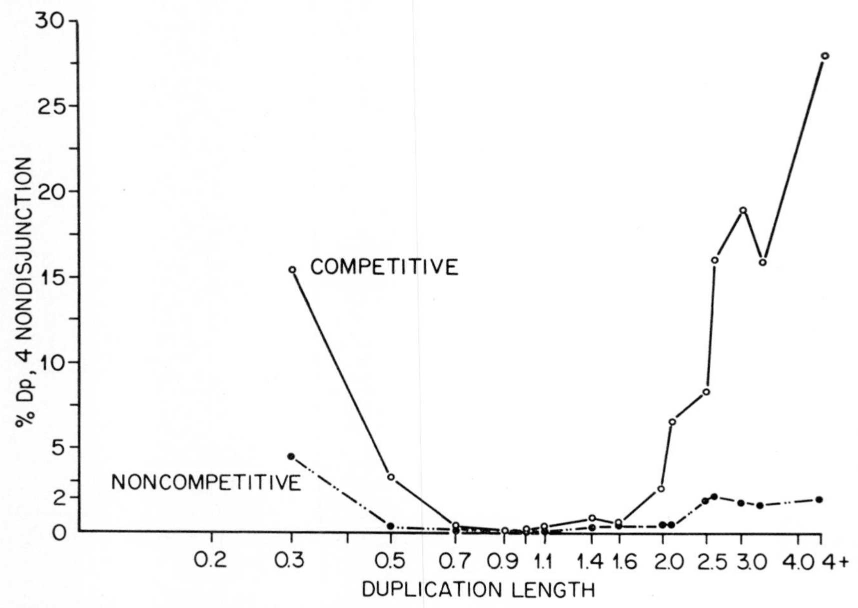

FIG. 16. Comparison of duplication, four nondisjunction frequencies under competitive and noncompetitive conditions. (From R. F. Grell, 1967.)

the X duplication and the four. With duplications close to the four in length (0.7 to 1.6), the presence of T_4 is not felt, since the two curves are practically coincident; with duplications smaller than 0.7 and larger than 1.6, however, T_4 is recognized, since nondisjunction values show conspicuous increases which become larger as the size discrepancy between the duplication and the four increases. The observations suggest that it is the ratio of chromosome size rather than absolute size which is important in recognition. When the size ratio of the duplication and the four is approximately 2/3 or greater (i.e., $\geqq 0.7/1$ and $1/\leqq 1.6$) a competitor of the size of T_4 is not recognized; when the ratio of the duplication and the four becomes less that 2/3 (i.e., $<0.7/1$ and $1/>1.6$) it is.

The effect of size on pairing preference is dramatically illustrated in Fig. 17, where the nondisjunction values for the three pairwise combinations of the X duplication (or Y), the four, and the T_4 are plotted as functions of chromosome length. The rightmost points of the three curves represent the nondisjunction values when the three competitors are the four, the T_4, and a Y chromosome (which is larger than any of the X

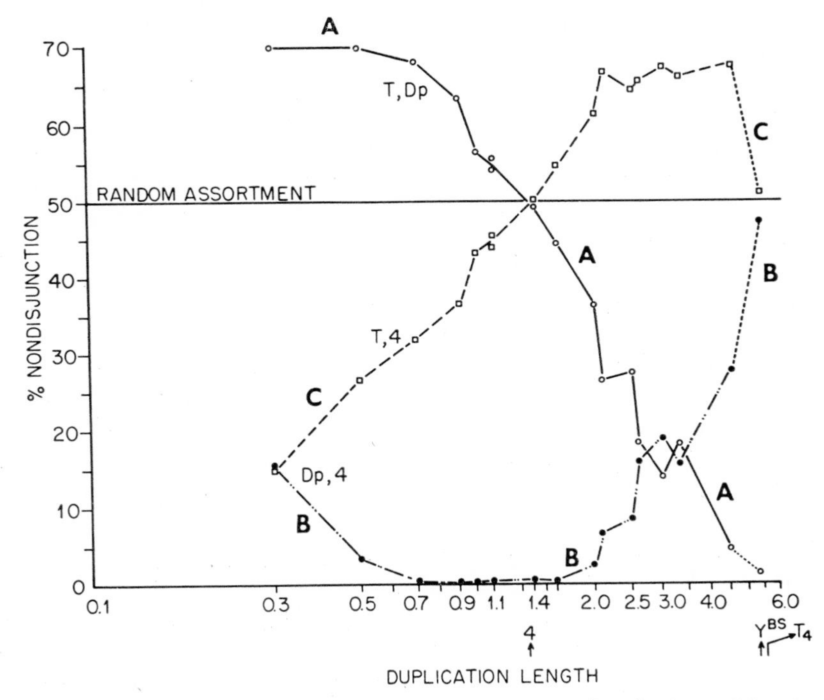

FIG. 17. Effect of size on chromosome preference at distributive pairing in the female of *Drosophila melanogaster*. (From R. F. Grell, 1965.)

duplications). Nondisjunction of the T_4 and Y is found to be very low (2%), but the small fourth chromosome shows ~ 50% nondisjunction (or random assortment) with both the T_4 and the Y. If it is true that preference in pairing is size-dependent, preferences should be alterable as size is altered. This is clearly the case; for as X duplications of smaller and smaller size are substituted for the Y, nondisjunction between the duplications and T_4 steadily increases (curve A), but at the same time nondisjunction between the X duplications and the four decreases (curve B). Finally when the duplications and four are very similar in size (~1.4 on abscissa), the large T_4 shows ~50% nondisjunction or random assortment with both the duplication and the four, while they, in turn, segregate ~98%.

An unexpected and interesting feature of the segregation pattern is the incidence of trivalents. As shown in Fig. 17, except for the two instances where random assortment of one chromosome occurs, one of the three pairwise combinations invariably shows nondisjunction frequencies significantly in excess of 50%. This means, first, that some trivalents are occurring and, second, that chromosome segregation from the trivalent is nonrandom. The type of nonrandomness which must be postulated to explain the data is such that the element of intermediate length directs the small and large chromosomes to the same pole while it moves to the opposite one. Thus with duplications smaller than chromosome four, the four is the directing element, and the small duplication and the large T_4 are recovered together in excess of 50%; with duplications larger than the four, the duplication is the directing element and the four and T_4 are recovered together in excess of 50%.

In addition to demonstrating that size is the critical factor in preference, the studies with the three elements show that the composition of the chromosomes is not a factor in recognition. The X duplications and the Y are primarily heterochromatic, the T_4 is predominantly euchromatic and the four carries its normal complement of heterochromatin and euchromatin; yet, the behavior of the chromosomes is correlated with their total length and is independent of the proportion of euchromatin and heterochromatin present.

iv. *Distributive pairing of the fourth chromosomes.* The fourth chromosomes of *Drosophila* are extremely small and do not undergo exchange in the diploid female. The mechanism underlying their regular segregation has long been a subject for conjecture. Darlington (1930) has proposed that localized double, but genetically undetectable, exchanges in the genetically inert segments are responsible for the regular segregation of small chromosomes such as the fours. The discovery of a second, presumably achiasmate kind of pairing in the *Drosophila* female, which provides a mechanism whereby noncrossover homologs may segregate regularly, immediately suggested the possibility that this type of pairing might also be responsible for the regular segregation of the fourth chromosomes. On the other hand, if the fourth chromosomes are normally members of the distributive pool, the failure of other chromosomes in the pool, such as noncrossover X's or the Y, to affect their regular segregation would need elucidation.

Two plausible alternatives could now be imagined. Either the fours are present in the pool but are too small to recognize other members; or a special device, divorced from distributive pairing, ensures regular segregation of the fours. A simple experiment sufficed to distinguish between the two possibilities (R. F. Grell, 1964b).

If the fours are present in the pool but are protected by their small size from heterologous involvement, the introduction of chromosomes

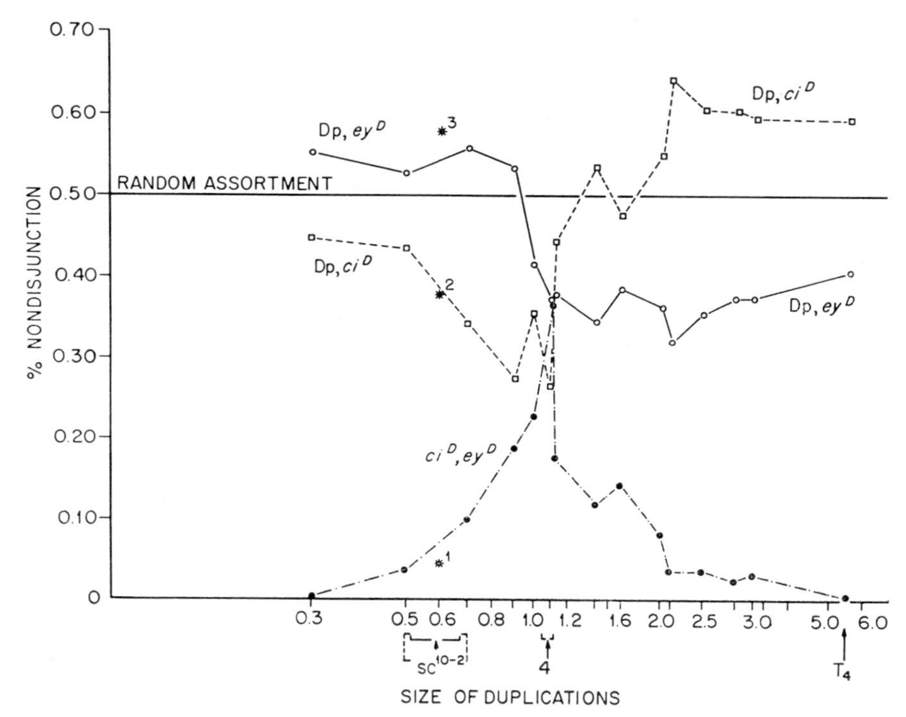

FIG. 18. Nondisjunction frequencies for the three chromosome sets: (1)ci^D, ey^D; (2) Dp, ci^D or T_4, ci^D; (3) Dp, ey^D or T_4, ey^D plotted as a function of duplication (and T_4) length. The asterisk indicates the nondisjunction frequencies obtained by Sturtevant with the sc^{10-2} duplication where $^{*1} = ci^D$, ey^D nondisjunction; $^{*2} =$ Dp, ci^D nondisjunction; $^{*3} =$ Dp, ey^D nondisjunction. (From R. F. Grell, 1964b).

of more or less equivalent length, such as the X duplications, should lead to recognition between the duplication and the four, to heterologous pairing, and to fourth chromosome nondisjunction. The results, shown in Fig. 18, clearly demonstrate that such is the case. Nondisjunction frequency of the two dominantly marked fours, ci^D and ey^D, which for the smallest duplication (0.3) is equivalent to the control value (0.2%), rises rapidly as the length of the duplication approaches 1, reaches a maximum of 36.6% with a duplication 1.1 in length, and then decreases as the length of the duplication exceeds that of chromosome four. Since the theoretical expectation of *ci*^{D}, *ey*^{D} nondisjunction in a triplo-4 female is only 33%, it is evident that recognition between a nonhomolog of the proper size and chromosome four is as efficient as recognition between two fourth chromosomes. As with the competitive situation previously described (Fig. 16), nondisjunction frequencies significantly in excess of 50% are found, indicating that trivalents between the two fours and the duplication, followed by their nonrandom segregation, must be occurring (Fig. 17). It should be noted that the *ey*^{D} four carries a sizable euchromatic duplication and is larger than the *ci*^{D} four. Again the pattern of segregation from the trivalent is such that the chromosome of intermediate size directs the small and large chromosomes to one pole while it moves to the other.

As the result of these studies it is possible to say that the fourth chromosomes are regular members of the pool, that the mechanism responsible for the orderly segregation of the fours is distributive pairing, and that homology is not a factor in recognition at this time.

Studies of distributive pairing of the fourth chromosome suggested the interesting possibility that the phenomenon of preferential segregation has as its basis a size difference between chromosomes (R. F. Grell, 1964b). Sturtevant (1934, 1936) found that in triplo-4 females, the three types of segregation giving one fourth chromosome to one pole and two to the other do not occur with equal frequency. An order of preference was established such that any particular fourth chromosome would pass to the diplo-4 pole more frequently than any chromosome lying below it in the series. This seriation was demonstrated most satisfactorily with the sc$^{10-2}$ duplication, which consists of a small segment of X attached to the four base. An examination of the order established by the sc$^{10-2}$ tests discloses that the *ey*^{D} chromosome, which carries a sizable euchromatic duplication, lies fourth from one end of the series of 26 chromosomes; that the *M-4* chromosome, which possesses a large deletion, lies next to last at the other end; and that *ci*^{D} lies close to the midpoint. The positions of the duplicated, the deleted, and the normal chromosomes in the sequence suggest that the seriation could be a reflection of size differences among the fours such that the *M-4* chromosome represents one of the smallest and the *ey*^{D} one of the largest of the fourth chromosomes. Preferential segregation would then be a consequence of slight size differences among the fourth chromosomes such that these differences would determine the association and/or disjunctive pattern of the fours at distributive pairing and would lead to their nonrandom segregation. Variation in chromosome size, as postulated here for every fourth chromosome which shows a distinctive segregation pattern, might be an inevitable consequence of the elimination of crossing-over and the concomitant retention and accumulation of small differences between fours.

e. *Compound Chromosomes at Distributive Pairing.* Compound chromosomes are a special class of chromosomes formed by the attachment of two homologous arms onto a single centromere. If the second arm is attached in reverse order, pairing between the two arms is easily achieved. With respect to exchange such a compound acts much like two independent homologs, since crossing-over values are essentially unaltered (Beadle and Emerson, 1935); but with respect to segregation a compound perforce functions as a single unit. This special behavioral aspect of compound chromosomes makes them a useful tool for investigating the role of exchange in excluding chromosomes from the distributive pool. If exchange, per se, is sufficient to disqualify a chromosome from membership, crossover compounds should not be available for nonhomologous associations. If, on the other hand, the special attribute of exchange resides in its ability to bind two units together until metaphase, crossover compounds, as single units, should be present in the pool and available for nonhomologous pairing.

E. H. Grell (1963) employed compound-X chromosomes ($\overline{XX}$), either with two isosequential arms and normal exchange or with one arm heterozygously inverted and greatly reduced exchange, to determine if exchange between the two arms affected the frequency with which they paired nonhomologously with an autosome. A value of 90% segregation of the X and autosome was obtained in the former case and 88% in the latter, demonstrating that exchange within a compound does not reduce its ability to engage in nonhomologous, distributive pairing. Apparently two centromeres are essential to satisfy the requirement for segregational pairing, presumably because they are needed for coorientation of the chromosomes on the spindle. Nevertheless, as was shown by the studies of chromosome size (Section II,B,5,d), the centromeres themselves possess no specificity in distributive pairing.

A Y chromosome is normally present in females carrying a compound-X, and the regularity of segregation between the compound-X and the Y (99.9%) had generally been attributed to the fact that the X and Y are quasihomologs and segregate regularly in the male. E. H. Grell's studies (1963) demonstrated that highly regular segregations can also occur between a compound-X and a second chromosome (90%) in the absence of a Y. He concluded that when a Y segregates from a compound-X, it does so by the same process of distributive pairing as occurs between a compound-X and a nonhomolog.

Distributive pairing between a compound-X and a Y can be made competitive by the addition of an autosome to the distributive pool. Although a preference remains for the compound-X and Y to go to opposite poles, the autosome can interfere with this process to the extent that 13–18% of the gametes now receive both or neither the compound-X and Y chromosomes as contrasted to 0.1% in the absence of the autosome (E. H. Grell, 1963). Similar studies by Ramel (1958) showed that the availability of either large autosome for nonhomologous pairing increased the nondisjunction of the compound-X and Y. In related experiments (1962b), Ramel was able to demonstrate that the frequency of nonhomologous association between heterozygously inverted second chromosomes and a compound-X can be made competitive by the addition of the Y chromosome, whereupon a decrease in second chromosome nondisjunction occurs. In other words, $\overline{XX}$,Y pairing reduces $\overline{XX}$,2 pairing and results in more regular segregation of the second chromosomes.

Ramel (1962b) examined the effect of heterozygous second chromosome inversions in increasing crossing-over in attached-X and in free X chromosomes. Exchange properties were found to be similar in the two situations despite the high frequency of nonhomologous X,2 pairing, which occurs in the former but not in the latter case. Ramel concluded that the increase in X crossing-over, induced by the heterologous inversions, occurs independently of the nonhomologous pairing induced by the same inversions.

A variety of compound autosomal chromosomes have been produced in the laboratory of E. B. Lewis (Rasmussen, 1960). These consist of the attachment of two left arms or of two right arms of chromosomes 2 or 3 to a common centromere to form an "attached-2L" or an "attached-2R" or an "attached-3L" or an "attached-3R." The two homologous arms now comprise a single unit and crossing-over occurs intrachromosomally instead of interchromosomally. Segregation results show that the behavior of compound autosomes is precisely that which would be predicted from the behavior of compound-X chromosomes and is entirely consistent with the distributive pairing model of meiosis. In a female, the attached-2L and attached-2R (or the attached-3L and attached-3R) segregate very regularly from one another, whereas in the male, where nonhomologous pairing fails to occur (R. F. Grell, 1957), the two elements assort independently with one another (E. H. Grell, 1965; Holm *et al.*, 1967). Furthermore, the introduction of a third element into the distributive pool, in the form of a Y chromosome or a compound-X chromosome greatly reduces the regularity of segregation between the compound autosomal elements and leads to the increased recovery of both autosomal components together. The involvement of one of the compound autosomes, e.g., attached-2R, in distributive pairing with attached-2L ceases, if each arm is provided with its own centromere to produce two independent 2R chromosomes (E. H. Grell, 1969). Should the other autosomal component remain in the distributive pool in conjunction with a nonhomolog, such as a Y chromosome, the attached-2L will now segregate from the Y with a very high frequency. As with the compound-X chromosome, it is predicted that the presence of a heterozygous inversion in a compound autosome, while reducing or eliminating exchange between the two homologous arms, will have no conspicuous effect on their segregation behavior.

f. *Translocations in Distributive Pairing.* A break in each of two nonhomologous chromosomes with an interchange of parts produces a reciprocal translocation. Pairing for exchange in translocation heterozygotes involves the two rearranged chromosomes and their two nontranslocated partners. The greatest opportunity for recovery of euploid gametes exists when a crossover is present in each of the four translocation arms and the members are associated as a ring at metaphase I. There is somewhat less chance if crossing-over occurs in three of the arms to produce a 4-membered chain at metaphase I. Exchange in two of the arms, depending on whether the affected arms are alternate or adjacent, will lead to either two pairs of bivalents with a 50% chance of recovery of balanced gametes or a trivalent and univalent with slightly less than a 50% chance of such recovery. Thus the number of exchanges required to produce balanced products (or to avoid univalent formation) is somewhat higher with a translocation heterozygote than with the same two pairs of chromosomes in their normal arrangement. In the latter case one exchange per homologous pair or a total of two exchanges ensures two pairs of bivalents at metaphase I and balanced gametes at the completion of meiosis.

The frequency of univalent formation with a translocation heterozygote may approach certainty if the locations of the breakpoints are such as to produce an asymmetrical rearrangement with one extremely small chromosome. Now, the chance of exchange in either of the translocation arms formed by this chromosome approaches zero. Jain and Basak (1963) studied the meiotic cytology of a translocation heterozygote of this type in *Delphinium*. They found that almost invariably a multiple association of three chromosomes and an extremely small unpaired chromosome were present at anaphase I.

In organisms which possess a *melanogaster* female type of distributive pairing mechanism, asymmetrical translocations constitute a source of univalents for the distributive pool. Chandley (1965) utilized a series of translocations between the X chromosome and the right arm of chromosome 2 to analyze the relationship between the locations of the breakpoints and the frequency of X nondisjunction. Chromosome behavior was consistent with that predicted by the distributive pairing hypothesis and was interpreted in this way. Exceptional progeny depend on a 3:1 segregation of the four chromosomes comprising the translocation heterozygote such that the nontranslocated second chromosome goes to one pole and the remaining chromosomes to the other pole. Segregations of this kind will not take place if two bivalents are present at metaphase I and should be rare with a ring or a chain of four. The configuration required for 3:1 segregation is a trivalent and a univalent. Furthermore, disjunction from the trivalent should be of a particular kind, namely, that which sends both X-bearing chromosomes to the same pole and the nontranslocated second chromosome to the other.

Chandley's results showed that the highest frequency of X nondisjunction occurred with the most asymmetrical translocation in which the probability of the smallest member's presence in the pool approached 1, the probability of a trivalent occurring between the three other members was high, and the chance of the two appropriate members of the trio moving to the same pole was $\sim$ 50%. The substitution of a multiply inverted X chromosome, FM6, and a multiply inverted second chromosome, SM5, either singly or in combination for the normal X and 2, served to suppress crossing-over in various arms and to make noncrossovers available for distributive pairing. A significant increase in X nondisjunction was observed when FM6 was the only member of the pool. On the other hand, the extremely asymmetrical translocation, one of whose members was normally present in the pool, showed a significant decrease in X nondisjunction when a second chromosome, FM6, was added to the pool. Apparently the two chromosomes paired distributively and segregated to opposite poles, in this way eliminating the possibility of the required 3:1 segregation. The introduction of SM5 presumably led to the formation of one crossover bivalent between the normal X and one part of the translocation plus one noncrossover bivalent formed distributively between SM5 and the other part of the translocation. As a result, X nondisjunction fell well below control levels in all cases. When FM6 and SM5 were both present, crossing-over was eliminated in all arms. Now all four chromosomes were present in the pool and pairing was presumed to occur according to size. The highly asymmetrical translocation possessed four elements of very diverse size, and trivalent-univalent formation was postulated to account for the significant increase in X nondisjunction. The four components of the other translocations more closely approximated two pairs with respect to size, and in these cases bivalent formation was considered responsible for significant decreases in X nondisjunction.

g. *X Duplications at Exchange and at Distributive Pairing.* The independence of the two types of meiotic pairing in *melanogaster* females, as well as the distinctive characteristics of each, are illustrated with particular clarity in studies of the exchange and segregational behavior of a series of free X duplications of different total length, each carrying different amounts of distal X euchromatin (Fig. 11) (R. F. Grell, 1967). At exchange pairing, the activity of each duplication is assayed by the decrease in crossing-over it causes between the two normal X chromosomes in the region from the tip to the locus of white (1.5 crossover units) where the duplication possesses homology (see Section II,A,3,d). At distributive pairing, the activity of each duplication is measured directly by the frequency with which it segregates from a fourth chromosome. The findings are given in Table V, where the data in the left half of the table are concerned with exchange pairing and those in the right half with distributive pairing.

In the case of exchange pairing, the reductions in exchange induced by the presence of the different duplications (Table V, column 4) are seen to be positively correlated with the physical extent of homologous X euchromatin carried by the duplications (Table V, column 3). They show no correlation with total duplication lengths (Table V, column 2). In the case of distributive pairing, the frequencies of nonhomologous associations between the X duplications and chromosome four (Table V, column 9) are positively correlated with the similarity between total duplication length (Table V, column 7) and the length of chromosome four (= 1). Duplication w^{m5}, which alone possesses homology for chromosome four, gives a value consistent with its total length and shows no enhancement of pairing because of its shared homology. Thus it is evident that different attributes of the duplication are utilized for the two effects that have been measured. For recognition at exchange pairing, homology is required; the degree of recognition is dependent on the extent of homology up to a certain minimal length, and total size is not a factor. For distributive pairing, recognition depends on total length and is independent of homology.

An appraisal of the frequencies with which the duplications must be involved in each function further emphasizes the discreteness of the two events. To cause a two-thirds reduction in exchange, the duplication

TABLE V

THE DUAL ROLE OF FREE DUPLICATIONS AT EXCHANGE PAIRING AND AT DISTRIBUTIVE PAIRING[a]

Exchange pairing				Distributive pairing				
Duplication	Total length (chromosome 4 = 1)[b]	*y-w* region (%)	Reduction crossing-over, *y-w* (%)	Duplication	*y-w* region (%)	Total length (chromosome 4 = 1)[b]	Duplication, four nondisjunction (%)	Duplication, four association (%)[c]
1144	1.1	6	11	1144	6	1.1	0.13	99.7
856	3.0	21	23	1337	44	1.4	0.38	99.2
1337	1.4	44	46	w^{m5}	100	1.9	0.45	99.1
w^{m5}	1.9	100	66	856	21	3.0	1.70	96.6
z^9	3.8	123	68	z^9	123	3.8	3.03	93.9

[a] From R. F. Grell (1967).

[b] Measured at mitotic metaphase.

[c] Calculated by use of the expression $a = 1 - 2n$, where a = association and n = nondisjunction.

must compete for exchange pairing close to 100%; to segregate from a fourth chromosome 97–99.9%, a duplication must participate in distributive pairing with the four 94–99.8%. Equally cogent evidence is the finding that the reduction in X exchange by the X duplication is the same whether chromosome four is available for nonhomologous pairing or is absent.

6. *Evidence of the* melanogaster *Female Type of Distributive Pairing in Other Organisms*

Nonhomologous pairing at meiotic metaphase may be regarded as misdirected distributive pairing. For genetic studies, its special value lies in its disclosure of the existence of a distributive mechanism which would not be detected if the second pairing were restricted to homologs. Since nonhomologous pairing at this time is recognized by a response to abnormal genetic conditions, special genetic tools are required for its induction. In most forms, such tools are lacking. In three forms where they are available, a search for nonhomologous pairing has been carried out. Pollard and Käfer (1967) investigated the influence of chromosomal aberrations on meiotic nondisjunction in *Aspergillus*. Rearrangements for linkage groups VI and VII were found to increase not only the frequencies of nondisjunction of the linkage groups involved, but to induce nondisjunction of structurally normal chromosomes belonging to other linkage groups as well. The results were interpreted according to the distributive pairing hypothesis. Cattanach's studies (1967) with the mouse did not disclose any evidence of this phenomenon, although XY pairing in this species is probably of the distributive type. Weber's studies (1967) with maize also gave negative results. Since both the mouse and maize lack very small chromosomes as well as inversion polymorphism, segregation of homologs, other than the sex bivalent, is most likely accomplished by means of chiasmata. Nevertheless, there are indications in several organisms, other than *D. melanogaster* and *Aspergillus*, that nonhomologs may undergo distributive pairing.

In *D. pseudoobscura* variability in chromosome structure is virtually restricted to chromosome 3, but within this chromosome inversion polymorphism is widespread. Terzhagi and Knapp (1960) studied the effect of introducing inversion heterozygosity into another pair of chromosomes as well. Egg viability, which is 95.2 and 93.1% with no inversion and one inversion, respectively, is markedly reduced to 78.8% with the introduction of a second, heterozygous inversion. The increase in dominant lethality is hypothesized by these authors to be a consequence of nonhomologous pairing leading to aneuploid gametes. The restriction of inversion polymorphism to chromosome 3 is, according to this interpretation, the result of a strong negative selection against the presence of inversions in chromosomes other than 3. If this explanation is correct, a distributive mechanism similar to that found in *melanogaster* would be operative in *pseudoobscura.* The possibility must be entertained that a distributive mechanism occurs in all of the *Drosophila* species which display a highly variable structure for only one pair of chromosomes, such as *D. nebulosa* (Pavan, 1946) and *D. athabasca* (Novitski, 1946). It is also tempting to speculate that the strong interchromosomal effect of heterozygous inversions on the frequency of exchange within heterologous chromosomes may serve a related function. The increased exchange in heterologs should decrease the possibility that they too will be noncrossovers and will enter the distributive pool.

A second kind of evidence suggestive of a distributive mechanism, perhaps of the *melanogaster* female variety, comes from studies of secondary nondisjunction. In *melanogaster* it has been shown that the role of the Y chromosome in secondary nondisjunction is to pair distributively with those X chromosomes that have failed to undergo exchange and to cause them to nondisjoin. The frequency of secondary nondisjunction depends then on the contribution of noncrossover X chromosomes to the distributive pool. Factors which decrease exchange between the X's, such as heterozygous X inversions, act to increase this contribution. On the other hand, increased exchange between the X's should generally decrease the frequency of secondaries. Sturtevant and Beadle (1936) noted that as the total map length of the X chromosome increased in different species of *Drosophila,* the frequency of secondary exceptions decreased. Thus, *melanogaster* and *simulans,* with map lengths of about 70 units, have 2–4% secondary nondisjunction; *willistoni,* with 84 units, about 2%; *virilis,* with 132 units, 0.5%; and *pseudoobscura,* with 170 units, 0%. To these observations may be added that of Philip *et al.* (1944) on *subobscura,* which has an X map length of more than 150 units and where the introduction of a Y causes no increase in X nondisjunction over the 0.01% found without a Y. Gregg and Day (1965) report that in *hydei* the map length of the X is 116 units and secondary nondisjunction is ~ 0.03%. They interpret the low rate in *hydei,* as compared to *melanogaster,* to the increased genetic length of the X and the corresponding reduction in the number of noncrossover X tetrads available for distributive pairing with the Y. In all of the species noted, the inverse relationship between the frequency of X exchange and the frequency of secondaries may be satisfactorily explained on the assumption of a distributive pairing mechanism of the *melanogaster* female type.

In *D. melanogaster* the regular segregation of the dot-like fourth chromosomes, which do not undergo exchange, has been shown to depend on a distributive mechanism. It is possible that in all *Drosophila* species which possess a pair of dot chromosomes, regular segregation of these small chromosomes also depends on such a mechanism.

New approaches to the study of human chromosomes have produced a voluminous amount of cytological and genetic data. A number of cases have been reported that are interpretable on the distributive pairing hypothesis and for which no adequate alternative interpretation is presently available. These are situations in which rare chromosomal abnormalities of different kinds occur within single families or kinships, such as an autosomal translocation present in one member of a family and aneuploidy for a sex chromosome in a second; or an autosomal translocation in one member and aneuploidy for an unrelated autosome in a second; or aneuploidy for an autosome in one member and aneuploidy for a sex chromosome in another. This evidence has led to the proposal (R. F. Grell and Valencia, 1964) that the causal event responsible for aneuploidy in all of these cases could be nonhomologous associations at the distributive phase of meiosis. If this interpretation should be correct, the frequency of aneuploidy in a human population would be expected to be related to the frequency of rearranged chromosomes that are present in the population.

In flowering plants it is frequently possible to recover haploid strains. The metaphase stage of meiosis I in monohaploid forms represents an ideal situation in which to search for cytological evidence of nonhomologous pairing, for it is assumed that these chromosomes lack homology for one another except for any interchromosomal duplications which might be present. This is in contrast to haploids derived from allo- or autopolyploids in which homologs are expected in the genome. Meiotic associations between nonhomologs at metaphase I have been reported for monohaploids of *Oenothera* (Catcheside, 1932), *Secale* (Levan, 1942), *Antirrhinum* (Rieger, 1957), *Oryza* (Hu, 1960), *Hordeum* (Tsuchiya, 1962), *Zea mays* (Snope, 1966), and others. Catcheside's studies of haploid *Oenothera blandina* ($n = 7$) have shown a decrease in the frequency of univalents from diakinesis to metaphase of 91.5 to 79%, suggesting the possibility of the initiation of pairing between univalents at late prometaphase. On the other hand, Catcheside considered all metaphase pairing to be chiasmate and to originate from crossing-over between reduplicated segments. A similar origin is proposed by Rieger for many of the metaphase associations in haploid *Antirrhinum* and by Levan for late associations in haploid *Secale*. John and Lewis (1965), however, after careful examination of multivalent associations of two to five chromosomes in slides of haploid *Secale* (prepared by Dr. Hubert Rees) were unable to detect any chiasmata (Fig. 19). These authors consider the appearance of the paired associations in haploid rye to be identical to the quasibivalents described by Östergren and Vigfusson (1953) in asynaptic rye. That pairing in haploids can have segregational consequences is indicated by Riegers' studies of chromosome distribution at anaphase I in haploid *Antirrhinum,* where a significant departure from the expected binomial distribution, with a bias toward a more equal sharing of chromosomes between the two poles, is observed.

Triploid plants possess an entire extra set of chromosomes, and for those chromosomes which fail to undergo exchange with either of their homologs an opportunity for nonhomologous pairing at metaphase I exists. Yarnell (1929) examined buds of triploid *Fragaria* plants ($3n = 21$) and reported that instead of forming 7 trisomes or 7 disomes plus 7 monosomes at diakinesis, the chromosomes usually formed 10 disomes plus an unpaired chromosome. The disomes involved chromosomes of different sizes, which supports the conclusion that some of the disomes represented two nonhomologous chromosomes. Chromosome counts at second metaphase most frequently revealed 10 or 11 chromosomes at the plate, indicating that the members of the disome had segregated at anaphase I. Yarnell points out that determination of relationships in species hybrids, which are based upon chromosome pairing, are valid only if pairing at reduction is determined by homology. In *Fragaria* triploids this is clearly not the case, nor would it appear to be so with two triploid *Oenotheras,* in both of which Gates (1909, 1923) has reported equal distribution of the 21 chromosomes at anaphase I. Longley (1924) reported for triploid *Rubus* species ($3n = 21$) that 10 or 11 bodies were observed at first meiotic metaphase; also for a triploid *Citrus* ($3N = 27$) mostly 13 bivalents plus one univalent were found at MI and 13 chromosomes at MII (Longley, 1926). Conclusions concerning the evolution of polyploid genera based upon pairing behavior alone, therefore appear to be inappropriate. In autotriploids of *Anthoxanthum* ($n = 5$) and *Dactylus* ($n = 7$), Carroll (1966) has observed multivalents involving as many as 6 chromosomes as well as a number of instances of more than the haploid number of associated chromosomes. Carroll believes that a single interchange linking two trisomes could account for the multivalents; but significantly, multivalents are not observed in the parental diploid material. An equally attractive hypothesis is that multivalents of more than 3 chromosomes represent nonhomologous pairing. As in the haploid situation in *Antirrhinum*, observations at anaphase I in triploid *Anthoxanthum* revealed that the actual assortment was biased toward a more equal sharing of chromosomes (7+8) between the two poles. King (1933) and Darlington (1929) have found that in *Tradescantia* triploids there is also a tendency for equal numbers of chromosomes to pass to opposite poles.

Even in haploids of *Triticum aestivum*, where the question of the occurrence of nonhomologous pairing is complicated by the existence of homeology, clear-cut evidence of a metaphase end-to-end pairing independent of homology, as well as side-by-side pairing presumably homologous but also achiasmate in nature, has been described by Person (1955). The distribution of achiasmate chromosomes to the poles was found to be nonrandom, and the deviation from randomness was considered to be due to preferential 1–1 distribution from achiasmate associations.

A number of authors have described achiasmate pairing sometimes occurring between nonhomologs in interspecific hybrids. Walters (1954), who reported associations of this kind in interspecific hybrids of *Bromus*, terms them pseudobivalents. The chromosomes involved may differ in size and morphology or be connected at noncorresponding regions. They are generally oriented on the equatorial plate at metaphase I and exhibit centromere movement toward the poles. In the pollen mother cells of F_1 plants of the intergeneric hybrid *Raphanus-Brassica*, 18 univalents are observed at diakinesis, but at metaphase I up to 9 bivalents were described by Richharia (1937). Sficas (1963) made a statistical study of chromosome distribution in three interspecific hybrids of *Nicotiana*. He found a nonrandom distribution and suggests an achiasmate pairing followed by segregation of the paired chromosomes to opposite poles to explain the deviation from randomness.

C. Anomalous Meiotic Pairing

A variety of chromosome associations, generally leading neither to the exchange nor to the segregation of the participating chromosomes, have been described as occurring during meiosis. In some cases, the associations occur normally between nonhomologs or portions of nonhomologous chromosomes, whereas in others, nonhomologous associations are induced as a consequence of abnormal genetic situations.

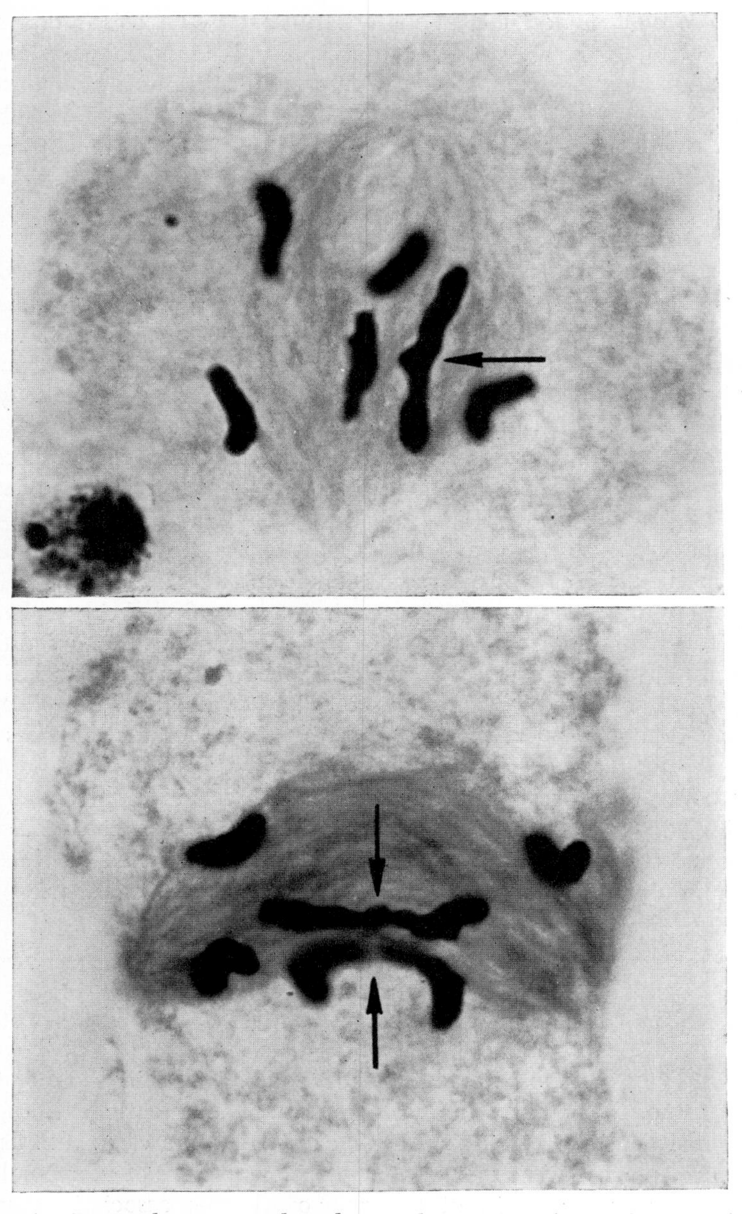

FIG. 19. Pairing between nonhomologous chromosomes (arrows) at metaphase I in haploid plants of *Secale*. (Courtesy H. Rees.)

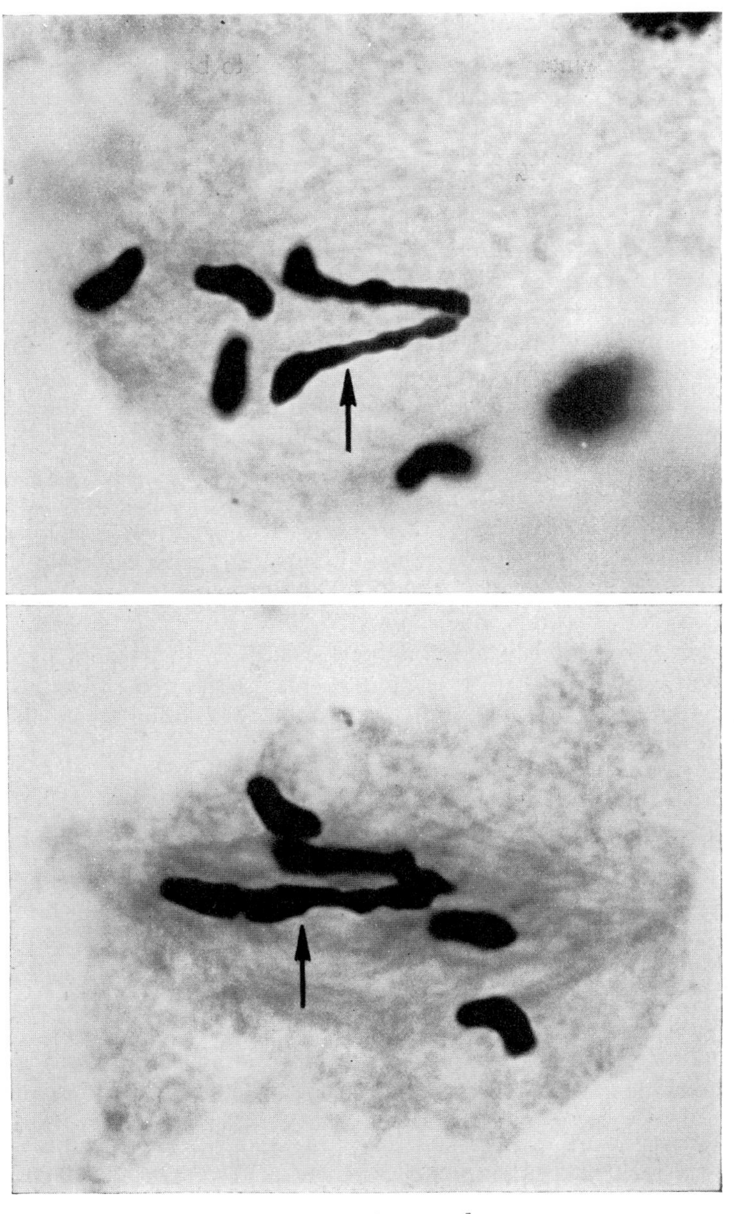

FIG. 19. *Continued.*

1. *Nonhomologous Pairing at Pachynema*

Pairing at pachynema is usually considered to be strictly homologous, chromomere for chromomere, but certain observations indicate that this is not invariably the case. Maguire (1960) reported for normal genomes that pachytene chromomere patterns do not always coincide in synapsed homologs of maize. With various abnormal genotypes, nonhomologous pairing at pachynema has been reported repeatedly. McClintock (1931, 1933) showed that reverse-loop formation (i.e., homologous pairing) does not always occur for heterozygously inverted regions in maize. The frequency of reverse loops is apparently correlated with the length of the inversion such that short inversions tend to be nonhomologously paired more frequently than long inversions. Maguire's studies (1966) of a small heterozygous inversion in chromosome 2 of maize showed that nonhomologous pairing, loop formation, and asynapsis occurred with frequencies of ~ 45, 35, and 20%, respectively. White and Morley (1955) have failed to find any evidence of reverse loops at pachynema in grasshopper chromosomes with heterozygous pericentric inversions. Similarly, McClintock (1931) has shown that in plants heterozygous for an internal deficiency, the position of the buckle in the nondeficient chromosome exhibits wide variability and an extensive amount of nonhomologous association is required to account for the latitude in pairing that is observed. Tabata's studies (1963) of intercrosses between stocks of maize interchanges involving the same chromosomes with different break points have shown that homologous terminal segments are usually paired but that the intercalary nonhomologous segments are often also associated. Tabata, like McClintock (1933), considers that associations initiated between two homologous regions proceed along the chromosome in a zipper-like manner, often bringing about an association of nonhomologous parts until prevented by a counter association from the opposite direction. Finally, in the haploid of maize, McClintock (1933) has described numerous cases of foldbacks within a chromosome as well as associations between chromosomes at pachynema. When supernumerary B chromosomes are present in the genome, they tend to pair at pachynema but are found more frequently as univalents at diakinesis. In general, McClintock found that associations of nonhomologous parts of chromosomes or of nonhomologs at pachynema lead to dissociations at diplonema and rarely continue into diakinesis. Often, however, two chromosomes are found loosely associated as a bivalent at metaphase I.

Perhaps the simplest interpretation of nonhomologous associations at pachynema is that such associations are initiated between duplicated regions and, once begun, association continues beyond the homologous segment into nonhomologous regions. In a monohaploid, the probability that small duplicated regions will recognize one another is enhanced by the absence of a competing homolog. The minimal length for recognition is unknown; but in view of the high frequency of recognition achieved under competitive conditions by duplicating fragments carrying well under one genetic unit of the chromosome in *Drosophila* (Section II,A,3,d), it might be suspected that the length required for recognition under noncompetitive conditions would be less than one map unit. With respect to exchange, the duplicating fragments rarely participate. If a similar situation prevails among monohaploids, chiasmata would be expected to be rare. A somewhat analogous case has been described in *E. coli* K12 by Berg and Curtiss (1967), who found that with haploid F′ two-ring genomes, the F′ segment integrates, though rarely, at a variety of sites, each of which presumably shares some homology with F′; whereas in the case of a partial diploid F′, integration is a much more frequent event but is virtually restricted to the region of diploidy.

2. *Nonhomologous Associations at Midprophase*

In certain forms possessing chromosomes with extensive, distally located, heterochromatic regions, the chromosome tips may be associated at pachynema, diplonema, and diakinesis. End-to-end adhesions between terminal segments have been described in the corixids *Sigara carinata, Corixa punctata,* and *Cymatia bonsdorfii* (Slack, 1938), and in the pentatomid *Edessa irrorata* (Schrader, 1941). The associations generally lapse before metaphase. Perhaps they reflect a tendency for metabolically inactive regions to adhere to each other at midprophase. The presence of the karyosome in the otherwise empty-looking nucleus of the *Drosophila* oocyte during the modified diplotene stages (3–12) might be interpreted in the same way. Labeling studies (Chandley, 1966) indicate that the karyosome is a dense aggregate of all of the chromatin in the nucleus; and according to the electron microscopic histochemical analysis of Koch *et al.* (1967), nucleolar synthesis of ribosomal RNA is suppressed in the oocyte and stimulated in the nurse cells concurrently with the appearance of the karyosome, suggesting a termination of synthetic activity in the oocyte at this time.

3. *Secondary Association at Prometaphase*

The terms secondary pairing and secondary associations were first used by Darlington and Moffett (1930) to describe the loose association of bivalents into groups at metaphase I. Pairing of this kind is not mediated by chiasmata nor does it affect the segregation of the bivalents involved, although the associations may be continued between pairs of daughter half-bivalents through anaphase I up to metaphase II. Secondary associations have been observed in plants considered to be allopolyploids, such as *Oryza, Dahlia, Brassica, Nicotiana, Digitalis, Rubus, Prunus, Pyrus, Hypericum, Viola, Aesculus, Betula, Corylus, Acer, Solanum, Taraxacum,* and *Carex.* They are apparently restricted to plants which possess small chromosomes (Lawrence, 1931). Darlington and Lawrence considered that the associations reflected a residual attraction between chromosomes related to one another but less closely than homologs. The occurrence of secondary pairing was, therefore, taken as an indication of a lower basic number of chromosomes for the plant. On this hypothesis, the "ur genomes" of a variety of plants, including *Brassica, Nicotiana, Dahlia, Oryza,* etc., have been inferred from the maximal number of bivalents present in a group.

On the other hand, Heilborn's studies (1936) with *Carex* led him to conclude that chromosomes grouped themselves according to size and irrespective of homology, but since chromosomes of equal size (or mass) are kept together, homologs, when present, will show association. Hirayashi (1957) also believed that secondary associations have no relationship to the true homology of chromosomes. His studies of *Oryza* ($2n = 24$) revealed as many as 12 bivalents associated together; and, similarly, Katayama (1965) observed chains of up to 7 bivalents in *Oryza.* The question of whether size or homology, or both, are the basis for associations of bivalents remains open.

D. Conclusions

Meiosis, as interpreted by the early cytologist, was solely a reductional process. For sexually reproducing forms it was the way of maintaining a stable chromosome number. Later, it was discovered that factors, known to be linked on the same chromosome at the onset of meiosis, need not be so at its termination, and crossing-over was recognized as a companion feature. Janssen's chiasma-type theory first equated the cytologically visible chiasma with the crossing-over event; but it was left to Darlington to elaborate the functional role of exchange, not only as a mechanism for reassorting genetic material between homologs, but as the equally important mechanism of holding them together and ensuring their proper segregation through the establishment of a physical bridge between them. Yet, the successful execution of meiosis in a number of forms, without the concomitant process of exchange and chiasma formation, corroborates in a sense the validity of the original view. While reduction constitutes the essence of the meiotic process and is indispensable to it, exchange may under certain conditions be eliminated.

Forms in which meiotic crossing-over is no longer retained appear to fall into two categories. In the first, crossing-over has disappeared throughout the genome in one sex or the other and regular segregation is achieved by the retention of an earlier pairing between homologs. The maintenance of this pairing in the absence of chiasmata seems to be accomplished through the telescoping or the abolition of the repulsion phase of meiosis, so that homologs, which normally fall apart except at loci of chiasmata, now remain associated. Presumably, the evolution of this kind of distributive mechanism must precede the elimination of exchange, for the presence of some segregational device would seem to be essential for survival. Exceptions to this category are those cases (e.g., *Lepidosiren*) in which desynapsis occurs for the entire genome only to be followed by a second, later pairing. Here it seems probable that exchange and chiasmata are completely absent but information on this point as well as on the recognition mechanism responsible for the second pairing is lacking.

In the second category, crossing-over is eliminated in certain specific members of the genome. Frequently these are small chromosomes or heteromorphic chromosomes. While meiosis remains normal for the remainder of the genome, the noncrossover members undergo a second kind of pairing which is often initiated at a late stage in meiosis I and involves greatly condensed chromosomes.

The necessity for invoking a second kind of pairing has been questioned by Novitski (1964). According to his view, it would seem inefficient for any organism to allow homologs, after achieving a synaptic configuration, to fall apart and then develop another system of distributive pairing to bring them together again. For this to be a valid objection, it would be necessary to establish that synapsis, per se, is sufficient to hold chromosomes together until metaphase. Although, as noted above, special adaptations of this kind may be made in cases of completely achiasmate meiosis, the repulsion phase which normally intervenes at diplonema and early diakinesis should cause homologs to fall apart if chiasmata are absent. Cytological evidence, particularly the description of chromosome behavior in interspecific and intergeneric hybrids, is fairly convincing on this point. The studies of Brown (1954, 1958) with species hybrids of *Gossypium*, of Menzel (1962) with intergeneric hybrids of *Lycopersicon* and *Solanum*, and of Tobgy (1943) with interspecific hybrids of *Crepis*, all agree that while synapsis appears intimate and complete at pachynema, univalents are frequent at metaphase. Moreover, those chromosomes that show the greatest reduction in chiasmata, as compared with the parental condition, are those that are most frequently found as univalents. Brown (1954) concluded that chiasma formation, and not pachytene synapsis, is the critical event for metaphase association. The conclusion seems inevitable that the development of a secondary pairing system to promote homologous pairing for segregational purposes would be of great value for those genomes in which noncrossover chromosomes normally form a fraction of the complement.

Furthermore, the existence of secondary pairing systems is well documented. Such systems have been described by E. B. Wilson (1928), who devotes a section of "The Cell in Development and Heredity" to late conjugation and diakinetic synaptic phenomena. Wilson states that in some cases the primary pairing occurs at the final stages of meiosis I, while in others the pairing seems *"to present a secondary coupling that has been preceded by a typical synapsis at the usual time and a subsequent deconjugation"* (italics mine). Muller (1941) recognized the occurrence of secondary pairing between condensed chromosomes and raised the question as to whether such pairing only superficially resembles real synapsis but has an entirely different basis.

Using a completely different approach, it has been possible to confirm the existence of such a secondary pairing system in the female of *D. melanogaster* and to learn something of its characteristics. Genetic analysis has shown that homologs that fail to undergo exchange, and thus lack chiasmata, such as the "dot-like" fourth chromosomes or the X chromosomes heterozygous for highly complex rearrangements, nevertheless segregate very regularly. Their segregation becomes irregular, however, when extra, nonhomologous chromosomes of the correct size are introduced into the genome. Under proper conditions, the nonhomologs are found to segregate from one another as regularly as the homologs. Does such nonhomologous pairing occur at the expense of homologous pairing? If so, it should reduce exchange between homologs. Our findings show that nonhomologs that segregate from one another have invariably failed to undergo exchange with an independent homolog. Yet, when conditions exist such that a particular chromosome is free to undergo exchange with its homolog and to segregate from a nonhomolog as well, the latter event in no way interferes with the former. The frequency of exchange between homologs remains constant whether nonhomologous pairing occurs or fails to occur. Thus, the two processes are noncompetitive and independent. To be so they must take place at different times.

Finally, we might speculate a bit on the conditions which would favor the inclusion of a secondary pairing system. As a first premise, we might assume that crossing-over, in the primitive organism in which the meiotic system arose, occurred more frequently than it does now in higher forms. Recombination per nucleotide pair has been calculated to be about 1000 times as frequent in phage T4, ~ 100 times as frequent in *E. coli,* and ~ 10 times as frequent in *Aspergillus* as it is in *Drosophila* (Pontecorvo, 1958). A tendency toward suppression of recombination, then, seems to have been one feature in the evolution of higher organisms. If this is true, a pair of homologs in the primitive form would have normally been bound together by a number of chiasmata, and regular segregation for the entire genome would have been assured by exchange. In many higher forms crossing-over may still be retained as the sole basis for orderly segregation, but in others it may not. Cooper (1945) recognized this premise many years ago when he wrote, "the chiasma proves only to be a sufficient—not a necessary—cause for segregation in some organisms, while in others, such as the lilies, of all possible mechanisms only chiasmata appear to be capable of guaranteeing segregation."

A variety of chromosomal alterations might be expected to interfere with the operation of the primitive system. First, a pair of homologs might diverge and eventually become so heteromorphic as to exclude the possibility of exchange. This situation has happened frequently with the sex chromosomes. Second, heterozygous inversions, which suppress recombination and keep portions of the genome intact may, because of selective value, become incorporated. Balanced inversion systems of this kind are well known in the genus *Drosophila.* Finally, a chromosome may become greatly reduced in size, so that small elements such as the dot chromosomes of *Drosophila* or the m chromosomes of Hemiptera may now be part of the genome. Darlington (1930), aware that chiasma frequency is correlated with chromosome length, also realized that failure of exchange in small chromosomes should prevent their perpetuation. To account for their continuance he postulated that localized chiasmata were present. In a number of instances the cytological evidence indicates that pairing between small chromosomes is initiated late in prophase and that localized chiasmata, if defined as the cytological equivalents of crossovers, do not exist (E. B. Wilson, 1928).

In all of the cases cited above, the primitive mechanism for ensuring regular segregation, via crossing-over, would no longer be operative. Since random assortment of univalent chromosomes leads to aneuploid gametes and generally to zygote lethality, any device that serves to circumvent this eventuality should have a high selective value. This, then, would appear to be the role of distributive pairing.

Under natural conditions, it is assumed that distributive pairing in *Drosophila* and elsewhere occurs only between homologs or heteromorphs and leads to balanced gametes. Thus, distributive pairing is considered to be a mechanism for promoting homologous pairing and regular segregation and, as such, to possess a selective advantage. Under natural conditions in the *melanogaster* female, only the fourth chromosomes and the X chromosomes are present together in the distributive pool with any frequency, and when this happens ($\sim$5% of the time) the difference in size between them ensures that they will not pair with one another. It is further surmised that the pairing of nonhomologs, encountered in the *melanogaster* female but not in the male of this species, is the consequence of highly abnormal genotypes in which the recognition mechanism, which is normally adequate to restrict distributive pairing to homologs, becomes inadequate to do so when grossly rearranged chromosomes of unusual lengths are introduced into the genome. In our view, then, nonhomologous pairing is considered to be largely a laboratory phenomenon which occurs only rarely in nature and which, when it does, acts as a selective device to prevent its own recurrence. In other words, it is those factors, whether chromosomal or genic, which permit nonhomologs to pair distributively that would be at a selective disadvantage and would be rapidly eliminated, while the mechanism itself would be at a great selective advantage and would be retained.

III. Somatic and Gonial Pairing

A. Discovery and Occurrence

Pairing of homologous chromosomes is generally considered to be a special property of the meiotic cell, with the conspicuous exception of the Diptera, where it apparently occurs in all cells. Yet, an examination of the cytological literature reveals that the forms in which somatic or gonial pairing has been observed include a wide variety of plant and animal species. The discovery that homologs may undergo pairing in nonmeiotic cells was first made in the gonial cells of certain Hemiptera by Montgomery (1901) and was followed by a similar report for the spermatogonia of the grasshopper *Brachystola magna* by Sutton (1902). These observations were extended to somatic cells by Strasburger (1905), who found association of chromosomes in pairs in the embryonic nuclei of the dicot *Galtonia candicans* and the monocot *Funkia sieboldiana*. Pairing in these forms as judged by the figures seldom appears intimate.

In the following decade the number of plant genera and species reported to exhibit somatic pairing was greatly increased and came to include *Pisum* (Strasburger, 1907), *Hydrocharis, Lychis* and *Bryonia* (Sykes, 1908), *Calycanthus floridus* (Overton, 1909), *Yucca* (Müller, 1909), *Melandrium rubrum, Mercurialis annua,* and *Cannabis sativa* (Strasburger, 1910), *Spinacea* (Stomps, 1910), *Ricinus* (Nemec, 1910), *Oryza sativa* (Kuwada, 1910), *Morus alba* and *M. indica* (Tahara, 1910), *Dahlia coronata* (Ishikawa, 1911), *Najas marina, Listera ovata, Albuca fastigiata, Aloe hanburyana, Eucomis lucolor, Bishornerea superba, Bullieni annua, Nerine rosea, Muscari botryodes, Scilla byolia, Chinodoxa luciliae, Hyacinthus oventalis* (Müller, 1912), and *Oenothera lata* (Gates, 1912).

Despite the fact that somatic pairing is apparently widespread in plants, it continues to be popularly associated with the order Diptera, where it was first described by Stevens (1908) and where it has been intensively investigated by Stevens (1908, 1910), Metz (1914, 1916), Whiting (1917), S. M. Grell (1946a), and others. Stevens, in the first report of somatic pairing in this group, wrote:

> Perhaps the most interesting point in the whole study is the pairing of chromosomes in cells somewhat removed from the sphere of the reduction process. This was first noticed in the oögonia of *Drosophila* and was also found to occur in the ovarian follicle cells, the spermatogonia and some embryonic cells. This is not an occasional phenomenon but one which belongs to every oögonial and spermatogonial mitosis.

Metz (1916) studied the chromosomes of about 80 species of Diptera, in many cases including somatic, spermatogonial, and oogonial cells, and found in all species that the chromosomes were uniformly associated in pairs in both somatic and germinal cells. The association in pairs was effected in early cleavage stages and continued through all stages of ontogeny from egg to adult. In cells that were polyploid, as many dipteran cells are, Metz found that the homologs associated in aggregates rather than pairs, so that the number of groups was equivalent to the haploid number of chromosomes, regardless of the degree of ploidy which in the gut cells of *Culex pipiens* may reach 64-ploid (Berger, 1941).

In certain cells—for example, those found in the salivary glands of many Diptera—reduplication occurs without separation of the sister chromatids, leading to a condition known as polyteny. Polytene homologs also exhibit somatic pairing, as do salivary gland chromosomes of triploid *Drosophila* larvae (Painter, 1934). An exception to this rule is found in a special type of giant cell present in the salivary glands of a Cecidomyid fly, *Lestodiplosis sp.* (White, 1946). This cell, which measures 150 $\times$ 220 μ, is simultaneously polyploid and polytene, for it contains 16 times the number of chromosomes found in the diploid cell and each chromosome is polytene. In this case chromosome pairing is restricted to 2's.

Thus far somatic pairing has been unambiguously reported for only one group of animals outside of the Insecta. Boss (1953, 1954) observed in both tissue culture fibroblasts and in tail tips of the newt *Triturus cristatus carnifex* an approximate halving of the chromosome number during anaphase. By following the mitotic process in living cells, Boss established that the reduction in number came about through chromosome pairing and that this process could be reversed by treatment of the cells with a hypotonic solution.

Recently, as more refined techniques for analyzing the spatial relationships of chromosomes have been utilized, somatic pairing has been reported for such cytogenetically well-known plant species as *Zea mays* (Maguire, 1967b), *Triticum aestivum* (Feldman *et al.*, 1966), *Haplopappus gracilis* (Mitra and Steward, 1961), *Crepis capillaris* and *Aegilops squarrosa* (Kitani, 1963), and *Beta vulgaris* (Butterfass, 1967). The possibility of homologous associations in human cells has been suggested by Barton and David (1962) and Schneiderman and Smith (1962). The latter workers analyzed statistically the distribution of four identifiable members of the genome at somatic metaphase and in each case found a tendency for homologous association which reached levels of significance with chromosomes 3 and 16. Analyses of this kind have led to positive reports so frequently as to suggest the possibility that homologous associations may be a universal cell phenomenon, differing only in intensity from one organism to another.

B. Stage of Cell Cycle

Most observations of somatic pairing have been made at metaphase, a time when condensed, well-spread chromosomes lend themselves most satisfactorily to analyses; but this phenomenon has also been reported for prophase, anaphase, telophase, and even interphase stages. One intriguing, but largely unanswered question concerns the normal distribution of homologs in interphase nuclei at the time they reach their maximal extension and when they are generally unidentifiable. The resolution of this problem would permit us to either affirm or deny the proposition that chromosome pairing during the metabolically active interphase is a normal event and that the retention of such pairing to a greater or lesser degree accounts for the extent of mitotic pairing that is typically observed in different forms.

Two types of observations bear on this issue. First, if somatic pairing achieves its greatest intimacy and functional significance during interphase, then mitotic pairing as either a relic or a prelude to this state should be greater during prophase and anaphase and minimal during metaphase. Unfortunately, reports on this subject are contradictory. Metz (1916) describes the pairing process in various dipteran species as continuing through all stages of cell division from earliest prophase to latest anaphase and being most intimate in the earliest and latest stages and least intimate at metaphase. On the other hand, Kaufmann (1934), studying the ganglion cells of *D. melanogaster*, believes that pairing reaches its maximal expression in late prophase, while T. Hinton (1946), using similar material, reported that pairing is not completed until late metaphase.

Kitani's studies (1963) of somatic pairing in *Crepis capillaris* and *Aegilops squarrosa* partially support the conclusion of Metz. He finds that at telophase homologous chromosomes that were arranged side-by-side at anaphase come together and lose their individuality under the light microscope. In *Crepis capillaris* ($n = 3$) only three chromosomes are observed at both telophase and at the beginning of interphase, although at later interphase stages no determination could be made. At early prophase, however, a 2N number was usually observed, but in certain exceptional cases intimately paired homologs were seen at midprophase. Somewhat similar results were obtained by Hiraoka (1958), who determined the number of chromosomal bodies present in the nuclei of a variety of tissues of *Daphne* ($n = 14$) at different mitotic stages. He found that $\sim$ 80% of the chromosomes are present as paired homologs at midprophase, whereas this number is reduced to $\sim$ 21% by metaphase. As the chromosomes move toward the poles at anaphase, their pairing tendency becomes more conspicuous, and by the time the nuclear membrane is formed, the chromosome bodies are again reduced to almost the haploid number.

A more direct approach to the problem is to study the interphase nucleus itself. Hiraoka's studies (1958) are again pertinent, for he finds in the interphase nuclei of *Daphne* a number of massive chromosomal bodies of irregular outline which are about twice as large as the metaphase chromosomes and which in some cases show a light line suggesting their double nature. A statistical investigation of their number revealed a variation from 11(3.3%) to 18(0.7%) with a mode at the haploid number of 14(51.0%). Hiraoka concludes that there is a strong pairing tendency at interphase.

Furthermore, if the salivary gland nucleus is interpreted correctly as an interphase nucleus which through multiplication of the number of chromosomal strands presents a magnified view of what normally exists at this time, then homologous pairing of a highly specific and intimate nature appears to take place.

When chromosomal condensation at interphase is absent, identifiable cytological markers may be used to determine the positions of homologs. Kitani (1963), using a distinctive satellite chromosome in *Aegilops squarrosa*, studied the distribution of three satellites in interphase nuclei of endosperm tissue (3N). He found that all three satellites were associated 36%, that two were associated 57%, and that no association occurred only 7%, thus suggesting a high frequency of interphase pairing. In a similar type of investigation Maguire (1967b) has used two maize stocks homozygous for chromosome "knobs" of various sizes to charter the positions of the knobs in interphase nuclei of tapetal, root tip, and premeiotic cells. In all cases very significant departures from random distribution were obtained, and Maguire concluded that some degree of homologous pairing existed during interphase. Since in many cases the heteropycnotic knobs show a tendency to be paired at a short distance, comparable to the type of somatic pairing of homologs found in Diptera, it is conceivable that the more extended regions of the chromosome are more closely aligned at this time. All traces of homologous pairing in maize are lost by prophase.

C. Specificity

Somatic and gonial chromosomes, much like meiotic chromosomes, exhibit a wide latitude in pairing specificity, depending on the type of cell and on the point in the cell cycle one chooses to study. Close to one extreme is the salivary gland chromosome, where pairing is so intimate as often to suggest that the two homologs are fused, and where the specificity lies in the band rather than in the chromosome. The consequence of band-to-band pairing leads to the characteristic configurations of reversed loops when heterozygously inverted sections of sufficient length are present or to asynapsis if the inverted portion is too short for loop formation. Similarly, heterozygously translocated segments or duplicated portions of the genome are generally found paired closely with their homologous regions of the normal chromosome. Although the specificity in salivary pairing resembles that normally found at pachynema of a meiotic cell, no restriction to two-by-two pairing applies to the salivary chromosome. On the other hand, the occurrence of 4-strand somatic and gonial exchange (see Section III,D,2,a) does imply that precise two-by-two exchange pairing can occur, though rarely, in somatic and gonial cells.

The degree of nonhomology that is compatible with good salivary pairing may not be very great. For example, the fourth chromosomes of *D. melanogaster* and *D. simulans* differ by an inversion of ~ 10 bands and by the presence of two extra bands in *melanogaster* and of one in *simulans*. These alterations apparently have little effect on viability, since the *simulans* fourth chromosomes may replace the *melanogaster* 4's in the *melanogaster* genome. Yet these divergencies, in conjunction with whatever cytologically undetectable differences may exist, are sufficient to eliminate completely pairing of the fourth chromosomes in the salivary gland nuclei of the hybrid. On the other hand, meiotic segregation of the *melanogaster* and *simulans* four is very regular, indicating good distributive pairing (R. F. Grell, 1968).

If the salivary gland nucleus is assumed to represent an interphase stage, then by comparison the intimacy of pairing during the mitotic stages in *Drosophila* appears to be considerably reduced. Although a side-by-side alignment of homologs is generally the rule, the homologs tend to be separated by a small distance over most of their length, but this is not invariably so since homologous segments do sometimes come into contact. Correlated with the decrease in pairing intimacy during mitosis, there appears to exist a decrease in the specificity of pairing. Oliver and Van Atta (1932) studied oogonial cells with heterozygous chromosomal rearrangements and found that in the case of a reciprocal translocation between chromosomes 2 and 3, with both breaks close to the centromere, pairing occurred arm-for-arm so that a quadrivalent cross formation was observed. Alternatively, if the fragment translocated was very small, no evidence for attraction between it and its homologous region in the normal chromosome was obtained. However, T. Hinton's observations (1946) on the giant cells of the larval ganglion of *Drosophila*, which carried heterozygously an insertion of a large interstitial region of X into the left arm of the second chromosome, indicated to him that there is an attraction of a specific nature between regions, as evidenced by the manner in which the inserted segment in chromosome two pulled out of position in an attempt to pair with its homolog. This interpretation has been seriously questioned by Cooper (1948b), who finds a similar extension of the constriction region of 2L in the normal cell when the heterozygous translocation is absent and which he attributes to distortions incident to the squashing of the nucleus.

The question of specificity in mitotic pairing has been reexamined by

Brosseau (1965), who used neuroblast cells of *Drosophila* larvae bearing three types of heterozygous rearrangements: (1) inversions alone, (2) inversions in one homolog and a transposition in the other, and (3) inversions in one homolog and a translocation in the other. In general, somatic synapsis was found to be fairly regular, being least so in the inversion, translocation heterozygote. The chromosomes paired as bivalents rather than multivalents, although a bivalent with one arm paired and the other unpaired was often seen. No evidence for the preferential involvement of the heterochromatic regions was found. In the case of the inversion heterozygote, pairing was good, indicating that nonhomologous pairing for large segments of the chromosome is the rule here.

Nonhomologous pairing of this variety may be looked upon as a transitional step between homologous pairing and pairing of completely nonhomologous chromosomes. A study of the latter type of pairing in the oogonial cells of *Drosophila melanogaster* has recently been carried out (R.F. Grell, 1967). The procedure involved the introduction of two extra, nonhomologous chromosomes (an X duplication and a fourth chromosome) into the genome. Cytological studies showed that pairing between the two nonhomologs occurred with a high frequency, reaching values approaching 50% (Fig. 20). Like distributive pairing in the oocyte of *melanogaster*, the frequency of nonhomologous pairing in the oogonia was found to be positively correlated with the similarity in size between the two participating nonhomologs. Unlike distributive pairing in the oocyte, the substitution of a fourth chromosome centromere and base into the X duplication increased the frequency with which it paired with the fourth chromosome, indicating a role for homology at this time.

Feldman *et al.* (1966) suggest that the centromere plays a part in somatic association. They analyzed the positions of identifiable telocentric chromosomes in the root tip cells of *Triticum aestivum*. Homologous telocentrics tended to lie near one another, whereas nonhomologous telocentrics were randomly distributed; but when the two arms of the same chromosome, each carrying an identical centromere, were present as telocentrics, they associated to the same degree as homologous telocentrics. Feldman's data do not exclude the possibility that size may be a factor in the associations, for while the homologous telocentrics were identical in length and the telocentrics derived from the two arms of the same chromosome were not markedly different in length (arm ratios 1.29:1 and 1.23:1), the lengths of the nonhomologous telocentrics for the one case shown were conspicuously different.

Nonhomologous pairing in somatic cells of vertebrates has been reported by Boss (1954). In haploid individuals of *Triturus cristatus*, as

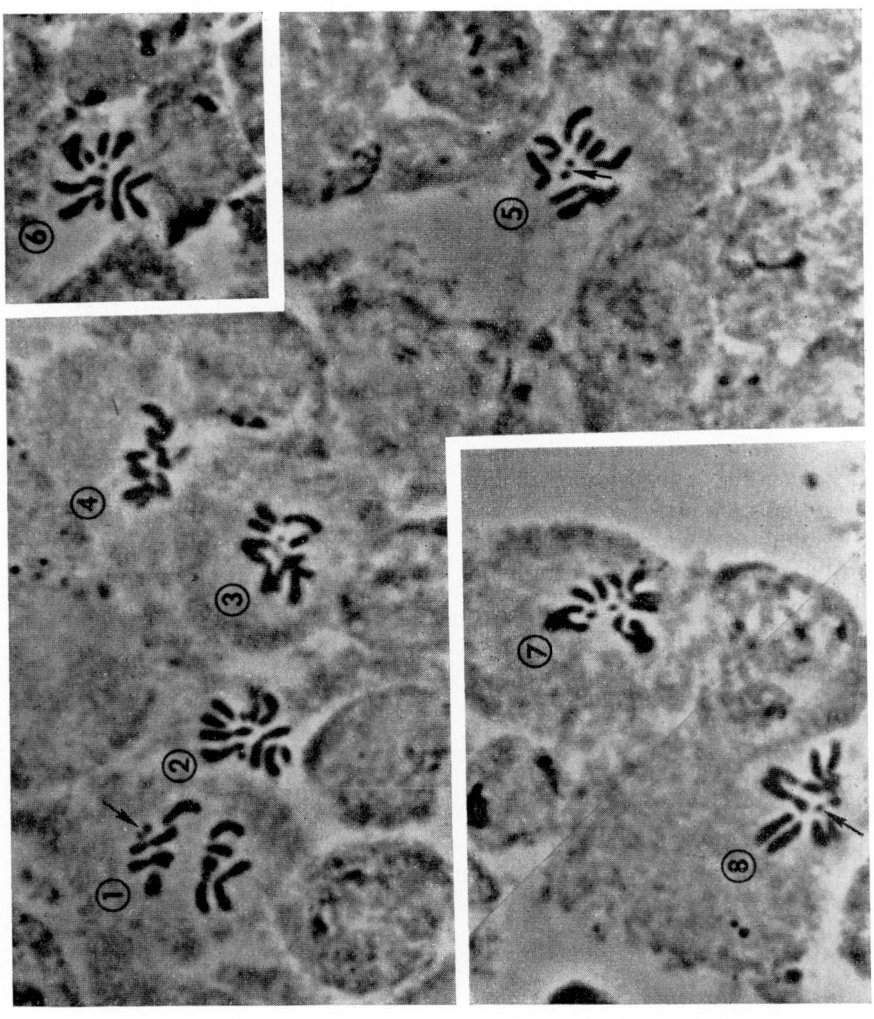

FIG. 20. Nonhomologous pairing between an X duplication and chromosome 4 in a cyst of eight oogonial cells at the final premeiotic metaphase. Arrows in cells numbered 1, 5, and 8 indicate paired nonhomologs. (From R. F. Grell, 1967.)

well as in the diploid, the chromosome number is halved during anaphase, indicating that pairing between nonhomologous chromosomes must occur here.

Perhaps the most consistent, although clearly speculative picture that can be drawn from these diverse observations is that intimacy and specificity of pairing go hand-in-hand and that both are a normal feature of the extended, metabolically active interphase chromosome; that pairing becomes less intimate and less specific when the chromosomes condense and the cell moves into its mitotic phase; and that the specificity requirement is so reduced with the approach of metaphase that nonhomologous pairing is permitted. Following metaphase and anaphase, as the cells progress again to the metabolically active, intermitotic stage, intimate, homologous pairing is resumed.

D. Functional Aspects

1. *Metabolic*

If active portions of the chromosome are normally paired during interphase, such pairing might play a role in certain known or inferred chromosomal functions such as replication, transcription, translation, regulation, and organization of the products of chromosomal activities. Evidence on this point is meager, but at least two established cases of position effects lend themselves to this interpretation. One such case is an unusual type of position effect, described by Lewis (1954), that is observed under special conditions when mutants of the bithorax region are present in *Drosophila*. Briefly, when two closely linked mutants *a* and *b*, both situated in the bithorax region, give a mutant phenotype in the *trans* but not in the *cis* configuration (i.e., a+/+b but not ab/++), the introduction of a chromosomal rearrangement may produce an enhancement of the mutant phenotype if the rearrangement transfers the bithorax region in one homolog of the double heterozygote, a+/+b, to a foreign position in the genome. Lewis (1954) has called this phenomenon "the transvection effect." Cytological studies of the salivary gland chromosomes indicate that the rearrangements which give rise to the transvection effect are those which disrupt somatic pairing. The effect apparently arises from a reduction in the degree of functional cooperation which normally exists between homologs in this region.

One of Lewis' interpretations (1954, 1963) of the transvection effect was a reasonable extension of his sequential reaction model, which he had originally proposed (1951) to explain the *cis-trans* effect. The bithorax region was assumed to consist of a series of functionally related genes or pseudoalleles acting in a polarized, sequential manner such that gene a^+ (but not a) makes a product A which is utilized in some way by gene b^+ (but not b) to make product B. In the *trans* heterozygote, a+/+b, the substance A would have to diffuse from the homolog carrying a^+ to the other carrying b^+ if it is to be transformed into B. If the substance A is effectively transported along its own chromosome but does not diffuse easily to its homolog, crossfeeding would be poor and B would be produced in lesser amounts than in the *cis* configuration ab/++, where crossfeeding is unnecessary. When the trans configuration is present in conjunction with a heterozygous rearrangement that reduces somatic pairing, crossfeeding would be expected to be further reduced; hence an intensification of the mutant phenotype would be expected (i.e., the transvection effect).

As pointed out by Lewis (1963), the sequential reaction model appeared to be inconsistent with biochemical studies which indicated that the translation process from RNA to protein in eucaryotes takes place in the cytoplasm, divorced and removed from chromosomal activities. However, more recent works (Byrne *et al.*, 1964; Shin and Moldave, 1966) have raised the possibility that transcription and translation may be transiently coupled in the nucleus through a DNA-RNA-ribosome-nascent protein complex. In this case, the above explanation becomes more feasible.

A second type of unusual position effect, which suggests that pairing of homologs in somatic cells may be a normal requirement for proper function, comes from the extensive studies of the R locus in maize by Brink and his co-workers. The R alleles, which control the formation and distribution of anthocyanin in plants and seeds, are divisible, with reference to paramutation, into three categories—one that is stable, a second that is paramutable, and a third that is paramutagenic. When a paramutable allele such as R^r is passed through a heterozygote with a paramutagenic allele such as R^{st}, the level of action of R^r is heritably altered, in that R shows a marked reduction in aleurone-pigmenting ability. This newly arisen paramutant form, now called R^{rl} is metastable, for it normally reverts partially back to R^r (Kermicle, 1963), although a further reduction in pigmentation can be induced if R^{rl} is kept heterozygous with a paramutagenic allele through subsequent generations. The change to the paramutant form R^{rl} has been shown to occur in somatic cells and not to require meiotic pairing (Brink, 1959). The pertinent information for the subject under consideration is that structural changes in chromosome 10, in which the R locus lies, when present heterozygously, reduce the sensitivity of the R^r allele in the translocated chromosome to the paramutagenic action of the R^{st} allele in the normal chromosome ten. It would thus appear that physical proximity of the two R alleles, possibly as a result of somatic pairing, may aid in the paramutation process.

2. *Genetic*

a. *Mitotic Exchange.* That crossing-over might occur in somatic cells was first postulated by Serebrovsky (1926) to account for the somatic segregation of genetically diverse tissue that he observed in the plumage of certain fowl hybrids. About the same time, mosaicism for X-linked genes in individuals of *Drosophila melanogaster* was described by Bridges (1925). In this case, a dominant factor on the X chromosome, M(1)n, appeared to be causally related to the mosaicism. Bridges proposed that the Minute factor caused the occasional elimination of the chromosome on which it was located, thus uncovering any recessive alleles that might be present in the remaining X, and permitting their expression in that tissue where elimination occurred. Similarly, mosaicism for autosomal recessives in flies heterozygous for both such recessives and for an autosomal Minute factor was reported by Stern (1927).

While subscribing to the Bridges' interpretation of chromosome elimination, Stern suggested that the loss was restricted to that portion of the chromosome which carried the Minute factor. Nine years later, in an elegant and comprehensive study of mosaicism, Stern (1936) repudiated the elimination hypotheses as the primary cause of mosaic formation in *Drosophila* and provided convincing evidence that mosaic areas, which appear in a varying percentage of flies whenever they are heterozygous for genes whose homozygous effect is recognizable in small spots, arise most frequently from somatic exchange.

The critical evidence in this study came from females which carried the recessive allele of yellow in one X chromosome and the recessive allele of singed in the other. Females of this doubly heterozygous genotype exhibited adjacent twin spots of approximately equal size, one spot showing a yellow phenotype and the other a singed phenotype. Equivalent twin spots are not an expected consequence of chromosome elimination but are readily explicable on the hypothesis of somatic crossing-over. As a result of Stern's studies it became clear that mitotic exchange differs in no essential features from meiotic exchange, for it too was found to be a reciprocal event, occurring at the four-strand stage and involving a precisely equivalent exchange of genetic material.

Stern's work and subsequent studies by Kaplan (1953) have revealed a number of intriguing aspects of somatic exchange which seem portentous but which still await final clarification. For instance, the enhancement of the frequency of somatic exchange in the presence of Minute factors, which are known to prolong development, is obscure; the occurrence of somatic but not meiotic crossing-over in male *Drosophila* is not understood; the tendency for somatic crossovers to occur in proximal heterochromatic regions, in contrast to the more distal location of meiotic crossovers, is curious. Judging by spot size, the proximal exchanges become less frequent, and the distal exchanges more frequent as development proceeds. It is tempting to speculate that this temporal pattern of exchange may reflect changes in activity in different chromosomal regions with time. Proximal heterochromatin may in fact constitute active euchromatic regions during embryonic and/or early larval stages. If crossing-over is restricted predominantly to extended, active regions, as is suggested by the virtual absence of meiotic exchange in regions designated heteropycnotic (Brown, 1940; Baker, 1958; Roberts, 1965), then the prolongation of the early stages by Minute factors could prolong the active period for proximal regions and in this way extend the time available here for somatic exchange.

There is a tendency to associate somatic crossing-over with somatic pairing in *Drosophila* on the assumption that such pairing would provide the opportunity for exchange. This correlation may be valid, but it should be remembered that the time of exchange may not coincide with the time of cytologically visible pairing. Furthermore, few organisms offer the genetic means for detection of somatic exchange, which may, in fact, be widespread. Among birds, Serebrovsky's early studies (1926) suggested that two recessive linked genes, present in the trans configuration, may undergo somatic exchange followed by somatic segregation to produce the mosaic condition of the plumage. A number of authors have tentatively proposed that somatic crossing-over occurs in the mouse. Carter (1952) described a mouse heterozygous for the semidominant color mutant W^v which showed patches of full-colored pigmentation (i.e., lacking W^v) and which bred as though germinal tissues were partly deficient for $+^w$. Somatic crossing-over offers the most probable explanation for the double alteration, although somatic reduction cannot be ruled out. Russell (1965) described three mice with coat pigmentation variations that might have originated through somatic crossing-over. In two of the mice, twin spotting was a possibility. Bateman (1967) has reported in the mouse a probable case of somatic crossing-over based on twin spotting. Grüneberg (1966), after reviewing a number of anomalous cases in the mammalian literature, many of which are inexplicable or poorly explicable, except on the assumption of somatic crossing-over, concluded that there is a strong case for the existence of somatic crossing-over in laboratory rodents.

No selective advantage would seem to accrue to most organisms from somatic crossing-over, and in the case of cell lethals or defective genes the results would be clearly deleterious. In certain fungi, on the other hand, mitotic crossing-over apparently plays a functional role as a mechanism for genetic recombination, supplementing or even replacing the sexual cycle. Pontecorvo and Roper (1952) have applied the term parasexual cycle to this method of providing for genetic variation. The parasexual cycle involves the fusion of two haploid nuclei (which must be genetically unlike if recombinational consequences are to ensue) in a heterocaryon to give a diploid, heterozygous nucleus. The frequency that diploid nuclei of spontaneous origin occur is estimated at 10^{-6}, but exposure to ultraviolet light raises the frequency to 10^{-2}. The diploid nuclei occasionally undergo mitotic exchange during their multiplication. They also undergo haploidization (see Section III,D,2,b) at a high rate to produce haploid nuclei which most often are different from the parental haploid in genotype.

The existence of mitotic recombination has been demonstrated for *Aspergillus nidulans* (Pontecorvo and Roper, 1952), *Aspergillus niger* (Pontecorvo *et al.*, 1953), *Aspergillus sojae* (Ishitani *et al.*, 1956), *Penicillium chrysogenum* (Pontecorvo and Sermonti, 1954), *Fusarium oxysporum* f. *pisi* (Buxton, 1956), and *Saccharomyces cerevisiae* (Roman, 1956a,b). Evidence for mitotic recombination in disomics of *Neurospora crassa* has been presented by Pittenger and Coyle (1963).

In *Aspergillus nidulans,* where the parasexual cycle supplements the sexual one, the frequency of mitotic recombination is roughly 2×10^{-3} that of meiotic recombination in the same species and should contribute only to a limited extent to gene recombination. In two species which lack the sexual cycle, *Aspergillus niger* and *Penicillium chrysogenum,* the parasexual cycle plays a more positive role, for diploidization occurs at a higher rate and mitotic exchange is apparently more frequent (Pontecorvo, 1958). As Pontecorvo points out, with a small shift in favor of diploids and an increase in the rate of mitotic recombination, the parasexual cycle can yield as high a frequency of recombinants as the sexual cycle.

b. *Mitotic Reduction.* In fungi, the process of haploidization from the diploid has been analyzed by Pontecorvo and Käfer (1956, 1958) and by Käfer (1961). Their results show that the haploids represent all possible genotypes that are expected from independent assortment of nonhomologous chromosomes without crossing-over. In other words, linked markers remain associated and only nonlinked markers recombine. The process of reduction appears, therefore, to be independent of the process of mitotic crossing-over, unlike the meiotic mechanism where crossing-over is most often a prerequisite for ensuring a reductional segregation. How, then, is reduction achieved in mitotic cells of fungi? According to Pontecorvo (1958) and Käfer (1961), haploidization is probably a multi-step process which utilizes a nondisjunctional mechanism. At first, non-disjunction of two sister chromatids leads to a monosomic and a trisomic for one or two members of the genome; this is followed by a reduction to the monosomic condition for all, through subsequent nondisjunctional events, with a strong selection in favor of a fully balanced haploid. This scheme presumes that a grossly unbalanced genome is not sufficiently deleterious as to be lost before full haploidization can be achieved.

Among higher plants, Brown (1947) has reported a case of spontaneous reduction of chromosome number in somatic tissue of cotton. Pollination of *Gossypium hirsutum* with pollen from *Hibiscus esculentus* produced a highly abnormal and sterile plant with ~ twice the normal chromosome number of cotton. After several seasons, the plant, which was propogated by grafting to a normal stock, was found to have developed a morphologically normal phenotype in the apical portion of a grafted branch, although the base was still abnormal. The normal portion possessed approximately the diploid number of chromosomes for cotton (2N — 1), but the method of reduction remains unknown. Christoff and Christoff (1948) have described a similar case in which a colchicine-treated seed of *Hieracium hoppeanum* gave rise to a plant with a doubled chromosome number but among whose progeny appeared several plants with a diploid number. The normal cells apparently arose from an unusual reduction division within the somatic integumental cells. Battaglia (1947) has found in *Sambucus ebulus* that the somatic cells at the base of the style undergo a somatic reduction to produce two haploid nuclei, and Gates (1912) reported the appearance of haploid cells in the nucellar tissue of *Oenothera*.

Huskins and Steinitz (1948) found that the differentiated root cells of *Rhoeo*, which ordinarily do not divide, can be stimulated to do so by treatment with indole acetic acid. Two haploid cells, presumably arising through somatic reduction, were found among the division products. Huskins (1948), using a 1–4% solution of sodium ribose nucleic acid on the root tips of *Allium cepa*, succeeded in inducing an increased frequency of somatic reductional divisions. The divisions resembled meiotic ones in that homologs underwent synapsis and paired homologs came to lie on the metaphase plate preceding their segregation to opposite poles at anaphase. Segregation of long, prophase-like chromosomes was observed even more frequently, both in treated plants and in untreated bulbs that were flaccid after several months' storage at room temperature. Huskins and Cheng (1950) found that the reductional type of mitosis could also be greatly increased in frequency by growing *Allium* bulbs at low temperatures. Spontaneous reduction of chromosomes of the leaf primordia cells and of the meristem cells of *Haplopappus gracilis* have been reported by Lima-de-Faria and Jaworska (1964) and by Ames and Mitra (1966), respectively.

Among higher animals, mitotic reduction has been found in certain of the insects. Hughes-Schrader (1925, 1927) reported a case of mitotic reduction in the coccid *Icerya purchasi*. In hermaphroditic embryos of this coccid, the germ cells were found to carry the diploid number of four chromosomes, but during the early part of the first nymphal instar, haploid nuclei were regularly observed in the gonad. The chromosomes of the haploid are recognizable by size and shape as comprising one member of each pair of the diploid complex. Rapid proliferation of these cells produces a mass of haploid tissue which gives rise to the primary spermatocytes. No cytological clue as to the mechanism underlying the reduction from diploid to haploid was obtained.

The most detailed studies of somatic reduction were carried out by Berger (1938, 1941) and by S. M. Grell (1946a) with the iliac epithelium of *Culex pipiens*. During larval life, the cells of the midgut of the mosquito increase from 3–4 μ to 10–17 μ as a consequence of chromosome reduplication without corresponding cell divisions. Thus, at the onset of metamorphosis, the ileum is composed of a few large, polyploid cells (48*n* or 96*n*). During metamorphosis these cells are reduced in strand number so that by its termination most of the cells are only 4*n* or 8*n*. The process of reduction involves the separation of the constituent strands of each of the six thick chromosomes during late prophase. Each single strand then pairs with a homologous strand so that at metaphase 24–48 pairs of homologs may be found on the plate. At anaphase a reductional division leads to two daughter cells, each with one half of the parental chromosome number. At anaphase, chromosomes come together again, and then at interphase, they enter a still closer association or somatic synapsis characterized by the absence of DNA synthesis. At the following prophase, thick chromosomes again appear in the diploid number, separate into their component strands, and repeat the process. In this way a progressive halving of the strand number takes place until the 4*n*–8*n* condition is reached.

A search of the literature has revealed relatively few instances of mitotic reduction of diploid tissue. This is not surprising, since in most organisms the process would be expected to uncover lethal or deleterious genes. If more instances of mitotic reduction are to be found, the search should probably be directed to forms in which detrimental genes are regularly eliminated by means of an independent, active haploid phase in the life cycle.

IV. Conclusions

Many facts and undoubtedly some fantasies about pairing have been reviewed in the preceding pages. Yet we have no definitive answer to the basic question: How do chromosomes pair? Specifically, what are the problems whose resolution might contribute ultimately to an understanding of the pairing phenomenon? Certainly we should like a final answer to the much-debated question of whether long-range attraction forces exist. Current thinking generally negates this idea, and envisions pairing to be initiated through chance contact between moving chromosomes. Chromosome movement might be promoted and regulated by concentration changes in the ionic composition of the medium within which the chromosomes lie, so that an attraction force at one time could be replaced by a repulsion force at another time (Yos *et al.*, 1957). Although this interpretation is consonant with present knowledge and although no physical forces are known which are capable of acting over the distances required by the alternative hypothesis, the possibility that long-range forces are responsible for some kinds of pairing cannot be unequivocally excluded.

We should like to know whether chromosomes need be in a condensed state for the initiation of pairing; and, if so, wherein pairing specificity lies. There is good cytological evidence for late pairing between condensed chromosomes at diakinesis or prometaphase. Studies of the *melanogaster* female indicate that specificity at this time is related to chromosome size and not to the sequence of nucleotides the chromosomes possess in common. This conclusion rests on the meiotic segregation ratios between nonhomologous chromosomes of varying size as well as on observations of pairing between nonhomologs in the oogonial cells of *D. melanogaster*. For pairing of condensed chromosomes at these times, chemical identity at the nucleoprotein level appears to suffice.

Is chromosome condensation obligatory for the initiation of all pairing? A rough approximation of condensed chromosomes that precedes the great extension that must take place, at least regionally, for exchange pairing, should help to eliminate entanglements and interlocking, which would seem to be an inevitable consequence of synapsis of greatly extended chromosomes, but which cytologists fail to observe. We know that before exchange, small segments of chromosomes, carried as duplications in the genome, find their homologous regions on the much larger normal chromosomes with remarkable fidelity. This must mean that if a rough pairing of condensed chromosomes precedes intimate exchange pairing, specificity resides at the nucleotide level and is independent of size, unlike the pairing of condensed chromosomes preceding segregation. Perhaps chromosome condensation is greater prior to metaphase than it is prior to exchange, when small extended portions of otherwise condensed chromosomes may provide the specificity needed for exchange pairing. It is also clear that there is a spatial constraint on pairing at this time, such that duplications placed close to their region of homology possess a pairing advantage, whereas those located distantly are at a disadvantage.

What is the minimal number of nucleotides required for exchange pairing? Is this number constant or does it vary within and between organisms? Is the requirement greater for exchange than for exchange pairing, as is suggested by the behavior of free duplications which grossly alter exchange values but rarely participate in exchange? Do heterochromatic regions fail to undergo exchange because they do not undergo exchange pairing, and does this mean that chromosome extension is a prerequisite for both processes?

What is the significance of the synaptinemal complex? This complex has been variously assigned roles in pairing, in exchange, and in holding sister chromatids together. The last possibility is a very appealing one, for it is evident that the maintenance of bivalents until metaphase requires not only the presence of a chiasma but also the retention of sister chromatid pairing. Were homologous chromatids rather than sister chromatids to become associated distal to the chiasma, the chiasma would be resolved, and the homologs would fall apart, as happens at anaphase. Could, then, the complex be a prerequisite for proper segregation and have nothing to do with exchange pairing or exchange? The presence of the complex in haploids would, in this view, become understandable.

Are homologs normally paired during mitotic interphase; and, if so, does pairing at this time aid in the regulation of the transcription and/or the translation processes? Are homologs paired during meiotic interphase; and, if so, does this pairing constitute exchange pairing?

For chromosomes to pair, they must not only come together, they must also remain together. Assuming that chance movements bring chromosomes into alignment with various degrees of precision, depending on their physical state, what then holds them together? In 1940, some twenty years before the structure of the DNA molecule was elaborated, two remarkable scientists considered the nature of the intermolecular forces operative in biological processes; and with extraordinary perception, they wrote,

> These interactions are such as to give stability to a system of two molecules with complementary structures in juxtaposition, rather than of two molecules with necessarily identical structures; we accordingly feel that complementariness should be given primary consideration in the discussion of the specific attraction between molecules. . . .
>
> A general argument regarding complementariness may be given. Attraction forces between molecules vary inversely with a power of the distance and the maximum stability of a complex is achieved by bringing the molecules as close together as possible in such a way that positively charged groups are brought near to negatively charged groups, electric dipoles are brought into suitable mutual orientations, etc. The minimum distances of approach of atoms are determined by their repulsive potential which may be expressed in terms of van der Waals radii; in order to achieve the maximum stability, the two molecules must have complementary surfaces, like die and coin, and also a complementary distribution of active groups (Pauling and Delbrück, 1940).

Ultimate complementariness would seem to be realized with two complementary strands of DNA, for here each base participates in the pairing process and in so doing becomes saturated. In a functional sense it is difficult to imagine any other way that the absolute precision that is required for exchange pairing could be achieved. Herein may also lie the answer to the provocative question of why intimate pairing at exchange is restricted to two strands, whereas intimate pairing between extended polytene strands shows no such restriction. If we assume that pairing for exchange involves two complementary strands, each with a single pairing face so constructed as to be saturated by the complementariness of the other, then complete saturation occurs when they lie front-to-front. Additional strands would find no complementariness and could not pair. By contrast, if pairing in polytene nuclei occurs between double-stranded DNA molecules embedded in protein, complementariness would probably be of a lower order of magnitude and could involve front-to-back rather than front-to-front pairing. Conceivably, front-to-back pairing would leave unsaturated charges on the front of one molecule and on the back of the other, so as to permit the involvement of an indefinite number of strands. Pairing between completely condensed chromosomes at metaphase may imply a still lower order of complementariness.

The key to the understanding of pairing, then, lies in the proper identification and compartmentalization of the different pairing processes. We have seen that the apparent contradiction between precise homologous pairing and nonhomologous pairing, occurring simultaneously during meiosis, eventually became resolved through the realization that each type represents a temporally and functionally distinct kind of pairing. So it will be necessary to recognize all of the other aspects of the pairing phenomenon, including the physical state that is characteristic of each and the chromosomal activity that each represents, if sense is to be made from the diverse array of facts that have been and will be accumulating.

The literature survey for this chapter was concluded in February, 1968.

REFERENCES

Abel, W. O. (1963). *Z. Vererbungslehre* **94**, 442–455.
Abel, W. O. (1964). *Z. Vererbungslehre* **95**, 306–317
Abel, W. O. (1965). *Z. Vererbungslehre* **96**, 228–233.
Agar, W. E. (1911). *Quart. J. Microscop. Sci.* **57**, Part 1, 1–44.
Ames, I. H., and Mitra, J. (1966). *Nature* **210**, 873–874.
Anderson, E. G. (1929). *Z. Induktive Abstammungs- Vererbungslehre* **51**, 397–441.
Baglioni, C. (1962). *Proc. Natl. Acad. Sci. U.S.* **48**, 1880–1886.
Baker, W. K. (1958). *Am. Naturalist* **92**, 59–60
Barton, D. E., and David, F. N. (1962). *Am. J. Human Genet.* **25**, 323–329.
Bateman, A. J. (1967). *Genet. Res.* **9**, 375
Battaglia, E. (1947). *Nuova Giorn. Botan. Ital.* [n. s.] **54**, 596–696.
Bautz, F. A., and Bautz, E. K. F. (1967), *Genetics* **57**, 887–895.
Beadle, G. W. (1933). *Cytologia (Tokyo)* **4**, 269–286.
Beadle, G. W. (1934). *J. Genet.* **29**, 277–309
Beadle, G. W. (1935). *Genetics* **20**, 179–191
Beadle, G. W., and Emerson, S. (1935). *Genetics* **20**, 192–206.
Beadle, G. W., and Ephrussi, B. (1937). *Proc. Natl. Acad. Sci. U.S.* **23**, 356–360.
Beermann, S. (1959). *Chromosoma* **10**, 504–514.
Beermann, W. (1954). *Chromosoma* **6**, 381–396.
Belling, J. (1933). *Genetics* **18**, 388–413.
Bennett, D., and Dunn, L. C. (1964). *Genetics* **49**, 949–958.
Berg, C., and Curtiss, R., III. (1967). *Genetics* **56**, 503–525.
Berger, C. A. (1938). *Carnegie Inst. Wash. Publ.* **476**, 209–232.
Berger, C. A. (1941). *Cold Spring Harbor Symp. Quant. Biol.* **9**, 19–21.
Blackman, M. W. (1903). *Biol. Bull.* **6**, 187–217.
Bodmer, W. F. (1965). *J. Mol. Biol.* **14**, 534–557.
Bodmer, W. F. (1966). *J. Gen. Physiol.* **49**, 233–258.
Boss, J. M. N. (1953). *J. Physiol. (London)* **120**, 32–35.
Boss, J. M. N. (1954). *Exptl. Cell Res.* **7**, 225–231.
Bridges, C. B. (1916). *Genetics* **1**, 1–52, 107–163.
Bridges, C. B. (1925). *Proc. Natl. Acad. Sci. U.S.* **11**, 701–706.
Bridges, C. B. (1935). *(Demonstr.) Records of Genet. Soc. Am.* **4**, 58.
Bridges, C. B. (1936). *Science* **83**, 210–211.

Bridges, C. B., and Anderson, E. G. (1925). *Genetics* **10**, 418–441.
Brink, R. A. (1959). *Proc. Natl. Acad. Sci. U.S.* **45**, 819–827.
Brinkley, B. R., and Bryan, J. H. D. (1964). *J. Cell Biol.* **23**, 14A.
Brosseau, G. E., Jr. (1965). *Drosophila Inform. Serv.* **40**, 71.
Brown, M. (1940). *Texas, Univ., Publ.* **4032**, 65–70.
Brown, M. (1947). *Am. J. Botany* **34**, 384–388.
Brown, M. (1954). *Genetics* **34**, 962–963.
Brown, M. (1958). *Genetics* **2**, 36–37.
Butterfass, T. (1967). *Chromosoma* **20**, 442–444.
Buxton, E. W. (1956). *J. Gen. Microbiol.* **15**, 133–139
Byrne, R., Levin, J. G., Bladen, H. A., and Nirenberg, M. W. (1964). *Proc. Natl. Acad. Sci. U.S.* **52**, 140–148.
Cameron, H. R., Hsu, K. S., and Perkins, D. D. (1966). *Genetics* **37**, 1–6
Carroll, C. P. (1966). *Chromosoma* **18**, 19–43
Carter, T. C. (1952). *J. Genet.* **51**, 1–6.
Catcheside, D. G. (1932). *Cytologia (Tokyo)* **4**, 68–113
Catcheside, D. G. (1947). *J. Genet.* **48**, 99–110.
Cattanach, B. M. (1967). *Cytogenetics* **6**, 67–77.
Chandley, A. C. (1965). *Genetics* **52**, 247–258.
Chandley, A. C. (1966). *Exptl. Cell Res.* **44**, 201–215
Chase, M., and Doermann, A. H. (1958). *Genetics* **43**, 332–353.
Chauhan, K. P. S., and Abel, W. O. (1968). *Chromosoma* **25**, 297–302.
Chen, K. C., and Ravin, A. W. (1966a). *J. Mol. Biol.* **22**, 109–121.
Chen, K. C., and Ravin, A. W. (1966b). *J. Mol. Biol.* **22**, 123–134.
Chiang, K. S., and Sueoka, N. (1967a). *J. Cellular Physiol.* **70**, Suppl. 1, 89–112.
Chiang, K. S., and Sueoka, N. (1967b). Personal communication.
Christensen, B. (1961). *Hereditas* **47**, 385–449.
Christoff, M., and Christoff, M. A. (1948). *Genetics* **33**, 36–42
Clark, A. J. (1964). *Z. Vererbungslehre* **95**, 368–373.
Cohen, A., Fisher, W. D., Curtiss, R., and Adler, H. I. (1968). *Cold Spring Harbor Symp. Quant. Biol.* **33** (in press).
Coleman, J. R., and Moses, M. J. (1964). *J. Cell Biol.* **23**, 63–78.
Comings, D. E., and Kakefuda, T. (1968). *J. Mol. Biol.* **33**, 225–229.
Cooper, K. W. (1944). *Genetics* **29**, 537–568.
Cooper, K. W. (1945). *Genetics* **30**, 472–484.
Cooper, K. W. (1948a). *Proc. Natl. Acad. Sci. U. S.* **34**, 179–187.
Cooper, K. W. (1948b). *J. Exptl. Zool.* **108**, 327–336.
Cooper, K. W. (1949). *J. Morphol.* **84**, 81–121.
Cooper, K. W. (1964). *Proc. Natl. Acad. Sci. U. S.* **52**, 1248–1255.
Cooper, K. W., Zimmering, S., and Krivshenko, J. (1955). *Proc. Natl. Acad. Sci. U.S.* **41**, 911–914.
Cronenwett, C., and Maguire, M. (1967). *Maize Genet. Cooperation Newsletter* **41**, 179–180.
Darlington, C. D. (1929). *J. Genet.* **21**, 207–286
Darlington, C. D. (1930). *Cytologia (Tokyo)* **2**, 37–55
Darlington, C. D. (1937). "Recent Advances in Cytology," 2nd ed. McGraw-Hill (Blakiston), New York.
Darlington, C. D., and Moffett, A. A. (1930). *J. Genet.* **22**, 130–151
Demerec, M. (1962). *Proc. Natl. Acad. Sci. U.S.* **48**, 1696–1704

Demerec, M. (1963). *Genetics* **48**, 1519–1531.
Demerec, M., Gillespie, D. H., and Mitzobuchi, K. (1963). *Genetics* **48**, 997–1009.
Dempsey, E. (1959). *Maize Genet. Cooperation Newsletter* **33**, 54–55.
de Serres, F. (1958). *Cold Spring Harbor Symp. Quant. Biol.* **23**, 111–118.
de Serres, F. (1962). *Genetics* **47**, 950–951.
de Serres, F. (1967). Personal communication.
Dobzhansky, T. (1931). *Am. Naturalist* **65**, 214–232.
Dobzhansky, T. (1932). *Z. Induktive Abstammungs- Vererbungslehre* **60**, 235–286.
Dobzhansky, T. (1933). *Z. Induktive Abstammungs- Vererbungslehre* **64**, 269–309.
Dobzhansky, T. (1934). *Z. Induktive Abstammungs- Vererbungslehre* **68**, 134–162.
Doermann, A. H. (1965). *Proc. 11th Intern. Congr. Genet., The Hague, 1963* pp. 69–80. Pergamon Press, Oxford.
Downing, E. R. (1905). *Zool. Jahrb.* **21**, 379–426.
Doyle, G. G. (1959). *Science* **130**, 1415.
Dublin, L. I. (1905). *Ann. N.Y. Acad. Sci.* **16**, 1–64.
Dunn, L. C. (1964). *Science* **144**, 260–263.
Falk, R. (1955). *Hereditas* **41**, 376–383.
Farmer, J. B., and Moore, J. E. S. (1905). *Quart. J. Microscop. Sci.* **48**.
Fawcett, D. W. (1956). *J. Biophys. Biochem. Cytol.* **2**, 402–406.
Feldman, M. (1966) *Proc. Natl. Acad. Sci. U.S.* **55**, 1447–1453.
Feldman, M., Mello-Sampayo, T., and Sears, E. R. (1966). *Proc. Natl. Acad. Sci. U.S.* **56**, 1192–1199.
Forbes, C. (1960). *Proc. Natl. Acad. Sci. U.S.* **46**, 222–225.
Fox, M. S. (1966). *J. Gen. Physiol.* **49**, 183–196.
Franchi, L. L., and Mandl, A. M. (1962). *Proc. Roy. Soc.* **B157**, 99–114.
Fraser, H. C. I. (1907). *Ann. Botany (London)* **21**, 307–308.
Fraser, H. C. I. (1908). *Ann. Botany (London)* **22**, 36–55.
Gall, J. G. (1961). *J. Biophys. Biochem. Cytol.* **10**, 163–193.
Gans, M. (1953). *Bull. Biol. France Belgi.* **38**, Suppl., 1–90.
Gassner, G. (1967). *J. Cell. Biol.* **35**, 166A–167A.
Gates, R. R. (1909). *Botan. Gaz.* **48**, 179–199.
Gates, R. R. (1912). *Ann. Botany (London)* **26**, 993–1010.
Gates, R. R. (1923). *Ann. Botany (London)* **37**, 565–569.
Gelei, J. (1921). *Arch. Zellforsch.* **16**, 1.
Gerstel, D. U. (1961). *Science* **133**, 579–580.
Gerstel, D. U. (1963). *Genetics* **48**, 677–689.
Gerstel, D. U., and Phillips, L. L. (1958). *Cold Spring Harbor Symp. Quant. Biol.* **23**, 225–237.
Giles, N. H. (1955). *Brookhaven Symp. Biol.* **8**, 103–125.
Green, M. M. (1959a). *Heredity* **13**, 303–315.
Green, M. M. (1959b). *Genetics* **44**, 1243–1256.
Green, M. M. (1959c). *Z. Vererbungslehre* **90**, 375–384.
Green, M. M. (1961). *Genetics* **46**, 1555–1560.
Green, M. M. (1962). *Genetics* **33**, 154–164.
Green, M. M. (1963). *Z. Vererbungslehre* **94**, 200–214.
Greenleaf, W. H. (1942). *J. Genet.* **43**, 69–96.
Gregg, T. G., and Day, J. W. (1965). *Genetica* 36, 172–182.
Grell, E. H. (1961). *Genetics* **46**, 1267–1271.
Grell, E. H. (1963). *Genetics* **48**, 1217–1229.

Grell, E. H. (1964). *Genetics* **50**, 251–252.
Grell, E. H. (1965). Unpublished data.
Grell, E. H. (1969). In preparation.
Grell, R. F. (1957). *Genetics* **42**, 374.
Grell, R. F. (1959a). *Genetics* **44**, 421–435.
Grell, R. F. (1959b). *Genetics* **44**, 514.
Grell, R. F. (1962a). *Proc. Natl. Acad. Sci. U.S.* **48**, 165–172.
Grell, R. F. (1962b). *Genetics* **47**, 1737–1754.
Grell, R. F. (1964a). *Genetics* **50**, 150–166.
Grell, R. F. (1964b). *Proc. Natl. Acad. Sci. U.S.* **52**, 226–232.
Grell, R. F. (1964c). Unpublished results.
Grell, R. F. (1965). *Natl. Cancer Inst. Monograph* **18**, 215–242.
Grell, R. F. (1966). *Genetics* **54**, 411–421.
Grell, R. F. (1967). *J. Cellular Physiol.* **70**, Suppl. 1, 119–145.
Grell, R. F. (1968a). Unpublished results.
Grell, R. F. (1968b). *Genetics* **60**, 184.
Grell, R. F., and Chandley, A. C. (1965). *Proc. Natl. Acad. Sci. U.S.* **53**, 1340–1346.
Grell, R. F., and Grell, E. H. (1959). *Drosophila Inform. Serv.* **33**, 137–139.
Grell, R. F., and Grell, E. H. (1960). *Proc. Natl. Acad. Sci. U.S.* **46**, 51–57.
Grell, R. F., and Valencia, J. I. (1964). *Science* **145**, 66–67.
Grell, S. M. (1946a). *Genetics* **31**, 60–76.
Grell, S. M. (1946b). *Genetics* **31**, 77–94.
Gross, J. (1904). *Zool. Jahrb., Abt. Anat. Ontog. Tiere* **20**.
Grüneberg, H. (1966). *Genet. Res.* **7**, 58–75.
Guillermond, M. A. (1904). *Compt. Rend. Soc. Biol.* **56**, 412.
Guyénot, E., and Naville, A. (1933–1934). *Cellule* **42**, 213–230.
Hayes, W. (1964). "The Genetics of Bacteria and their Viruses." Blackwell, Oxford.
Heilborn, O. (1936). *Hereditas* **22**, 167–188.
Henderson, S. A. (1963). *Heredity* **18**, 77–94.
Henderson, S. A. (1966). *Nature* **211**, 1043–1047.
Henking, H. (1891). *Z. Wiss. Zool.* **51**.
Hinton, C. W. (1962). *Genetics* **47**, 959.
Hinton, C. W. (1965). *Genetics* **51**, 971–982.
Hinton, C. W. (1966). *Genetics* **53**, 157–164.
Hinton, T. (1946). *J. Exptl. Zool.* **102**, 237–251.
Hiraoka, T. (1941). *Cytologia (Tokyo)* **11**, 473–492.
Hiraoka, T. (1958). *Proc. Japan Acad.* **34**, 700–705.
Hirayashi, I. (1957). *Proc. Intern. Genet. Symp., Tokyo Kyoto, 1956* pp. 293–297. Sci. Council Japan, Tokyo.
Holliday, R. (1964). *Genet. Res.* **5**, 282–304.
Holm, D. G., Deland, M., and Chovnick, A. (1967). *Genetics 56*, 565–566.
Hotta, Y., Ito, M., and Stern, H. (1966). *Proc. Natl. Acad. Sci. U.S. 56*, 1184–1191.
Hu, C. H. (1960). *Cytologia (Tokyo)* **25**, 437–449.
Hughes-Schrader, S. (1925). *Z. Zellforsch. Mikroskop. Anat.* **2**, 264–292.
Hughes-Schrader, S. (1927). *Z. Zellforsch. Mikroskop. Anat.* **6**, 509–540.
Huskins, C. L. (1948). *J. Heredity* **39**, 310–325.
Huskins, C. L., and Cheng, K. C. (1950). *J. Heredity* **41**, 13–18.
Huskins, C. L., and Steinitz, L. N. (1948). *J. Heredity* **39**, 34–43 and 66–77.

Illert, G. (1956). *Chromosoma* **7**, 608–619.
Ishikawa, M. (1911). *Botan. Mag. (Tokyo)* **25**.
Ishitani, C., Ikeda, Y., and Sakaguchi, K. (1956). *J. Gen. Appl. Microbiol. (Tokyo)* **2**, 401–430.
Jain, H. K., and Basak, S. L. (1963). *Genetics* **48**, 329–339.
John, B., and Lewis, K. R. (1965). "Protoplasmatologia." Springer, Vienna.
Judd, B. H. (1961a). *Genetics* **46**, 1687–1697.
Judd, B. H. (1961b). *Proc. Natl. Acad. Sci. U.S.* **47**, 545–550.
Käfer, E. (1961). *Genetics* **46**, 1581–1609.
Kaplan, W. D. (1953). *Genetics* **38**, 630–631.
Katayama, T. (1965). *Japan. J. Genet.* **40**, 21–32.
Kaufmann, B. P. (1934). *J. Morphol.* **56**, 125–155.
Kermicle, J. (1963). Ph.D. Thesis, University of Wisconsin, Madison, Wisconsin.
Keyl, H. G. (1957). *Chromosoma* **8**, 719–729.
King, E. (1933). *J. Heredity* **24**, 253–256.
Kiritani, K. (1962). *Japan. J. Genet.* **37**, 42–56.
Kitani, Y. (1963). *Japan. J. Genet.* **38**, 244–256.
Koch, E. A., Smith, P. A., and King, R. C. (1967). *J. Morphol.* **121**, 55–70.
Kozhevnikov, B. T. (1940). *Bull. Biol. Med. Exptl. URSS* **9**, 13–14.
Kuwada, Y. (1910). *Botan. Mag. (Tokyo)* **39**.
Lamm, R. (1936). *Hereditas* **31**, 217–240.
Lawrence, W. J. C. (1931). *Cytologia (Tokyo)* **2**, 352–384.
Le Calvez, J. (1950). *Arch. Zool. Exptl. Gen.* **87**, 211–214.
Levan, A. (1942). *Hereditas* **28**, 177–211.
Levine, R. P., and Levine, E. E. (1954). *Genetics* **39**, 677–691.
Lewis, E. B. (1951). *Cold Spring Harbor Symp. Quant. Biol.* **16**, 159–174.
Lewis, E. B. (1954). *Am. Naturalist* **88**, 225–239.
Lewis, E. B. (1963). *Am. Zoologist* **3**, 33–56.
Lima-de-Faria, A., and Jaworska, H. (1964). *Hereditas* **52**, 119.
Linnert, G. (1955). *Chromosoma* **7**, 90–128.
Longley, A. E. (1924). *Am. J. Botany* **11**, 249–282.
Longley, A. E. (1926). *J. Wash. Acad. Sci.* **16**, 543–545.
McClintock, B. (1931). *Missouri, Univ., Agr. Expt. Sta., Res. Bull.* **163**, 1–30.
McClintock, B. (1933). *Z. Zellforsch. Mikroskop. Anat.* **19**, 191–237.
McClintock, B. (1945). *Am. J. Botany* **32**, 671–678.
McClung, C. E. (1927). *J. Morphol. Physiol.* **43**, 181–266.
McNelly-Ingle, C., Lamb, B. C., and Frost, L. C. (1966). *Genet. Res.* **7**, 169–183.
Maeda, T. (1939). *Japan. J. Genet.* **15**, 118–127.
Magni, G. E., and von Borstel, R. C. (1962). *Genetics* **47**, 1097–1108.
Maguire, M. P. (1960). *Genetics* **45**, 651–664.
Maguire, M. P. (1961). *Exptl. Cell Res.* **24**, 21–36.
Maguire, M. P. (1965). *Genetics* **51**, 23–40.
Maguire, M. P. (1966). *Genetics* **53**, 1071–1077.
Maguire, M. P. (1967a). *Maize Genet. Cooperation Newsletter* **41**, 177–179.
Maguire, M. P. (1967b). *Chromosoma* **21**, 221–231.
Maguire, M. P. (1968). *Genetics* **60**, 353–362.
Maillet, P. L., and Folliot, R. (1965). *Compt. Rend.* **260**, 3486–3489.
Menzel, M. Y. (1962). *Am. J. Botany* **49**, 605–615.
Menzel, M. Y. (1965). Personal communication.

Menzel, M. Y., and Price, J. M. (1966). *Am. J. Botany* **53**, 1079–1086.
Menzel, M. Y., and Price, J. M. (1967). *J. Cell Biol.* **35**, 189.
Merriam, J. R. (1967). *Genetics* **57**, 409–425.
Metz, C. W. (1914). *J. Exptl. Zool.* **17**, 45–59.
Metz, C. W. (1916). *J. Exptl. Zool.* **21**, 213–280.
Metz, C. W. (1926). *Anat. Anz.* **4**, 1–28.
Metz, C. W., and Nonidez, J. F. (1921). *J. Exptl. Zool.* **32**, 165–185.
Meyer, G. F. (1961). *Proc. 2nd Reg. Conf. (Eur.) Electron Microscopy, Delft, 1960* pp. 951–954. Almqvist & Wiksell, Uppsala.
Meyer, G. F. (1964). *Proc. 3rd Reg. Conf. (Eur.) Electron Microscopy, Prague, 1964,* pp. 461–462. Czech. Acad. Sci., Prague.
Miller, B. A., and Grell, R. F. (1963). *Drosophila Inform. Serv.* **38**, 56–66.
Miller, O. L. (1963). *Genetics* **48**, 1445–1466.
Mitchell, M. B. (1956). *Compt. Rend. Trav. Lab. Carlsberg* **26**, 285–298.
Mitra, J., and Steward, F. C. (1961). *Am. J. Botany* **48**, 358–368.
Miyake, K. (1905). *Jahrb. Wiss. Botan.* **42**, 83–120.
Moens, P. B. (1964). *Chromosoma* **15**, 231–242.
Montgomery, T. H. (1900). *Zool. Jahrb.* **14**.
Montgomery, T. H. (1901). *Trans. Am. Phil. Soc.* **20**.
Moore, J. E. S., and Embleton, A. L. (1905). *Proc. Roy. Soc.* **B77**, 555–562.
Morgan, L. V. (1938). *Genetics* **23**, 423–462.
Moses, M. J. (1956). *J. Biophys. Biochem. Cytol.* **2**, 215–218.
Moses, M. J. (1958). *J. Biophys. Biochem. Cytol.* **4**, 633–638.
Moses, M. J. (1964). *In* "Cytology and Cell Physiology" (G. H. Bourne, ed.), 3rd ed., p. 424. Academic Press, New York.
Mückenthaler, F. A. (1964). *Exptl. Cell Res.* **35**, 531–547.
Müller, H. A. C. (1909). *Jahrb. Wiss. Botan.* **47**.
Müller, H. A. C. (1912). *Arch. Zellforsch* **8**, 1–51.
Muller, H. J. (1941). *Cold Spring Harbor Symp. Quant. Biol.* **9**, 290–308.
Nebel, B. R., and Coulon, E. M. (1962). *Chromosoma* **13**, 272–291.
Nemec, B. (1910). "Das Problem der Befruchtungsvorgange und andere Zytologische Fragen." Berlin.
Nichols, M. L. (1902). *Proc. Am. Phil. Soc.* **41**, 77–112.
Nonidez, J. F. (1920). *J. Morphol.* **34**, 69–117.
Notani, N., and Goodgal, S. H. (1966). *J. Gen. Physiol.* **49**, 197–209.
Novitski, E. (1946). *Genetics* **31**, 508–524.
Novitski, E. (1964). *Genetics* **50**, 1449–1451.
Oehlkers, F., and Eberle, P. (1957). *Chromosoma* **8**, 351–363.
Ohno, S., and Atkins, N. B. (1966). *Chromosoma* **18**, 455–466.
Okamoto, M., and Sears, E. R. (1962). *Can. J. Genet. Cytol.* **4**, 24–30.
Oksala, T. (1958). *Cold Spring Harbor Symp. Quant. Biol.* **23**, 197–210.
Oksala, T., and Therman, E. (1958). *Chromosoma* **9**, 505–513.
Oliver, C. P., and Van Atta, E. W. (1932). *Proc. 6th Intern. Congr. Genet., Ithaca, 1932* Vol. **2**, pp. 145–147, Brooklyn Botanic Garden.
Östergren, G., and Vigfusson, E. (1953). *Hereditas* **39**, 33–50.
Overton, J. B. (1905). *Jahrb. Wiss. Botan.* **42**, 121–153.
Overton, J. B. (1909). *Ann. Botany (London)* **23**, 19–57.
Painter, T. S. (1934). *Genetics* **19**, 175–188.
Pauling, L., and Delbrück, M. (1940). *Science* **92**, 77–79.

Paulmier, F. C. (1899). *J. Morphol.* **15**, Suppl.
Pavan, C. (1946). *Genetics* **31**, 546–557.
Person, C. (1955). *Can. J. Botany* **33**, 11–30.
Philip, U., Rendel, J. M., Spurway, H., and Haldane, J. B. S. (1944). *Nature* **154**, 260–262.
Pittenger, T. H., and Coyle, M. B. (1963). *Proc. Natl. Acad. Sci. U.S.* **49**, 445–451.
Piza, S. De T. (1943). *Rev. Agr. (Sao Paulo)* **18**, 249–276.
Plough, H. H. (1917). *J. Exptl. Zool.* **24**, 147–209.
Pollard, D. R., and Käfer, E. (1967). *Can. J. Genet. Cytol.* **9**, 662.
Pontecorvo, G. (1958). "Trends in Genetic Analysis." Columbia Univ. Press, New York.
Pontecorvo, G., and Käfer, E. (1956). *Proc. Roy. Soc. Edinburgh* **B25**, 16–20.
Pontecorvo, G., and Käfer, E. (1958). *Advan. Genet.* **9**, 71–104.
Pontecorvo, G., and Roper, J. A. (1952). *J. Gen. Microbiol.* **6**, vii.
Pontecorvo, G., and Sermonti, G. (1954). *J. Gen. Microbiol.* **11**, 94–104.
Pontecorvo, G., Roper, J. A., and Forbes, E. (1953). *J. Gen. Microbiol.* **8**, 198–210.
Pritchard, R. H. (1955). *Heredity* **9**, 343–371.
Pritchard, R. H. (1960a). *Genet. Res.* **1**, 1–24.
Pritchard, R. H. (1960b). *In* "Microbial Genetics" (W. Hayes and R. C. Clowes, eds.), pp. 155–180. Oxford Univ. Press (Univ. London), London and New York.
Ramel, C. (1958). *Drosophila Inform. Serv.* **32**, 150–151.
Ramel, C. (1962a). *Hereditas* **48**, 1–58.
Ramel, C. (1962b). *Hereditas* **48**, 59–82.
Rasmussen, I. E. (1960). *Drosophila Inform. Serv.* **34**, 53.
Ravin, A. W., and Chen, K.-C. (1967). *Genetics* **57**, 851–864.
Redfield, H. (1930). *Genetics* **15**, 205–252.
Redfield, H. (1932). *Genetics* **17**, 137–152.
Redfield, H. (1957). *Genetics* **42**, 712–728.
Rees, H., and Thompson, J. B. (1956). *Heredity* **10**, 409–424.
Rhoades, M. M. (1931). *Genetics* **16**, 490–504.
Rhoades, M. M. (1947). *Genetics* **32**, 101.
Rhoades, M. M. (1957). *Maize Genet. Cooperation Newsletter* **31**, 75–76.
Rhoades, M. M. and Dempsey, E. (1949). *Maize Genet. Cooperation Newsletter* **23**, 56–57.
Richharia, R. H. (1937). *J. Genet.* **34**, 19–44.
Rieger, R. (1957). *Chromosoma* **9**, 1–38.
Riley, R. (1965). *In* "Essays on Crop Plant Evolution" (J. B. Hutchinson, ed.), pp. 103–122. Cambridge Univ. Press, London and New York.
Riley, R., and Chapman, V. (1958). *Nature* **182**, 713–715.
Roberts, P. A. (1962). *Genetics* **47**, 1691–1709.
Roberts, P. A. (1965). *Nature* **205**, 725–726.
Roberts, P. A. (1966a). *Genetics* **54**, 969–979.
Roberts, P. A. (1966b). Personal communication.
Roman, H. (1956a). *Compt. Rend. Trav. Lab. Carlsberg* **26**, 299–314.
Roman, H. (1956b). *Cold Spring Harbor Symp. Quant. Biol.* **21**, 175–185.
Rossen, J. M., and Westergaard, M. (1966). *Compt. Rend. Trav. Lab. Carlsberg* **35**, 233–260.
Roth, T. F. (1966). *Protoplasma* **61**, 346–386.

Rothfels, K. H. (1952). *Heredity* **9**, 343.
Rückert, J. (1892). *Anat. Anz.* **8**.
Rückert, J. (1893). "Merkel and Bonnet," Erg. III.
Russell, L. B. (1965). *In* "The Role of Chromosomes in Development" (M. Locke, ed.), pp. 153–181. Academic Press, New York.
Sandler, L., and Novitski, E. (1956). *Genetics* **41**, 189–193.
Sarvella, P. (1958). *Genetics* **43**, 601–619.
Sax, K. (1931). *Proc. Natl. Acad. Sci. U.S.* **17**, 601–603.
Schin, K. (1965). *Chromosoma* **16**, 436–452.
Schneiderman, L. J., and Smith, C. A. B. (1962). *Nature* **195**, 1229–1230.
Schrader, F. (1941). *J. Morphol.* **69**, 587–604.
Sears, E. R., and Okamoto, M. (1959). *Proc. 10th Intern. Congr. Genet., Montreal, 1958,* Vol. 2, pp. 258–259. Univ. of Toronto Press, Toronto.
Serebrovsky, A. S. (1926). *J. Genet.* **16**, 33–42.
Sficas, A. G. (1963). *Genet. Res.* **4**, 266–275.
Shaver, D. L. (1962). *Genetics* **47**, 984.
Shin, D. H., and Moldane, K. (1966). *J. Mol. Biol.* **21**, 231–245.
Singh, R. M., and Wallace, A. T. (1967). *Can. J. Genet. Cytol.* **9**, 87–96.
Singleton, J. R. (1953). *Am. J. Botany* **40**, 124–144.
Slack, H. D. (1938). *Proc. Roy. Soc. Edinburgh* **B58**, 192–212.
Smith, S. G. (1942). *Can. J. Res.* **20**, 221–229.
Smithies, O., Connell, G. E., and Dixon, G. D. (1962). *Nature* **196**, 232–236.
Snope, A. J. (1966). Doctoral Thesis, Indiana University, Bloomington, Indiana.
Sotelo, J. R., and Trujillo-Cenoz, O. (1958). *Exptl. Cell Res.* **14**, 1–8.
Sotelo, J. R., and Trujillo-Cenoz, O. (1960). *Z. Zellforsch. Mikroskop. Anat.* **51**, 243–277.
Sotelo, J. R., and Wettstein, R. (1965). *Natl. Cancer Inst. Monograph* **18**, 133–152.
Stack, S. M., and Brown, W. V. (1968). *Bull. Torr. Bot. Club* **95**, 369–378.
Stadler, D. R., and Towe, A. M. (1962). *Genetics* **47**, 839–856.
Stern, C. (1927). Verh. V. int. Kongr. Vererbungsw.: 1403–1404. *Z. Induktive Abstammungs-Vererbungslehre* Supp. II, 1928.
Stern, C. (1936). *Genetics* **21**, 625–730.
Stevens, N. M. (1908). *J. Exptl. Zool.* **5**, 359–374.
Stevens, N. M. (1910). *J. Exptl. Zool.* **8**, 207–241.
Stomps, T. J. (1910). *Biol. Zentra.* **31**.
Strasburger, E. (1905). *Jahrb. Wiss. Botan.* **42**, 1–71.
Strasburger, E. (1907). *Jahrb. Wiss. Botan.* **49**.
Strasburger, E. (1910). *Ann. Botany (London)* **7**.
Streisinger, G., and Franklin, N. C. (1956). *Cold Spring Harbor Symp. Quant. Biol.* **21**, 103–111.
Strickland, W. N. (1958). *Proc. Roy. Soc.* **B149**, 82–101.
Sturtevant, A. H. (1919). *Carnegie Inst. Wash. Publ.* **278**, 305–341.
Sturtevant, A. H. (1925). *Genetics* **10**, 117–147.
Sturtevant, A. H. (1934). *Proc. Natl. Acad. Sci. U.S.* **20**, 515–518.
Sturtevant, A. H. (1936). *Genetics* **21**, 444–466.
Sturtevant, A. H. (1944). *Carnegie Inst. Wash. Year Book* **43**, 164–165.
Sturtevant, A. H., and Beadle, G. W. (1936). *Genetics* **21**, 554–604.
Sturtevant, A. H., and Morgan, T. H. (1923). *Science* **57**, 746–747.
Suomalainen, E. (1946). *Ann. Acad. Sci. Fennicae, Ser. A IV* **4**, 1–60.
Suomalainen, E. (1947). *Ann. Acad. Sci. Fennicae, Ser. A IV* **A13**, 1–65.

Sutton, W. S. (1902). *Biol. Bull.* **4**, 24–39.
Swift, H. (1950). *Physiol. Zool.* **23**, 169–198.
Sykes, M. G. (1908). *Arch. Zellforsch.* **1**, 380–398.
Tabata, M. (1963). *Cytologia (Tokyo)* **28**, 278–292.
Tahara, M. (1910). *Botan. Mag. (Tokyo)* **24**.
Taylor, J. H. (1953). *Exptl. Cell Res.* **4**, 164–173.
Taylor, J. H. (1967). *In* "Molecular Genetics" (J. H. Taylor, ed.), Part 2, pp. 95–132. Academic Press, New York.
Taylor, J. H., and McMaster, R. D. (1954). *Chromosoma* **6**, 489–521.
Taylor, J. H., Wood, P. S., and Hughes, W. L. (1957). *Proc. Natl. Acad. Sci. U.S.* **43**, 122–128.
Terzhagi, E., and Knapp, D. (1960). *Evolution* **14**, 347–350.
Therman, E. (1953). *Ann. Botan. Soc. Zool. Botan. Fennicae, "Vanamo"* **25**, 1–26.
Therman, E. (1956). *Am. J. Botany* **43**, 134–142.
Tobgy, H. A. (1943). *J. Genetics* **45**, 67–111.
Towe, A. M., and Stadler, D. R. (1964). *Genetics* **49**, 577–583.
Tsuchiya, T. (1962). *Chiba Ikg. Z. (Tokyo)* **3**, 14–15.
Ullerich, F. H. (1961). *Chromosoma* **12**, 215–232.
Upcott, M. (1936). *Cytologia (Tokyo)* **7**, 118–130.
Virkki, N. (1967). *Hereditas* **57**, 275–288.
Virkki, N. (1968). *Caryologia* **21**, 47–51.
Walters, M. S. (1954). *Am. J. Botany* **41**, 160–171.
Weber, D. F. (1967). Ph.D. Thesis, Indiana University, Bloomington, Indiana.
Wenrich, D. H. (1917). *J. Morphol.* **29**, 471.
White, M. J. D. (1938). *Proc. Roy. Soc.* **B125**, 516–523.
White, M. J. D. (1946). *J. Morphol.* **78**, 201–219.
White, M. J. D. (1954). "Animal Cytology and Evolution." Cambridge Univ. Press., London and New York.
White, M. J. D. (1965a). *Chromosoma* **16**, 271–307.
White, M. J. D. (1965b). *Chromosoma* **16**, 521–547.
White, M. J. D., and Morley, F. H. W. (1955). *Genetics* **40**, 604–619.
Whitehouse, H. L. K. (1963). *Nature* **199**, 1034.
Whitehouse, H. L. K., and Hastings, P. J. (1965). *Genet. Res.* **6**, 27–92.
Whiting, P. W. (1917). *J. Morphol.* **28**, 2.
Whittinghill, M. (1955). *J. Cellular Comp. Physiol.* **45**, Suppl. 2, 189–220.
Wilson, E. B. (1905a). *J. Exptl. Zool.* **2**, 371–405.
Wilson, E. B. (1905b). *J. Exptl. Zool.* **2**, 507–545.
Wilson, E. B. (1928). "The Cell in Development and Heredity." Macmillan, New York.
Wilson, I. M. (1937). *Ann. Botany (London)* **1**, 655–671.
Wimber, D. E., and Prensky, W. (1963). *Genetics* **48**, 1731–1738.
Winiwarter, H. de (1901). *Arch. Biol. (Liege)* **17**.
Wolstenholme, D. R., and Meyer, G. F. (1966). *Chromosoma* **18**, 272–286.
Woolam, D. H. M., Ford, E. H. R., and Millen, J. W. (1966). *Exptl. Cell Res.* **42**, 657–661.
Yarnell, S. H. (1929). *Proc. Natl. Acad. Sci. U.S.* **15**, 843–844.
Yos, J. M., Bade, W. L., and Jehle, H. (1957). *In* "Molecular Structure and Biological Specificity" (L. Pauling, ed.), pp. 28–60. Am. Inst. Biol. Sci., Washington, D.C.

AUTHOR INDEX

Numbers in italics refer to pages on which the references are listed.

C

N

O

P

Y

Z

SUBJECT INDEX

E

H

I

N

O

P

T

U

V